Catalysis for Global Development. Contributions around the Iberoamerican Federation of Catalysis

Catalysis for Global Development. Contributions around the Iberoamerican Federation of Catalysis

Special Issue Editors

Helder T. Gomes
Joaquim Luís Faria

MDPI • Basel • Beijing • Wuhan • Barcelona • Belgrade • Manchester • Tokyo • Cluj • Tianjin

Special Issue Editors
Helder T. Gomes
Instituto Politécnico de Bragança (IPB)
Portugal

Joaquim Luís Faria
Associate Laboratory LSRE-LCM
Faculdade de Engenharia da
Universidade do Porto (FEUP)
Portugal

Editorial Office
MDPI
St. Alban-Anlage 66
4052 Basel, Switzerland

This is a reprint of articles from the Special Issue published online in the open access journal *Catalysts* (ISSN 2073-4344) (available at: https://www.mdpi.com/journal/catalysts/special_issues/Global_Development).

For citation purposes, cite each article independently as indicated on the article page online and as indicated below:

LastName, A.A.; LastName, B.B.; LastName, C.C. Article Title. *Journal Name* **Year**, *Article Number*, Page Range.

ISBN 978-3-03928-891-5 (Hbk)
ISBN 978-3-03928-892-2 (PDF)

Contents

About the Special Issue Editors

Helder Gomes is currently Coordinator Professor at the Polytechnic Institute of Bragança and the Product and Process Engineering topic leader at the Mountain Research Centre (CIMO). He graduated with a degree in Chemical Engineering from the Faculty of Engineering of the University of Porto (FEUP) in 1997. In 2002, also from FEUP, he completed his Ph.D. in Chemical Engineering in the area of Catalytic Wet Air Oxidation of Organic Pollutants. He joined the Department of Chemical and Biological Technology (DTQB) at the Polytechnic Institute of Bragança (IPB) in 2001. Between 2016 and 2018 he was President of the Division of Catalysis and Porous Materials of the Portuguese Chemical Society and, since 2018, he is Vice-President of the Iberoamerican Federation of Catalysis Societies. His main research interests are focused on the synthesis and characterization of heterogeneous carbon-based materials for environmental, biomedical and bioenergy applications, the valorization of industrial and agro-industrial wastes into materials and fuels, and the development of water/wastewater treatment solutions based on advanced oxidation processes. He has published 63 ISI papers and more than 200 communications in international/national congresses. He has participated in more than 30 projects and 8 networks as coordinator.

Joaquim L Faria joined the Laboratory of Catalysis and Materials (University of Porto) in 1994 and founded a research line on Heterogeneous Photocatalysis. He is currently Associate Professor in the Faculty of Engineering of the University of Porto (FEUP), Board Member of the Department of Chemical Engineering of the same Faculty, member of the Scientific Committee of the MSc in Chemical Engineering, and member of the Scientific Council of the Associate Laboratory LSRE-LCM (Laboratory of Separation and Reaction Engineering – Laboratory of Catalysis and Materials—a distinguished research unit devoted to specific objectives of the scientific and technological policy laid down by the government. He has chaired and collaborated on the organization of several national and international meetings, and has acted as a member of several scientific boards of national and international meetings. His work focuses on chemical emergent systems (including catalytic) for sustainability and development in the areas of environmental protection, energy, and fine chemical synthesis. He has published over 180 articles in international ISI journals and more than 300 communications in international congresses and symposia (h index of 48, more than 6616 times cited without self-citations). He has written book chapters, been a guest editor of volumes on specific collections, and acted as a reviewer of scientific journals and other non-periodic publications. Prof. Faria has also performed public demonstrations of popular science for young people.

catalysts

Editorial

Catalysis for Global Development. Contributions around the Iberoamerican Federation of Catalysis

Helder Gomes [1,2,*] and Joaquim Faria [2]

1 Centro de Investigação de Montanha (CIMO), Instituto Politécnico de Bragança, 5300-253 Bragança, Portugal
2 Laboratory of Separation and Reaction Engineering—Laboratory of Catalysis and Materials (LSRE-LCM), Faculdade de Engenharia, Universidade do Porto, 4200-465 Porto, Portugal; jlfaria@fe.up.pt
* Correspondence: htgomes@ipb.pt

Received: 28 February 2020; Accepted: 11 March 2020; Published: 19 March 2020

Following biennial meetings held since 1968, the Iberoamerican Federation of Catalysis Societies (FISoCat), the Portuguese Chemical Society (SPQ) and the University of Coimbra jointly organized the XXVI Iberoamerican Congress on Catalysis (CICat 2018), which took place in the historic city of Coimbra, Portugal, between the 9th and 14th of September 2018. CICat 2018 was of particular importance in the history of these events, as it marked the 50th anniversary since the beginning of this series of meetings—by far the most important in the field of catalysis in the Iberoamerican region. Associated with the commemoration of this event, this Special Issue, Catalysis for Global Development: Contributions around the Iberoamerican Federation of Catalysis, emerged to feature selected works presented at CICat 2018. Other possible additional contributions promoting linkages among catalytic science, technology, education, and culture plans and processes involved in cooperation programs and projects among the Iberoamerican Member States, as well as states and institutions of other regions, were also envisaged.

The topics of the conference covered various aspects of catalysis in all its diversity (environmental catalysis, industrial catalysis, oil refining, natural gas conversion and petrochemistry, catalyst design, preparation and characterization, sustainable processes and clean energies, fine chemistry, and other topics on biocatalysis, homogeneous and heterogeneous catalysis), as well other related areas. The diversity of the topics covered was evidenced by the 442 delegates from 20 countries, mostly Iberoamerican countries, attending the conference, and by the presentation of 510 works, together with five plenary lessons—one lesson alluding to the history of CICat to celebrate their 50 years—eight keynotes, and six awards given during the event.

Its extensive scope and interdisciplinarity affirm catalysis to be an essential part of the process and the chemical industry. Research in catalysis supports several strategic industrial sectors in Iberoamerica development through products and processes from energy to the manufacture of materials, and has implications for the development of digital applications and devices. Catalysis plays a crucial role in environmental protection, whether by recycling waste or by reducing gas emissions that contribute to increasing global warming—thereby opening new routes for eco-friendly processes and products that are sustainable and ecologically correct. It is on these topics and directly-related subjects that sixteen selected contributions from the Iberoamerican Federation of Catalysis are gathered in this Special Issue [1–16].

We believe that the contributions published will serve as a source of inspiration and guidance to all those involved in the exciting field of catalysis, particularly for the young researchers and students taking their first steps into research on catalysis. We would like to thank the authors for their enthusiasm since the call for papers was opened, showing from the very earliest stages of the production of the Special Issue their motivation to contribute to this collection. Finally, we acknowledge the unmeasurable help of the assistant editors and reviewers involved, which allowed speeding up the production process and promoting the quality of the manuscripts presented in this Special Issue, making it, so far, the most successful conference issue in Catalysts.

Conflicts of Interest: The authors declare no conflict of interest

References

1. Carriel Schmitt, C.; Gagliardi Reolon, M.; Zimmermann, M.; Raffelt, K.; Grunwaldt, J.; Dahmen, N. Synthesis and Regeneration of Nickel-Based Catalysts for Hydrodeoxygenation of Beech Wood Fast Pyrolysis Bio-Oil. *Catalysts* **2018**, *8*, 449. [CrossRef]
2. Tomé, V.; Calvete, M.; Vinagreiro, C.; Aroso, R.; Pereira, M. A New Tool in the Quest for Biocompatible Phthalocyanines: Palladium Catalyzed Aminocarbonylation for Amide Substituted Phthalonitriles and Illustrative Phthalocyanines Thereof. *Catalysts* **2018**, *8*, 480. [CrossRef]
3. Hurtado, B.; Posadillo, A.; Luna, D.; Bautista, F.; Hidalgo, J.; Luna, C.; Calero, J.; Romero, A.; Estevez, R. Synthesis, Performance and Emission Quality Assessment of Ecodiesel from Castor Oil in Diesel/Biofuel/Alcohol Triple Blends in a Diesel Engine. *Catalysts* **2019**, *9*, 40. [CrossRef]
4. Parejas, A.; Cosano, D.; Hidalgo-Carrillo, J.; Ruiz, J.; Marinas, A.; Jiménez-Sanchidrián, C.; Urbano, F. Aldol Condensation of Furfural with Acetone Over Mg/Al Mixed Oxides. Influence of Water and Synthesis Method. *Catalysts* **2019**, *9*, 203. [CrossRef]
5. Sánchez-Bayo, A.; Morales, V.; Rodríguez, R.; Vicente, G.; Bautista, L. Biodiesel Production (FAEEs) by Heterogeneous Combi-Lipase Biocatalysts Using Wet Extracted Lipids from Microalgae. *Catalysts* **2019**, *9*, 296. [CrossRef]
6. Montaña, M.; Leguizamón Aparicio, M.; Ocsachoque, M.; Navas, M.; de CL Barros, I.; Rodriguez-Castellón, E.; Casella, M.; Lick, I. Zirconia-Supported Silver Nanoparticles for the Catalytic Combustion of Pollutants Originating from Mobile Sources. *Catalysts* **2019**, *9*, 297. [CrossRef]
7. Silva, L.; Neves, V.; Ramos, V.; Silva, R.; Campos, J.; Silva, A.; Malta, L.; Senra, J. Layered Double Hydroxides as Bifunctional Catalysts for the Aryl Borylation under Ligand-Free Conditions. *Catalysts* **2019**, *9*, 302. [CrossRef]
8. Fidelis, M.; Abreu, E.; Dos Santos, O.; Chaves, E.; Brackmann, R.; Dias, D.; Lenzi, G. Experimental Design and Optimization of Triclosan and 2.8-Diclorodibenzeno-p-dioxina Degradation by the Fe/Nb2O5/UV System. *Catalysts* **2019**, *9*, 343. [CrossRef]
9. Escobar, J.; Barrera, M.; Valente, J.; Solís-Casados, D.; Santes, V.; Terrazas, J.; Fouconnier, B. Dibenzothiophene Hydrodesulfurization over P-CoMo on Sol-Gel Alumina Modified by La Addition. Effect of Rare-Earth Content. *Catalysts* **2019**, *9*, 359. [CrossRef]
10. Aiube, C.; Oliveira, K.; Macedo, J. Effect of Cerium Precursor in the Synthesis of Ce-MCM-41 and in the Efficiency for Liquid-Phase Oxidation of Benzyl Alcohol. *Catalysts* **2019**, *9*, 377. [CrossRef]
11. Dominguez, C.; Romero, A.; Santos, A. Improved Etherification of Glycerol with Tert-Butyl Alcohol by the Addition of Dibutyl Ether as Solvent. *Catalysts* **2019**, *9*, 378. [CrossRef]
12. Gutiérrez-Ortega, N.; Ramos-Ramírez, E.; Serafín-Muñoz, A.; Zamorategui-Molina, A.; Monjaraz-Vallejo, J. Use of Co/Fe-Mixed Oxides as Heterogeneous Catalysts in Obtaining Biodiesel. *Catalysts* **2019**, *9*, 403. [CrossRef]
13. Freitas, C.; Pereira, M.; Souza, D.; Fonseca, N.; Sales, E.; Frety, R.; Felix, C.; Azevedo, A.; Brandao, S. Thermal and Catalytic Pyrolysis of Dodecanoic Acid on SAPO-5 and Al-MCM-41 Catalysts. *Catalysts* **2019**, *9*, 418. [CrossRef]
14. Ramos-Ramírez, E.; Tzompantzi-Morales, F.; Gutiérrez-Ortega, N.; Mojica-Calvillo, H.; Castillo-Rodríguez, J. Photocatalytic Degradation of 2,4,6-Trichlorophenol by MgO–MgFe2O4 Derived from Layered Double Hydroxide Structures. *Catalysts* **2019**, *9*, 454. [CrossRef]
15. Tavizón-Pozos, J.; Santolalla-Vargas, C.; Valdés-Martínez, O.; de los Reyes Heredia, J. Effect of Metal Loading in Unpromoted and Promoted CoMo/Al2O3–TiO2 Catalysts for the Hydrodeoxygenation of Phenol. *Catalysts* **2019**, *9*, 550. [CrossRef]
16. V. Sales, R.; Moura, H.; Câmara, A.; Rodríguez-Castellón, E.; Silva, J.; Pergher, S.; Campos, L.; Urbina, M.; Bicudo, T.; de Carvalho, L. Assessment of Ag Nanoparticles Interaction over Low-Cost Mesoporous Silica in Deep Desulfurization of Diesel. *Catalysts* **2019**, *9*, 651. [CrossRef]

catalysts

MDPI

Article

Photocatalytic Degradation of 2,4,6-Trichlorophenol by MgO–$MgFe_2O_4$ Derived from Layered Double Hydroxide Structures

Esthela Ramos-Ramírez [1,*], Francisco Tzompantzi-Morales [2], Norma Gutiérrez-Ortega [3,*], Héctor G. Mojica-Calvillo [1] and Julio Castillo-Rodríguez [2]

1 Laboratory of Advanced Materials and Processes, Department of Chemistry, Division of Natural and Exact Sciences, University of Guanajuato, Guanajuato, Gto. 36050, Mexico; jhg.mojicacalvillo@ugto.mx

2 Laboratory of Ecocatalysis, Department of Chemistry, Metropolitan Autonomous University, México City 09340, Mexico; fjtz@xanum.uam.mx (F.T.-M.); jarry-rasec@hotmail.com (J.C.-R.)

3 Laboratory of Environmental Engineering, Department of Civil Engineering, Division of Engineering, University of Guanajuato, Guanajuato, Gto. 36000, Mexico

* Correspondence: ramosre@ugto.mx (E.R.-R.); normagut@ugto.mx (N.G.-O.); Tel.: +52-473-732-0006 (ext. 1457) (E.R.-R.); +52-473-732-0006 (ext. 2227) (N.G.-O.)

Received: 5 April 2019; Accepted: 14 May 2019; Published: 17 May 2019

Abstract: In recent years, the search for solutions for the treatment of water pollution by toxic compounds such as phenols and chlorophenols has been increasing. Phenols and their derivatives are widely used in the manufacture of pesticides, insecticides, paper, and wood preservers, among other things. Chlorophenols are partially biodegradable but not directly photodegradable by sunlight and are extremely toxic—especially 2,4,6-trichlorophenol, which is considered to be potentially carcinogenic. As a viable proposal to be applied in the treatment of water contaminated with 2,4,6-trichlorophenol, this paper presents an application study of the thermally activated Mg/Fe layered double hydroxides as photocatalysts for the mineralization of this contaminant. Activated Mg/Fe layered double hydroxides were characterized by X-ray diffraction, thermal analysis, N_2 physisorption, and scanning electron microscopy with X-ray dispersive energy. The results of the photocatalytic degradation of 2,4,6-trichlorophenol in aqueous solution showed good photocatalytic activity, with an efficiency of degradation of up to 93% and mineralization of 82%; degradation values which are higher than that of TiO_2-P25, which only reached 18% degradation. The degradation capacity is attributed to the structure of the MgO–$MgFe_2O_4$ oxides derived from double laminate hydroxide Mg/Fe. A path of degradation based on a mechanism of superoxide and hollow radicals is proposed.

Keywords: photocatalysis; Mg/Fe layered double hydroxides; coprecipitation; chlorophenols; mixed oxides; elimination; degradation

1. Introduction

Layered double hydroxides (LDH) or hydrotalcite-type compounds are a large class of natural and synthetic compounds of the anionic clay type [1,2]. These compounds are characterized by having a laminar structure with octahedral arrays of double metal hydroxides that generate a positive residual charge, which is neutralized by the presence of hydrated interlaminar anions [3–5]. Its structural formula is represented in a general way as $[M(II)_{1-x}M(III)_x(OH)_2]^{x+}[A_{x/n}{}^{n-}]\cdot mH_2O$, where M (II) and M (III) can be any divalent cation (Mg^{2+}, Fe^{2+}, Co^{2+}, Zn^{2+}, Cu^{2+}, and Ni^{2+}) or trivalent cation (Al^{3+}, Fe^{3+}, Cr^{3+}, In^{3+}, Ga^{3+}, and Mn^{3+}), respectively; A is any anion ($CO_3{}^{2-}$, $NO_3{}^{-}$, $SO_4{}^{2-}$, Cl^-, $CrO_4{}^{2-}$, etc.) of n charge; x is the fraction of the trivalent cation (at the ratio of x = M(III)/M (III)+M(II)]);

and m is the number of water molecules in the interlaminar space [6,7]. The properties of an LDH will depend on its structural characteristics and will determine the applications that can be given to them. A characteristic that is determinant for the use of an LDH is its profile of evolution of phases by thermal treatment, which allows dehydration, deanionization, and dehydroxylation, with the subsequent formation of a variety of double and simple mixed metal oxides [8–10] Among the main applications that have been found to thermally decompose the products of LDHs is heterogeneous catalysis [11–14]. More specifically, heterogeneous catalysis assisted by irradiation of ultraviolet and visible light as an advanced process of oxidation of recalcitrant and/or persistent organic molecules has become of interest as a potential application of LDHs [15,16]. Several advanced oxidation processes, such as O_3/Ultraviolet (UV), O_3/H_2O_2, UV/H_2O_2, Fenton, UV/Fenton, and UV/TiO_2, have been applied in wastewater as a treatment to mineralize many organic chemicals [17–21]. Within the persistent molecules that exist, interest in their degradation tends to increase by their toxicity and damage to the environment, as is the case for chlorophenols. Chlorophenols are aromatic compounds that are released into the environment in wastewater generated by the wood and petroleum processing industries, as well as the production of drugs, weapons, paper, textiles, and pesticides [22,23]. Most chlorophenols are considered highly toxic, depending on the nature and degree of ring substitution by chlorine. In general, the toxicity increases directly with the degree of chlorination. Specifically, 2,4,6-trichlorophenol tends to accumulate in the lipid tissues of several organisms, is mutagenic or co-mutagenic, and has been linked to cancer in animals, producing lymphomas and leukemia after consumption of contaminated food and water for long periods of time, and in high concentrations [24,25]. Specific studies on the degradation of 2,4,6-trichlorophenol using various technologies have included biological and physicochemical treatments. Specifically, biological treatments have been inefficient because 2,4,6-trichlorophenol is a molecule resistant to biodegradation, as well as being toxic to microorganisms. Physicochemical treatments such as thermal treatment and adsorption have the disadvantage of generating other dangerous compounds as a result of decomposition or generating residues with high concentrations of the contaminant, respectively [26–30]. Therefore, it is feasible to think of advanced oxidation processes as a more effective alternative for the destruction of this pollutant, such as ozone, hydrogen peroxide, photocatalysts, and even combinations of these. Among the semiconductors most used as photocatalysts in the degradation of 2,4,6-trichlorophenol are TiO_2, Fe_2O_3, CeO_2, CuO, ZnO, ZrO, and Al_2O_3, among others; they have been used either alone and mixed or doped, preferably with metals such as Ag, Au, Fe, Co, and Ni, to degrade the pollutant in a significant way [31–36]. The present work reports MgO–$MgFe_2O_4$ derived from layered double hydroxides Mg/Fe with band energy in the range of 2.28–2.47 eV, and its application in the degradation of 2,4,6-trichlorophenol in the aqueous phase using ultraviolet radiation as a source of light.

2. Results and Discussion

2.1. X-Ray Diffraction (XRD)

The X-ray diffraction patterns of the LDH are shown in Figure 1a. The three patterns are similar in relation to the position of the signals, but it is observed that crystallinity increases when the Mg/Fe ratio increases, with a tendency of LDHM1F < LDHM2F < LDHM3F, which is associated with a greater crystallinity produced by the decrease of Fe^{3+} within the brucite type network [Mg $(OH)_2$]. The characteristic signals are in the 2θ angles of 11.2 (003), 23.0 (006), 34.0 (012), 38.0 (015), 45 (018), 59.5 (110), and 61 (013), corresponding to the pyroaurite phase with the PDF card 25-0521 [2,37].

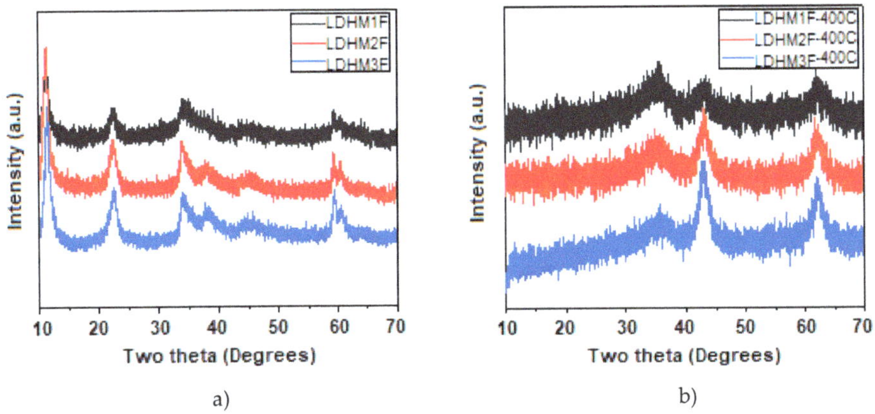

Figure 1. X-ray diffraction patterns of Mg/Fe layered double hydroxides. (**a**) Fresh and (**b**) thermally active at 400 °C.

In Figure 1b, the X-ray diffraction pattern of the layered double hydroxides, calcined at 400 °C, is shown. The presence of the crystalline phase of periclase (MgO) can be observed by characteristic signals in the 2θ angles of 36.80 (111), 42.85 (200), and 62.23 (220) (JCPDS-4-0829), showing a tendency to increase the crystallinity with respect to the Mg/Fe ratio. In the case of the solid LDHM1F, it is observed that at 400 °C the periclase phase has been formed but with a lower crystallinity, which is attributed to the fact that at these temperatures phase changes have occurred, although a higher temperature is required to favor crystallization. For the LDHM2F and LDHM3F solids, a better definition of the peaks is observed, which is associated with a better crystallinity, attributed to the fact that—having a lower amount of Fe—the MgO is more rapidly segregated. The above is associated with the thermal profile described in the following section, in which, in addition to the crystalline phases of MgO, the spinel phase ($MgFe_2O_4$) is segregated, which at 400 °C is amorphous. It is not identifiable by XRD.

2.2. Thermal Analysis: DTA and TGA

Figure 2 shows the differential thermal and thermogravimetric analysis of laminar double hydroxides. In the curve of the DTA (Figure 2a), in all cases an endothermic reaction is observed centered at 145 °C and associated with the elimination of interlaminar water molecules, followed by endothermic reactions centered at 290 °C and 350 °C, due to the dehydroxylation of the sheets and decomposition of the interlaminar carbonate. It can be observed that in the case of solid LDHM1F, the evolution to periclase (MgO) and to amorphous spinel ($MgFe_2O_4$) occurs at 330 °C. In the case of the solid LDHM2F and LDHM3F, the same profile of thermal decomposition is observed; however, it has a stability up to 350 °C, which is attributed to the greater crystallinity that they present as this favors the activation of the catalyst. The reactions associated with the thermal decomposition of LDH are shown in Equations (1)–(3). Thermal reactions of the LDH can be represented as:

$$Mg_6Fe_2(OH)_{16}CO_3{\cdot}4H_2O \rightarrow Mg_6Fe_2(OH)_{16}CO_3 + 4H_2O\uparrow \quad 20\text{–}220\ ^\circ C \tag{1}$$

$$Mg_6Fe_2(OH)_{16}CO_3 \rightarrow Mg_6Fe_2O_8(OH)_2 + 7H_2O\uparrow + CO_2\uparrow \quad 220\text{–}400\ ^\circ C \tag{2}$$

$$Mg_6Fe_2O_8(OH)_2 \rightarrow MgFe_2O_4 + 5MgO + H_2O\uparrow \quad 400\text{–}900\ ^\circ C \tag{3}$$

Figure 2b shows the weight loss profiles associated with the thermal decomposition of LDH, which are similar for all solids. In the first thermal decomposition reaction the solid LDHM1F reaches a weight loss of 21%, the solid LDHM2F a loss of 12.3%, and the solid LDHM3F a loss of 14.6%, which can be attributed to the desorption of water adsorbed on the porous surface of the solid and the water occluded in the interlaminar space. The second weight loss is attributed to the partial dehydroxylation of the lamellar structure and the elimination of interlaminar carbonate ions; 16.86% for LDHM1F, 17.7% for LDHM2F, and 19.4% for LDHM3F. Finally, the final weight loss occurs when the solid is dehydroxylated completely, resulting in the collapse of the laminar structure with the subsequent segregation of the oxide phases, with losses of 7.72% for LDHM1F, 6.86% for LDHM2F, and of 9.69% for LDHM3F. The cumulative loss of transition from LDH precursors to the MgO and $MgFe_2O_4$ oxides was 45.58%, 36.86%, and 43.69% for LDHM1F, LDHM2F, and LDHM3F, respectively.

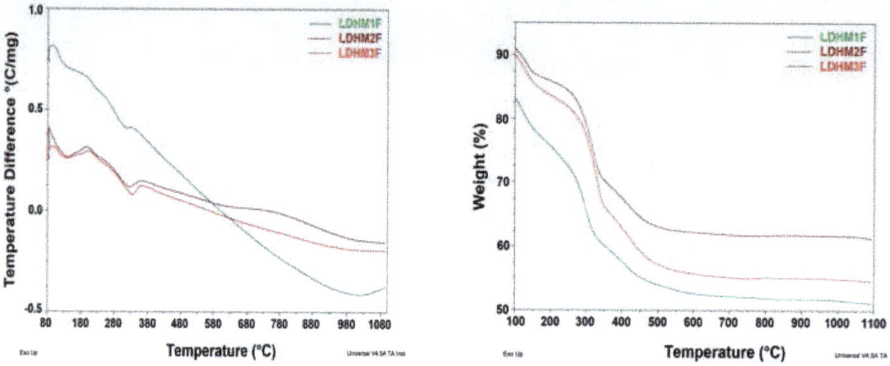

a) b)

Figure 2. Thermal analysis of Mg/Fe layered double hydroxides. (**a**) Differential thermal analysis (DTA) and (**b**) thermogravimetric analysis (TGA).

2.3. *Fourier Transformed Infrared Spectrocopy (FTIR)*

The FTIR spectra of the LDH calcined at 400 °C are shown in Figure 3. The signals are similar for all solids. They present a band centered on 3550 cm^{-1} that corresponds to the hydrogen bridge vibrations of the OH–OH_2 and H_2O–OH_2 types of the hydroxyl and water molecules remaining. A signal at 1640 cm^{-1} corresponds to the H–OH vibration of the water. The band at 1360 cm^{-1} is attributed to the carbonate ions remaining in the structure. The bands in the range of 750 to 500 cm^{-1} are attributed to the metal-oxygen-metal stretch; specifically, the vibration frequency of 590 cm^{-1} is attributed to the Fe–OH bond, the band at 630 cm^{-1} corresponds to the vibration of O–Fe–O, and the band at 648 cm^{-1} to the Mg–OH vibration.

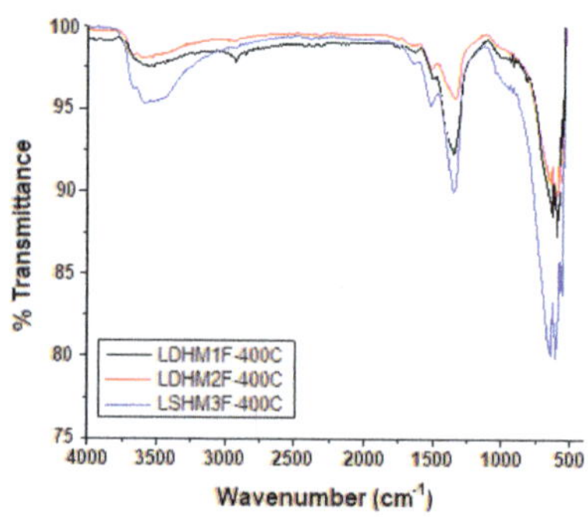

Figure 3. Fourier transformed infrared spectra of activated Mg/Fe layered double hydroxides.

2.4. Textural Analysis

The N_2 adsorption-desorption isotherms of LDH calcined at 400 °C are shown in Figure 4, showing that for all cases, type IV isotherms corresponding to mesoporous materials are presented according to the IUPAC classification. The isotherms at high values of relative pressure (P/Po) did not show a horizontal tendency, which indicates that the nitrogen physisorption took place between the aggregates of particles that have a laminar morphology. Complementarily, the hysteresis cycles of the type H3 can be observed, which indicate the presence of pores of asymmetric size and asymmetrical shape.

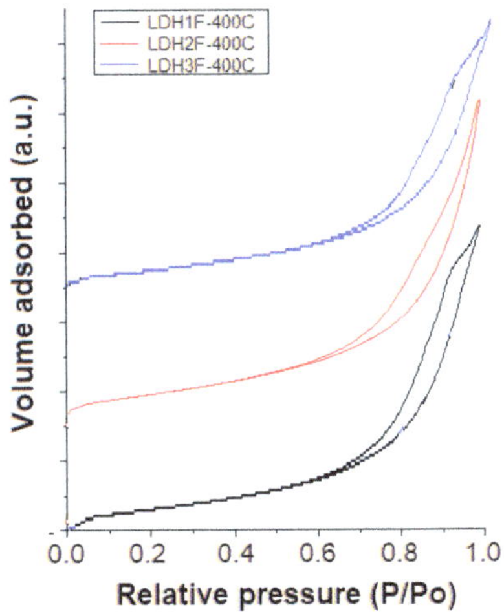

Figure 4. N_2 physisorption isotherms of activated Mg/Fe layered double hydroxides.

Table 1 shows the values of the specific areas determined by the Brunauer–Emmett–Teller (BET) method for the physisorption isotherms of N_2 of activated Mg/Fe layered double hydroxides, as well as the average pore diameter calculated by the Barrett, Joyner, and Halenda (BJH) method and the volume of the pores. It is observed that when the Mg/Fe ratio increases, there is a tendency to decrease the surface area, which is associated with the increase in the crystallinity of the solids and with the respective decrease in the pore size of the solids. The increase of area in the solids is because when the materials are calcined, the double structures collapse, and the pores of the tubular material expand (shape of the isotherm). The LDH with the highest specific area is the HTM1F-400 °C catalyst, which reaches specific areas close to 300 m^2/g, which will favor the contact area of the photocatalyst with the pollutant.

Table 1. Textural properties of activated Mg/Fe layered double hydroxides.

Catalyst	BET Area (m^2/g)	Pore Diameter (nm)	Pore Volume (cm^3/g)
LDHMIF-400 °C	282.2	8.72	0.0178
LDHM2F-400 °C	253.1	7.34	0.0391
LDHM3F-400 °C	248.9	6.30	0.0136

2.5. Scanning Electron Microscopy (SEM) with Energy-Dispersive X-Ray Spectroscopy (EDS)

Figure 5 shows the scanning microphotographs of activated layered double hydroxides. In the case of the solids LDHM2F-400 °C and LDHM3F-400 °C, crystals of heterogeneous size with numerous edges can be observed, in which the structure of stacked sheets can be seen, with crystallinity and very similar particle size, which will favor catalytic capacity for both solids. For the solid LDHM1F-400 °C, the morphology looks similar, however, the particles are larger with the formation of aggregates of larger crystals, which hinders access to the active sites of the photocatalyst.

c)

Figure 5. Scanning micrograph of activated layered double hydroxides at 5000x. (**a**) LDHM1F-400 °C, (**b**) LDHM2F-400 °C, (**c**) LDHM3F-400 °C.

The elemental composition of the surface of the activated LDH particles is shown in Figure 6, which contains the X-ray scattering energy spectra. As can be seen, the three solids have the same elements: Mg, Fe, and O, corresponding to the MgO and the spinel $MgFe_2O_4$, but with different elemental molar ratios. Only in the case of solid LDHM1F-400 °C is a small amount of Na present, which was trapped in the LDH network at the time of synthesis. Regarding the metal molar ratio of Mg/Fe on the surface, this was 1, 1.97, and 2.1 for LDHM1F-400 °C, LDHM2 F-400 °C, and LDHM3F-400 °C, respectively, which for the first two solids corresponds to the theoretical molar ratio, while for the third solid a significant decrease of the same is observed.

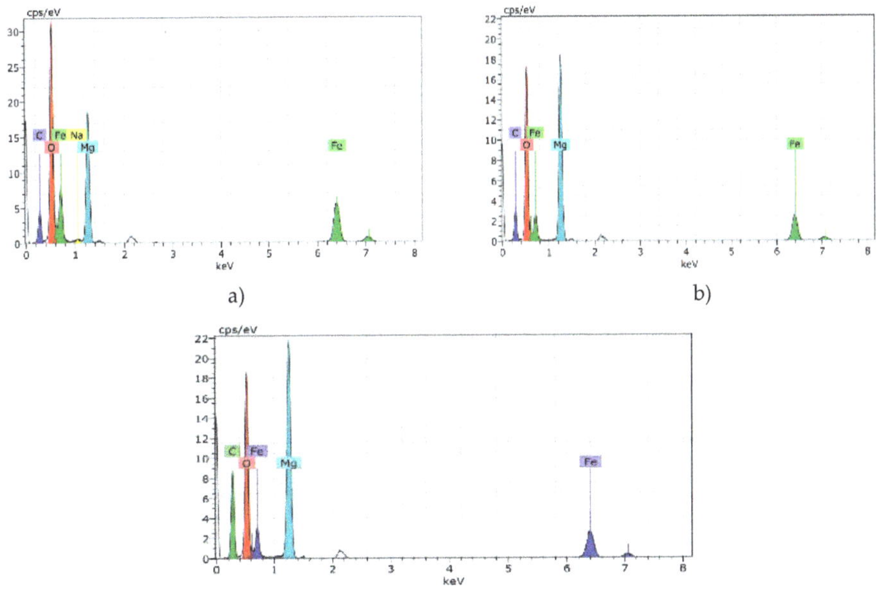

c)

Figure 6. Energy-dispersive X-ray spectroscopy of activated layered double hydroxides at 5000x. (**a**) LDHM1F-400 °C, (**b**) LDHM2F-400 °C, (**c**) LDHM3F-400 °C.

2.6. *Diffuse Reflectance Spectroscopy (DRS)*

The evaluation of the band gap energy (Eg) for activated LDH was calculated using the Kubelka–Munk equation [$F(R) = (1-R)^2/2R$], where R is the converted reflectance (%) of the UV adsorption spectra, and these are reported in Figure 7. It can be observed that the values increase from 2.28 eV to 2.46 eV as a function of the Mg/Fe ratio, associated with the Fe^{3+} content. These results show that the content of Fe^{3+} in activated LDH materials modifies the semiconductor properties of solids due to a decrease in bandgap values by increasing the amount of Fe^{3+}, reaching values lower than 3.20 eV, corresponding to TiO_2-P25.

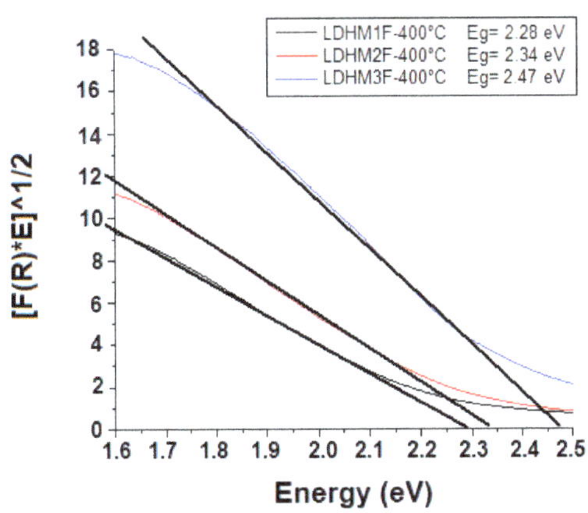

Figure 7. UV-vis-KM (Kubelka–Munk Method) spectra of activated Mg/Fe layered double hydroxides.

2.7. Photocatalytic Degradation of 2,4,6-Trichlorophenol

Figure 8 shows the ultraviolet-visible spectra of the degradation of a solution with a concentration of 80 parts per million (ppm), equivalent to 80 mg/L, of 2,4,6-trichlorophenol as a function of time, using activated LDH catalysts and TiO_2 at 400 °C as a reference control. The TiO_2-P25 catalyst is used as a reference, since at the level of research and industrial application, it is used for the degradation of water polluting compounds due to its excellent stability, non-toxicity, being a semiconductor material with high photo-oxidation power, and low cost [19,36,38]. The 2,4,6-trichlorophenol spectra show three characteristic absorption bands, with the primary transition $\pi \rightarrow \pi^*$ assigned to the aromatic group between 208 and 220 nm, the secondary transition $\pi \rightarrow \pi^*$ at 243 nm due to the aromatic group, and the transition $n \rightarrow \pi^*$ that is attributed to the C–Cl link located at 311 nm [33,39]. In the case of activated LDH (Figure 8a–c), it can be observed that as time passes, the intensity of the three bands decreases, which is associated with the degradation of the 2,4,6-trichlorophenol molecule, until the complete disappearance of the bands of the secondary transition $\pi \rightarrow \pi^*$ at 243 nm occurs due to the aromatic group and the transition $\pi \rightarrow \pi^*$ of the C–Cl bond, as well as the decrease until almost the disappearance of the primary transition $n \rightarrow \pi^*$ assigned to the aromatic group at 240 min.

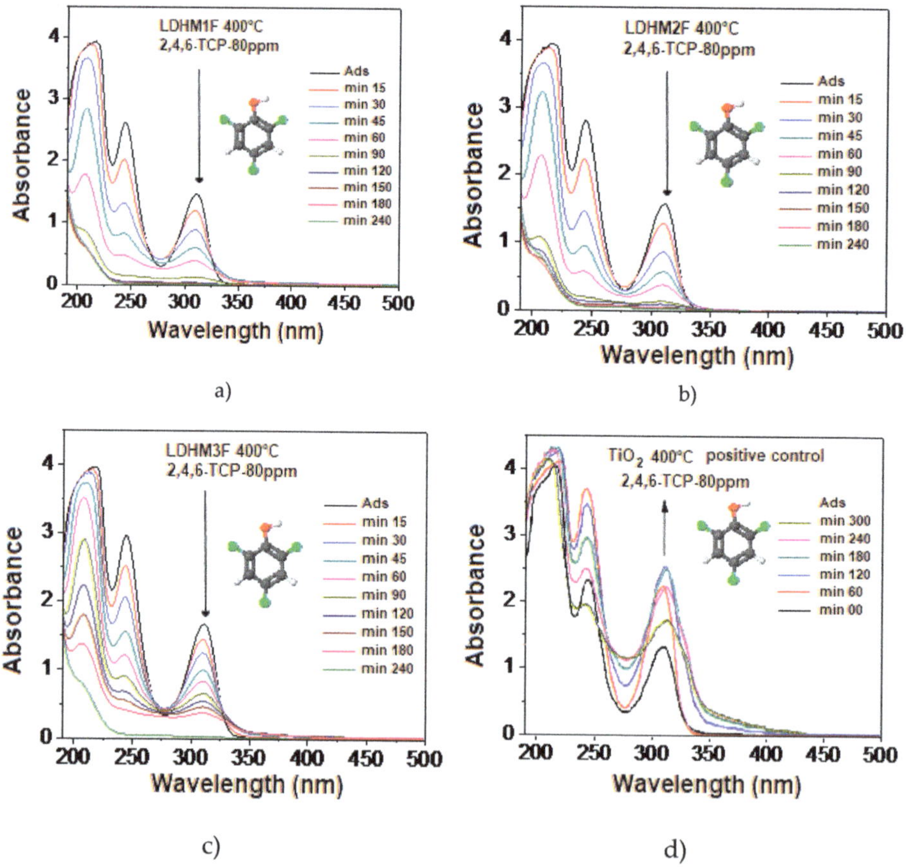

Figure 8. UV-vis spectra of the photocatalytic degradation of 2,4,6-TCP using Activated Lamellar Double Hydroxides (**a**) LDHM1F-400 °C, (**b**) LDHM2F-400 °C, (**c**) LDHM3F-400 °C, and (**d**) TiO_2 positive control.

In the case of the LDHM1F and LDHM2F catalysts, the decrease of the bands is very similar as a function of time, reaching degradation values higher than 90% at 90 min, while the LDHM3F catalyst at 90 min has only about 50% degradation, which is attributed to its band gap values of 2.34 and 2.47 eV. This is favored by the presence of Fe in the catalysts, requiring 240 min to reach values greater than 90% degradation of the entire organochlorine molecule. For TiO_2 used as a reference catalyst as a photocatalyst in the degradation of 2,4,6-trichlorophenol (Figure 8d), the intensity of the absorbance increases with the irradiation time, which is the opposite of the behavior of activated Mg/Fe LDH catalysts. The increase in the intensity of the signals is because the degradation of TiO_2 produces intermediates, which have higher absorptivity coefficients (This coefficient is proportional to the intensity in the absorbance), modifying the characteristic signals of 2,4,6-tetrachlorophenol at 245 and 310.5 nm. The intermediaries that are formed are mainly catechols, as well as benzoquinones and hydroxyquinones; all of them are compounds that preserve the aromatic ring increasing the signals between 200 and 225 nm.

Figure 9 shows the graph of the relative degradation rate of 80 ppm of 2,4,6-trichlorophenol with different catalysts. It can be corroborated that the photocatalysts show a good degradation of 2,4,6-trichlorophenol at 90 min, reaching 93% for LDHM2F-400 °C, 92% for LDHM1F-400 °C, and 55% for LDHM3F-400 °C—values higher than the 18% degradation of TiO_2-P25-400 °C. For the catalysts LDHM1F-400 °C and LDHM2F-400 °C, the degradation behavior as a function of time is very similar, only showing a small difference at the beginning of the degradation process, whereas the catalyst of Mg/Fe = 1 ratio at 15 min has degraded the pollutant more quickly, but at 30 min, both catalysts reach the same speed, which at 90 min is slightly higher for the ratio catalyst Mg/Fe = 2.

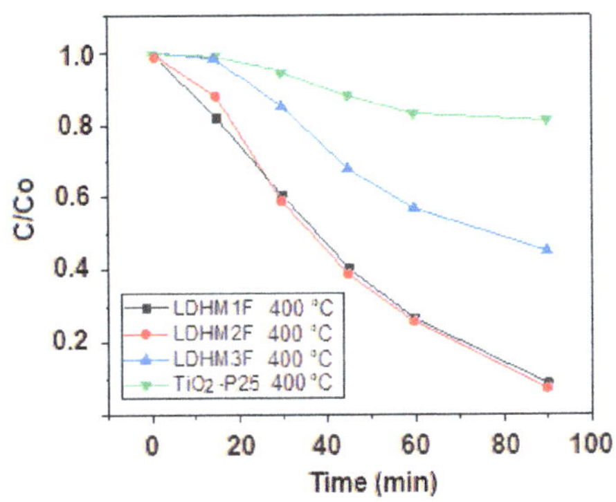

Figure 9. Relative photodegradation rate of 80 ppm of 2,4,6-trichlorophenol.

On the other hand, TiO_2 shows zero activity during the first 15 min, and then the degradation of the pollutant begins at a low speed, which allows it to reach 14% degradation at 60 min after the degradation process has begun, and only 18% at 90 min, showing a decrease in rate.

2.8. Kinetic Model Adjustment Study

Figure 10 shows the adjustment of the kinetic model of the relative degradation rate of 80 ppm of 2,4,6-trichlorophenol with the different catalysts, which corresponds to a pseudo-first-order degradation behavior in all cases—corresponding to the kinetic model of Langmuir–Hinshelwood, which is used to describe many photocatalytic reactions where the rate follows the kinetics of the pseudo-first-order considering steady state conditions [38,40].

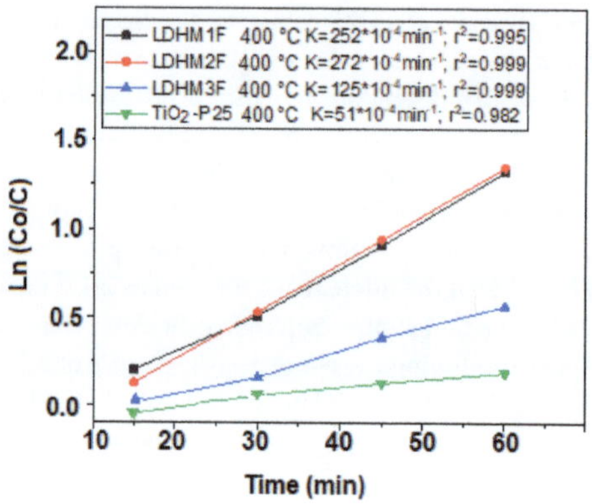

Figure 10. Pseudo-first-order degradation kinetics for 80 ppm of 2,4,6-trichlorophenol with activated layered double hydroxides and TiO_2-P25.

As seen in the Figure 10, the velocity constantly increases when the Mg/Fe molar ratio decreases, which is attributed to the role that Fe^{3+} plays in decreasing the bandgap value and allowing the photodegradation process to be carried out faster for the catalysts LDHM1F-400 °C and LDHM2F-400 °C, with the constant of apparent speed greater than 250×10^{-4} min^{-1} and half-lives of 25 min. These are much higher than those of TiO_2 of 51×10^{-4} min^{-1}, with average lifetimes of 136 min.

2.9. Mineralization Study

The analysis of organic carbon (TOC) is shown in Figure 11, where it can be seen that the degradation of 2,4,6-trichlorophenol reaches values close to 80% for the catalysts LDHM1F-400 °C and LDHM2F-400 °C, reaching a greater mineralization using the LDHM2F-400 °C catalyst at 120 min, with values of 82%, which are higher than those reached by LDHM3F-400 °C and TiO_2-P25 of 61 and 29% at 120 min, respectively.

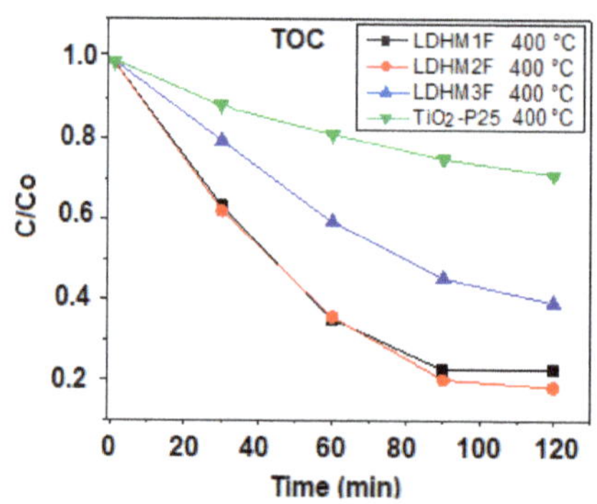

Figure 11. TOC removal during photodegradation of a 2,4,6-trichlorophenol solution.

Table 2 shows the values of the band gap, apparent speed constant, average flow time, percentage of degradation, and percentage of mineralization of the catalysts used in the degradation of 80 ppm of 2,4,6-trichlorophenol.

Table 2. Bandgap energy, specific area, apparent kinetic constant, half-life, percentage of degradation, and percentage of mineralization of activated layered double hydroxides.

Catalyst	Bandgap (eV)	$K_{app}*10^{-4}$ (min^{-1})	$t_{1/2}$ (min)	%Degradation (UV-Vis)	%Mineralization (TOC)
LDHMIF-400 °C	2.28	252	28	92	78
LDHM2F-400 °C	2.34	272	25	93	82
LDHM3F-400 °C	2.47	125	56	55	61
TiO_2-P25-400 °C	3.20	51	136	18	29

As can be seen, the LDHM1F and LDHM2 catalysts have very similar behaviors, however, the LDHM2F catalyst has a greater apparent speed constant, as well as a half-life of less than 25 min, 93% degradation and mineralization of 82%, which is why it can be considered as the catalyst with better photodegradation capacity for 2,4,6-trichlorophenol. This allowed LDHM2F-400 °C to be chosen to carry out studies on the possible mechanism of this photocatalytic reaction.

2.10. Detection of Hydroxyl Radicals ($OH^{\bullet}$)

In order to explore the mechanism of photocatalytic degradation, the possible formation of radicals was analyzed by fluorescence spectroscopy. For the detection of said radicals, coumarin was used as a molecule that captures the hydroxyl radicals, producing the hydroxyproduct 7-hydroxycoumarin [41], as shown in Figure 12.

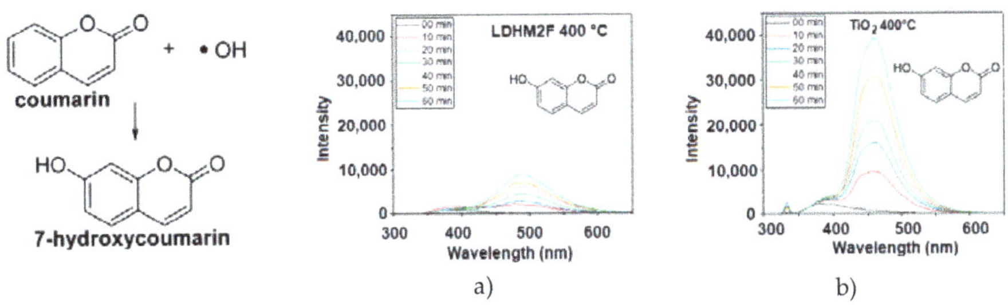

Figure 12. Fluorescence spectra of coumarin employed in the detection of $OH^{\bullet}$ generated by a photocatalyst without air bubbling: (**a**) LDHM2F-400 °C and (**b**) TiO_2-P25-400 °C.

The fluorescence spectra obtained with hydroxycoumarin exhibit a characteristic band of emission between 400 and 600 nm. In the case of the activated LDHM2F catalyst, the intensity of the signal is low, which implies that there is little presence of $OH^{\bullet}$ radicals associated with the scarce formation of these in 60 min of reaction. On the other hand, in the case of the reference catalyst TiO_2-P25, an intense band associated with the production of hydroxycoumarin can be observed, due to the ability of the reference catalyst to produce hydroxyl radicals [42,43]. The above suggests that the activated LDHM2F catalyst does not favor the formation of $OH^{\bullet}$ radicals.

2.11. Photocatalytic Evaluation of Superoxide Radicals ($O_2^{-\bullet}$)

As observed in Equation (4), when a semiconductor material is irradiated with energy greater than or equal to that of its forbidden bandwidth, the electrons are photoexcited from the valence band to the conduction band, with the probability of migrating to the surface of the semiconductor and reacting with bubbled oxygen (Equation (5)) during the reaction to produce superoxide radicals, which, like the hydroxyl radicals, have great oxidizing power for degrading the organic compound [44].

$$h\upsilon + \text{photocatalyst} \rightarrow e^- + h^+ \tag{4}$$

$$O_2 + e^- \rightarrow O_2^{-\bullet} \quad (5)$$

Using this photocatalytic foundation as a model to determine the formation of superoxide radicals in the process, the UV-Vis spectra were obtained in the photodegradation of 2,4,6-trichlorophenol. How the process develops in the presence of bubbled oxygen can be compared with a process with an absence of bubbling oxygen, where instead nitrogen gas is emitted, which is inert in the processes of photodegradation [45], as shown in Figure 13.

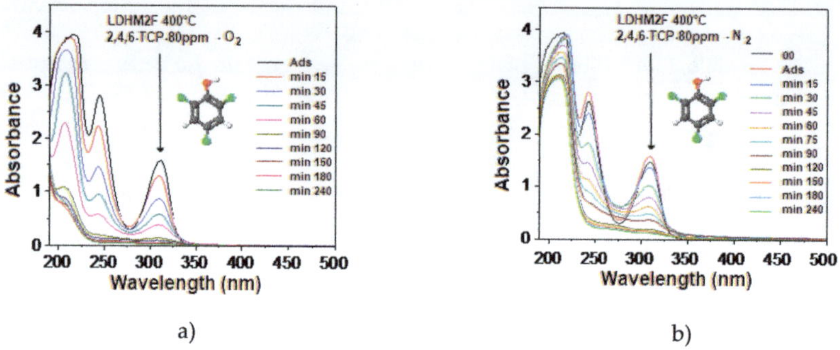

Figure 13. UV-Vis spectra in the photodegradation of 246-trichlorophenol in (**a**) the presence and (**b**) the absence of oxygen.

As can be seen, in the presence of O_2 the catalyst shows a good degradation capacity of 2,4,6-trichlorophenol, achieving the elimination of the signals of the vibrations of the functional groups C–Cl and C–OH, as well as a decrease of the CC and CH vibrations of the aromatic ring at 240 min. On the other hand, in the presence of N_2 it is observed that the catalyst is active in the partial degradation of the 2,4,6-trichlorophenol molecule, showing an almost total decrease of the vibrations of the functional groups C–Cl and C–OH, but only a small decrease in the CC and CH vibrations of the aromatic ring at 240 min.

As seen in Figure 14, the photocatalytic activity decreases by about 35% for the elimination of the OH and Cl functional groups in the absence of bubbled oxygen, according to the values of the apparent kinetic constants. The above confirms that the generation of superoxide radicals is determinant for the elimination of the functional groups, as well as for the degradation of the aromatic ring for the complete photodegradation of 2,4,6-trichlorophenol. The partial degradation of the 2,4,6-trichlorophenol molecule in the absence of oxygen suggests the participation of photogenerated voids (h^+) in the catalyst.

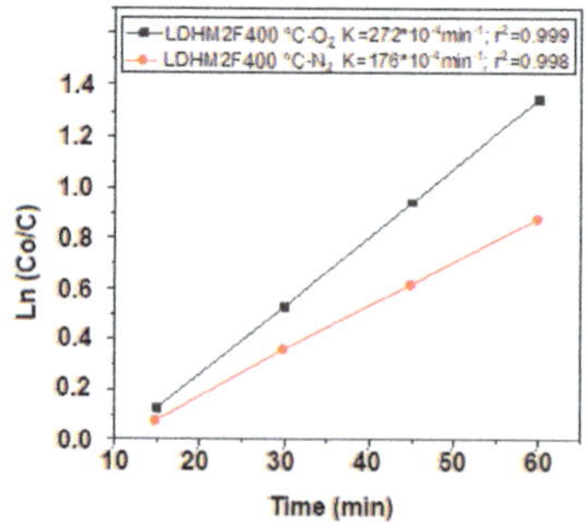

Figure 14. Kinetic constants in the photodegradation of 2,4,6-trichlorophenol in the presence and absence of oxygen.

2.12. Photocatalytic Evaluation of the Hole Trap (h^+)

The evaluation of the photocatalytic activity by gauging holes was evaluated using a model molecule of ammonium oxalate as a sacrificial agent [46–49]. Figure 15 shows the UV-Vis absorption spectra of the photodegradation of 2,4,6-trichlorophenol in the presence of ammonium oxalate with oxygen and nitrogen flow.

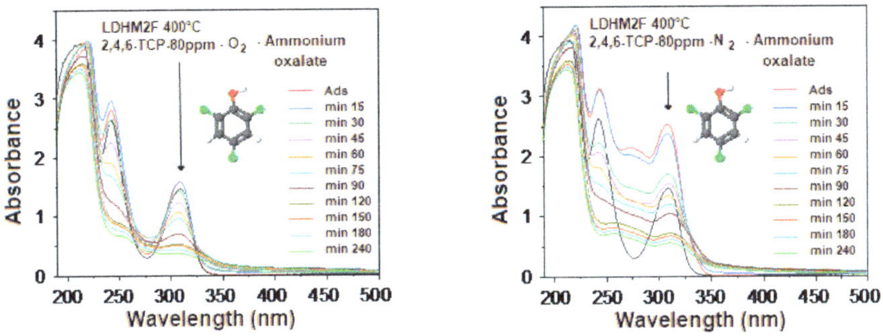

a) b)

Figure 15. Absorption of UV-Vis for the photodegradation of 2,4,6-trichlorophenol in the presence of ammonium oxalate as a hole capture agent: (**a**) in the presence of oxygen; (**b**) in the presence of nitrogen.

According to the adsorption spectra, in the case of the photodegradation of 2,4,6-trichlorophenol in the presence of oxygen and ammonium oxalate, a reduction in the degradation of both the functional groups and the aromatic ring is observed. This confirms that the gaps that are formed in the LDHM2F-400 °C catalyst contribute in a significant way to the catalytic activity. On the other hand, in the case of the process of photodegradation in the absence of oxygen and in the presence of ammonium oxalate, it can be observed that the photodegradation process is completely modified, with the formation of a new band at the center at 275 nm. This is attributed to the formation of intermediaries which are identified by the signal of 2-chlorophenol, as well as a phase shift of the aromatic band attributed to the formation of phenol [27]. The above confirms that the catalytic activity of LDHM2F depends on the presence of voids, as well as the formation of superoxide radicals.

Table 3 shows the comparison of the apparent kinetic constants, percentage of degradation, and percentage of mineralization of the evaluation of the active species involved in the photodegradation of 2,4,6-trichlorophenol.

Table 3. Apparent kinetic constant, percentage of degradation, and percentage of mineralization of activated layered double hydroxides.

Catalyst	K_{app}*10^{-4} (min^{-1})	%Degradation (UV-Vis)	%Mineralization (TOC)
LDHM2F-400 °C with O_2	252	91	77
LDHM2F-400 °C with N_2	176	76	66
LDHM2F-400 °C-O_2 with AO	87	47	20
LDHM2F-400 °C with N_2-AO	83	44	18

The catalytic activity related to the degradation of 2,4,6-trichlorophenol is decreased by more than 50% in the presence of ammonium oxalate and oxygen but is even more decreased in the absence of oxygen. As for mineralization, this is reduced by 77% in the absence of oxygen and in the presence of ammonium oxalate, which confirms that the degradation is partial, with the formation of intermediaries and not until mineralization has occurred. This behavior indicates that both the superoxide radicals and the holes (h^+) are the determinants, and both species contribute to the mechanism of photodegradation until the mineralization of 2,4,6-trichlorophenol using the LDHM2F-400 °C catalyst. The results in Table 3 show that the oxidation of 2,4,6-trichlorophenol occurs first in the holes (h^+) photogenerated in the double-layered hydroxides under reconstruction and subsequently the intermediates are removed by further oxidation with superoxide radicals or other oxygen oxidizing species.

2.13. *Possible 2,4,6-Trichlorophenol Degradation Mechanisms*

Figure 16 shows a diagram of the proposed mechanism for the degradation of 2,4,6-trichlorophenol based on the properties of the catalysts and the different types of free radicals involved. As could be demonstrated in the fluorescence test for the detection of the generation of superoxide radicals, this is the main route involved in the degradation of the pollutant and, to a lesser extent, the holes. As observed in coumarin tests, photocatalysts do not favor the formation of hydroxyl radicals, so these free radicals do not contribute to the mechanism. Magnesium has a Pauling electronegativity of 1.31 eV and iron of 1.83 eV, so iron has a greater tendency to attract electrons to itself and transfer them to O_2 to generate the superoxide radical.

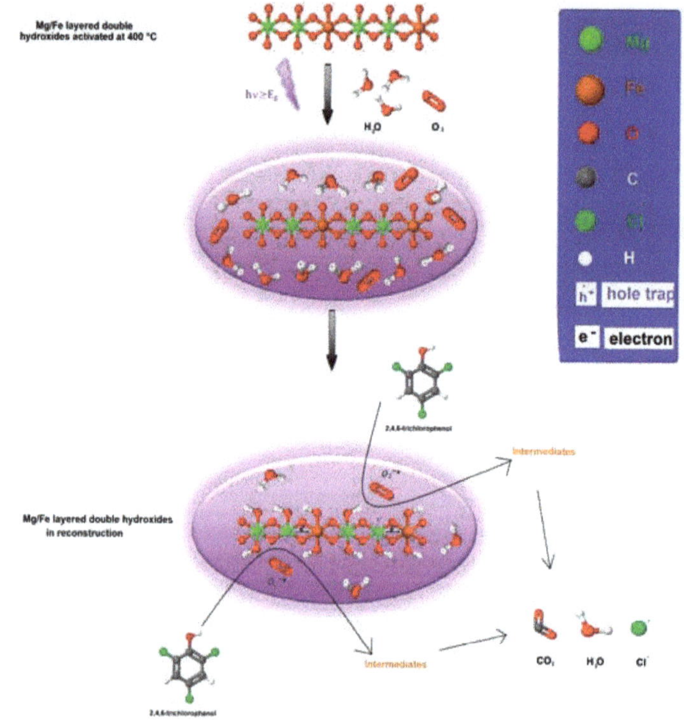

Figure 16. Possible routes for the degradation/mineralization of 2,4,6-trichlorophenol by activated layered double hydroxides.

As proposed in the possible degradation routes, this occurs when the activated layered double hydroxides, which are in the form of simple and mixed oxides, meet the aqueous medium with oxygen, which when irradiated produce an excess of electrons in the conduction band and positive gaps in the valence band, as shown in Equation (6).

$$LDHMF\text{-}400\ °C + hv \rightarrow \mathrm{h}^+ + e^- \quad (6)$$

On the other hand, electrons over the conduction band react with molecular oxygen, which acts as an electron acceptor to form superoxide radicals (Equation (7)).

$$e^- + O_2 \rightarrow O_2^{-\bullet} \quad (7)$$

The generated superoxide radicals can attack and oxidize 2,4,6-trichlorophenol (Equation (8)).

$$O_2^{-\bullet} + \mathit{2,4,6\text{-}trichlorophenol} \rightarrow \text{intermediates} \rightarrow CO_2 + H_2O \quad (8)$$

The final LDHMF catalytic precursor is recovered thanks to the memory effect, a property characteristic of the laminar double hydroxide materials [50–52]. Said precursor can be reactivated for a second cycle as a catalyst for the degradation of 2,4,6-trichlorophenol.

The degradation to the mineralization takes place thanks to the presence of superoxide radicals accompanied by the holes until the disappearance of the total organic carbon [32,53–56], implying a mechanism in which the superoxide radicals are carrying out the degradation of the aromatic ring simultaneously with the elimination of chlorine, since the formation of monochlorinated intermediates or phenols is not observed by the ultraviolet spectrum.

The catalyst with the highest photocatalytic activity is LDHM2F, whose pore volume value is higher compared to the LDHM1F-400 °C and LDHM3F-400 °C catalysts. This may indicate that there is enough separation between the laminar structures of the catalysts, promoting an improvement in the transfer of charges (e^- and h^+) by decreasing the recombination and, therefore, increasing the photocatalytic activity.

In the analysis by SEM, it is observed that the particle size of the LDHM2F-400 °C catalysts is lower than that of LDHM1F-400 °C, which indicates that, by the synthesis method used, the nucleation is different with the addition of Fe to the LDH. Being smaller particles, a lower percentage of conglomerates is obtained, suggesting that Fe increases directly in the LDH by modifying its electronegativity in such a way that a better transfer of charges is obtained, and with this the photocatalytic activity increases.

The main advantages of the activated layered double hydroxides are that they are not toxic, they can regenerate for continuous use, and they can be synthesized in an easy way, and without needing to be doped.

3. Materials and Methods

3.1. Obtaining Photocatalysts

3.1.1. Synthesis of Mg/Fe-Layered Double Hydroxide Catalytic Pre-Cursors of Ratios 1, 2, and 3

Layered double hydroxides with different Mg/Fe molar ratios were obtained by the coprecipitation method. The stoichiometric amounts of magnesium nitrate [Mg $(NO_3)\cdot 6H_2O$], (Fermont, Lopez Mateos, Mexico), and iron nitrate [Fe $(NO_3)_3\cdot 9H_2O$] (Fermont, Lopez Mateos, Mexico)were dissolved in water and subsequently coprecipitated at a pH of 11.5 with sodium hydroxide (NaOH) (Fermont, Lopez Mateos, Mexico). The coprecipitate was left and aged for 24 h, washed until it reached a pH of 9, and dried at 100 °C. The materials were identified as LDHM1F, LDHM2F, and LDHM3F, corresponding to the Mg/Fe = 1, 2, and 3 ratios.

3.1.2. Activation of Layered Double Hydroxides

The LDHs that were synthesized were thermally treated at 400 °C in an air atmosphere for 4 h for the formation of the mixed Mg/Fe oxides and were identified as LDHM1F-400 °C, LDHM2F-400 °C, and LDHM3F-400 °C.

3.2. *Physicochemical Characterization of Catalytic Pre-Cursors and Catalysts*

The materials were characterized by physicochemically fresh and calcined. To corroborate the present crystalline phase, X-ray diffraction was performed on a Siemens D500 powder diffractometer (University of Guanajuato, Guanajuato, Mexico) with CuKα radiation, a graphite crystal monochromator with a step of 2θ equal to 0.02 s^{-1}, and a counting time of 9 s per point. The thermal analyses were carried out in a TA Instruments Thermoanalyzer (University of Guanajuato, Guanajuato, Mexico) at a heating rate of 10 °C/min in an air atmosphere, at a speed of 100 mL/min and using α-alumina as a reference. The infrared spectra with Fourier transform were obtained in a Bruker model Tensor 27 Hyperion 7000 (University of Guanajuato, Guanajuato, Mexico) with a resolution of 4 cm^{-1} and 20 scans, from 400 to 4000 cm^{-1}, and the samples were prepared by dilution with KBr. The nitrogen adsorption-desorption isotherms were determined in TriStar Micromeritics equipment (University of Guanajuato, Guanajuato, Mexico), with the samples previously dried at 100 ° C for 24 h, with a degassing at 150 °C, and a vacuum pressure of up to 0.5 mmHg. The BET model was used to obtain the specific area. Scanning electron microscopy analysis was performed on a Seizz 1550VP microscope with field emission, with Oxford EDS equipment (University of Guanajuato, Guanajuato, Mexico) with an opening of 30.0 μm and a width of 3.1 mm. The band gap energy of the photocatalysts was calculated from the ultraviolet absorption spectra coupled with an integrating sphere (diffuse reflectance) in a UV-Vis Recording S UV2401 Shimadzu instrument (University of Guanajuato, Guanajuato, Mexico), and using the Kubelka–Munk theory, which consists of plotting the energy of the photon against the square root of the Kubelka–Munk function multiplied by the energy of the photon and extrapolating the linear part with the abscissa axis.

3.3. *Evaluation of the Photocatalytic Activity in the Degradation of 2,4,6-Trichlorophenol*

3.3.1. Degradation of 2,4,6-Trichlorophenol

The photocatalytic degradation capacity of 2,4,6-trichlorophenol using activated LDH was determined under the following conditions: standard 200 ml solution at a concentration of 80 ppm of 2,4,6-trichlorophenol (Sigma-Aldrich, St. Louis, MO, USA), in a Batch reactor at a controlled temperature of 25 °C with constant magnetic stirring of 700 rpm, with a flow of air of 2 mL/s and a UV light irradiation of 254 nm, and an emission of 2.5 mW/cm^2 generated by a Pen-Ray UV lamp inserted in a tube of quartz. The photoactivity of all materials was determined by using 1 g/L of photocatalyst. To determine the adsorption phenomena effect, the suspension was stirred at 800 rpm under airflow in dark conditions for 1 h; after that, an aliquot was collected, and the reaction was started by turning on the UV lamp that was kept on for 240 min while stirring. Additionally, a sample of 2,4,6-trichlorophenol solution was subjected to photolysis in the absence of any catalyst to determine the effect of the radiation on the contaminant. In addition, for the comparison against a reference photocatalyst, commercial TiO_2-P25 was used. In all cases, the process of degradation of 2,4,6-trichlorophenol was monitored using an aliquot sample of the reactor, with the subsequent quantification of this in a UV-Vis Cary 100 spectrophotometer (Metropolitan Autonomous University, Mexico City, Mexico) at a wavelength of 310.5 nm. The aliquots were collected at 15-minute intervals for 240 min, and in order to separate the solid catalyst from the aqueous solution, the aliquots were filtered using a nitrocellulose membrane with a pore size of 0.22 μm (Millipore Corporation, Burlington, Massachusetts, Estados Unidos). The amount of total organic carbon present in the irradiated solution was determined in a TOC-V-CSH/CSN Shimadzu 5000 TOC (Metropolitan Autonomous University, Mexico City, Mexico) to corroborate the degradation of the molecule until mineralization.

3.3.2. Detection of Hydroxyl and Superoxide Radicals and Study of the Hole Trap

A dilute aqueous solution (2×10^{-3} M) of coumarin (Sigma-Aldrich, St. Louis, MO, USA) was prepared, in which 200 mL of solution and 200 mg of photocatalyst to be analyzed were poured into a Batch reactor. TiO_2-P25 was used as the reference material. As above, the study was carried out in the absence of a photocatalyst (Photolysis) to observe the possible production of hydroxyl radicals with and without a photocatalyst. Once this was done, the solution was irradiated with UV light for 1 h with a Pen-Ray lamp (λ = 254 nm and I0 = 4.4 mWcm^{-2}) under conditions of constant stirring, oxygen bubbling, and temperature (800 rpm, 1 mL/s, and 25 °C, respectively), extracting 3 mL aliquots at 10 min intervals. Finally, the fluorescence emission spectra in the irradiated solution were analyzed by photoluminescence in a Fluorescence spectroscopy Scinco FS-2 spectrometer (Metropolitan Autonomous University, Mexico City, Mexico), a wavelength of 320 nm, and the results obtained with the synthesized materials and photolysis were compared with those obtained using the reference solid TiO_2-P25.

3.3.3. Photocatalytic Evaluation of Hole Capture with Ammonium Oxalate as a Sacrificial Agent

The photocatalytic reaction was monitored with the material with the highest photodegradation (HTM2F-400 °C) in the presence of ammonium oxalate (Sigma-Aldrich, St. Louis, MO, USA), and with oxygen and nitrogen flow. An aqueous solution of 246-TCF at 80 ppm was used, and 1 g/L of the photocatalyst with higher activity and a concentration of 0.004 M of ammonium oxalate were poured into a reactor. The solution was irradiated with UV light, maintaining the conditions of constant stirring, bubbling of nitrogen/oxygen, and temperature (800 rpm, 1 mL/s, and 25 °C, respectively), while extracting aliquots of 3 mL to be analyzed by UV-Vis spectroscopy.

4. Conclusions

The results of the present work show that layered double hydroxides of different Mg/Fe molar ratio can be synthesized by the coprecipitation method, which have a hydrotalcite crystalline structure, and which when calcined at 400 °C evolve to a crystalline phase of periclase and an amorphous phase of spinel, with Type IV isotherms characteristic of mesoporous materials and specific areas between 273.9 and 248.9 m^2/g. Layered double hydroxides activated at 400 °C have Eg values ranging from 2.28 to 2.47, which are lower values than TiO_2, so they have better semiconductor properties. These photocatalysts have good photocatalytic activity, with degradation efficiencies for 2,4,6-trichlorophenol of 93% LDHM2F-400 °C > 92% LDHM1F-400 °C > 55% LDHM3F-400 °C > 18% TiO_2-P25- 400 °C in less than 90 minutes. Therefore, it is concluded that the most active catalysts are LDHM2FI-400 °C and LDHM1F-400 °C, which present a greater amount of iron in the structure and which reach mineralization values of up to 82%. The degradation mechanism proposed for the photodegradation of 2,4,6-trichlorophenol considers that the degradation occurs first in the photogenerated holes in the layered double hydroxides under reconstruction and then the intermediates are removed by further oxidation with the superoxide radicals or other oxidizing species of oxygen. These photocatalysts obtained from Mg/Fe layered double hydroxides are promising for the photodegradation of recalcitrant chlorinated compounds.

Author Contributions: Conceptualization, E.R.-R. and N.G.-O.; methodology, E.R.-R., H.G.M.-C., and J.C.-R.; software, J.C.-R.; validation, F.T.-M.; formal analysis, F.T.-M., H.G.M.-C., and J.C.-R.; investigation, F.T.-M.; resources, H.G.M.-C.; data curation, N.G.-O.; writing—original draft preparation, E.R.-R.; writing—review and editing, N.G.-O. and E.R.-R.; funding acquisition, E.R.-R.

Funding: This research received no external funding. Funding was received from the resources of the University of Guanajuato.

Acknowledgments: We would like to especially thank the Directorate for Research and Postgraduate Support (DAIP) at the University of Guanajuato for their support in developing this project. Also, we thank the University of Guanajuato-CONACyT National Laboratory SEM-EDS. Additionally, we thank the CONACyT for the graduate scholarship and UAM-I for technical support.

Conflicts of Interest: The authors declare no conflict of interest.

References

1. Miyata, S. Physico-chemical properties of synthetic hydrotalcites in relation to composition. *Clays Clay Miner.* **1980**, *28*, 50–56. [CrossRef]
2. Mills, S.J.; Christy, A.G.; Génin, J.M.R.; Kameda, T.; Colombo, F. Nomenclature of the hydrotalcite supergroup: Natural layered double hydroxides. *Mineral. Mag.* **2012**, *76*, 1289–1336. [CrossRef]
3. Roelofs, J.C.A.A.; Bokhoven, J.A.; Dillen, A.J.; Geus, J.W.; Jong, K.P. The Thermal Decomposition of Mg±Al Hydrotalcites: Effects of Interlayer Anions and Characteristics of the Final Structure. *Chem. Eur. J.* **2002**, *8*, 5571–5579. [CrossRef]
4. Nguyen, H.K.D.; Nguyen, H.V.; Nguyen, V.A. Effect of synthetic conditions on the structure of mesoporous Mg-Al-Co hydrotalcite. *J. Mol. Struct.* **2018**, *1171*, 25–32. [CrossRef]
5. Sikander, U.; Sufian, S.; Salam, M.A. A review of hydrotalcite based catalysts for hydrogen production systems. *Int. J. Hydrogen Energy* **2017**, *42*, 19851–19868. [CrossRef]
6. Cavani, F.; Trifiro, F.; Vaccari, A. Hydrotalcite–type anionic clays: Preparation, properties and applications. *Catal Today* **1992**, *11*, 173–301. [CrossRef]
7. Othman, M.R.; Helwani, Z.; Martunus, W.J.N. Synthetic hydrotalcites from different routes and their application as catalysts and gas adsorbents: A review. *Appl. Organomet. Chem.* **2009**, *23*, 335–346. [CrossRef]
8. Yahyaoui, R.; Sanchez, P.E.; Pérez, L.A.; Nahdi, K.; Criado, J.M. Synthesis, characterization and combined kinetic analysis of thermal decomposition of hydrotalcite ($Mg_6A_{12}(OH)_{16}CO_3{\cdot}4H_2O$). *Thermochim. Acta* **2018**, *667*, 177–184. [CrossRef]
9. Vágvölgyi, V.; Palmer, S.J.; Kristóf, J.; Frost, R.L.; Horváth, E. Mechanism for hydrotalcite decomposition: A controlled rate thermal analysis study. *J. Colloid Interface Sci.* **2008**, *318*, 302–308. [CrossRef]
10. Mokhtar, M.; Inayat, A.; Ofili, J.; Schwieger, W. Thermal decomposition, gas phase hydration and liquid phase reconstruction in the system Mg/Al hydrotalcite/mixed oxide: A comparative study. *Appl. Clay Sci.* **2010**, *50*, 176–181. [CrossRef]
11. Parida, K.M.; Sahoo, M.; Singha, S. Synthesis and characterization of a Fe(III)-Schiff base complex in a Zn-Al LDH host for cyclohexane oxidation. *J. Mol. Catal A Chem.* **2010**, *329*, 7–12. [CrossRef]
12. Védrine, J.C. Heterogeneous Catalysis on Metal Oxides. *Catalysts* **2017**, *7*, 341. [CrossRef]
13. Baloyi, J.; Ntho, T.; Mom, J. Synthesis and application of pillared clay heterogeneous catalysts for wastewater treatment: A review. *RSC Adv.* **2018**, *8*, 5197. [CrossRef]
14. Prasad, C.; Tang, H.; Liu, W. Magnetic Fe_3O_4 based layered double hydroxides (LDHs) nanocomposites (Fe_3O_4/LDHs): Recent review of progress in synthesis, properties and applications. *J. Nanostruct. Chem.* **2018**, *8*, 393–412. [CrossRef]
15. Moma, J.; Baloyi, J.; Ntho, T. Synthesis and characterization of an efficient and stable Al/Fe pillared clay catalyst for the catalytic wet air oxidation of phenol. *RSC Adv.* **2018**, *8*, 30115. [CrossRef]
16. Valencia-Lopez, C.D.; Zafra-Calvo, M.; Martín de Vidales, M.J.; Blanco-Gutierrez, V.; Atanes-Sanchez, E.; Merayo, N.; Fernandez-Martinez, F.; Nieto-Marquez, A.; Dos santos-Garcia, J. Synthesis of $NiFe_2O_4$-LDH Composites with High Adsorption and Photocatalytic Activity for Methyl Orange Degradation. *Inorganics* **2018**, *6*, 98. [CrossRef]
17. Li, Z.; Chen, H.; Liu, W. Full-Spectrum Photocatalytic Activity of ZnO/CuO/$ZnFe_2O_4$ Nanocomposite as a PhotoFenton-Like Catalyst. *Catalysts* **2018**, *8*, 557. [CrossRef]
18. Descorme, C. Catalytic wastewater treatment: Oxidation and reduction processes. Recent studies on chlorophenols, Claude Descorme. *Catal Today* **2017**, *297*, 324–334. [CrossRef]
19. Lazar, M.A.; Varghese, S.; Nair, S.S. Photocatalytic Water Treatment by Titanium Dioxide: Recent Updates. *Catalysts* **2012**, *2*, 572–601. [CrossRef]
20. Pi, Y.; Wang, J. Pathway of the ozonation of 2,4,6-trichlorophenol in aqueous solution. *Front. Environ. Sci. Eng. China* **2007**, *1*, 179–183. [CrossRef]
21. Yan, Y.; Wu, X.; Zhang, H. Catalytic wet peroxide oxidation of phenol over Fe_2O_3/MCM-41 in a fixed bed reactor. *Sep. Purif. Technol.* **2016**, *171*, 52–61. [CrossRef]
22. Honda, M.; Kannan, K. Biomonitoring of chlorophenols in human urine from several Asian countries, Greece and the United States. *Environ. Pollut.* **2018**, *232*, 487–493. [CrossRef]
23. Badanthadka, M.; Mehendale, H.M. Chlorophenols. In *Encyclopedia of Toxicology*, 3rd ed.; Elsiever: Amsterdam, The Netherlands, 2014; pp. 896–899. [CrossRef]

24. Mostafalou, S.; Abdollahi, M. Pesticides: An update of human exposure and toxicity. *Arch. Toxicol.* **2017**, *91*, 549–599. [CrossRef]
25. Michałowicz, J. Pentachlorophenol and its derivatives induce oxidative damage and morphological changes in human lymphocytes (in vitro). *Arch. Toxicol.* **2010**, *84*, 379–387. [CrossRef]
26. Khorsandi, H.; Ghochlavi, N.; Aghapour, A.A. Biological Degradation of 2,4,6-Trichlorophenol by a Sequencing Batch Reactor. *Environ. Process.* **2018**, *5*, 907–917. [CrossRef]
27. Yi, A.; Feng, Y.; Du, Z.; Li, H. Mechanism of 2, 4, 6-Trichlorophenol Degradation in Microbial Fuel Cells System with Microbe Isolated from Submarine Sediment. *Int. J. Electrochem. Sci.* **2015**, *10*, 1459–1468.
28. Krishnaiah, D.; Anisuzzaman, S.M.; Bono, A.; Sarbatly, R. Adsorption of 2,4,6-trichlorophenol (TCP) onto activated carbon. *J. King Saud Univ. Sci.* **2013**, *25*, 251–255. [CrossRef]
29. Zhao, P.; Liu, X.; Tian, W.; Yan, D.; Sun, X.; Lei, X. Adsolubilization of 2,4,6-trichlorophenol from aqueous solution by surfactant intercalated ZnAl layered double hydroxides. *Chem. Eng. J.* **2015**, *279*, 597–604. [CrossRef]
30. Saritha, P.; Suman, D.S.; Aparna, C.; Vijaya, P.N.; Himabindu, V.; Anjaneyulu, Y. Degradative Oxidation of 2,4,6 Trichlorophenol Using Advanced Oxidation Processes—A Comparative Study. *Water Air Soil Pollut.* **2009**, *200*, 169–179. [CrossRef]
31. Kashyap, J.; Riaz, U. Facile synthesis of novel polypyrrole dispersed $AgFeO_2$ nanohybrid with highly efficient photocatalytic activity towards 2,4,6-trichlorophenol degradation. *RSC Adv.* **2018**, *8*, 13218–13225. [CrossRef]
32. Benbachir, H.; Gaffour, H.; Mokhtari, M. Photodegradation of 2,4,6-trichlorophenol using natural hematite modified with chloride of zirconium oxide. *React. Kinet. Mech. Catal.* **2017**, *122*, 635–653. [CrossRef]
33. Pino-Chamorro, J.Á.; Ditrói, T.; Lente, G.; Fábián, I. A detailed kinetic study of the direct photooxidation of 2,4,6-trichlorophenol. *J. Photochem. Photobiol. A* **2016**, *330*, 71–78. [CrossRef]
34. Chaliha, S.; Bhattacharyya, K.G. Wet oxidative method for removal of 2,4,6-trichlorophenol in water using Fe(III), Co(II), Ni(II) supported MCM41 catalysts. *J. Hazard. Mater.* **2008**, *150*, 728–736. [CrossRef] [PubMed]
35. Yang, J.; Chen, H.; Gao, J.; Yan, T.; Zhou, F.; Cuin, S.; Bi, W. Synthesis of Fe_3O_4/g-C_3N_4 nanocomposites and their application in the photodegradation of 2,4,6-trichlorophenol under visible light. *Mater. Lett.* **2016**, *164*, 183–189. [CrossRef]
36. Hee-Chan, K.; Sang-Hyup, L.; Dong-Ju, K.; Jae-Woo, C. Photocatalytic Activity of 2,4,6-Trichlorophenol by TiO_2 Mesostructures: Effects of Surface Modification, Calcination Temperature and Initial pH. *Water Air Soil Pollut.* **2013**, *224*, 1459–1467. [CrossRef]
37. Rozov, K.; Berner, U.R.; Kulik, D.A.; Diamond, L.W. Solubility and thermodynamic properties of carbonate-bearing hydrotalcite-pyroaurite solid solutions with a 3:1 Mg/(Al+Fe) mole ratio. *Clays Clay Miner.* **2011**, *59*, 215–232. [CrossRef]
38. Lavand, A.B.; Malghe, Y.S. Nano sized C-doped TiO_2 as a visible-light photocatalyst for the degradation of 2,4,6- trichlorophenol. *Adv. Mater. Lett.* **2015**, *6*, 695–700. [CrossRef]
39. Skrabal, P.M. *Spectroscopy: An Interdisciplinary Integral Description of Spectroscopy from UV to NMR*, 1st ed.; vdf Hochschulverlag AG an der ETZ Zurich: Zurich, Switzerland, 2012; Volume 1, pp. 216–224. ISBN 978-3-7281-3385-4.
40. Turchi, C.S.; Ollis, D.F. Mixed Reactant photocatalysis: Intermediates and Mutual Rate Inhibition. *J. Catal.* **1989**, *119*, 483–496. [CrossRef]
41. Lin, Z.R.; Zhao, L.; Dong, Y.H. Quantitative characterization of hydroxyl radical generation in a goethite-catalyzed Fenton-like reaction Author links open overlay panel. *Chemosphere* **2015**, *141*, 7–12. [CrossRef]
42. Fu, S.; Deng, B.; Ma, D.; Cheng, H.; Dong, S. Visible-Light-Driven Photocatalytic Fuel Cell with an Ag-TiO_2 Carbon Foam Anode for Simultaneous 4-Chlorophenol Degradation and Energy Recovery. *ChemEngineering* **2018**, *2*, 20. [CrossRef]
43. An, X.; Liu, H.; Qu, J.; Moniz, S.J.A.; Tang, J. Photocatalytic mineralisation of herbicide 2,4,5-trichlorophenoxyacetic acid: Enhanced performance by triple junction Cu–TiO_2–Cu_2O and the underlying reaction mechanism. *New J. Chem.* **2015**, *39*, 314–320. [CrossRef]
44. Nosaka, Y.; Nosaka, A.Y. Generation and Detection of Reactive Oxygen Species in Photocatalysis. *Chem. Rev.* **2017**, *117*, 11302–11336. [CrossRef]
45. Chen, F.; Yang, Q.; Pehkonen, S.O.; Ray, M.B. Modeling of Gas-Phase Photodegradation of Chloroform and Carbon Tetrachloride. *J. Air Waste Manag. Assoc.* **2004**, *54*, 1281–1292. [CrossRef] [PubMed]

46. Rodríguez, E.M.; Márquez, G.; Tena, M.; Álvarez, P.M.; Beltrán, F.J. Determination of main species involved in the first steps of TiO_2 photocatalytic degradation of organics with the use of scavengers: The case of ofloxacin. *Appl. Catal. B* **2015**, *178*, 44–53. [CrossRef]
47. Wu, W.; Liang, S.; Chen, Y.; Shen, L.; Zheng, H.; Wu, L. High efficient photocatalytic reduction of 4-nitroaniline to p-phenylenediamine over microcrystalline $SrBi_2Nb_2O_9$. *Catal. Commun.* **2012**, *17*, 39–42. [CrossRef]
48. Xian, T.; Yang, H.; Di, L.; Ma, J.; Zhang, H.; Dai, J. Photocatalytic reduction synthesis of $SrTiO_3$-graphene nanocomposites and their enhanced photocatalytic activity. *Nanoscale Res. Lett.* **2014**, *9*, 1–9. [CrossRef] [PubMed]
49. Liu, W.; Wnag, M.; Xu, C.; Chen, S.; Fu, X. Significantly enhanced visible-light photocatalytic activity of g-C3N4 via ZnO modification and the mechanism study. *J. Mol. Catal. A Chem.* **2013**, *368*, 9–15. [CrossRef]
50. Mascolo, G.; Mascolo, M.C. On the synthesis of layered double hydroxides (LDHs) by reconstruction method based on the "memory effect". *Microporous Mesoporous Mater.* **2015**, *214*, 246–248. [CrossRef]
51. Kim, B.K.; Gwak, G.H.; Okada, T.; Oh, J.M. Effect of particle size and local disorder on specific surface area of layered double hydroxides upon calcination-reconstruction. *J. Solid State Chem.* **2018**, *263*, 60–64. [CrossRef]
52. Teodorescu, F.; Paladuta, A.M.; Pavel, O.D. Memory effect of hydrotalcites and its impact on cyanoethylation reaction. *Mater. Res. Bull.* **2013**, *48*, 2055–2059. [CrossRef]
53. Czaplicka, M. Photo-degradation of chlorophenols in the aqueous solution. *J. Hazard Mater.* **2006**, *134*, 45–59. [CrossRef]
54. Xu, L.; Wang, J. Degradation of 2,4,6-trichlorophenol using magnetic nanoscaled Fe_3O_4/CeO_2 composite as a heterogeneous Fenton-like catalyst. *Sep. Purif. Technol.* **2015**, *149*, 255–264. [CrossRef]
55. Odling, G.; Robertson, N. Bridging the gap between laboratory and application in photocatalytic water purification. *Catal. Sci. Technol.* **2019**, *9*, 533–545. [CrossRef]
56. Guo, Q.; Zhou, C.; Ma, Z.; Ren, Z.; Fan, H.; Yang, X. Elementary Chemical Reactions in Surface Photocatalysis. *Annu. Rev. Phys. Chem.* **2018**, *69*, 451–472. [CrossRef]

catalysts

MDPI

Article

Experimental Design and Optimization of Triclosan and 2.8-Diclorodibenzeno-p-dioxina Degradation by the Fe/Nb_2O_5/UV System

Michel Z. Fidelis [1,*], Eduardo Abreu [1], Onélia A. A. Dos Santos [2], Eduardo S. Chaves [3], Rodrigo Brackmann [4], Daniele T. Dias [5] and Giane G. Lenzi [1]

1 Departamento de Engenharia Química, Universidade Tecnológica Federal do Paraná, Av Monteiro Lobato, s/n—Km 04, Ponta Grossa, PR 84016-210, Brazil; eduardo_abreu@live.com (E.A.); gianeg@utfpr.edu.br (G.G.L.)
2 Departamento de Engenharia Química, Universidade Estadual de Maringá, Avenida Colombo, 5790, Maringá, PR 87020-900, Brazil; oaasantos@uem.br
3 Departamento de Química, Universidade Federal de Santa Catarina, Campus Reitor João David Ferreira Lima, R. Eng. Agronômico Andrei Cristian Ferreira, s/n, Florianópolis, SC 88040-900, Brazil; eschaves@hotmail.com
4 Departamento de Química, Universidade Tecnológica Federal do Paraná, Via do Conhecimento, s/n—Km 01, Pato Branco, PR 85503-390, Brazil; rodrigobrackmann@utfpr.edu.br
5 Departamento Acadêmico de Física, Universidade Tecnológica Federal do Paraná, Av Monteiro Lobato, s/n—Km 04, Ponta Grossa, PR 84016-210, Brazil; danieletdias@utfpr.edu.br
* Correspondence: michelmzzf@gmail.com; Tel.: +55-42-3220-4800

Received: 11 February 2019; Accepted: 27 March 2019; Published: 8 April 2019

Abstract: This study describes the experimental design and optimization of the photocatalytic reaction using the immobilized catalyst Fe/Nb_2O_5 in the degradation of Triclosan and 2.8-DCDD. The techniques employed to characterize the photocatalysts were: specific surface area, average pore volume, average pore diameter, photo-acoustic spectroscopy (PAS), X-ray diffraction (XRD), and scanning electron microscopy (SEM/EDS). The reaction parameters studied were pH, catalyst concentration, catalyst calcination temperature, and nominal metallic charge. The results indicated that the immobilized Fe/Nb_2O_5 catalysts were efficient in the degradation of Triclosan and 2.8-dichlorodibenzene-p-dioxin. The catalysts with nominal metal loading of 1.5% Fe calcined at 873 K showed the highest constant reaction rate and the lowest half-life 0.069 min^{-1} and 10.04 min. Tests in different matrices indicated that the photocatalytic reaction using aqueous solution containing Cl^- is faster when compared with the ultrapure water matrix.

Keywords: Fe/Nb_2O_5 immobilized catalyst; emerging pollutants; degradation

1. Introduction

The chemical compounds progressive use, coupled with the low efficiency of effluent collection and treatment systems, have contributed to the significant increase in contaminants of emerging concern found in surface waters. This was considered to be of low risk to the environment, and these contaminants were detected only a few years ago.

Among the contaminants of emerging concern is the Triclosan (TCS) and 5-chloro-2-(2,4-dichlorophenoxy)-5-phenol—an antimicrobial agent that is extensively used as a preservative in a variety of consumer products such as toothpaste, antiseptic soaps, detergents, cosmetics, plastic kitchenware, socks, rugs, and toys [1–3]. The TCS presence can be harmful to both human and animal health since it tends to undergo bio-magnification and can be degraded by photolysis, which generates by-products of greater toxicity, such as chlorophenols and dioxins [4–6]. Among the dioxins that may be formed, 2,8-dichlorodibenzene-p-dioxin (2.8-DCDD) stands out as being a chemically stable compound and it is known to suppress antibody responses. In addition, exposure to radiation, mainly in an environment with free chlorine, favors the reaction of transformation of the TCS in this compound [6,7].

Conventional water treatment methods are ineffective for removing substances with biological activity, such as Triclosan, and extremely stable substances such as dioxins. Therefore, in recent decades, Advanced Oxidation Processes (AOPs) have received attention as an alternative for removing these compounds. They are based on the hydroxyl radicals generation, which has a strong oxidizing character and has the capacity to degrade compounds through oxidation. The formation of free radicals that react with dissolved oxygen is an example of such [8,9].

Many semiconductors have been studied in the heterogeneous photo catalysis in order to degrade pollutants in water bodies (ZnO, CdO, TiO_2, Fe_2O_3, among others). Among these, TiO_2 stands out as the most used photo catalyst due to its highly chemical stability, nontoxic, band gap of 3.2 eV for UV radiation (natural/artificial), and relatively low cost [10–13]. However, since its band gap is 3.2 eV, the highest efficiency of the process is limited by the absorption of the TiO_2 semiconductor with a radiation of over 385 nm, which corresponds to approximately 3% of the solar spectrum at sea level.

Thus, the search for new catalysts becomes essential. The use of niobium pentoxide (Nb_2O_5) as a semiconductor for application in heterogeneous photocatalysis has few literature references. This compound has chemical characteristics of low toxicity and band gap (3.4 eV), which is very similar to TiO_2 [14,15]. It is a stable semiconductor and insoluble in water, which can be attacked only by concentrated hydrofluoric acid. Compared with vanadium oxide, it is more stable and difficult to reduce. In contrast to titanium oxide, it presents an advantage in some types of photo degradation [16].

Due to the characteristics of the oxide, it has been used in catalytic reactions such as dehydrogenation, alkylation, hydrolysis, and photocatalytic degradation. The study of Nb_2O_5 shows that it has a combination of energy when in contact with ultraviolet radiation, and, also, when together with another semiconductor, can greatly increase the catalytic activity and the life of a catalyst [16,17].

One of the negative aspects of the photocatalytic process is the difficulty to separate the catalyst after its treatment. In this direction, the material immobilization allows the separation of the catalyst by avoiding the need of a later step in the process and increases the porosity, the adsorption capacity, and thermal stability [18]. On the other hand, a catalytic activity reduction occurs [19,20]. In order to improve the photoreaction efficiency, many authors have metals such as Ag [21], Fe [22], and Cu [23] added to it in order to prevent the process of electron/role recombination in semiconductors.

In this context, this study describes the optimization of the photocatalytic degradation of TCS and 2.8-DCDD using the immobilized Fe/Nb_2O_5 as the photo-catalyst. It was also analyzed using different parameters such as pH, catalyst concentration, and the catalyst calcination temperature.

2. Results and Discussion

2.1. Catalysts Characterization

2.1.1. Pore Properties (BET Method)

The results of the textural properties for the calcined Fe/Nb_2O_5 catalysts at different temperatures are shown in Table 1.

A variation of the specific surface area for the studied catalysts ranges from 20.26 to 149.63 $m^2\ g^{-1}$. The lowest area obtained for the samples was 0.5% Fe/Nb_2O_5 calcined at 673 K, and the highest for the sample was 1.0% Fe/Nb_2O_5 at 773 K. These results indicate the influence of the calcination temperature and the addition of Fe. The 1.0% Fe/Nb_2O_5 catalysts and 0.5% Fe/Nb_2O_5 calcined at 773 K and 873K practically obtained the same results of approximately 150 $m^2\ g^{-1}$. The catalysts are mesoporous materials (pores of internal width between 20 Å and 50 Å). When comparing the results of the pore diameter, it is possible to verify that the samples did not have significant variations in this property. Regarding the micro-pore volumes, there was considerable variation of

0.00518 cm^3 g^{-1} (1.5% Fe/Nb_2O_5 calcined at 873 K) and 0.01064 cm^3 g^{-1} (1.7% Fe/Nb_2O_5 calcined at 773 K). Fontana et al. (2018) [24] profiled a surface area of 1.50 m^2 g^{-1}, pore volume of 0.049 cm^3 g^{-1}, and pore diameter of 15.60 Å for non-calcined Nb_2O_5 samples. Nakajima et al. (2011) [25] found that Nb_2O_5 samples calcined at 573 K had a pore volume of 0.21 cm^3 g^{-1} and a diameter of 18 Å. When compared with the results obtained, it is noted that the surface areas of the immobilized Fe/Nb_2O_5 catalysts are larger than those of the non-calcined and non-immobilized Nb_2O_5 catalysts, as well as their pore diameter.

Table 1. Summarizes the surface textural properties catalysts (Specific surface area (So), Micropore Volume (Vm), and average pore diameter (Dm).

Catalyst Fe/Nb_2O_5 wt %-T Calcination (K)	So (m^2 g^{-1})	Vm (cm^3 g^{-1})	Dm (Å)
0.3%-773	49.51	0.00972	33.03
0.5%-673	20.26	0.00589	33.25
0.5%-873	148.83	0.00716	33.05
1.0%-633	29.55	0.00772	33.16
1.0%-773	149.63	0.00816	33.34
1.0%-913	25.42	0.00635	33.10
1.5%-673	64.90	0.00769	33.07
1.5%-873	21.24	0.00518	33.34
1.7%-773	43.44	0.01064	33.25

2.1.2. X-ray Diffraction (XRD)

The XRD analysis of the calcined Fe/Nb_2O_5 immobilized samples are presented in Figure 1. The peaks obtained were compared with the standards published by ICDD (International Center for Diffraction Data).

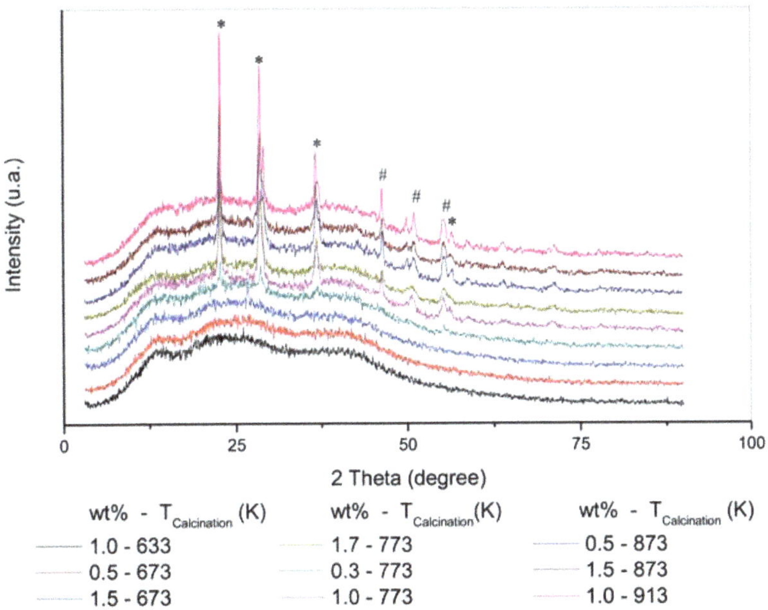

Figure 1. XRD pattern of Fe/Nb_2O_5 sample where (*) TT phase (pseudo-hexagonal form) (#) T phase (orthorhombic form).

By analyzing the diffractograms, it is possible to verify that the catalyst 1.0% Fe/Nb_2O_5, 0.5% Fe/Nb_2O_5, and 1.5% Fe/Nb_2O_5 samples calcined at 673K have a non-crystalline (amorphous) structure. Similar results were found for niobium samples at the same catalyst calcination temperatures from research conducted by Morais et al. (2017) [26]. The diffractograms indicate that the *TT*-Nb_2O_5 (ICDD00-028-0317) with the cell parameters a = b = 3.6070 Å, c = 3.9250 Å, and *T*-Nb_2O_5 (ICDD00-027-1313) forms with the cell parameters a = 6.1680 Å, b = 29.3120 Å, and c = 3.9380 Å were obtained for the catalysts calcined at temperatures higher than 673 K. It was also observed that 1.0% Fe/Nb_2O_5, 1.7% Fe/Nb_2O_5 calcined at 773 K and 0.5% Fe/Nb_2O_5, 1.5% Fe/Nb_2O_5 calcined at 873 K and 1.0% Fe/Nb_2O_5 calcined at 913 K showed a semi-crystalline structure, with crystallite sizes of 30, 26, 35, 35, and 40 nm, respectively. The sample 0.3% Fe/Nb_2O_5 773 K showed a semi-crystalline structure, but it was not possible to calculate the size of the crystallites and to determine a hexagonal shape due to it being very amorphous. Sreethawonget al. (2012) [27] found similar results for Nb_2O_5 sol-gel samples calcined at 773 K with an orthorhombic form (T) with crystallites of approximate size of 18.5 nm. A result was also found by Liu et. al. (2011) [28] in Nb_2O_5 samples of an orthorhombic form (T). However, the crystallite sizes were 0.39 nm.

2.1.3. Photoacoustic Spectroscopy (PAS)

The band gap results and the absorption threshold for the Nb_2O_5 catalysts are shown in Table 2. The values found ranged from 3.05 to 3.94 eV. According to Greenwood and Eranshaw (2003) [29], the band gap of Nb_2O_5 has a range of 3.1 to 4.0 eV. Analyzing the values obtained in relation to the nominal metallic load of Fe, it was observed that the band gap is likely to decrease with an increase in the Fe quantity. This also happens with samples of iron-doped TiO_2 catalysts [30,31]. It obtained the following values for the calcination temperature at 673 K (0.5% Fe-4.75 eV and 1.5% Fe-3.28 eV). It also occurred for the catalysts calcined at 873 K. An approximate value was found for the catalyst 0.3% and 1.0% when calcined at 773 K. It was also found that, when the calcination temperature increases, the band gap tends to increase as well. In particular, this was observed for the catalysts with a low metal nominal loading. On the other hand, Yoshimura et. al. (1996) [32] indicated band gap energies of 3.41 eV for samples of Nb_2O_5 calcined at 773 K, and 3.45 eV for non-calcined samples. The values found were very similar to those presented by the other authors. The samples 1.0% Fe/Nb_2O_5 913 K and 0.5% Fe/Nb_2O_5 873 K showed the highest band gap energy values of 3.94 eV, while 1.5% Fe/Nb_2O_5 873 K presented the lowest band gap energy values of 3.05 eV. This latter result (3.05 eV) suggests that this catalyst is absorbing in regions closer to the visible region of the electromagnetic spectrum. This means that it requires less energy for electronic transactions to occur. In this context, Weibin et al. (2013) [33] reported a value of 3.70 eV, while Scheidt et al. (2014) [34] presented a band gap energy of 3.61 eV for non-calcined Nb_2O_5 samples. Other authors such as Hamaguchi, (2011) [35], Prado et. al. (2008) [36], Liu et. al. (2011) [28] and Gallo (2016) [37] found band gap energies for Nb_2O_5 of 3.61, 3.40, 3.72, and 3.60 eV, respectively.

Table 2. Result of band gaps UV–VIS.

Catalyst Fe/Nb_2O_5 wt %-$T_{Calcination}$ (K)	Band Gap (eV)	Absorption Threshold (nm)
0.3%-773	3.88	281
0.5%-673	3.75	281
0.5%-873	3.94	279
1.0%-633	3.92	281
1.0%-773	3.90	281
1.0%-913	3.94	281
1.5%-673	3.28	292
1.5%-873	3.05	300
1.7%-773	3.85	281

2.1.4. Scanning Electron Microscopy (SEM/EDS)

Figure 2 shows the images obtained by the scanning electron microscopy with an EDS detector (SEM/EDS), for 0.3 and 1.7% Fe/Nb_2O_5 samples calcined at 773 K, respectively.

A

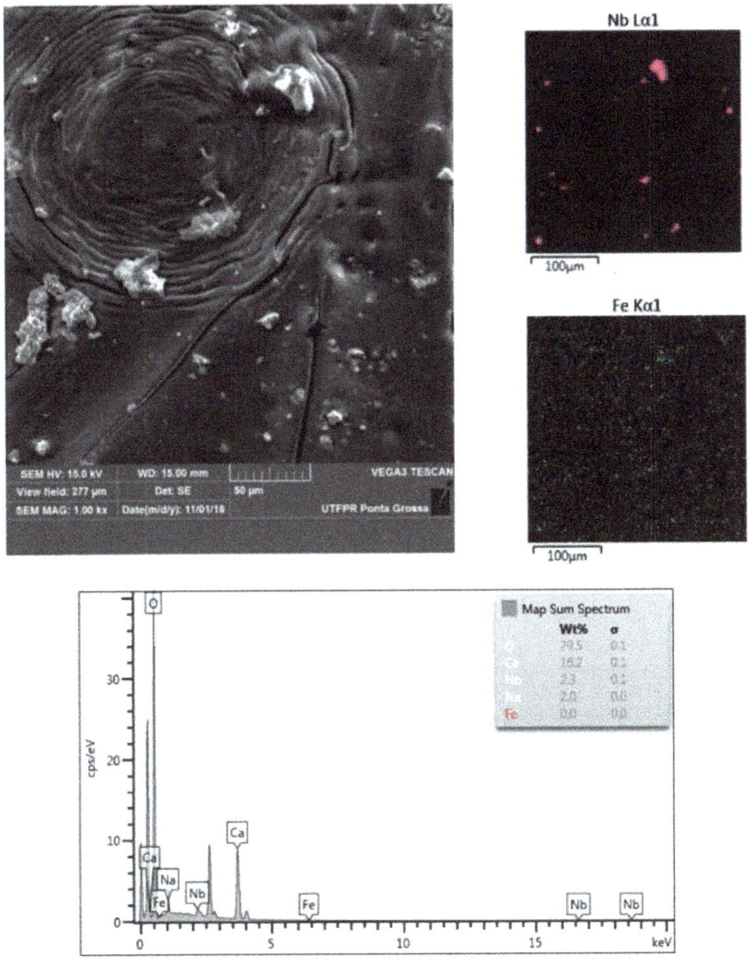

Figure 2. *Cont.*

B

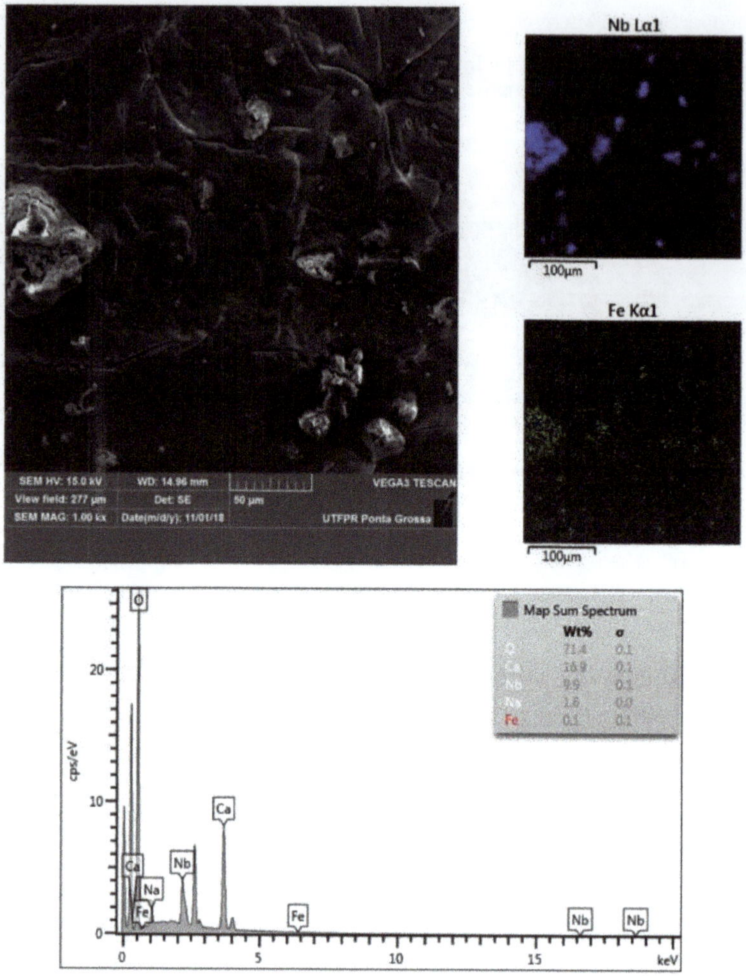

Figure 2. MEV/EDS of sample (**A**) 0.3% Fe/Nb_2O_5 773 K, (**B**) 1.5% Fe/Nb_2O_5 at 873 K, increase 500×.

Using the SEM/EDS analysis for both samples, it is possible to observe that the dispersion on the surface catalysts was homogeneous since the formation of Fe/Nb_2O_5 agglomerates is distributed on the alginate sphere surface. The surface presents some rugosity, which can favor the interaction between the adsorbate and adsorbent. This benefits the catalytic activity. Similar structures on alginate spheres immobilized catalysts were reported by References [38,39]. Thus, it is verified that there were no major structural changes on the catalysts, and, subsequently, in the catalytic activity as well.

2.2. Photocatalytic Reaction

2.2.1. Catalyst Influence

The effect of the photocatalysis process in the TCS degradation was evaluated using TiO_2 and Nb_2O_5 as catalysts (calcined at 773 K). The results are shown in Figure 3.

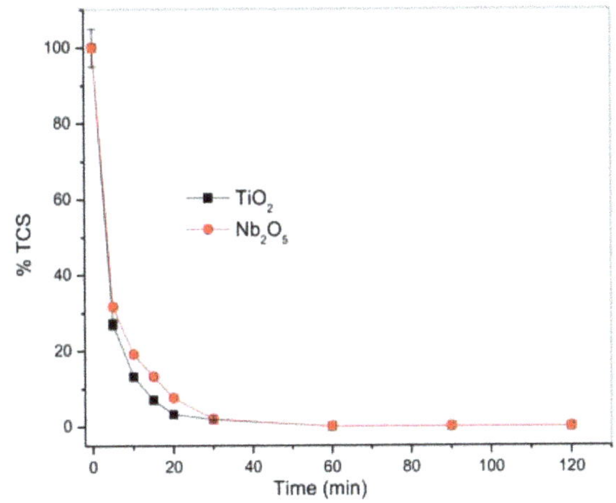

Figure 3. Influence of photo-catalysts using TiO_2 and Nb_2O_5 (calcined at 773K).

The results presented in Figure 3 showed that there is no significant difference between the photo-catalysts TiO_2 and Nb_2O_5 for TCS degradation. Therefore, in the next steps, the Fe/Nb_2O_5 catalyst was used.

2.2.2. pH Influence and Mass Catalitic

In order to analyze the pH and catalyst mass influence in the TCS degradation, the conditions were verified at pH 5.5, 7.0, and 8.5 using the 1.0% Fe/Nb_2O_5 catalyst calcined at 773 K, 1 g L^{-1} (Figure 4A) and 0.5, 1.0, and 2.0 g L^{-1} catalyst (Figure 4B).

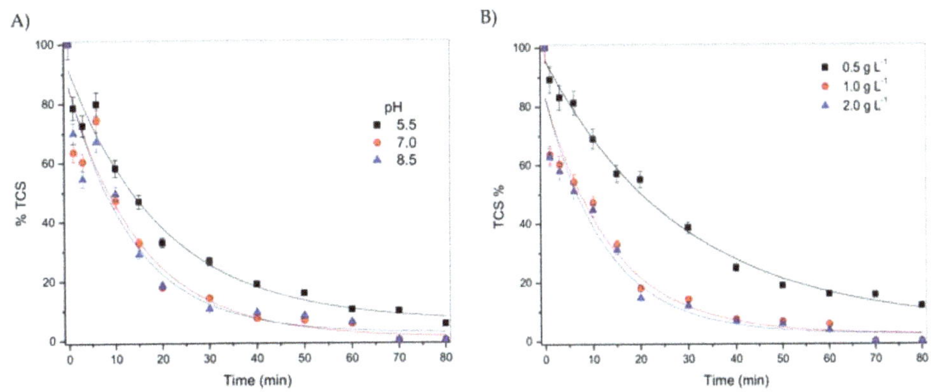

Figure 4. Influence of (**A**) pH (1% Fe/Nb_2O_5 calcined 773 K, 1.0 g L^{-1}) and (**B**) catalyst mass (pH 7.0, 1.0% Fe/Nb_2O_5 calcined 773 K).

The results indicated that, in the neutral and alkaline pH, the TCS degradation kinetics obtained were higher than in an acidic pH. As described in the literature, the photocatalytic efficiency of Triclosan did not change at a neutral pH, but significantly decreased in the acidic condition by inhibiting the generation of OH radicals [40]. On the other hand, the degradation in pH 7.0 and 8.5 was practically the same. A concentration of 0.5, 1.0, and 2.0 g L^{-1} was used to verify the catalyst mass influence. It was observed that the TCS total degradation does not occur within 80 min at the lowest catalyst concentration (0.5 g L^{-1}). However, concentrations of 1.0 and 2.0 g L^{-1} showed very similar results

(Figure 4B). Therefore, a pH 7.0 and a catalyst mass concentration of 1.0 g L^{-1} were maintained, in the following tests.

2.2.3. Adsorption, Photolysis, and Photo Catalyst Tests

The results of adsorption and photocatalytic for the TCS degradation using the Fe/Nb_2O_5 immobilized catalysts are shown in Figure 5. The adsorption was performed in the first 30 min before the photo catalysis.

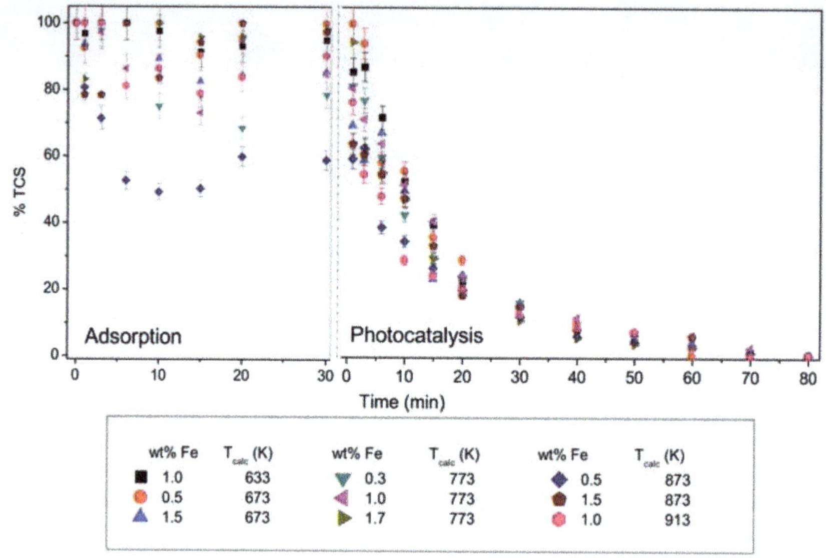

Figure 5. Adsorption and photo catalysis results (pH = 7.0, 1 g L^{-1}).

It was observed that there was an instability in the adsorption process (adsorption and desorption of TCS on catalytic surface), which indicates that only the adsorption process is not adequate for the TCS removal. In addition, it was verified that a greater adsorption does not directly imply a higher rate of degradation. For example, 0.5 wt % catalyst calcined at 873 K obtained approximately 40% adsorption. However, other catalysts saw a similar level of degradation of TCS, without the same adsorption capacity. This catalyst had a high specific surface area (~149 m^2 g^{-1}). The photocatalytic degradation presented an exponential profile for all catalysts studied. The data obtained indicated that all the photo catalysts degraded the TCS after 80 min of the reaction.

It is also possible to verify that catalysts with higher calcination temperatures of 873 K and 913 K have a higher photocatalytic activity in the first 15 min of the reaction, when compared to the other catalysts used.

These values show that the photocatalytic degradation of TCS in an aqueous solution can be described by a pseudo-first order kinetic model. In addition, to compare the results obtained, the half-life ($t_{1/2}$) was evaluated, which indicates the approximate time required for the contaminant concentration to be halved.

Its half-life was calculated according to Equation (1), where *k* is the constant reaction rate. The results are shown in Table 3.

$$t_{\left(\frac{1}{2}\right)} = -\frac{\ln\left(\frac{1}{2}\right)}{k} \tag{1}$$

Table 3. Half-life $t_{1/2}$, first-order reaction rate constant results.

wt %-$T_{calcination}$ (K)	k (min^{-1})	$t_{1/2}$ (min)	R^2
0.3%-773	0.0570	12.16048	0.97926
0.5%-673	0.0666	10.40605	0.98392
0.5%-873	0.0595	11.64953	0.98398
1.0%-633	0.0623	11.12596	0.99583
1.0%-773	0.0500	13.86294	0.97886
1.0%-913	0.0538	12.89336	0.93194
1.5%-673	0.0514	13.48535	0.96524
1.5%-873	0.0690	10.04561	0.93221
1.7%-773	0.0670	10.34548	0.99106

R^2 = Coefficient of Determination.

Among all the catalysts, the 1.5% Fe/Nb_2O_5 calcined at 873 K was the one with the highest constant reaction rate and the lowest half-life of 0.069 min^{-1} and 10.04 min. Another relevant characteristic is that it has the smallest band gap (3.05 eV). Figure 5 shows a comparison degradation kinetics of TCS between photo catalysis (1.5% Fe/Nb_2O_5calcined at 873 K) and photolysis. The experiments were performed in different matrices, ultrapure water, and water containing Cl^-. The reaction orders, rate constant, and $t_{1/2}$ are shown in the subtitle of Figure 6.

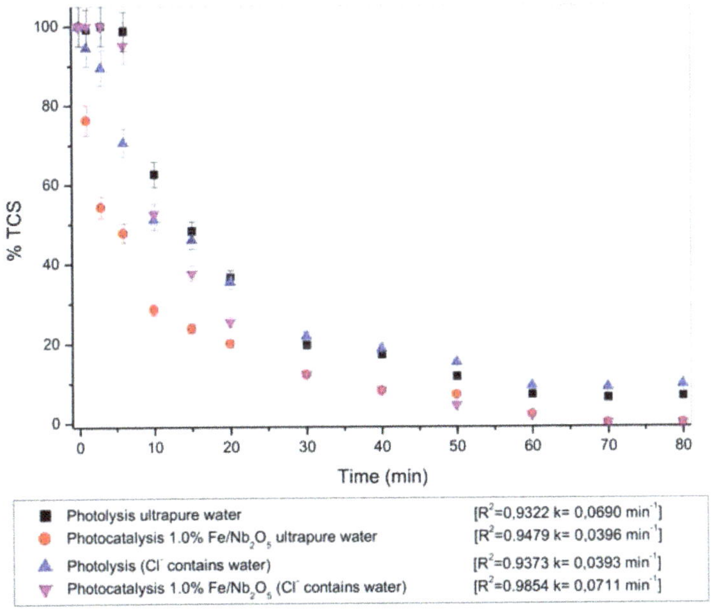

Figure 6. Photo catalysis and photolysis in different matrices.

In Figure 7, it can be verified that the photo catalysis is more efficient in the 100% degradation of TCS, which does not occur for photolysis. In both matrices used, the photo catalysis was efficient.

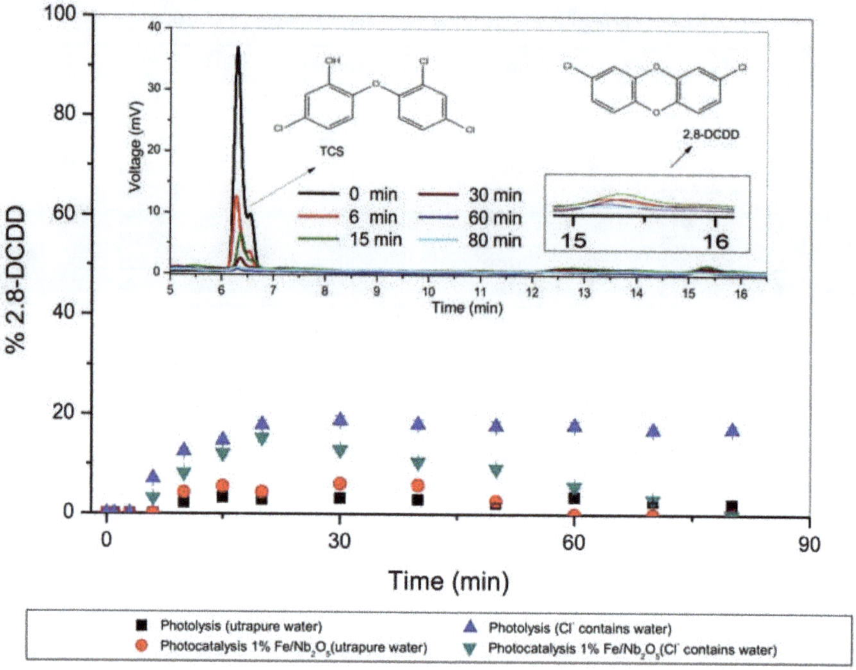

Figure 7. Generation of 2.8-DCDD from photocatalytic and photolytic processes in different matrices.

At the same time, an analysis on the generation/degradation of 2.8-DCDD was performed, where the TCS in the presence of Cl^- and radiation, is transformed into this compound.

Figure 7 shows that 2.8-DCDD has a kinetics of higher formation in matrices containing Cl^-, which was previously described by Reference [6]. It is also noted that, in these environments, the transformation of TCS into 2,8-DCDD occurs extremely fast in approximately 20 min. It is observed that the photolysis is not efficient in the 2.8-DCDD degradation. However, the photocatalysis is able to degrade it completely after 80 min of reaction.

Then the formation of hydroxyl radicals (·OH) occur. According to Reference [5], two conditions need to be satisfied for the formation of DCDDs and DCHDFs (dichlorohydroxydibenzofurans). First, ·OH radicals are present for the cleavage and oxidation. Second, the chlorinated phenoxyphenols are not to be de-chlorinated into phenoxyphenols before they react with ·OH.

Iron leaching tests were performed in triplicate. The concentrations of iron were determined by Flame atomic absorption spectrometry (F AAS) using a spectrometer (Perkin Elmer AAnalyst 700, Lübeck, Germany). The results indicated that leaching of Fe does not occur.

2.2.4. Photo Stability

To evaluate photo stability, tests in successive cycles (80 min each) were performed (Figure 8) and 1.5% Fe/Nb_2O_5 calcined at 873 K were used. In the first cycle, a TCS degradation of about 91% occurs, decreasing to 89% in the first reuse cycle, 68% in the second reuse, and reaching only 50% in the third reuse cycle. This time is sufficient to TCS and 2.8-DCDD degradation. However, the catalytic deactivation may be related with surface adsorbed compounds and by-products formed in the reaction.

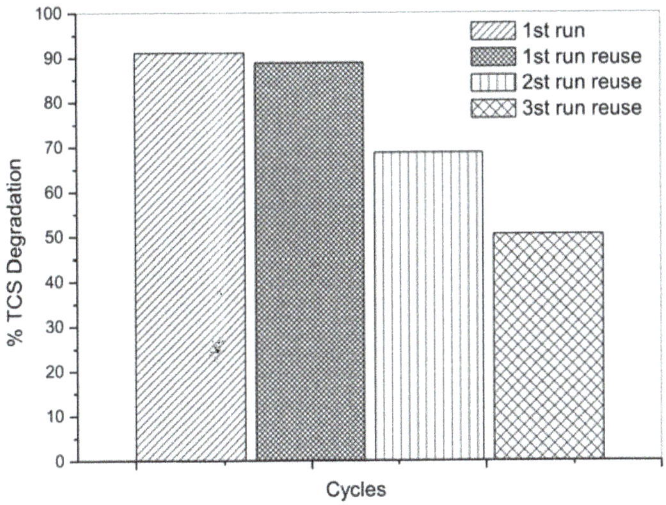

Figure 8. Photocatalytic reuse cycle (1.5%Fe/Nb_2O_5 calcined at 873 K).

2.3. Experimental Design and Optimization

The effects on the parameters of the nominal load of metal (%Fe) and calcination temperature on the TCS degradation were investigated.

According to the results obtained in the Response surface and Surface Contour (Figure 9A,B) and the e Pareto chart (Figure 10), all variables studied were significant for the process.

It was mainly observed that the effects are more accentuated at higher calcination temperatures and higher iron percentages. These are two conditions that significantly affect the photocatalytic process.

From the data obtained, it was possible to table a new construct with R^2 = 0.98204 presented in Table 4.

Table 4. Analysis of variance.

Source of Variation	Sum of Squares (SS)	Degrees of Freedom (df)	Mean Square (MS)	F	Prob > F
Regression	115.4253	5	23.08506	54.66544	5.05
Resisual	2.1115	5	0.422297		
Lack of Fit	2.0356	3	0.678545	17.891521	19.16
Pure erro	0.0759	2	0.037925		
Total	117.5368	10	-	-	-

In order to verify if the generated model was significant, the statistical test F was used as a basis. For $\alpha = 0.05$, we have F (Reg, Res) 5.5 > F (Tab) 5.5 and F (Err) 3.2 < F (Tab) 3.2. Subsequently, it can confidently be concluded to a percentage of 95% that the generated model is significant and there is no evidence of significant differences between treatments.

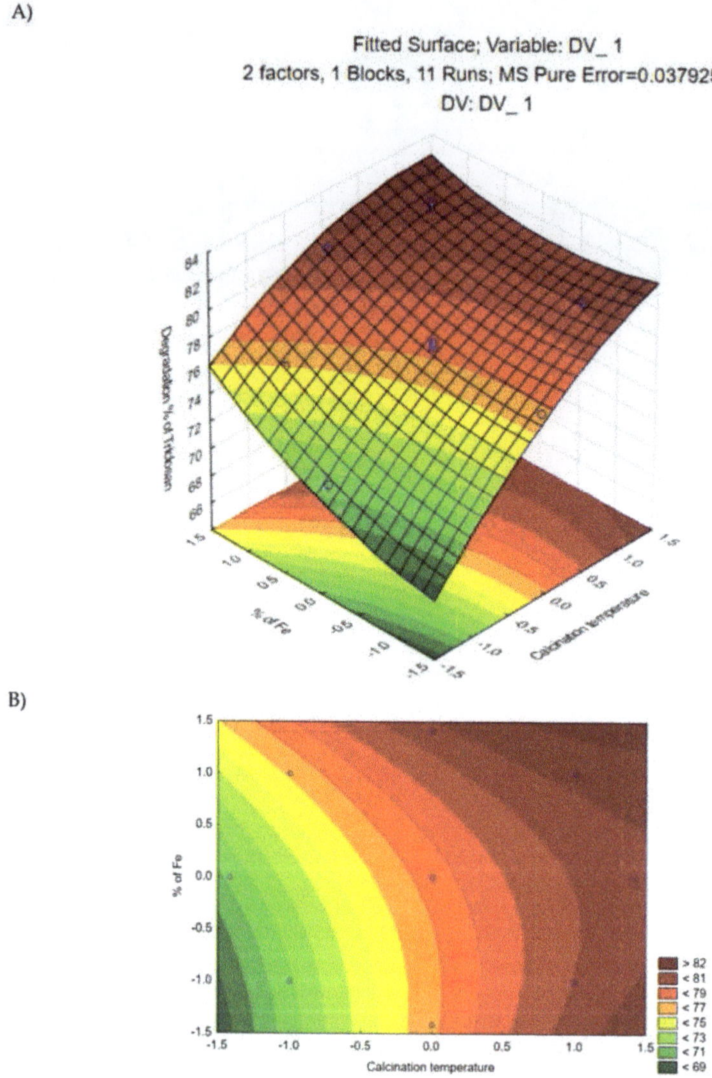

Figure 9. (**A**) Response surface plot for degradation TCS by Nb_2O_5 versus wt % Fe and Temperature Calcination. (**B**) Surface contour obtained in the experimental design. (Coded variable Table 5).

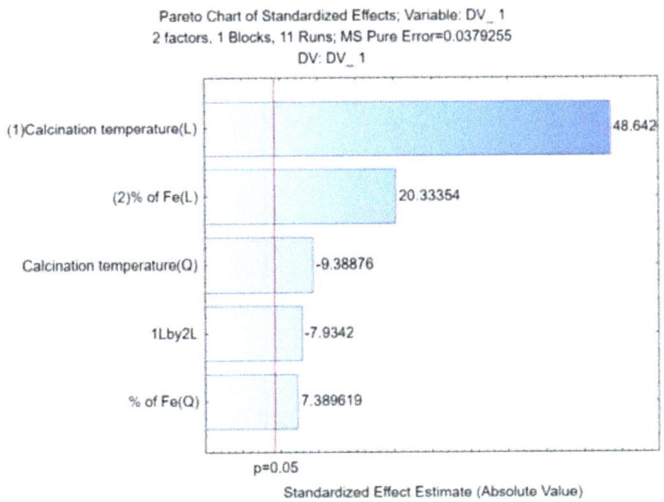

Figure 10. Pareto chart (pH = 7.0, 1 g L^{-1}, and 20 min of reaction).

3. Materials and Methods

Chemicals: Acetonitrile (HPLC–supplied by J.T.Barker, Ciudad de México, México) PA-ACS–CH3CN. Panreac-AppliChem., Triclosan–(supplied by Merck, Darmstadt, Germany); Niobium pentoxide hydrated (Nb2O5) (supplied by CBMM–Companhia Brasileira de Matalurgia e Mineração, Araxá, Minas Gerais, Brasil), Iron Nitrate III (ICO) nonohydrate P.A. ACS ($Fe(NO_3)\cdot 3.9\cdot H_2O$) (supplied by SYNTH, São Paulo, Brasil), Sodium Alginate (P.A) (supplied by Sigma Aldrich, St Louis, MO, USA), and Calcium Chloride-(P.A.) (supplied by Nuclear, São Paulo, Brasil).

3.1. *Fe/Nb$_2$O$_5$ Immobilized Synthesis*

The catalyst synthesis was performed in two stages: (i) Impregnation method–used for supported iron on niobium, with solvent in excess, with different nominal metal loads (0.3%, 0.5%, 1.0%, 1.5% and 1.7%), as described by Reference [33,41]. It was then dried in a rotary evaporator and stove to remove the solvent in excess. Afterward, it was then calcined at different temperatures (633, 673, 773, 873, and 913 K) with heating ramps. (ii) Catalytic immobilization—for the catalyst immobilization, it used a methodology adapted from Reference [31], which incorporated sodium alginate as the polymer matrix. The concentrations used were: sodium alginate 2% (w/v) and photocatalyst 1g L^{-1}. The catalyst was agitated with sodium alginate in a solution and then dripped in a 2% (w/v) calcium chloride solution ($CaCl_2$) in a 288 K ultra-thermostatic bath to form the sodium alginate beads. After 24 hours, the beads were washed with ultrapure water and oven dried at 333 K for 8 hours.

3.2. *Characterization*

Textural Properties: Porous properties such as specific surface area, mean pore diameter, and pore volume were established using a QUANTACHROME Analyzer—Model Novatouch 2 LX with N^2 adsorption at 77 K. The samples were previously submitted to a thermal treatment at 333K for 12 hours, which puts them under vacuum to eliminate any existing water within the pores of the solids. Photoacoustic spectroscopy (PAS): The photoacoustic spectroscopy measurements in the UV–VIS spectral regions were performed using a lab-made experimental setup.

It obtained a monochromatic light through a 1000-Watt xenon arc lamp (66926, Newport Corporation/Oriel) and a monochromator (74100, Newport Corporation by Oriel Instruments, São Paulo, São Paulo, Brasil. The light beam was modulated using a mechanical chopper

(SR540—Stanford Research Systems). A lab-made photoacoustic cell was designed to have a very low volume and was made of aluminum block. It was machined to hold samples with dimensions of up to 10 mm in diameter and 1.5 mm in thickness. This allowed light to enter through a high transparent quartz window of a 6-mm diameter and 2-mm thickness. The distance of the microphone chamber was 15 mm. It was connected to the sample holder room through a 1-mm diameter duct. The capacitive microphone (4953, Brüel&Kjaer) used was a 12-mm diameter. It was very sensitive and presented a gain of 50 mV/Pa, and a flat frequency response between 3 to 10 kHz. It also used a lock-in amplifier (SR830, Stanford Research System). All the photoacoustic spectra were obtained at a 23 Hz modulation frequency and recorded between 225 to 700 nm. Data was acquired using a personal computer, and the PAS spectra were normalized regarding the carbon black signal. The band gap energies were established through Equation (2):

$$\lambda = \frac{hc}{E_{gap}} = \frac{1240}{E_{gap}} \tag{2}$$

where E_{gap} is the *band gap* energy in eV units. The direct method was applied to obtain the values, i.e., m = 2.

X-ray diffraction: The samples were measured through a Rigaku-Denki Diffractometer with Cu-Kα radiation (λ = 1.5406 Å) at 40 kV voltage and 40 mA current. Subsequently, the obtained patterns were compared with the diffraction dataset cards from the ICDD.

Scanning electron microscopy (SEM/EDS): For the analysis, a scanning electron microscope (model VEGA 3 LMU, brand TESCAN) was used, complete with W 30kV filament, 3.0 nm resolution, SE detectors, and retractable BSE. It also featured a low vacuum mode (500 Pa), chamber with internal diameter of 230 mm, and a door width of 148 mm. The microscope also included compucentric rotation, 5-axis fully motorized, with X movement: 80 mm, Y: 60 mm and Z: 47mm, CCD camera for viewing samples, and "chamberview" software, VegaTC operational software, data processing system, and track-ball. The Microscope is also equipped with the EDS detector, dry, model AZTec Energy X-Act, resolution 130 eV, brand Oxford.

3.3. Photocatalytic Tests

The reaction mixture was transferred to a reactor consisting of a cylindrical Pyrex cell (5.0×10^{-4} m^3) surrounded by a water jacket and a magnetic stirrer. This was used to ensure a constant temperature (298 K) with ultra-thermostatic bath and solution homogeneity throughout the experiment. The tests were performed for 80 min and opened to the air. Oxygen was pumped at 8.3×10^{-9} m^3 s^{-1} into the solution containing 10 ppm of TCS and a photo catalyst. The TCS 10 mg L^{-1} solution was prepared with 70% of acetonitrile and 30% of ultrapure water. A total of 500 mL was inserted into the reactor, along with the catalyst mass. UV light was irradiated by using a 125-W medium pressure mercury lamp, the radiation incidence was measured (2.48 mW/cm^2). Aliquots were collected at regular intervals of time. Chromatographic analysis of Triclosan was performed with a high performance liquid chromatograph YL Clarity model 9100 equipped with a pre-column, reverse phase C-18 column, and visible ultraviolet detector (UV-VIS). It used the standard acetonitrile HPLC for chemical analysis and the pH control of sodium hydroxide and nitric acid.

Different conditions were studied: pH 5.5, 7.0, and 8.5, catalyst mass 0.5, 1.0, and 2.0 g L^{-1} and calcination temperature of the catalyst between 633 and 913 K. The adsorption tests were performed by applying the same procedures of the previous photocatalytic test, but without the presence of light. The photolysis experiments were conducted using no catalysts and followed the same procedures of the photocatalytic tests. The tests were completed twice.

Tests about photo stability were performed in four successive cycles. Between the cycles, the catalyst was filtered, washed (ultrapure water), and dried.

3.4. Optimization of Experimental Parameters

The effects of operating parameters of TCS degradation using the Fe/Nb_2O_5 catalyst were analyzed by using a central composite factorial. The general form of this factorial design is 2n plus a star configuration ($\pm\alpha = 2^{n^{(1/4)}}$), with a central point, where n and 2 represent the number of factors and the two levels of work [33], respectively. The levels are defined as low level (−1), high level (+1), central point (0), and two outer points. The independent variables were the Calcination Temperature of catalysts and nominal metal load (%Fe), while the dependent variable Y (variable response) was the TCS removal percentage. The settings for the independent variables were as follows (low/high value): calcination Temperature: 633 to 913 K, and nominal metal load: 0.3 to 1.7 %Fe. Table 5 shows experimental results of the independent variables. The values for coded and decoded variables are also shown.

Table 5. Design matrix and experimental results for the central composite factorial design.

Run	Calcination Temperature (K)	%Fe and (Coded Variable)
1	873 (+1)	1.5 (+1)
2	873 (+1)	0.5 (−1)
3	673 (−1)	1.5 (+1)
4	673 (−1)	0.5 (−1)
5	913 (+α)	1.0 (0)
6	633 (−α)	1.0 (0)
7	773 (0)	0.3 (−α)
8	773 (0)	1.7 (+α)
9	773 (0)	1.0 (0)
10	773 (0)	1.0 (0)
11	773 (0)	1.0 (0)

4. Conclusions

The Fe/Nb_2O_5 immobilized catalysts were efficient in the Triclosan and 2,8-dichlorodibenzene-p-dioxin degradation. The characterization indicated that a variation in the specific surface area occurs for the catalysts studied. This is influenced by the calcination temperature and nominal metal loading.

In the XRD analysis, catalysts calcined at 673 K or at lower temperatures have a non-crystalline structure. Above this calcination temperature, structures in the form *T* and *TT* were found. The band gap results, in general, see a decrease of the band gap in accordance with an increase in the nominal metallic charge of Fe. On the other hand, when the calcination temperature increases, the band gap also tends to increase. The SEM/EDS analysis showed the formation of agglomerates throughout the sphere surface and rugosity. The photocatalytic tests indicated that the operating optimum conditions were pH 7.0 and catalyst mass of 1.0 g L^{-1}. The adsorption process is not efficient for the TCS removal and a higher adsorption on the catalytic surface does not imply a faster photocatalytic process. All catalysts were efficient in the TCS degradation after 80 min of reaction. The catalysts 1.5% Fe/Nb_2O_5 calcined at 873 K showed the highest constant reaction rate and the lowest half-life—0.069 min^{-1} and 10.04 min. This was, consequently, the best catalyst studied.

Photo catalysis was efficient (100% degradation) in both matrices studied (ultrapure and water containing Cl^-). On the other hand, this did not occur with photolysis. In the process, the generation and subsequent degradation of 2.8-DCDD occurred. This degradation was larger in matrices containing Cl^-, since it was totally degraded after 80 min of reaction. The parameters studied (nominal metal load (%Fe) and calcination temperature) in the experimental design and optimization were significant for the process. From the results obtained, it was concluded that the greater the amount of Fe percentage and calcination temperature, the better the result.

Author Contributions: M.Z.F, E.A., and E.S.C. carried out the experiment, conceived, and planned the experiments. R.B. and D.T.D. Characterization Analysis. G.G.L. and O.A.A.D.S. original draft preparation. All authors discussed the results and contributed to the final manuscript.

Acknowledgments: The authors thank the CBMM-CompanhiaBrasileira de Matalurgia e Mineração by the Niobium, and CNPq agency.

Conflicts of Interest: The authors declare no conflict of interest.

References

1. Bedoux, G.; Roig, B.; Thomas, O.; Dupont, V.; Le Bot, B. Occurrence and toxicity of antimicrobial triclosan and by-products in the environment. *Environ. Sci. Pollut. Res.* **2012**, *19*, 1044–1065. [CrossRef]
2. Reiss, R.; Mackay, N.; Habig, C.; Griffin, J. An ecological risk assessment for triclosan in lotic systems following discharge from wastewater treatment plants in the United States. Environ. *Toxicol. Chem.* **2002**, *21*, 2483–2492. [CrossRef]
3. Singer, H.; Müller, S.; Tixier, C.; Pillonel, L. Triclosan: Occurrence and Fate of a Widely Used Biocide in the Aquatic Environment: Field Measurements in Wastewater Treatment Plants, Surface Waters, and Lake Sediments. *Environ. Sci. Technol.* **2002**, *36*, 4998–5004. [CrossRef] [PubMed]
4. Wang, J.L.; XU, L.J. Advanced oxidation processes for wastewater treatment: formation of hydroxyl radical and application. *Crit. Rev. Environ. Sci. Technol.* **2012**, *42*, 251–325. [CrossRef]
5. Niu, J.; Dai, Y.; Yin, L.; Shang, J.; Crittenden, J.C. Photocatalytic reduction of triclosan on Au-Cu_2O nanowire arrays as plasmonic photoatalysts under visible light irradiation. *Phys. Chem. Chem. Phys.* **2015**, *17*, 17421–17428. [CrossRef] [PubMed]
6. Buth, J.M.; Steen, P.O.; Sueper, C.; Blumentritt, D.; Vikesland, P.J.; Arnold, W.A.; McNeill, K. Dioxin Photoproducts of Triclosan and Its Chlorinated Derivatives in Sediment Cores. *Environ. Sci. Technol.* **2010**, *44*, 4545–4551. [CrossRef]
7. UNEP—United Nations Environment Programme. Persistent Organic Pollutants. 2013. Available online: http://www.chem.unep.ch/pops/ (accessed on 22 January 2018).
8. Chen, J.; Huang, Y.; Li, G.; An, T.; Hu, Y.; Li, Y. VOCs elimination and health risk reduction in e-waste dismantling workshop using integrated techniques of electrostatic precipitation with advanced oxidation technologies. *J. Hazard. Mater.* **2016**, *302*, 395–403. [CrossRef] [PubMed]
9. Zeng, Z.-Q.; Wang, J.-F.; Li, Z.-H.; Sun, B.-C.; Shao, L.; Li, W.-J.; Zou, H.-K. The Advanced Oxidation Process of Phenol Solution by O3/H2O2in a Rotating Packed Bed. *Ozone Sci. Eng.* **2013**, *35*, 101–108. [CrossRef]
10. Abdullah, A.M.; Al-Thani, N.J.; Tawbi, K.; Al-Kandari, H. Carbon/nitrogen-doped TiO_2: New synthesis route, characterization and application for phenol degradation. *Arab. J. Chem.* **2016**, *9*, 229–237. [CrossRef]
11. Pelaez, M.; Nolan, N.T.; Pillai, S.C.; Seery, M.K.; Falaras, P.; Kontos, A.G. A review on the visible light active titanium dioxide photocatalysts for environmental applications. *Appl. Catal. B Environ.* **2012**, *125*, 331–349. [CrossRef]
12. Espino-Estévez, M.R.; Fernández-Rodríguez, C.; González-Díaz, O.M. Effect of TiO_2–Pd and TiO_2–Ag on the photocatalytic oxidation of diclofenac, isoproturon and phenol. *Chem. Eng. J.* **2016**, *298*, 82–95. [CrossRef]
13. Léonard, G.L.; Pàez, C.A.; Ramírez, A.E.; Mahy, J.G.; Heinrichs, B. Interactions between Zn^{2+} or ZnO with TiO_2 to produce an efficient photocatalytic, superhydrophilic and aesthetic glass. *J. Photochem. Photobiol. A Chem.* **2018**, *350*, 32–43. [CrossRef]
14. Yan, J.; Wu, G.; Guan, N.; Li, L. Nb_2O_5/TiO_2 heterojunctions: Synthesis strategy and photocatalytic activity. *Appl. Catal. B* **2014**, *152–153*, 280–288. [CrossRef]
15. Kalan, R.E.; Yaparatne, S.; Amirbahman, A.; Tripp, C.P. P25 titanium dioxide coated magnetic particles: Preparation, characterization and photocatalytic activity. *Appl. Catal. B* **2016**, *187*, 249–258. [CrossRef]
16. Castro, D.C. Síntese de Nb_2O_5 mesoporo para aplicação em fotocatálise heterogênea. Master's Degree, Universidade Federal de Mato Grosso do Sul, Campo Grande, MS, Brazil, 2014.
17. Zhao, Y.; Zhou, X.; Ye, L.; Chi Edman Tsang, S. Nanostructured Nb_2O_5 catalysts. *Nano Rev.* **2012**, *3*, 17631. [CrossRef]
18. Dong, Y.; Tang, D.; Li, C. Photocatalytic oxidation of methyl orange in water phase by immobilized TiO_2-carbon nanotube nanocomposite photocatalyst. *Appl. Surf. Sci.* **2014**, *296*, 1–7. [CrossRef]
19. Katsaros, G.; Stergiopoulos, T.; Arabatzis, I.; Papadokostaki, K.; Falaras, P. A solvent-free composite polymer/inorganic oxide electrolyte for high efficiency solid-state dye-sensitized solar cells. *J. Photochem. Photobiol.* **2002**, *149*, 191–198. [CrossRef]

20. Sakkas, V.; Arabatzis, I.; Konstantinou, I.; Dimou, A.; Albanis, T.; Falaras, P. Metolachlor photocatalytic degradation using TiO_2 photocatalysts. *Appl. Catal. B Envioron.* **2004**, *49*, 195–205. [CrossRef]
21. Santos, L.M.; Machado, W.A.; França, M.D.; Borges, K.A.; Paniago, R.M.; Patrocinio, A.O.T.; Machado, A.E.H. Structural characterization of Ag-doped TiO2 with enhanced photocatalytic activity. *RSC Adv.* **2015**, *5*, 103752–103759. [CrossRef]
22. Arcanjo, G.S.; Mounteer, A.H.; Bellato, C.R.; da Silva, L.M.M.; Brant Dias, S.H.; da Silva, P.R. Heterogeneous photocatalysis using TiO_2 modified with hydrotalcite and iron oxide under UV–visible irradiation for color and toxicity reduction in secondary textile mill effluent. *J. Environ. Manage.* **2018**, *211*, 154–163. [CrossRef]
23. Yadav, H.M.; Otari, S.V.; Koli, V.B.; Mali, S.S.; Hong, C.K.; Pawar, S.H.; Delekar, S.D. Preparation and characterization of copper-doped anatase TiO_2 nanoparticles with visible light photocatalytic antibacterial activity. *J. Photochem. Photobiol.* **2014**, *280*, 32–38. [CrossRef]
24. Fontana, K.B.; Chaves, E.S.; Kosera, V.S.; Lenzi, G.G. Barium removal by photocatalytic process: An alternative for water treatment. *J. Water Process Eng.* **2018**, *22*, 163–171. [CrossRef]
25. Nakajima, K.; Baba, Y.; Noma, R.; Kitano, M.N.; Kondo, J.; Hayashi, S.; Hara, M. $Nb_2O_5{\cdot}nH_2O$ as a Heterogeneous Catalyst with Water-Tolerant Lewis Acid Sites. *J. Am. Chem. Soc.* **2011**, *133*, 4224–4227. [CrossRef]
26. Morais, L.A.; Adán, C.; Araujo, A.S.; Guedes, A.P.M.A.; Marugán, J. Synthesis, Characterization, and Photonic Efficiency of Novel Photocatalytic Niobium Oxide Materials. *Glob. Chall.* **2017**, *1*, 1700066. [CrossRef]
27. Sreethawong, T.; Ngamsinlapasathian, S.; Yoshikawa, S. Crystalline mesoporous Nb_2O_5 nanoparticles synthesized via a surfactant-modified sol–gel process. *Mater. Lett.* **2012**, *78*, 135–138. [CrossRef]
28. Liu, J.; Zhang, T.; Wang, Z.; Dawson, G.; Chen, W. Simple pyrolysis of urea into graphitic carbon nitride with recyclable adsorption and photocatalytic activity. *J. Mater. Chem.* **2011**, *21*, 14398. [CrossRef]
29. Greenwood, N.N.; Earnshaw, A. *Chemistry of the Elements*, 2nd ed; Butterworth-Heinemann: Oxford, UK, 1998.
30. Yamashita, H.; Harada, M.; Misaka, J.; Takeuchi, M.; Neppolian, B.; Anpo, M. Photocatalytic degradation of organic compounds diluted in water using visible light-responsive metal ion-implanted TiO_2 catalysts: Fe ion-implanted TiO_2. *Catal. Today* **2003**, *84*, 191–196. [CrossRef]
31. Malengreaux, C.M.; Pirard, S.L.; Léonard, G.; Mahy, J.G.; Klobes, B.; Herlitschke, M.; Hermann, R.; Heinrichs, B.; Bartlett, J.R. Study of the photocatalytic activity of Fe^{3+}, Cr^{3+}, La^{3+} and Eu^{3+} single- doped and co-doped TiO_2 catalysts produced by aqueous sol-gel processing. *J. Alloys Compd.* **2017**, *691*, 726–738. [CrossRef]
32. Yoshimura, K.; Miki, T.; Iwama, S.; Tanemura, S. Characterization of niobium oxide electrochromic thin films prepared by reactive d.c. magnetron sputtering. *Thin Solid Films* **1996**, *281–282*, 235–238. [CrossRef]
33. Weibin, Z.; Weidong, W.; Xueming, W.; Xinlu, C.; Dawei, Y.; Changle, S.; Li, B. The investigation of NbO_2 and Nb_2O_5 electronic structure by XPS, UPS and first principles methods. *Surf. Interface Anal.* **2013**, *45*, 1206–1210. [CrossRef]
34. Scheidt, G. Caracterização óptica de filmes finos de NbO_x obtidos por sputtering reativo. Master's Degree in Nuclear Technology-Materials-Nuclear and Energy Research Institute, University of São Paulo, São Paulo, Brazil, 2014.
35. Hamaguchi, R. Preparação, caracterização e estudo das propriedades fotocatalíticas de catalisadores obtidos de dopagem de TiO_2 e Nb_2O_5. Master's Degree, DQ-FFCLRP-USP, Ribeirão Preto, Brazil, 2011.
36. Prado, A.G.S.; Bolzon, L.B.; Pedroso, C.P.; Moura, A.O.; Costa, L.L. Nb_2O_5 as efficient and recyclable photocatalyst for indigo carmine degradation. *Appl. Catal. B* **2008**, *82*, 219–224. [CrossRef]
37. Gallo, I.F.L. Preparação e caracterização de fotocatalisadores heterogêneos de titânio e nióbio e avaliação do potencial de fotodegradação. Master's Degree, FFCL-USP, Ribeirão Preto, Brazil, 2016.
38. Dalponte, I.; Mathias, A.L.; Jorge, R.M.M.; Weinschutz, R. Degradação fotocatalítica de tartrazina com TiO_2 imobilizado em esferas de alginato. *Quim. Nova.* **2016**, *39*, 1165–1169. [CrossRef]
39. Kosera, V.S.; Cruz, T.M.; Chaves, E.S.; Tiburtius, E.R.L. Triclosan degradation by heterogeneous photocatalysis using ZnO immobilized in biopolymer as catalyst. *J. Photochem. Photobiol.* **2017**, *344*, 184–191. [CrossRef]

40. Son, H.-S.; Lee, S.-J.; Cho, I.-H.; Zoh, K.-D. Kinetics and mechanism of TNT degradation in TiO_2 photocatalysis. *Chemosphere* **2004**, *57*, 309–317. [CrossRef]
41. Lenzi, G.G.; Fávero, C.V.B.; Colpini, L.M.S.; Bernabe, H.; Baesso, M.L.; Specchia, S.; Santos, O.A.A. Photocatalytic reduction of Hg(II) on TiO_2 and Ag/TiO_2 prepared by the sol–gel and impregnation methods. *Desalination* **2011**, *270*, 241–247. [CrossRef]

catalysts

Article

Effect of Cerium Precursor in the Synthesis of Ce-MCM-41 and in the Efficiency for Liquid-Phase Oxidation of Benzyl Alcohol

Carlos M. Aiube [1,*], Karolyne V. de Oliveira [2] and Julio L. de Macedo [1,*]

1 Group of New Materials for Sustainable Chemical Catalysis, Institute of Chemistry, Campus Darcy Ribeiro, University of Brasília, Brasília-DF 70910-900, Brazil
2 Laboratory of Materials and Fuels, Institute of Chemistry, Campus Darcy Ribeiro, University of Brasília, Brasília-DF 70910-900, Brazil; karolyne.vilela@gmail.com
* Correspondence: carlos.unb12@gmail.com (C.M.A.); julio@unb.br (J.L.d.M.); Tel.: +55-61-3107-3877 (J.L.d.M.)

Received: 14 March 2019; Accepted: 17 April 2019; Published: 23 April 2019

Abstract: Understanding the effects of synthetic parameters in the catalytic activity of heterogeneous catalysts is of utmost importance when aiming for optimal reaction conditions. Hence, we disclose in this work the synthesis and characterization of cerium-modified MCM-41 materials. In addition, it was observed for the first time, differences in catalytic activity when using different cerium synthetic precursors: $CeCl_3{\cdot}7H_2O$ and $Ce(NO_3)_3{\cdot}6H_2O$ (Ce-MCM-Cl and Ce-MCM-NO_3, respectively). A mechanism for cerium incorporation in MCM-41 was proposed, where $[Ce(OH)_3]$ species were hydrogen bonded to silicate anions, forming framework Ce-O-Si bonds during condensation and, consequently, causing distortion of the typical hexagonal mesophase. It was also observed that $Ce(OH)_3$ formed aggregated layers with template assemblies during synthesis, resulting in non-framework CeO_2 species on the MCM-41 surface after calcination. These CeO_2 species were preferentially formed for Ce-MCM-NO_3 and were attributed to the nitrate ions' strong binding to template molecules. In the solvent free liquid-phase oxidation of benzyl alcohol (BzOH), Ce-MCM-Cl achieved better BzOH conversions and benzaldehyde (BzD) yields, while Ce-MCM-NO_3 offered increased BzD selectivity. The catalysts' reusability was also studied over three catalytic runs, where Ce-MCM-NO_3 was more resistant than Ce-MCM-Cl towards deactivation. The observed catalytic behavior shows the importance of metal precursors in the obtainment of materials with desirable final properties.

Keywords: MCM-41; cerium; benzyl alcohol; oxidation; benzaldehyde

1. Introduction

The need for starting materials in industrial processes makes fine chemistry one of the most relevant sectors in the modern industry, having an expressive economic impact in our society. By definition, fine chemicals are economically relevant products manufactured in a limited content (less than 5000 tons per year) with high added-value associated to its production process and end use [1]. In Brazil, more than 90% of the income from this sector belongs to the production of important starting materials for pharmaceutical, agricultural, and livestock industries [2]. According to data collected by the Brazilian Association of Chemical Industries (ABIQUIM), the cited industries ended 2017 with a net billing of approximately US$25.7 billion altogether, which corresponds to 21.4% of the national chemical industry total income [3]. It is noteworthy that the sustainability of industrial actions must be a major motive to investigate alternative routes for established chemical processes. This concern arises from the main disadvantage of fine chemistry: The generation of about 5 to 50 kg of waste per kg of product due to the use of homogeneous catalysis, stoichiometric reagents, and harsh reaction conditions [4,5].

Benzaldehyde (BzD) is an essential fine chemical material for several industrial sectors, being used as a flavoring agent, as a denaturant, as a fragrance, and as a precursor to the manufacture of important products, such as, cinnamaldehyde, mandelic acid, and malachite green [6]. According to an economic report from Transparent Market Research [7], the global revenue of the benzaldehyde market surpassed US$234 million in 2014 and the demand of approximately 80 Mt is expected to increase 3.06% until 2023. Since the last century, benzaldehyde is produced following two main industrial processes [6,8]: (i) The alkali hydrolysis of benzal chloride; and (ii) the aerobic oxidation of toluene. Both reaction pathways present environmental and procedural drawbacks: High basicity of effluents and a large amount of residues in the alkali hydrolysis; and high pressure and temperature at the expense of low conversions in the aerobic oxidation [6,8]. Thus, the development of safer, more efficient, and environmentally friendly technologies to improve the production of benzaldehyde is of great interest to the fine chemical industry. The liquid-phase oxidation of benzyl alcohol (BzOH) is an interesting alternative due to its intrinsic benefits: (i) Generation of low levels of effluents with little or no negative environmental effects [9]; (ii) use of milder conditions than the cited industrial processes, with temperatures and partial pressures (for oxidants, like molecular oxygen) not higher than 150 °C and 3 atm, respectively [9,10]; and (iii) the protective effect of BzOH towards over-oxidation of benzaldehyde to benzoic acid (HBz), mainly due to the low dissociation energy of the methylene group C-H bond [11]. Hence, the study of this catalytic oxidation process offers good opportunities for the renewal of the knowledge in catalytic oxidative processes, and in the preparation and use of heterogeneous catalysts as well.

The application of numerous heterogeneous catalysts for both gas- and liquid-phase oxidation of BzOH has been reported in the literature [12–16] and microstructural/physicochemical aspects of these materials are decisive in their efficient application. For example, Della Pina and co-workers [12] reported the gas-phase oxidation of BzOH by air using 1 wt.% of gold and/or copper supported on silica as catalysts. The monometallic materials exhibited lower conversion (75% for Au/SiO_2) or selectivity (78% for Cu/SiO_2) when compared to the bimetallic combination. Among these, the material Au-Cu/SiO_2 with an Au/Cu weight ratio = 4 was the best catalyst, showing 98% of BzOH conversion and >99% of selectivity towards BzD at 313 °C. Similarly, Enache and co-workers [13] used Au-Pd/TiO_2 catalysts in the liquid-phase oxidation of BzOH under O_2 constant pressure. The material containing 2.5 wt.% of each metal showed good performance under mild conditions (pO_2 = 0.2 MPa and T = 100 °C) when compared to their monometallic analogs, exhibiting 75% of BzOH conversion and 92% of selectivity towards BzD after 8 h. Even though the good redox activity of metal nanoparticles in the activation of oxygenated species to form peroxo/hydroperoxo intermediates via hydride transfer and dehydrogenative pathways is well known [17], the combination with other metallic nanoparticles seeks the enhancement of the catalyst conversion, selectivity, and durability. Jia and co-workers [14] modified ZSM-5 zeolites with sodium hydroxide to investigate their activity in the liquid-phase oxidation of BzOH with hydrogen peroxide (H_2O_2/BzOH ratio = 1.3) at reflux conditions. All obtained materials presented a higher Lewis acidity and external surface area after the alkali-treatment. The best catalyst was obtained after 0.5 h of the latter process, showing 53% of BzOH conversion and 86% of selectivity towards BzD. Both Lewis and Brønsted acid-base sites can play important roles in the efficiency of this process by performing a dehydrogenative step in the methylene carbon that allows BzOH oxidation [14,18]. Chen and co-workers [16] synthesized a mesoporous hybrid material based on the interaction between the heteropolyacid $H_3PW_{12}O_{40}$ and *N″*-(3-amino-2-hydroxypropyl)-*N,N,N′,N′*-tetramethylguanidinium chloride (TMGHACl), an ionic liquid. The catalyst, $[TMGHA]_{2.4}H_{0.6}PW_{12}O_{40}$, was tested in the water-mediated liquid-phase oxidation of BzOH with H_2O_2 to test the reaction viability in a triphasic system. The catalyst gave 98% of BzOH conversion and 94% of selectivity towards BzD at 90 °C after 6 h. The interaction between the heteropolyanion $PW_{12}O_{40}{}^{3-}$ in the catalyst and hydrogen peroxide forms $\{PO_4[WO(O_2)_2]_4\}^{3-}$ species that are active in oxidation reactions [19,20]. Also, the catalyst was efficient in a great variety of pure and mixed solvents due to the polar groups of the ionic liquid, providing a range of triphasic systems that allow better catalyst recovery and benzaldehyde separation.

MCM-41 (mobile composition of matter) is one of the most studied mesoporous silica-based molecular sieves [21]. These materials have desirable properties for heterogeneous catalysis due to: (i) Long-range ordering of cylindrical mesopores in an hexagonal array; (ii) specific surface areas higher than 700 $m^2 g^{-1}$; (iii) tunable pore volumes between 0.7 and 1.2 $cm^3 g^{-1}$; and (iv) pore diameter between 2.5 and 4.5 nm. Such properties allow a high dispersion of catalytic species, a better accessibility to these active sites, and increase the mass diffusion of reactants than those observed for microporous materials [22]. This singular structure is possible because the synthesis procedure employs long-chain alkyl-ammonium surfactants that arrange themselves in rod-like micelles in water [23]. Such supramolecular arrangements act as structural directing agents and define the manner that silicate oligomers will interact with the cationic head group before crystallizing in the final material [24]. Therefore, disturbances in this molecular assembly might lead to microstructural changes in MCM-41 solids, affecting their final properties as well [25,26].

In the context of benzyl alcohol oxidation, studies have shown the potential of MCM-41 as support for active catalysts, including heteropolyacids [27,28], metal complexes [29,30], and metallic nanoparticles [31]. Few reports in the literature have dedicated attention to the catalytic activity of metal-incorporated MCM-41 (i.e., isomorphically substituted) in this subject [32,33], although examples towards the oxidation of other substrates have been demonstrated over the years [34,35]. Specifically, cerium-incorporated MCM-41 (Ce-MCM-41) has been explored towards oxidation reactions with positive results [36,37], mainly because of the active Ce^{4+}/Ce^{3+} redox pair that allows oxygen mobility in the structure of these materials. This feature refers to the ability of oxide-based Ce^{4+} materials to generate and relinquish oxygen vacancies, easily generating reactive sites for the activation of oxygenated species that can intermediate redox chemical processes. The insertion of cerium ions in the inorganic framework of MCM-41 could propitiate the formation of distorted coordinative environments not found on bulk materials and enhance the redox activity inherent to the element. So far, only Pal and co-workers [32] has explored this characteristic in the liquid-phase oxidation of BzOH. The material obtained by the researchers, named HSCS-20, presented 28.1 wt.% of the cerium content and showed 53.5% of BzOH conversion and 100% of selectivity towards BzD at room temperature for 24 h using *tert*-butyl hydroperoxide (TBHP) as the oxidant. Even though this result encourages the investigation of mesoporous cerium-silica catalysts in oxidation reactions, synthetic aspects involving these materials must be taken into account. For example, little has been done to unravel the effect that cerium precursors have in the catalytic activity of Ce-MCM-41 in any oxidation reaction whatsoever.

This work studies the preparation of Ce-MCM-41 materials by using different metal precursors, $CeCl_3{\cdot}7H_2O$ and $Ce(NO_3)_3{\cdot}6H_2O$ (generating the catalysts, Ce-MCM-Cl and Ce-MCM-NO_3, respectively). The structural, morphological, and textural differences between the two materials were demonstrated through characterization techniques (XRD, XRF, FTIR, UV-Vis, TG/DTG, DSC, N_2 physisorption, solid state NMR, TEM, and Raman) and correlated with disturbances in the assembly of surfactant molecules and the stability of cerium species in the synthesis medium. In addition, the catalytic activity of the materials was evaluated in the liquid-phase oxidation of BzOH with TBHP.

2. Results and Discussion

2.1. Catalysts' Characterization

The results described below are related to the calcined materials, except for the Raman spectroscopy and thermal analysis also done for the as-synthesized materials. Also, the Ce^{3+} ions used as a precursor in the synthesis procedure were oxidized to Ce^{4+} during the two step calcination to remove the template [38] (see Materials and Methods).

2.1.1. X-ray Fluorescence (XRF) and X-ray Diffraction (XRD)

The XRF elemental analysis of cerium-incorporated samples is compiled in Table 1. As can be seen, both materials have a similar cerium content, 5.89 and 5.24 wt.% for Ce-MCM-Cl and Ce-MCM-NO_3, respectively. In both cases, the Ce content is lower than the expected value, which led to Si/Ce molar ratios higher than the one used in the synthesis gel (Si/Ce = 25), i.e., Si/Ce = 31 for Ce-MCM-Cl and Si/Ce = 36 for Ce-MCM-NO_3. This result is not uncommon since the incorporation of heteroatom ions in silica matrices is strongly dependent on the synthetic procedure and precursors used [39–42].

Table 1. XRD and XRF data of calcined MCM-41 materials.

Sample	d_{200} (nm)	d_{100} (nm) [1]	a (nm) [2]	Ce content (wt.%)	Si/Ce Molar Ratio
Si-MCM-41	1.94	3.88	4.48	-	∞
Ce-MCM-Cl	2.00	4.00	4.61	5.89	31
Ce-MCM-NO_3	2.15	4.30	4.97	5.24	36

[1] Calculated from Equation (2). [2] Calculated from Equation (1).

Haller and Cesteros [39] studied the effects of several synthesis parameters during the preparation of Al-MCM-41 materials with Si/Al = 24, including different aluminum and silicon sources. Some samples showed large deviations from the desired Si/Al ratio, reaching values between 27.48 and 37.61, an error of 15% and 57%, respectively. Also, not all heteroatom ions are incorporated into the framework of silica materials, even when the silicon-to-metal molar ratio obtained in the solid is the same as the one used in the synthesis gel [43,44]. For example, Matsumoto and co-workers [43] studied Al-MCM-41 with different Si/Al ratios by solid state ^{27}Al MAS-NMR and observed a signal at 53 ppm, assigned to tetrahedrally coordinated framework aluminum, and another at 0 ppm, associated with octahedrally coordinated non-framework aluminum.

The slightly higher Si/Ce ratio obtained for Ce-MCM-NO_3, when compared to Ce-MCM-Cl, suggests that cerium(III) precursors with different anions might influence the outcome in the silica-based molecular sieves' synthesis. Indeed, XRD patterns of the calcined Ce-MCM-41 materials in the low angle region (Figure 1) showed not only differences towards Si-MCM-41, but also among themselves. Si-MCM-41 displayed reflections at 2.30, 3.95, and 4.55°, corresponding to structural planes 100, 110, and 200 of the MCM-41 hexagonal mesophase (planar space group p6m) [45]. For the prepared Ce-MCM-41, the 100 reflection was noticeably shifted towards lower 2θ values and the 110 and 200 reflections overlap with each other, forming a broad signal between 3 and 5°. The first indicates an expansion of the unit cell due to cerium incorporation in the MCM-41 framework, which is expected since the ionic radii of the Ce^{4+} ions (87 pm for 6-coordination [46]) is higher than Si^{4+} (40 pm for 6-coordination [46]). The second observation is characteristic of hexagonal disordered mesoporous materials [47] and has been observed in the literature for Ce-MCM-41 [48–51]. This loss of mesophase ordering can be attributed to the partial substitution of Si^{4+} for Ce^{4+} ions due to changes in the Si-O-Si bond angles that lead to the emergence of local structural defects and bond strain [37,47]. However, metal precursor salts in the synthesis gel could cause changes in the assembly of CTA^+ molecules, leading to less organized materials when compared to Si-MCM-41. As Barkam and co-workers [52] demonstrated, even the use of different anions for the same metal precursor can extensively affect the physicochemical properties of particles in solution [52].

Since MCM-41 is a molecular sieve with a bidimensional structure, the calculation of the hexagonal unit cell parameter (a) could be determined by using Equation (1) (see Materials and Methods). Unfortunately, the 100 reflection cannot be observed in the XRD pattern for the cerium-incorporated materials (Figure 1), as a consequence of the experimental limitations (i.e., the signal of the beam source was only obtainable starting at 2θ = 2°). As mentioned before, the 110 and 200 reflections could be distinguished, but are overlapped. If the sole 200 reflection could be observed in the modified materials, d_{100} could be calculated by a relationship between the interplanar spacing of the 200 and 100 reflections (Equation (2), see Materials and Methods). Therefore, deconvolution methods [53] were applied in the overlapped signal to gain information towards the 2θ position of both the 110 and 200 reflections to find d_{100} and, consequently, the unit cell, a, for the modified materials (see Figure 2).

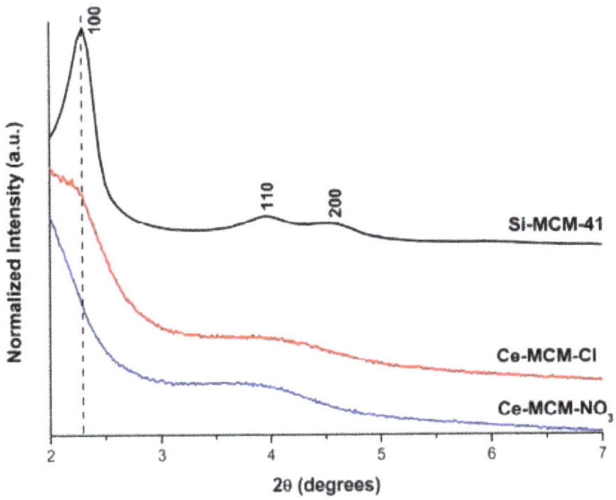

Figure 1. XRD patterns of the calcined MCM-41 materials at the low angle region.

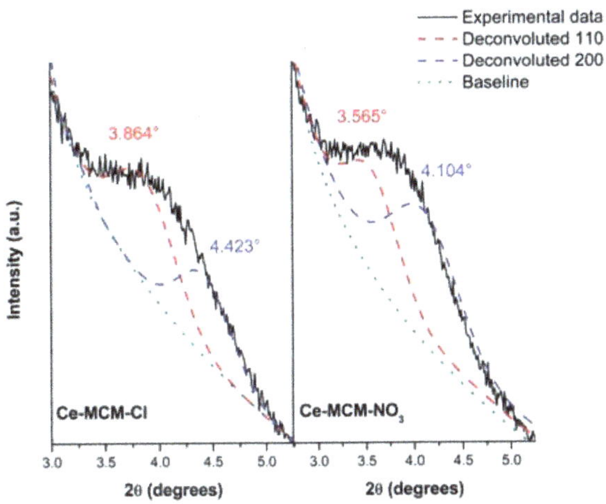

Figure 2. Deconvolution of the overlapped XRD reflections and their 2θ positions for Ce-MCM-Cl and Ce-MCM-NO_3 materials.

The 110 and 200 reflections were found in 3.86 and 4.42° for Ce-MCM-Cl and in 3.56 and 4.10° for Ce-MCM-NO_3, respectively. The values obtained for d_{100} and a using the information of the 200 deconvoluted reflections are compiled in Table 1. For Si-MCM-41, the d_{100} value calculated from the 200 reflection (3.88 nm) was close to the one observed experimentally (3.84 nm), validating our data obtained from Equations (1) and (2). As can be observed in Table 1, the values of a calculated for the cerium materials confirmed the unit cell expansion mentioned before and this effect was more expressive towards Ce-MCM-NO_3.

Differences between both cerium-modified materials could also be observed at the high angle region (Figure S1 in Supplementary Material). While the Ce-MCM-Cl pattern only exhibited the usual broad signal around 25° common to amorphous silica in the MCM-41 walls, Ce-MCM-NO_3 presented three reflections around 28, 47, and 57° that suggests the formation of CeO_2 (reflections 110, 220, and 311 according to JCPDS 34-0394, respectively [54]). The surfacing of non-framework species is generally associated with the synthetic procedure [37,55] and the influence of a particular metal precursor is little discussed in the literature. However, the absence of peaks from segregated CeO_2 in the particles of the Ce-MCM-Cl pattern does not mean there are not any; they might be highly dispersed on the MCM-41 surface. Nonetheless, the XRD analysis demonstrated the clear presence of CeO_2 in one of the materials.

2.1.2. Spectroscopy Studies (FTIR, UV-Vis, Raman, and ^{29}Si MAS-NMR)

To investigate the origin of non-framework CeO_2 species, as well as the consequences of using different cerium precursors, Raman spectroscopy was performed in both as-synthesized and calcined MCM-41 materials. The Raman spectra of as-synthesized materials can give information on how an assembly of surfactants is affected by other ionic species in the synthetic medium. Raman signals of organic molecules are dominant in this situation and the occluded CTA^+ cations can be in the same arrangement from the synthetic medium, counter-balancing the negative charges from deprotonated silanol groups. Also, the use of CTA^+ as a probe to access information towards microstructural characteristics from the framework of as-synthesized MCM-41 has been little explored in the literature [51,56,57], encouraging the use of conformational-sensible spectroscopic techniques, such as Raman spectroscopy, in the unraveling of microstructural changes.

Figure 3 displays the spectra of the as-synthesized materials and, for means of comparison, the spectra of template molecules with two different molecular orderings were acquired: A 23 wt.% aqueous solution of cetyltrimethylammonium bromide (CTAB) and solid CTAB (Figure S2 in Supplementary Material). The concentration of 23 wt.% was chosen because it is in the range of the concentration where surfactant molecules form an hexagonal liquid-crystal phase in solution [58]. The comparison between the spectra of the as-synthesized materials and the two template standards clearly shows that CTA^+ cations are organized as cylindrical micelles inside the pores of as-synthesized materials, as it is expected based on the mechanism for MCM-41 crystallization [59].

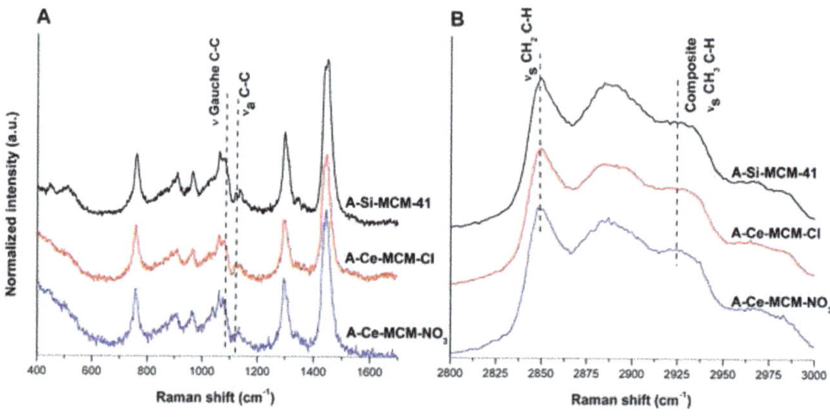

Figure 3. Raman spectra of the as-synthesized materials between 400 and 1700 cm^{-1} (**A**) and 2800 and 3000 cm^{-1} (**B**). The signals used for the determination of the disorder parameters are indicated.

The assignment of each vibrational mode observable in the spectra of the as-synthesized materials and template standards is summarized in Table S1 (see Supplementary Material). For the solid CTAB, two of the most important Raman bands that contain information towards molecular configuration are those in 1061 and 1126 cm^{-1}, attributed to C-C symmetric and asymmetric stretching, respectively [57,60,61]. These are characteristic of extended hydrocarbon chains with all-*trans* ordered structures in which alternate carbon atoms move in the opposite direction. Other important vibrational modes at 2848/2880 and 2941/2958 cm^{-1} were assigned to the C-H symmetric/asymmetric stretching of

CH_2 and CH_3, respectively. These bands are the strongest in the Raman spectra and highly sensitive to molecular conformation [57,62]. In comparison with solid CTAB, the Raman spectra of CTAB solution and occluded CTA^+ in as-synthesized materials showed less and broader signals, as well as other relevant differences. First, a new signal appeared at 1076 cm^{-1} for both as-synthesized materials and CTAB solution. This signal, which was not observed for solid CTAB, was assigned to the C-C stretching mode of hydrocarbon chains in the gauche configuration and is expected in more disordered systems [57,60]. Second, common signals found for solid CTAB were absent or became overlapped for CTA^+ molecules in as-synthesized materials or in solution, especially those at 1175 and 1464 cm^{-1}, associated to the dynamic motion of molecules in supramolecular assemblies [57]. This result indicates that occluded CTA^+ molecules in as-synthesized materials act as it were in a liquid-like state [57]. Finally, intensity and shift changes in conformational-sensitive peaks between 2850 and 3000 cm^{-1} were also observed for CTA^+ occluded and in solution. The asymmetric methylene C-H mode around 2885 cm^{-1} is less intense than in the solid CTAB and the signals assigned to methyl C-H modes were considerably displaced, both evidence of higher statistical chances for attaining the gauche configuration, i.e., a more disordered state [57,61,62]. This was even more noticeable for the broad band around 2920 to 2928 cm^{-1} for CTA^+ occluded and in solution. This band is a composite of two overlapping modes, one from the polymethylene chain (usually around 2920 cm^{-1}) and the other from the methyl group (at 2941 cm^{-1} for solid CTAB), making it sensitive to conformational disorder of the long methylene chain, since the former contributes more than the latter [61,62].

As mentioned before, changes in the intensity of C-H and C-C stretching modes can reveal effects towards disordering since liquid-like states permit an increase in the number of molecules that can attain the gauche conformation. In the literature, it was demonstrated that the ratio between the intensities of certain long alkyl chain Raman bands can reflect their conformational state [61–63]. Therefore, we employed such parameters to understand the influence of different cerium precursors in the assembly of the template molecules. Two of the most employed ratios are those between the C-C gauche stretching around 1080 cm^{-1} and the C-C asymmetric stretching around 1120 cm^{-1} [57,62]; and the C-H stretchings around 2920 and 2850 cm^{-1} [61,63]. Higher values for these parameters are related to an increase of feasibility to attain the gauche conformation and, consequently, disordered the template assembly more in the synthetic medium that was retained in MCM-41 pores. Both ratios for the as-synthesized materials were calculated and are compiled in Table 2.

Table 2. Template assembly disorder for the as-synthesized MCM-41 materials.

As-Synthesized Sample	I_{1080}/I_{1120} [1]	I_{2920}/I_{2850} [2]
A-Si-MCM-41	1.994	0.731
A-Ce-MCM-Cl	2.310	0.760
A-Ce-MCM-NO_3	2.024	0.747

[1] Ratio between the intensities of bands around 1080 and 1120 cm^{-1}. [2] Ratio between intensities of bands around 2920 and 2850 cm^{-1}.

The I_{2920}/I_{2850} ratio demonstrates small changes in its value when compared to the I_{1080}/I_{1120} ratio, which indicates that the latter is a more sensible parameter towards conformational disordering than the former [62]. Nonetheless, the values presented for both ratios allowed the materials to be organized in increasing disorder of template assemblies: A-Si-MCM-41 < A-Ce-MCM-NO_3 < A-Ce-MCM-Cl. Thus, it was possible to perceive that: (i) The presence of cerium species altered the supramolecular arrangement of CTA^+ cations; (ii) the disordering was more prominent when utilizing cerium(III) chloride as the precursor; and (iii) the degree of disordering obtained for A-Ce-MCM-NO_3 was close to A-Si-MCM-41. To explain these observations, it is worth recalling the chemistry of cerium species in high pH aqueous mediums [64–66] and the synthetic procedure used in this work to prepare Ce-MCM-41 materials (see Materials and Methods). Under basic conditions, the hydrolysis of Ce(III) generates $[Ce(OH)_y]^{(3-y)}$ species, where y goes from 0 to 3. During our synthesis, the dropwise addition of cerium(III) precursor turned the clear solution into a dark slurry before it became a light brown/purple solid. This could be addressed by assuming that $[Ce(OH)]^{2+}$, $[Ce(OH)_2]^+$, and $[Ce(OH)_3]$ formation occurred progressively and continuously [65]. In addition, the oxidation of Ce^{3+} to Ce^{4+} is not thermodynamically favored under the conditions used in this work (i.e., non-aerated solution and pH = 11.2) [64] and $[Ce(OH)_3]$ can be regarded as the main cerium compound in solution [67].

Based on the above discussion and the well-known self-assembled systems used for synthesizing MCM-41 [59], this work proposes the following mechanism for cerium incorporation: (i) CTA^+ cations are counterbalanced by basic silicate anions; (ii) the latter interacts through hydrogen bonds with $[Ce(OH)_3]$ species; and (iii) the gradual condensation under basic conditions leads to silicate oligomers containing Si-O-Si and Ce-O-Si bonds. The higher disorder detected by Raman for occluded CTA^+ indicates that cerium incorporation into silica causes different electrostatic interactions between CTA^+ cations and the anionic surface. Since both cerium(III) precursors are highly soluble salts with similar stability constants [68], the differences observed between Ce-MCM-Cl and Ce-MCM-NO_3 by the techniques used in this work must be related to the different anions. Berr and co-workers [69] studied the effect of anions on alkyltrimethylammonium micelles and observed that micelle aggregation increases in the following order: $OH^- << Cl^- < CH_3SO_4^- < Br^- < NO_3^-$. For that reason, nitrate anions induce longer rods of CTA^+ cations in solution than chloride [70]. In addition, the molar ratio used in the synthesis of Ce-MCM-41 materials presented a concentration of nitrate or chloride anions close to CTAB, exhibiting 1TEOS:0.04Ce:0.12A:0.152CTAB:2.8NH_4OH:141.2H_2O, where A = NO_3^- or Cl^-. Since NO_3^- ions have a higher ionic exchange constant on CTAB micelles than Cl^- [71], they are expected to present a stronger binding to CTA^+ rods. For Ce-MCM-NO_3, nitrate ions bonded to CTA^+ could also be hydrogen bonded to $[Ce(OH)_3]$. This layer of concentrated cerium(III) species will aggregate during calcination to form non-framework CeO_2 species on the surface of the MCM-41 structure. The higher disordering level observed for Ce-MCM-Cl can be associated to a higher degree of framework Ce-O-Si due to the lower binding between chloride ions and CTA^+ molecules. Thus, the physicochemical differences among both cerium modified samples may arise fundamentally from an anion effect in the disturbance of CTA^+ molecule assemblies, especially the formation of different cerium species in the calcined materials.

Further effects in the formation of cerium species can be perceived in the calcined materials. Figure 4 displays the Raman spectra of the calcined materials between 650 and 150 cm^{-1}. In general, they all present signals common to the amorphous nature of the silica present in MCM-41. At the studied range, the Si-MCM-41 spectrum presents a signal at 483 cm^{-1}, typical of the stretching mode of Si-O bonds for fourfold cyclosiloxane groups [72–74]. For Ce-MCM-Cl, a weak peak appears aside the cyclosiloxane group mode at 460 cm^{-1} and this signal becomes dominant over the 483 cm^{-1} mode in the spectrum of Ce-MCM-NO_3. This feature is commonly attributed in the literature to the symmetric stretching of Ce-O bonds of the fluorite crystal structure of CeO_2 [75] and was assigned as the F_{2g} mode of CeO_2. The spectrum of crystalline CeO_2 was obtained for means of comparison (Figure S3 in Supplementary Material) and the F_{2g} mode presents itself as a strong signal centered in 463 cm^{-1}. In the case of the modified materials, the presence of this extra signal in both samples indicates the existence of non-framework cerium and that Ce-MCM-NO_3 possess a higher content of these species due to a much stronger intensity for this vibrational mode, which is in agreement with the XRD data presented earlier. A similar observation in the F_{2g} mode appearance was made by Zhan and co-workers [48] for Ce-MCM-48 materials, where the mode increased in intensity with the increase of cerium content. However, such a difference between two cerium-modified MCM-41 materials with similar Si/Ce ratios as it is presented in this work has not been observed in the literature so far.

Correlating with the information extracted for the as-synthesized materials, it is possible to ascertain the mechanism of crystallization proposed earlier. Thus, the disordering observed in the assembly of CTA^+ molecules of Ce-MCM-Cl was driven by isomorphic substitution. As for Ce-MCM-NO_3, part of the cerium added in the synthetic medium does not react with silicate oligomers and precipitates to form non-framework $[Ce(OH)_3]$, which converts completely to CeO_2 after calcination.

Such a hypothesis is scarcely pointed out to explain the difference in the reactivity of cerium precursors in the synthesis of oxides, but there is an extensive literature regarding the differences in the use of precursors that leads to singular microstructural properties of these materials [52,76,77]. These works clearly suggest that the anions in cerium precursors have major roles in defining the properties of the final cerium-based materials more than just the mere electrostatic counterparts in the synthetic medium.

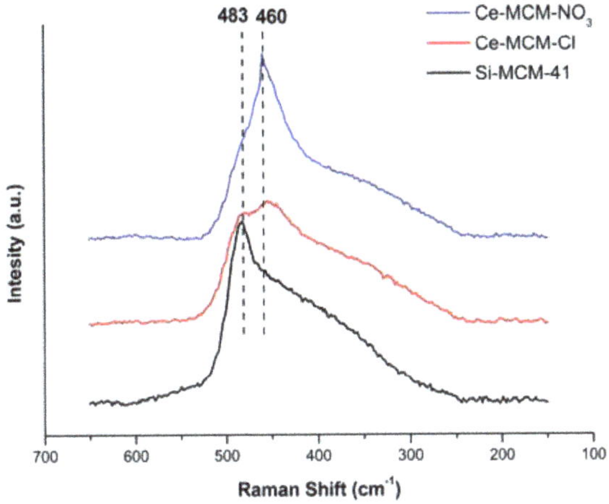

Figure 4. Raman spectra of the calcined materials between 150 and 650 cm^{-1}.

Figure 5 displays the Fourier transform infrared spectroscopy (FTIR) spectra for all calcined MCM-41 materials.

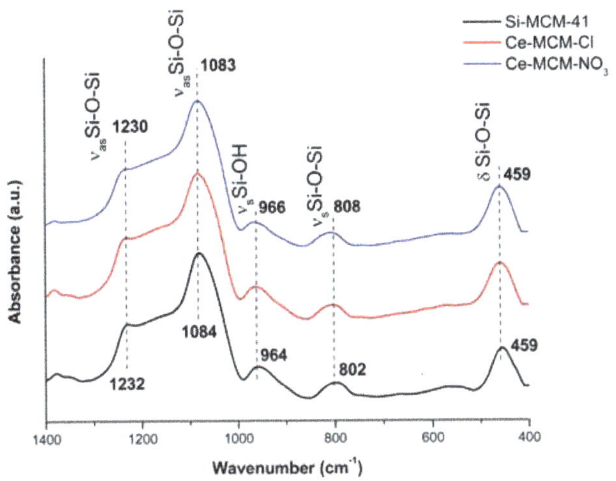

Figure 5. FTIR spectra of the calcined MCM-41 materials.

The vibrational modes between 1400 and 400 cm^{-1} are typical of Si-O-Si bonds [78]. In the case of Si-MCM-41: (i) Asymmetric stretching of Si-O-Si bonds at 1232 and 1084 cm^{-1}; (ii) symmetric stretching of Si-OH bonds at 964 cm^{-1}; (iii) symmetric stretching of Si-O-Si at 802 cm^{-1}; and (iv) the Si-O-Si bending mode at 459 cm^{-1}. Changes in the wavenumber of these modes are frequently associated to isomorphic substitution of heteroelements in the framework of silica-based materials, because the average length of T-O bonds (T corresponds to any heteroatom) in the walls of the MCM-41 is increased with the insertion of a larger atom than silicon [37,49,51]. However, the shifts observed for the cerium-incorporated materials were too small to make any assumption of isomorphic substitution. In addition, no vibrational modes attributed to Ce-O bonds, between 1600 and 900 cm^{-1} and below 700 cm^{-1}, were evidenced. These observations (i.e., small shifts and absence of Ce-O bonds) could be respectively attributed to the low concentration of framework Ce^{4+} ions and non-framework CeO_2 when compared to the amount of silica in MCM-41.

Diffuse reflectance UV-Vis (DRS UV-Vis) spectra for the calcined materials are displayed in Figure 6A. When compared to Si-MCM-41, cerium-incorporated materials presented stronger absorption in the ultraviolet range and a red shift of the absorption edge due to $O^{2-} \rightarrow Ce^{4+}$ charge transfer [32]. These experimental observations are proof of cerium incorporation in the materials, as already described by Pal and co-workers [32]. Moreover, the greater intensity of the Ce-MCM-Cl spectrum can be due to its slightly higher cerium content (see Table 1).

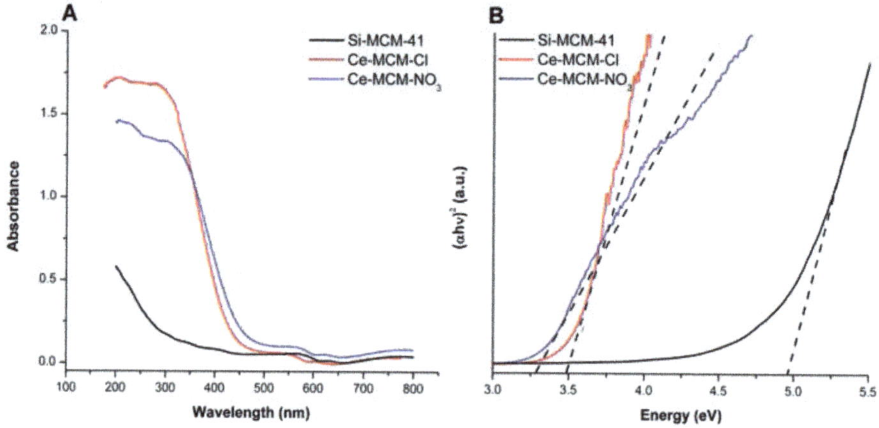

Figure 6. Absorption spectra (**A**) and Davis-Mott plots (**B**) for the calcined MCM-41 materials.

When compared to a standard sample of CeO_2 (Figure S4 in Supplementary Material), it is possible to gain information towards the coordination environment of the heteroatom, since the charge transfer transition is symmetry-sensible [37,51,79]. The CeO_2 spectrum presented more intense bands at 380 and 250 nm, owing to Ce^{4+} ions in octahedral/polymeric and tetrahedral coordination environments, respectively. The octahedral/polymeric environment is the main coordination type for cerium in the CeO_2 fluorite crystal structure. Its non-stoichiometric lattice [80] gives the dioxide a high oxygen mobility and absorptions near or above 380 nm is strong evidence of non-framework species. The spectra of cerium-incorporated materials in Figure 6A showed that UV-absorption finishes around 310 nm for Ce-MCM-Cl with a steep absorption edge until 450 nm, while for Ce-MCM-NO_3, it ends at 355 nm with a less steep edge until 500 nm. These results are evidence that the heteroatoms in Ce-MCM-Cl have more tetrahedral coordination sites than Ce-MCM-NO_3, which absorbs more radiation in the visible range of the spectrum due to non-framework species [37,51,79].

The different positions found for the absorption edge in Ce-MCM-41 samples can be used to obtain important information towards the heteroatom by determining the optical band gap. The latter was calculated by Equation (3) (see Materials and Methods) using $n = \frac{1}{2}$, due to direct allowed transitions, and Davis-Mott plots with suitable linear ranges were obtained ($R^2 > 0.99$). CeO_2 has an optical band gap of 3.02 eV, which is near to the values described in the literature for the bulk material [81]. Ce-MCM-Cl and Ce-MCM-NO_3 showed band gap values of 3.50 and 3.33 eV, respectively, which were lower than Si-MCM-41 (4.97 eV). As studied by Casas-Orozco and co-workers [73], higher band gap values of metal-incorporated MCM-41 than the metal oxide counterpart indicates high dispersion of the heteroatom in the siliceous framework. Also, the lower band gap value observed for Ce-MCM-NO_3

could be associated with the higher content of non-framework species identified by XRD and Raman, resulting in less dispersed CeO_2 particles.

^{29}Si MAS-NMR spectra of the calcined materials are available in Figure 7. As can be seen, the signals obtained for all samples were broad and overlapped in the range between −80 and −120 ppm. This interval is typical for the Q^n-groups with a general formula of $Si(OSi)_n(OH)_{4-n}$ [82]: (i) Geminal silanols or Q^2 groups (resonating around −90 ppm); (ii) terminal silanols or Q^3 groups (resonating around −100 ppm); and (iii) siloxane or Q^4 groups (resonating around −110 ppm). The overlapping effect is typical in systems based on amorphous silica due to the wide distribution of Si-O-Si bond angles and Si-O bond lengths [82]. It was possible to perceive that, in the case of Ce-MCM-Cl, there was a slight diminishment in the signal of Q^3 groups, indicating a decrease in its content. This type of change was even more noticeable for Ce-MCM-NO_3, where all defect-related environments (Q^2 and Q^3) were significantly reduced when compared to Si-MCM-41. Such qualitative observations indicate that the cerium incorporation is modifying the condensation of Si-O-Si bonds in the structure of these materials. To access more information towards the latter, deconvolution methods were applied and are denoted in Figure 7.

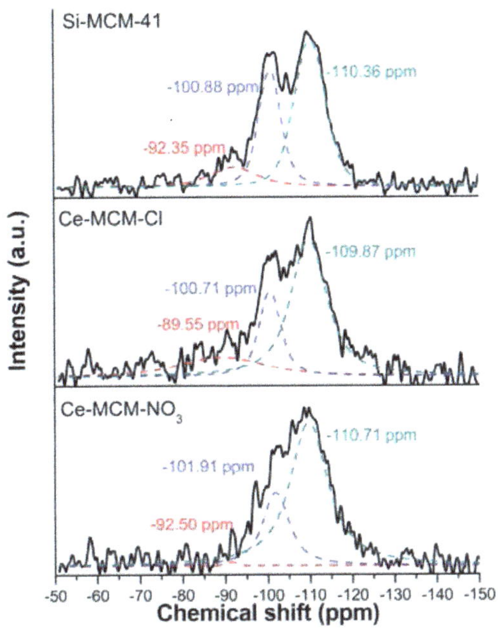

Figure 7. ^{29}Si MAS-NMR spectra for the calcined MCM-41 materials, highlighting the position of each chemical environment detected by deconvolution.

No additional peaks to those mentioned earlier were found after applying the deconvolution methods. Therefore, the traditional parameters associated to the condensation of Si-O-Si bonds were applied to further analyze the changes perceived for cerium modified materials. Among them, the percentage of each siliceous chemical environment was calculated based on the area of each peak and they were used to determine $Q^4/(Q^2 + Q^3)$ ratios and the molar percentage of silanol groups (Si molar%) using Equation (4) (see Materials and Methods). Higher values for the mentioned ratio and lower ones for Si molar% indicates more condensed Si-O-Si bonds in the inorganic framework. These quantitative data were compiled in Table 3.

Table 3. Content of Q^n-groups and related parameters obtained from ^{29}Si MAS-NMR spectra of the calcined MCM-41 materials.

Sample	Q^2 (%) [1]	Q^3 (%) [1]	Q^4 (%) [1]	$Q^4/(Q^2 + Q^3)$	Silanol (Si Molar%) [2]
Si-MCM-41	12.4	33.1	54.5	1.20	57.9
Ce-MCM-Cl	19.3	21.2	59.5	1.47	59.7
Ce-MCM-NO_3	0.4	25.6	74.0	2.85	26.4

[1] Calculated from normalization of the area of each peak. [2] Calculated from Equation (4).

From the values displayed in Table 3, it was possible to infer that: (i) For Ce-MCM-Cl, even though the content of terminal silanols (Q^3) decreases, the siloxane group (Q^4) does not increase significantly when compared to geminal silanols (Q^2), suggesting that even if the condensation of Si-O-Si bonds slightly increases, the percentage of silanol groups (Si molar%) remains relatively constant if Q^2 increases; and (ii) for Ce-MCM-NO_3, the modification decreases drastically the content of silanol groups, indicating that the non-framework cerium species might be located in the mesopores' wall or external surface, but Raman data suggests the former. Once again, there is little contribution from the literature to explain effectively the reason for differences in ^{29}Si MAS NMR arising solely with varying the counterion in cerium precursor. Even though this is a scarce topic, it is well known that such effects in siliceous molecular sieves are dependent more in the synthetic procedure than in any other aspect and the fact that cerium incorporation can lead to a more or less condensed degree of chemical bonds has already been reported [32,51]. Thus, the MAS-NMR data is in good agreement with the observations made earlier due to the crystallization mechanism proposed: (i) The smaller aggregation effect of chloride anions on CTA^+ assemblies and the formation of framework Ce-O-Si bonds led to the emergence of silanol defects (Q^2); and (ii) the precipitation of $Ce(OH)_3$ and conversion to non-framework CeO_2 resulted in a coating effect in the MCM-41 surface, leading to the decrease of Q^2.

2.1.3. Morphological (TEM and HRTEM), Textural (Surface Area and Pore Volume), and Thermal Analyses (TG/DTG/DSC)

The morphology of the calcined MCM-41 materials was accessed through low-resolution transmission electron microscopy (TEM). Representative images are selected and displayed in Figure 8.

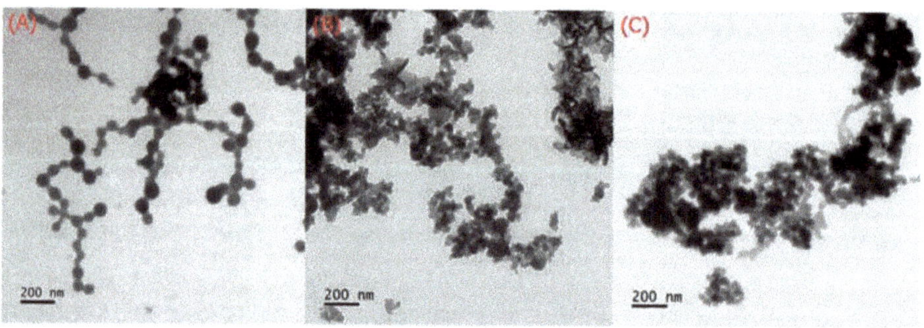

Figure 8. TEM images for the calcined MCM-41 materials: Si-MCM-41 (**A**); Ce-MCM-Cl (**B**); and Ce-MCM-NO_3 (**C**).

For Si-MCM-41, perfectly spherical nanoparticles were obtained (mean size of 61.8 ± 15.9 nm). There is little aggregation between the particles, possibly a synthetic consequence of the rapid formation of ethanol as a side product during TEOS hydrolysis [83]. On the other hand, the modified materials present smaller and spherical nanoparticles, with a mean size of around 34.2 ± 14.5 nm and 38.7 ± 9.1 nm for Ce-MCM-Cl and Ce-MCM-NO_3, respectively. Also, the nanoparticles have a more aggregated aspect when compared with Si-MCM-41, exhibiting more irregular shapes. These observations are possibly an evidence of a faster nucleation process, a consequence of the presence of cerium precursors in the synthetic medium. This supports the loss of long-range ordering observed previously by XRD, as well as the disordering effect of the supramolecular assembly of CTA^+ cations evidenced by Raman spectroscopy.

To further study the modified materials, high-resolution transmission electron microscopy (HRTEM) was used to access the formation of CeO_2 species on Ce-MCM-41 materials and representative images are displayed in Figure 9. As one can see, some of the irregular entities observed in TEM for both materials showed dark nanorods on their surface that were not observable on Si-MCM-41. High magnifications of these nanorods (Figure S5 in Supplementary Material) displayed lattices fringes around 0.33 nm, consistent with the interplanar distance characteristic to 110 XRD reflection of CeO_2, thus confirming their identity. For Ce-MCM-Cl, these nanorods displayed lengths between 22.0 and 70.0 nm and widths between 1.75 and 2.50 nm, while for Ce-MCM-NO_3, these entities were bigger and thicker, with lengths between 100 and 250 nm and widths between 5.0 and 6.7 nm. The differences observed in their measures were consistent with the crystallization mechanism proposed for non-framework cerium species, where longer and thicker rods were expected for Ce-MCM-NO_3 due to the aggregation effect of micelles attributed to the nitrate anion. Also, the formation of these nanorods was consistent with the mechanism proposed for non-framework cerium species. The formation of ceria nanorods is widely studied in the literature [84,85] and includes surfactant-assisted synthetic procedures using CTAB. However, these methodologies are normally performed under hydrothermal conditions with high temperatures (80–160 °C), longer time periods (8–96 h), and low Ce/CTAB molar ratios (0.33–0.66) [84,85]. The results observed in this work suggest that nanorods can also form under milder conditions (30–35 °C and 4 h) and high Ce/CTAB molar ratios (3.8).

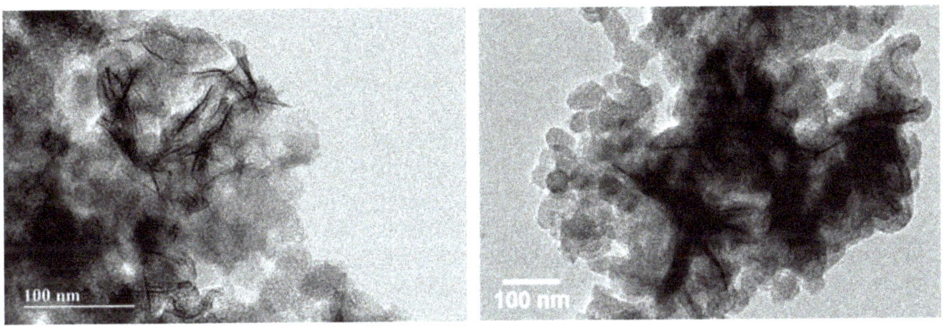

(**a**) (**b**)

Figure 9. HRTEM images for the calcined Ce-MCM-Cl (**a**) and Ce-MCM-NO_3 (**b**) materials.

The textural characterization of the calcined MCM-41 materials was performed by N_2 physisorption using t-plot curves obtained from analytical functions for each portion of the P/P_0 domain (see Equations (5)–(8) in Materials and Methods) [86]. The initial part of the curve was used to calculate the total specific surface area (S_{total}) of the materials using the slope of the linear fit going through the origin (red line in Figure 10a). On the other hand, the final part of the curve was used to calculate the external specific surface area (S_{ext}), indicated in Figure 10a by the blue line.

Figure 10b exhibits the N_2 adsorption/desorption isotherms obtained for Si-MCM-41, Ce-MCM-Cl, and Ce-MCM-NO_3, respectively. According to the IUPAC classification [87], all curves were type IV isotherms, presenting a characteristic hysteresis loop associated with capillary condensation in mesopores. This hysteresis was type H1 according to the IUPAC classification [87] and could be associated with a narrow and uniform pore size distribution. This result implies that the MCM-41 pore system in cerium-incorporated materials was preserved and that the long-range disorder was responsible for the loss of the mesophase observed by XRD. This is not uncommon and even in highly disordered materials, such as titanosilicate ETS-10, short-range ordering could not be affected by long-range disordering [88].

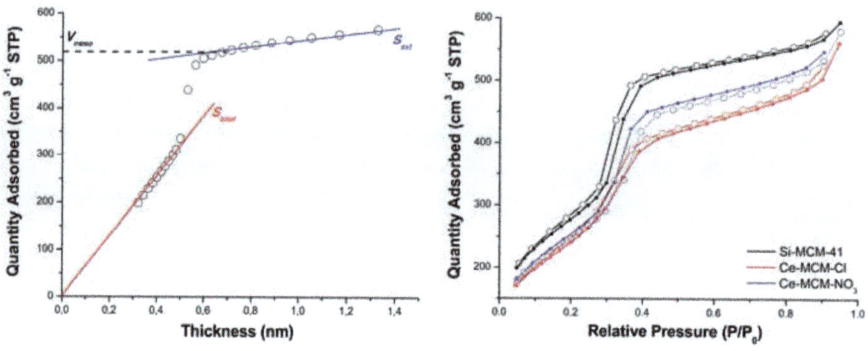

(a) (b)

Figure 10. Thickness or t-plot curve for Si-MCM-41 (**a**) where the blue line slope corresponds to S_{total}, the red line slope corresponds to S_{ext}, and the dashed black line intercept with the *y*-axis corresponds to V_{meso}; N_2 adsorption/desorption isotherm for Si-MCM-41, Ce-MCM-Cl, and Ce-MCM-NO_3 (**b**) where filled circles correspond to adsorption and hollow circles corresponds to desorption.

The initial part of the isotherm, before capillary condensation, is associated to monolayer-multilayer adsorption on both mesopores and external surface, and consequently, was used to calculate S_{total}, as described above [86]. Similarly, in the final part of the isotherm, after capillary condensation, adsorption occurs only on the external surface and, for that reason, was used to determine S_{ext}. The capillary condensation is identified by an increase of adsorbed N_2 at the relative pressure range of $P/P_0 = 0.30$ to 0.45, due to the filling of mesopores.

The data obtained by N_2 physisorption experiments (Table 4) showed a reduction of S_{total}, mesoporous surface area (S_{meso}), and mesopore volume (V_{meso}) for the cerium-incorporated samples, which could be attributed to the loss of long-range order. The sharp increase observed in all isotherms at the relative pressure range of $P/P_0 = 0.9$ to 1.0 is related to the presence of interparticle porosity [32] and its effect can be observed in a higher value of the total pore volume (V_{total}) than V_{meso}. For that reason, the average pore diameter (*PD*) was calculated using the penultimate point from the adsorption isotherm (see Materials and Methods) to avoid contributions from interparticle porosities, since the materials showed different particle sizes. The values obtained for *PD* were almost identical, which is in agreement with the narrow and uniform pore size distribution expected for the type H1 hysteresis loop observed for all materials and discussed above. However, the sample of Ce-MCM-NO_3 showed a small increase that could be attributed to its different mechanism of pore formation, possibly due to the coating effect of non-framework CeO_2. Additionally, *PD* values were used to determine the wall thickness (*WT*) of MCM-41 materials, evidencing that cerium incorporation affects this parameter due to framework cerium and non-framework CeO_2. In the case of Ce-MCM-Cl, the increase in *WT* was consistent with the incorporation of Ce in the inorganic framework, causing an expansion of the unit cell (*a*) and, ultimately, of the pore walls. However, in the case of Ce-MCM-NO_3, not only isomorphic substitution could cause such an increase, but also the existence of non-framework CeO_2 and the higher value of *WT* corroborates the possible coating effect mentioned previously, which can make the pore walls appear thicker. These results are in accordance with the higher value of *a* found for Ce-MCM-NO_3 (4.97 nm) when compared to Ce-MCM-Cl (4.61 nm), explaining the cell expansion observed by XRD. Finally, the higher S_{ext} value obtained for cerium-incorporated materials are linked to their smaller particle size when compared to Si-MCM-41, as observed by TEM.

Table 4. Textural properties for the calcined MCM-41 materials.

Properties	Si-MCM-41	Ce-MCM-Cl	Ce-MCM-NO_3
S_{total} ($m^2\ g^{-1}$) [1]	1032.1	909.7	907.0
S_{ext} ($m^2\ g^{-1}$) [2]	158.1	267.6	252.4
S_{meso} ($m^2\ g^{-1}$) [3]	874.0	642.1	654.6
V_{total} ($cm^3\ g^{-1}$) [4]	0.92	0.87	0.90
V_{meso} ($cm^3\ g^{-1}$) [5]	0.80	0.68	0.71
PD (nm) [6]	3.40	3.43	3.64
WT (nm) [7]	1.08	1.18	1.33

[1] Total specific surface area. [2] External specific surface area. [3] Mesoporous surface area. [4] Total pore volume. [5] Mesopore volume. [6] Average pore diameter. [7] Wall thickness.

Thermal analyses were performed for the as-synthesized and calcined MCM-41 materials to further analyze the consequences of texture changes towards microstructural aspects. The thermogravimetry (TG), differential thermogravimetry (DTG), and differential scanning calorimetry (DSC) curves for the as-synthesized materials are displayed in Figures 11–13. The values of mass loss, temperature range, and DTG maximum were collected in Table S2 (see Supplementary Material).

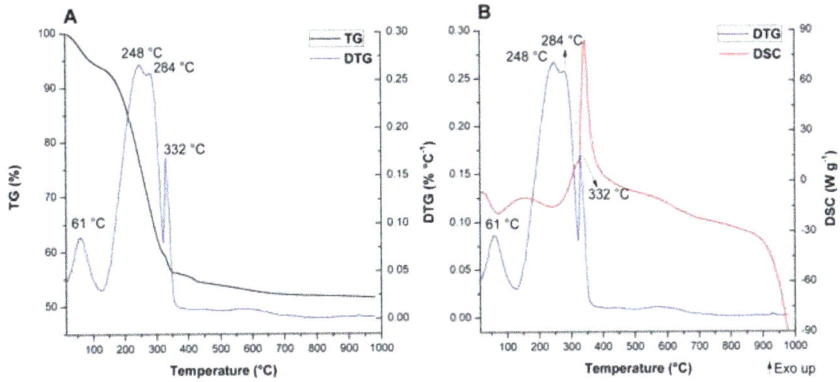

Figure 11. Thermal analysis obtained under synthetic air for A-Si-MCM-41: TG and DTG curves (**A**); DTG and DSC curves (**B**). The temperatures indicated refers to the DTG maximums of observable peaks.

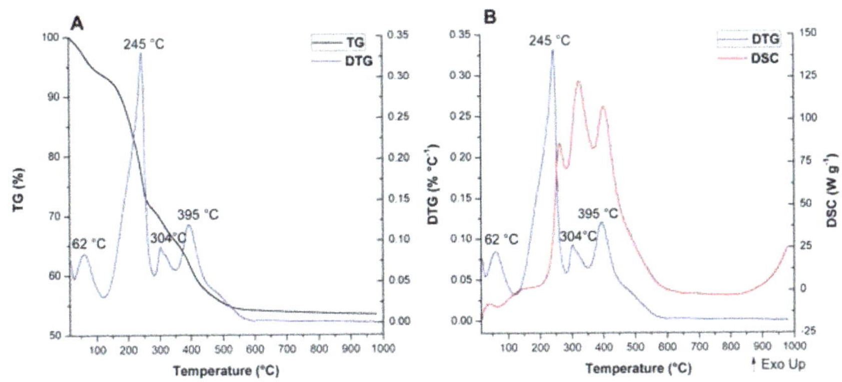

Figure 12. Thermal analysis obtained under synthetic air for A-Ce-MCM-Cl: TG and DTG curves (**A**); DTG and DSC curves (**B**). The temperatures indicated refers to DTG maximums of observable peaks.

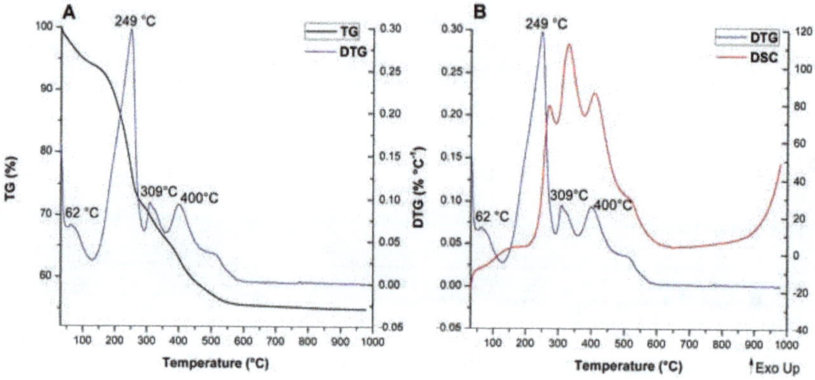

Figure 13. Thermal analysis obtained under synthetic air for A-Ce-MCM-NO_3: TG and DTG curves (**A**); DTG and DSC curves (**B**). The temperatures indicated refers to DTG maximums of observable peaks.

For A-Si-MCM-41, five thermal events were observed and are consistent with previously reported events for similar as-synthesized MCM-41 materials [89]: (i) Evaporation of physically adsorbed water (endothermic, from room temperature to 129 °C); (ii) initial decomposition of occluded CTA^+ molecules through Hofmann elimination (endothermic, from 129 to 270 °C), generating triethylamine and alkenes; (iii) further fragmentation of carbon chains that resisted Hofmann elimination (exothermic; from 270 to 322 °C); (iv) oxidation of reminiscent organic residues from previous thermal events, like coke (exothermic; from 322 to 482 °C); and (v) condensation of silanol groups (with almost undetectable DTG and DSC maximums, from 482 to 1000 °C).

The comparison between the modified as-synthesized materials and A-Si-MCM-41 showed differences in the DTG and DSC curves, even though the TG curves display similar profiles and CTA^+/SiO_2 ratios. The different thermal behavior for both cerium incorporated materials, especially after the Hofmann elimination, revealed that: (i) In the temperature range associated to CTA^+ decomposition, one extra endothermic event occurs at 304 °C for A-Ce-MCM-Cl and 309 °C for A-Ce-MCM-NO_3; (ii) these extra events were overlapped with the first oxidation process, now shifted in the range of 315 to 360 °C; (iii) increase in the temperature of the DTG maximum for the final oxidation process, occurring around 70 °C higher than for A-Si-MCM-41; and (iv) increase in mass loss for the final oxidation processes, 5.59% for A-Si-MCM-41 and above 8.50% for both A-Ce-MCM-41 samples. Overall, this evidence demonstrates clearly that template molecules in the modified materials are thermally more resistant than in the as-synthesized Si-MCM-41. The extra thermal event might be due to an extra Hofmann elimination of molecules that survived the first decomposition process, delaying both exothermic events. Such phenomena can be either related to an increase of the interaction stability between CTA^+ and the inorganic framework or to diffusion limitations of products from thermal processes, since there was a decrease in V_{meso} after cerium incorporation. Similar observations have been made by Khalil [55] and de Souza and co-workers [90], especially as further evidence of isomorphic substitution of heteroelements in the inorganic framework.

The TG curves of the calcined MCM-41 materials (Figure 14) and the mass loss associated to all processes (Table S3) are displayed in the Supplementary Material. These curves provided significant information to differentiate the materials among themselves, where a difference in total mass loss from room temperature to 1000 °C was observed in all solids. The two thermal events were associated to the evaporation of adsorbed water (endothermic event from room temperature to temperatures around 115 to 150 °C) and the condensation of silanol groups (water evaporation until 1000 °C). The major mass loss was associated to the initial evaporation of water: 4.58%, 14.23%, and 12.29% for Si-MCM-41, Ce-MCM-Cl, and Ce-MCM-NO_3, respectively. Also, this event ends in higher temperatures for the cerium-modified samples than for Si-MCM-41 (usually above 130 °C).

Altogether, these observations indicate that cerium incorporation increases the surface polarity of the materials, somewhat expected since cerium is less electronegative than silicon. Also, the mass loss from silanol condensation was different for the three materials: 1.70%, 3.50%, and 2.67% for Si-MCM-41, Ce-MCM-Cl, and Ce-MCM-NO_3, respectively. Thus, it is noticeable that the isomorphic substitution leads to less thermally stable materials, highlighting the fact that Ce-MCM-Cl is less stable than Ce-MCM-NO_3, evidence of its higher degree of silicon substitution in the framework. This is expected once the insertion of heteroelements causes bond angle tension and loss of the crystalline mesophase as indicated by XRD data, which is in accordance with the disordering effect proposed for the crystallization of the materials and corroborated previously with other characterization techniques.

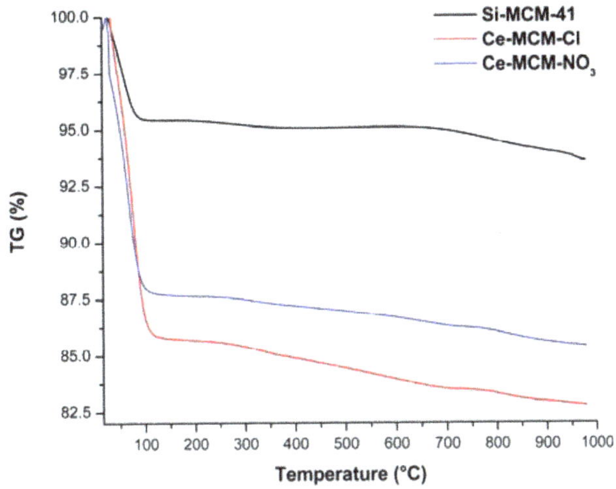

Figure 14. TG curves for the calcined MCM-41 materials.

2.2. Catalytic Tests

The liquid-phase oxidation of benzyl alcohol (BzOH) was used to further analyze the consequences of cerium precursors in the generation of catalytic active species. In all reaction analyses, the only products detected were benzaldehyde (BzD) and benzoic acid (HBz), both oxidation products. The other possible product, benzyl benzoate, was not observed. The reaction parameters were varied to achieve an optimal condition (see Materials and Methods for reaction details), where the maximum BzD yield was used as a reference. The results of all catalytic essays were compiled in Tables 5 and 6 for Ce-MCM-Cl and Ce-MCM-NO_3, respectively.

The first parameter studied was the type of oxidant: Hydrogen peroxide (H_2O_2, 30 wt.%); *tert*-butyl hydroperoxide (TBHP, 70 wt.%); and di-*tert*-butyl peroxide (DTBP), runs 1 to 3. As can be perceived in Tables 5 and 6, the only peroxide that produced significant values of BzOH conversion (*C*, %) and BzD yield (*Y*, %) was TBHP. Other studies involving the use of cerium-modified molecular sieves in different oxidation reactions suggest the use of H_2O_2 as a milder oxidant [91,92]. However, the catalysts' behavior in this work were consistent with the reactive conditions reported by Pal and co-workers [32], which performed liquid-phase BzOH oxidation with cerium-modified mesoporous materials. The values of *Y* and *TON* with TBHP as the oxidant were different for both cerium materials: 22.0% and 74.0 mol mol_{cat}^{-1} for Ce-MCM-Cl; 17.5% and 49.5 mol mol_{cat}^{-1} for Ce-MCM-NO_3, respectively. Thus, in these conditions, Ce-MCM-Cl presented better *Y* and was more active than Ce-MCM-NO_3. This result was consistent with previous literature indicating that non-framework CeO_2 species (observed mostly in Ce-MCM-NO_3) are less active and selective as catalytic sites [48,51].

Table 5. Catalytic performance optimization of Ce-MCM-Cl for benzyl alcohol oxidation.

Runs	Oxidant	BzOH:Oxidant Molar Ratio	T (°C)	Catalyst Amount (wt.%)	*C* (%) [1]	*S* (%) [2]	*Y* (%) [3]	*TON* [4]
run 1	H_2O_2	1:1	70	10	0.80	50.0	0.40	1.85
run 2	TBHP	1:1	70	10	33.6	65.4	22.0	74.0
run 3	DTBP	1:1	70	10	2.00	70.0	1.40	4.39
run 4 [5]	TBHP	1:1	70	10	15.1	89.0	13.4	27.3
run 5 [6]	TBHP	1:1	70	10	7.00	99.0	6.93	15.4
run 6	TBHP	2:1	70	10	25.8	74.9	19.4	56.8
run 7	TBHP	1:2	70	10	34.6	63.5	22.0	76.1
run 8	TBHP	1:1	80	10	54.3	48.7	26.4	119.4
run 9	TBHP	1:1	90	10	60.2	41.0	24.7	132.3
run 10	TBHP	1:1	80	5	35.9	57.1	20.5	151.8
run 11	TBHP	1:1	80	15	43.0	50.0	21.5	62.2
run 12 [7]	TBHP	1:1	80	10	0.50	100.0	0.50	n.a. [8]

[1] Conversion, calculated from Equation (11). [2] BzD selectivity, calculated from Equation (12). [3] BzD yield, calculated from Equation (13). [4] Turnover number, calculated from Equation (14). [5] Acetonitrile as solvent. [6] Isopropanol as solvent. [7] Si-MCM-41 as catalyst. [8] Not available.

Table 6. Catalytic performance optimization of Ce-MCM-NO_3 for benzyl alcohol oxidation.

Runs	Oxidant	BzOH:Oxidant Molar Ratio	T (°C)	Catalyst Amount (wt.%)	*C* (%) [1]	*S* (%) [2]	*Y* (%) [3]	*TON* [4]
run 1	H_2O_2	1:1	70	10	1.30	30.8	0.40	3.23
run 2	TBHP	1:1	70	10	23.2	75.6	17.5	49.5
run 3	DTBP	1:1	70	10	1.30	32.0	0.42	3.17
run 4 [5]	TBHP	1:1	70	10	14.1	92.0	13.0	30.1
run 5 [6]	TBHP	1:1	70	10	14.5	59.7	8.60	35.8
run 6	TBHP	2:1	70	10	19.7	80.3	15.9	48.8
run 7	TBHP	1:2	70	10	27.7	64.5	17.8	68.4
run 8	TBHP	1:1	80	10	38.5	56.5	21.7	95.1
run 9	TBHP	1:1	90	10	52.5	40.8	21.4	129.6
run 10	TBHP	1:1	80	5	33.7	64.3	21.7	160.3
run 11	TBHP	1:1	80	15	63.7	31.9	20.3	103.5
run 12 [7]	TBHP	1:1	80	5	0.35	100.0	0.35	n.a. [8]

[1] Conversion, calculated from Equation (11). [2] BzD selectivity, calculated from Equation (12). [3] BzD yield, calculated from Equation (13). [4] Turnover number, calculated from Equation (14). [5] Acetonitrile as solvent. [6] Isopropanol as solvent. [7] Si-MCM-41 as catalyst. [8] Not available.

To further optimize the system, the use of solvents was tested to observe if the changes in hydrophobicity/hydrophilicity of the medium can affect positively the reaction towards better BzD yields. Acetonitrile (CH_3CN) and isopropanol (*i*-PrOH) were selected to observe this influence (runs 4 and 5, respectively). For both cerium materials, the use of CH_3CN and *i*-PrOH resulted in less efficient conditions towards BzD yield and catalyst activity, even though BzD selectivity improved significantly (up to 99%). The lowering of *C* was the main issue, and this parameter was more affected using *i*-PrOH, possibly due to adsorptive competition of the protic solvent molecules [93].

The next parameter studied was the BzOH:TBHP molar ratio, since reactions so far were done in stoichiometric quantities. Changes in the chemical equilibrium can possibly affect the catalytic output and two conditions were tested: Excess of alcohol (2:1, run 6) and excess of TBHP (1:2, run 7). For Ce-MCM-Cl and Ce-MCM-NO_3, excess of alcohol decreased *C*, 25.8% and 19.7%, and *TON*, 56.8 and 48.8 mol mol_{cat}^{-1}, respectively. Even though there was a better peroxide efficiency in the molar ratio of 2:1, lower *Y* values were obtained (19.4% and 15.9% for Ce-MCM-Cl and Ce-MCM-NO_3). With the excess of peroxide, the main improvement was in the catalytic efficiency (*TON*), exhibiting 76.1 mol mol_{cat}^{-1} for Ce-MCM-Cl and 68.4 mol mol_{cat}^{-1} for Ce-MCM-NO_3. However, the little improvements observed in *Y* for both materials do not justify such an increase in the oxidant amount. Thus, the following essays were done in stoichiometric proportions of alcohol:oxidant.

Next, the effect of temperature was studied, increasing it from 70 to 80 and 90 °C (runs 8 and 9). With increasing temperature, significant enhancements in *C* were obtained (up to 60.2% and 52.5% for Ce-MCM-Cl and Ce-MCM-NO_3, respectively), but with concomitant decreases in *S* (close to 41% for both materials at 90 °C). In addition, improvements in *Y* and *TON* were significant, especially at 80 °C (26.4% and 119.4 mol mol_{cat}^{-1} for Ce-MCM-Cl, and 21.7% and 95.1 mol mol_{cat}^{-1} for Ce-MCM-NO_3). At higher temperatures, a better activation of oxidants and generation of reactive species were obtained, but it was only beneficial until a certain point, since a loss of selectivity was visible at 90 °C for both materials when compared to 80 °C. Therefore, the following essays were performed at 80 °C.

The final parameter evaluated was the catalyst amount, 5 and 15 wt.%, runs 10 and 11, respectively. In the case of Ce-MCM-Cl, both changes led to less efficient catalytic situations in terms of *C* and *Y*, where decreases of up to 20.0% for the latter and around 40.0% of the former were perceived when compared to run 8 (10 wt.%). This can be associated to a more exacerbated decomposition of TBHP in the case of 15 wt.% and a less efficient activation of the oxidant at 5 wt.%, achieving the optimal mass and active sites' amount around 10 wt.% for this catalyst. Surprisingly, a different trend was observed for Ce-MCM-NO_3, where the catalyst amount of 5 wt.% showed a negligible decrease of *C* and a considerable increase of *S* (33.7% and 64.3%, respectively). This combination resulted in the same *Y* and better catalytic performance (TON = 160.3 mol mol_{cat}^{-1}) than from run 8 (10 wt.%), while the use of 15 wt.% catalyst increased *C* to 63.7% with concomitant decrease in *S* to 31.9%, resulting in a decrease in *Y* (20.3%) and catalytic performance (TON = 103.5 mol mol_{cat}^{-1}). This can be associated to the mass transfer in Ce-MCM-NO_3 mesopores, while 5 wt.% offered an optimal condition of catalytic site usage and diffusion of reactants and products; 10 or 15 wt.% seems to lead to conditions where over-oxidation was favored due to an excess of active sites. These results are consistent with the fact that mainly non-framework CeO_2 (less selective) is present in the material's constitution as mentioned before. Thus, the amount of catalyst plays a significant role in attaining better *Y*, but also in differentiating the applicability of both active materials in this study. Also, for comparison means, Si-MCM-41 was tested in the optimal conditions found for Ce-MCM-Cl and Ce-MCM-NO_3 (run 12), and negligible values of *C* and *Y* (less than 1%) were obtained, proving that the activity of MCM-41 materials is due to the presence of cerium species.

In order to evaluate the reusability of the catalysts, the solids from the optimized conditions were recovered and studied for further cycles, see Materials and Methods. After each cycle, the cerium content and coordination state were analyzed by XRF and Raman to understand possible changes in the catalytic performance. The recycled results are compiled in Figure 15.

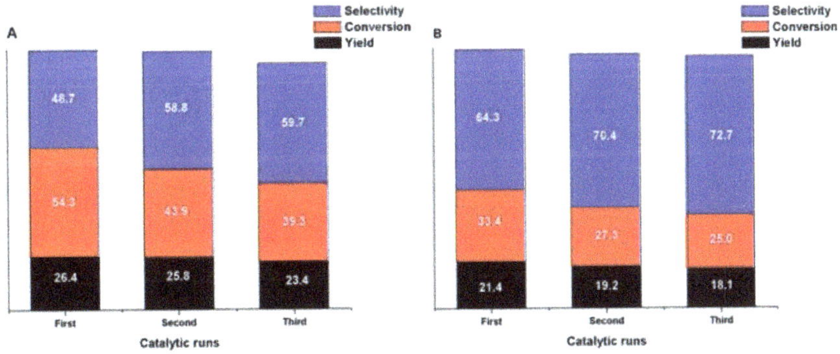

Figure 15. Performance of recovered solids after three catalytic cycles: Ce-MCM-Cl (**A**) and Ce-MCM-NO_3 (**B**).

As can be perceived from Figure 15, BzD *Y* of both catalysts maintains a stable state after three catalytic cycles, but the conversion of both materials decreases. For Ce-MCM-Cl, the conversion decreases from 54.3% to 43.9% in the second cycle and 39.3% in the third cycle. Meanwhile, the conversion of Ce-MCM-NO_3 decreases from 33.4% to 27.3% in the second cycle and 25.0% in the third cycle. Also, according to XRF analyses of the recovered solids, the amount of cerium in Ce-MCM-Cl diminishes from 5.89% to 5.77% in the last cycle, while for Ce-MCM-NO_3, the diminishment goes from 5.24% to 5.16%. Thus, it is possible to conclude from the previous data that: (i) Ce-MCM-Cl deactivates in a higher rate than Ce-MCM-NO_3; and (ii) the main reason for deactivation can be associated to microstructural changes in the recovered materials, since cerium leaching was negligible.

According to Timofeeva and coworkers [92], changes in the Ce coordination environment can occur after oxidation processes due to etching effects of peroxides in the surface of cerium-incorporated silica materials, where framework sites (more active) can turn to non-framework species (less active). In our study, the Raman spectra of the recovered materials (Figure 16) proved such a point, where an increase in the F_{2g} mode intensity of CeO_2 after each cycle confirmed the change in the cerium coordination state towards non-framework species. It was also noticeable in the Ce-MCM-Cl spectrum that the intensity of this band doubles itself after the second cycle, while for Ce-MCM-NO_3, it happens to a lesser degree. Such an observation is consistent with the steadier loss in catalytic performance observed in Ce-MCM-NO_3, indicating that this catalyst is more resistant towards deactivation. This effect is possibly a consequence of the CeO_2 coating mentioned earlier in this work, since pore walls were thicker for Ce-MCM-NO_3 (3.64 nm) than for Ce-MCM-Cl (3.43 nm), making the former less susceptible towards possible etching effects from TBHP. The protective effect described is not usually highlighted in the literature for metal-incorporated MCM-41 materials and shows how relevant the choice of metal precursors is to attain not only active materials, but more stable ones towards unfavorable processes inherent of catalytic applications.

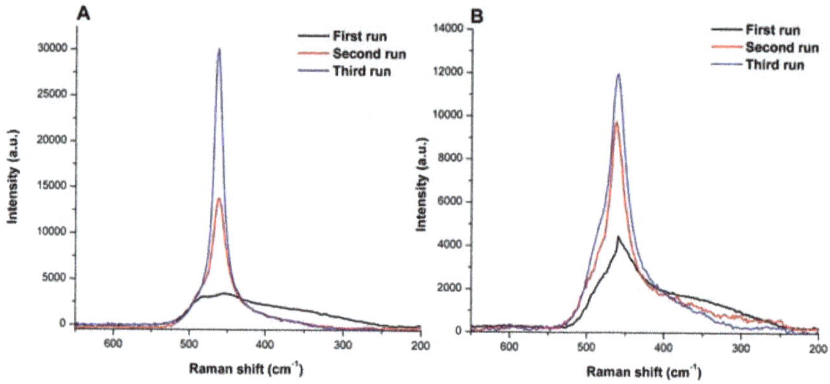

Figure 16. Raman spectra of recovered solids: Ce-MCM-Cl (**A**) and Ce-MCM-NO_3 (**B**).

Overall, the best conditions for the BzOH liquid-phase oxidation were the ones in run 8 for Ce-MCM-Cl (TBHP, molar ratio of 1:1, 10 wt.% of catalyst, and 80 °C), while the conditions of run 10 were best for Ce-MCM-NO_3 (TBHP, molar ratio of 1:1, 5 wt.% of catalyst, and 80 °C). In situations where better BzOH conversions, BzD yields, and heteroelement incorporation are aimed for, cerium(III) chloride precursor affords the ideal material. In situations where BzD selectivity and catalytic endurance is the objective, cerium(III) nitrate precursor might be the best choice of synthetic precursor.

3. Materials and Methods

3.1. Materials

Cetyltrimethylammonium bromide (CTAB, Sigma-Aldrich, >99.9%, Darmstadt, Germany), ammonium hydroxide (NH_4OH, Sigma-Aldrich, 28–30 wt.% aqueous solution, Darmstadt, Germany), ethyl orthosilicate (TEOS, Sigma-Aldrich, 98%, Darmstadt, Germany), silver nitrate ($AgNO_3$, Honeywell Fluka, ≥99.5%, Seelze, Germany), cerium chloride heptahydrate ($CeCl_3 \cdot 7H_2O$, Sigma-Aldrich, >98%,

Darmstadt, Germany), cerium nitrate hexahydrate ($Ce(NO_3)_3{\cdot}6H_2O$, Vetec, 99%, Duque de Caxias, RJ, Brazil), benzyl alcohol (BzOH, Tedia®, >99%, Fairfield, OH, USA), *tert*-butyl hydroperoxide (TBHP, Sigma-Aldrich, 70 wt.% aqueous solution, Darmstadt, Germany), hydrogen peroxide (H_2O_2, Vetec, 30 wt.% aqueous solution, Duque de Caxias, RJ, Brazil), di-*tert*-butyl peroxide (DTBP, Sigma-Aldrich, 98%, Darmstadt, Germany), deuterated chloroform ($CDCl_3$, Sigma-Aldrich, ≥ 99.8% D atom and 0.03% v/v of tetramethylsilane, TMS, Darmstadt, Germany), acetonitrile (CH_3CN, J.T.Baker, 99.99%, Phillipsburg, NJ, USA), and isopropanol (*i*-PrOH, Cromoline, 99.5%, Diadema, SP, Brazil) were used as supplied.

3.2. Synthesis of MCM-41 Materials

The catalysts were synthesized following an adaptation of the precipitation method proposed by Grün and co-workers [94]. In the synthesis of Si-MCM-41, 55 mL of deionized water and 1.24 g of CTAB were added in a round bottom flask. The mixture was stirred at 400 rpm and heated at 30 to 35 °C until homogenization. Then, 4.1 mL of concentrated NH_4OH and 5 mL of TEOS were added, the latter being dropwise with a Pasteur pipette. The synthetic mixture contained the following molar ratio of the reactants: $1TEOS{:}0.152CTAB{:}2.8NH_4OH{:}141.2H_2O$. After 4 h of reaction, the resulting white solid was filtered and washed with distilled water until it was free of halide anions (tested with a 1 mol L^{-1} aqueous solution of $AgNO_3$). The material was dried at room temperature for 3 days and then calcined in two steps: 300 °C for 3 h and 550 °C for 3 h. Ce-MCM-41 samples were prepared by adding, simultaneously, 5 mL of an aqueous solution of $CeCl_3{\cdot}7H_2O$ or $Ce(NO_3)_3{\cdot}6H_2O$ and TEOS in a dropwise manner. The synthetic mixture contained a Si/Ce molar ratio = 25, generating the materials of Ce-MCM-Cl and Ce-MCM-NO_3, respectively. The resulting light brown/purple solids were filtered, washed, dried, and calcined under the same conditions as Si-MCM-41. After the calcination procedure, both cerium-incorporated materials showed a light-yellow appearance. The as-synthesized materials (obtained after the drying step) were designated as A-Si-MCM-41, A-Ce-MCM-Cl, and A-Ce-MCM-NO_3.

3.3. Characterization of the Catalysts

X-ray fluorescence (XRF) analyses were made under vacuum on a Shimadzu EDX-720 (Tokyo, Japan) equipped with a rhodium tube as the X-ray source.

X-ray diffraction (XRD) patterns were obtained on a Rigaku Ultima IV diffractometer (Tokyo, Japan) with CuKα radiation (λ = 0.15409 nm) at 1° min^{-1}, 2θ range between 2 and 80°, and a step of 0.01°. For peak deconvolution, Gaussian functions were fitted with an exponential decay baseline using the Peak Analyzer tool in OriginPro 2019. The calculation of the cell lattice parameter (a) was made using crystallographic knowledge of hexagonal crystal lattices and Bragg's law [95] according to Equations (1) and (2):

$$a = (2/3^{1/2})d_{100} \tag{1}$$

$$d_{100} = 2d_{200} \tag{2}$$

where d_{100} and d_{200} are interplanar spaces of the 100 and 200 reflections, respectively.

Fourier transform infrared spectroscopy (FTIR) spectra were obtained on a Varian 640-IR (Palo Alto, CA, USA) using a 4 cm^{-1} resolution, 32 scans, and 1 wt.% of sample in KBr pellets.

The Raman spectroscopy measurements were performed on an In-Via^{TM} Raman system from Renishaw (Wotton-under-Edge, Glos., England) equipped with a thermoelectric cooled CCD detector and a 1200 lines/groove diffraction grating. A Leica microscope (Wetzlar, Germany) with a 50× objective lens was used to focus the radiation of the excitation laser on the sample and to collect the scattered photons. The wavenumbers were calibrated using the Raman phonon of a silicon wafer at 520.5 cm^{-1}. Each spectrum was the accumulation of 10 scans with a 20 s integration time at 632.8 nm (HeNe laser). All Raman spectra were recorded at a laser power of 20 mW and the nominal spectra resolution was 6 cm^{-1}. A 1 mm optical path quartz cuvette was used to perform Raman measurements of the solutions. All Raman measurements were performed at room temperature.

Diffuse reflectance UV-Vis spectroscopy (UV-Vis-DRS) spectra were obtained on a Varian Cary 5000 UV-Vis-NIR spectrophotometer (Palo Alto, CA, USA) equipped with a Harrick diffuse reflectance accessory and using $BaSO_4$ as a background standard. The optical band gap was calculated using the Davis-Mott method [96], in which a plot is done according to Equation (3). After obtaining the Davis-Mott plot, an extrapolation to zero of the linear portion where the curvature of the graph first changes gives the optical band gap value:

$$[\alpha h\upsilon]^{1/n} = A(E_g - h\upsilon) \tag{3}$$

where α is the absorption coefficient, replaced in this work by the Kubelka-Munk function, $F(R)$; E_g is the optical band gap; $h\upsilon$ is energy; A is a proportionality constant; and $1/n$ is a coefficient dependent on the type of transition responsible for the absorption edge observed.

Nuclear magnetic resonance (NMR) experiments were conducted in the solid state at 14 T with a Bruker Ascend^{TM} 600 (Kontich, AN, Belgium) equipped with a 4 mm magic-angle spinning (MAS) probe and zirconia rotors. ^{29}Si MAS-NMR spectra (119.237 MHz) were obtained with spinning rate at 10 kHz, pulse length of 4.25 μs (π/2), 512 scans, and a recycle delay of 20 s. The spectra were referenced to TMS (δ = 0.0 ppm). The molar percentage of silanol groups (Si molar%) was calculated using Equation (4) [97]:

$$\text{Si molar\%} = [(2Q^2 + Q^3)/(Q^2 + Q^3 + Q^4)]100 \tag{4}$$

where Q^2, Q^3, and Q^4 correspond to the relative proportions in the percentage of geminal silanols, terminal silanols, and siloxane groups, respectively. Peak areas were obtained after deconvolution by fitting the signals to Voigt peaks in the range between −110 and −90 ppm.

Thermogravimetry (TG), derivative thermogravimetry (DTG), and differential scanning calorimetry (DSC) curves were obtained on a TA Instruments SDT Q600 (New Castle, DE, USA) from room temperature to 1000 °C (20 °C min^{-1}), using synthetic air (99.999%) as the purge gas (flow rate of 60 mL min^{-1}). The analysis was performed on alumina crucibles and the external standard employed for calibration was sapphire.

N_2 physisorption measurements were performed at −196 °C on a Quantachrome Instruments NOVA 4200e apparatus (Boynton Beach, FL, USA). All samples were outgassed at 200 °C under vacuum for 24 h prior to the analysis. Surface area and porosity data were obtained from N_2 adsorption/desorption isotherms. Total specific surface area (S_{total}), external specific surface area (S_{ext}), and mesopore volume (V_{meso}) were determined using a t-plot curve obtained from analytical functions for each portion of the P/P_0 domain (see Equations (5)–(8) and Figure 10a) [86]. Mesoporous surface area (S_{meso}) was calculated by subtracting S_{ext} from S_{total}. Total pore volume (V_{total}) was calculated using the penultimate point from the adsorption isotherm. Average pore diameter (PD) was calculated according to Equation (9). Wall thickness (WT) was calculated according to Equation (10):

$$t\ (nm) = A(P/P_0)^a + B(P/P_0)^b \text{ for } 0.03 \leq P/P_0 < 0.25 \tag{5}$$

$$t\ (nm) = \{C/[D - log(P/P_0)]\}^c \text{ for } 0.25 \leq P/P_0 < 0.60 \tag{6}$$

$$t\ (nm) = \{E/[F - log(P/P_0)]\}^e \text{ for } 0.60 \leq P/P_0 < 0.90 \tag{7}$$

$$t\ (nm) = G(P/P_0)^g + H(P/P_0)^h \text{ for } 0.90 \leq P/P_0 \tag{8}$$

$$PD\ (nm) = 4V_{total}/S_{total} \tag{9}$$

$$WT\ (nm) = a - PD \tag{10}$$

where $A = 0.5675647$, $a = 0.199735$, $B = 0.4116168$, $b = 2.00834$, $C = 0.1423566$, $D = 0.1078$, $c = 0.4888$, $E = 0.08309076$, $F = 0.02995$, $e = 0.369$, $G = 1.268066$, $g = 1.931$, $H = 0.76934$, and $h = 51.09$.

Low-resolution transmission electron microscopy (TEM) images were obtained on a JEOL-1011 microscope (Tokyo, Japan) at 120 kV. Before analysis, each sample was dispersed in ethanol (0.1 mg mL^{-1})

through sonication. Subsequently, 5 μL of this suspension was deposited to a carbon-coated copper grid and dried overnight at room temperature. The mean particle size was determined using the software, ImageJ (version 1.52a), counting at least 100 particles for each sample.

High-resolution transmission electron microscopy (HRTEM) images were obtained on a JEOL JEM-2100 microscope (Tokyo, Japan) at 200 kV under vacuum and N_2 purge. Before analysis, each sample was dispersed in ethanol and subsequently deposited to a carbon-coated copper grid and dried at room temperature.

3.4. Liquid-Phase Oxidation of Benzyl Alcohol

In a typical essay, the desired catalyst amount (5, 10, or 15 wt.% of BzOH) was transferred to a 10 mL glass reactor and dried for 30 min at 70 °C. Then, 240 μL of BzOH and the required amount of TBHP were added for reactions with 1:1, 1:2, and 2:1 (alcohol:oxidant) molar ratios. The glass reactor was closed with a cover and subjected to magnetic stirring and heating at the desired temperature (70, 80, or 90 °C) for 24 h in an aluminum dry bath block/hot plate stirrer system. In addition, H_2O_2 and DTBP were also studied as oxidant agents, while acetonitrile and isopropanol were tested as solvents. After the reaction, the system was cooled at room temperature, centrifuged at 2500 rpm, and an aliquot removed for analysis by ^{1}H NMR (see the schematic representation of these procedures in Figure 17). For the reusability studies, the catalysts used in the first cycle of the optimized conditions were washed with ethanol twice, dried at room temperature, and calcined at 300 °C for 2 h in order to eliminate the residual organic content.

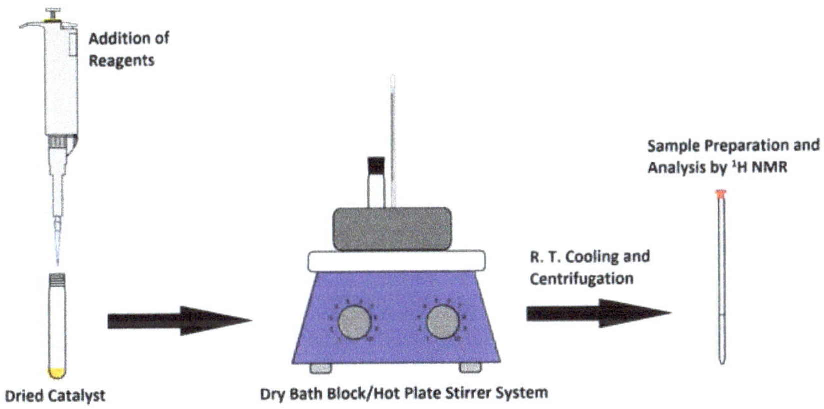

Figure 17. Schematic description for liquid-phase oxidation of BzOH and catalytic measurements.

The reaction analysis was performed by NMR on a Bruker Magneto Oxford YH300 using $CDCl_3$ as the solvent and TMS as internal standard for spectra referencing (δ = 0.0 ppm). ^{1}H NMR spectra (300 MHz) were obtained with a spinning rate at 20 Hz, pulse length of 4.90 μs (π/4), 16 scans, and a delay recycle of 1.36 s. Identification of the substrate and main products was performed by comparison with known standards. The BzOH conversion (*C*, %), BzD selectivity (*S*, %), BzD yield (*Y*, %), and turnover number (*TON*, mol mol_{cat}^{-1}) were determined using Equations (11)–(14), respectively:

$$C, \% = [(2A_{BzD} + A_{HBz})/(2A_{BzD} + A_{HBz} + A_{BzOH})]100 \quad (11)$$

$$S, \% = [2A_{BzD}/(2A_{BzD} + A_{HBz})]100 \quad (12)$$

$$Y, \% = (CS)/100 \quad (13)$$

$$TON, \text{mol mol}_{\text{cat}}^{-1} = n/n_{cat} \quad (14)$$

where A_{BzOH}, A_{BzD}, and A_{HBz} are the integrated areas of the methylene singlet at 4.69 ppm, the aldehyde hydrogen singlet at 10.02 ppm, and the deshielded aromatic double multiplet centered at 8.10 ppm, respectively; n = moles of BzOH converted; and n_{cat} = moles of Ce in the catalyst.

4. Conclusions

For the first time, the effect of different cerium(III) precursors was reported in the synthesis of Ce-MCM-41 materials. Cerium incorporation into the MCM-41 framework led to long-range disordered materials, but the results obtained by several characterization techniques showed the influence of different anions in the microstructural properties and catalytic activity of Ce-MCM-Cl and Ce-MCM-NO_3. A mechanism for cerium incorporation in the MCM-41 structure was proposed, where [$Ce(OH)_3$] species were hydrogen bonded to silicate anions, forming framework Ce-O-Si bonds during condensation. It was also observed that the synthesis of Ce-MCM-NO_3 follows a different path when compared to Ce-MCM-Cl. The former induced $Ce(OH)_3$ monomers to aggregate, forming non-framework CeO_2 during calcination on the MCM-41 surface, while the latter preferentially formed framework Ce-O-Si bonds. The preferential formation of CeO_2 species on Ce-MCM-NO_3 was attributed to the nitrate ions' strong binding to template molecules. The prepared materials were tested in the solvent free oxidation of benzyl alcohol to produce benzaldehyde, an important fine chemical compound for several industrial sectors, and the reaction parameters were optimized for each catalyst. Overall, Ce-MCM-Cl achieved better BzOH conversions and benzaldehyde (BzD) yields, while Ce-MCM-NO_3 offered increased BzD selectivity. The catalysts' reusability was also studied over three cycles. Ce-MCM-Cl and Ce-MCM-NO_3 showed a reduction in conversion, but increase in selectivity, resulting in negligible changes in BzD yield values. Also, it was observed that the different crystallization mechanism differentiates both catalysts' endurance towards deactivation, where Ce-MCM-NO_3 was more resistant than Ce-MCM-Cl. This effect was attributed to the CeO_2 coating on the Ce-MCM-NO_3 surface, making the latter less prone to possible etching effects from TBHP. Thus, the manuscript shows the importance of the selection of suitable heteroelement precursors in the synthesis of materials. Even more, there is scarce literature regarding differences in the same metal-incorporated silicas that can achieve different conditions for liquid-phase reactions, an aspect very important when seeking the best choices for the synthetic preparation of catalysts with the desired activity/selectivity in industrial processes.

Supplementary Materials: The following are available online at http://www.mdpi.com/2073-4344/9/4/377/s1, Figure S1: XRD patterns of Ce-MCM-Cl and Ce-MCM-NO_3 materials at high angle region, Figure S2. Raman spectra of CTA^+ systems between 400–1700 cm^{-1} (A) and 2800–3000 cm^{-1} (B), Figure S3. Raman spectrum of crystalline CeO_2, Figure S4. Absorption spectra (A) and Davis-Mott plot (B) for of CeO_2, Figure S5. High magnification in HRTEM images highlighting CeO_2 110 interplanar spacing for: Ce-MCM-Cl (A) and Ce-MCM-NO_3 (B), Table S1. Raman band assignments for CTA^+ cations in as-synthesized MCM-41 materials and template standards, Table S2. TGA results for the as-synthesized MCM-41 materials, Table S3. TGA results for the calcined MCM-41 materials.

Author Contributions: Conceptualization, formal analysis, methodology, project administration, visualization, writing—Review and editing, C.M.A. and J.L.M.; investigation, C.M.A. and K.V.d.O.; writing—Original draft preparation, C.M.A., K.V.d.O. and J.L.M.; validation, C.M.A.; funding acquisition, resources, supervision, J.L.M.

Funding: This research was funded by Conselho Nacional de Desenvolvimento Científico e Tecnológico—Brasil (CNPq), grant number 308061/2015-2; Fundação de Apoio à Pesquisa do Distrito Federal (FAPDF)—Brasil, grant number 0193001347/2016; and Coordenação de Aperfeiçoamento de Pessoal de Nível Superior—Brasil (CAPES), finance code 001.

Acknowledgments: The authors would like to thank Financiadora de Inovação e Pesquisa—Brasil (Finep), Instituto de Química da Universidade de Brasília—Brasil, Ministério da Ciência, Tecnologia e Inovação—Brasil (MCTI), Linconl Araújo Teixeira and Sandra Maria da Luz (Faculdade Gama da Universidade de Brasília—Brasil) for TG/DTG/DSC measurements, Rafael Brasil Figueras (Instituto de Geociências da Universidade de Brasília—Brasil) for XRD measurements, Tatiane Oliveira dos Santos (Laboratório Multiusuário de Alta Resolução da Universidade Federal do Goiás—Brasil) for high resolution TEM measurements, Ingrid Gracielle Martins da Silva (Laboratório de Microscopia Eletrônica e Microanálise do Instituto de Ciências Biológicas da Universidade de Brasília—Brasil) for low resolution TEM measurements, Maryene Alves Camargo for theoretical discussion and Claudia Cristina Gatto for the support during this manuscripts writing.

Conflicts of Interest: The authors declare no conflict of interest. The funders had no role in the design of the study; in the collection, analyses, or interpretation of data; in the writing of the manuscript, or in the decision to publish the results.

References

1. Ciriminna, R.; Pagliaro, M. Green chemistry in the fine chemicals and pharmaceutical industries. *Org. Process Res. Dev.* **2013**, *17*, 1479–1484. [CrossRef]
2. de Oliveira, N.B. Inovação e produção na química fina. *Quim. Nova* **2005**, *28*, S79–S85. [CrossRef]
3. O desempenho da indústria quimica em 2017. Available online: http://www.webcitation.org/75xqkk7e9 (accessed on 5 February 2019).
4. Anastas, P.; Eghbali, N. Green chemistry: Principles and practice. *Chem. Soc. Rev.* **2010**, *39*, 301–312. [CrossRef]
5. Sheldon, R.A. The E Factor: Fifteen years on. *Green Chem.* **2007**, *9*, 1273–1283. [CrossRef]
6. Brühne, F.; Wright, E. Benzaldehyde. In *Ullmann's Encyclopedia of Industrial Chemistry*, 7th ed.; Elvers, B., Ed.; Wiley-VCH Verlag GmbH & Co. KGaA: Weinheim, Germany, 2011; Volume 5, pp. 223–235.
7. Benzaldehyde Market for Aroma Chemicals, Pharmaceuticals, Agriculture, Coatings and Other End-Users—Global Industry Analysis, Size, Share, Growth, Trends and Forecast 2015–2023. Available online: http://www.webcitation.org/75xscDvAV (accessed on 5 February 2019).
8. Shearon, W.H.; Hall, H.E.; Stevens, J.E. Fine chemicals from coal. *Ind. Eng. Chem.* **1949**, *41*, 1812–1820. [CrossRef]
9. Sheldon, R.A. Recent advances in green catalytic oxidations of alcohols in aqueous media. *Catal. Today* **2015**, *247*, 4–13. [CrossRef]
10. Kopylovich, M.N.; Ribeiro, A.P.C.; Alegria, E.C.B.A.; Martins, N.M.R.; Martins, L.M.D.R.S.; Pombeiro, A.J.L. Catalytic oxidation of alcohols: Recent advances. *Adv. Organomet. Chem.* **2015**, *63*, 91–174.
11. Sankar, M.; Nowicka, E.; Carter, E.; Murphy, D.M.; Knight, D.W.; Bethell, D.; Hutchings, G.J. The benzaldehyde oxidation paradox explained by the interception of peroxy radical by benzyl alcohol. *Nat. Commun.* **2014**, *5*, 3332–3337. [CrossRef]
12. Della Pina, C.; Falletta, E.; Rossi, M. Highly selective oxidation of benzyl alcohol to benzaldehyde catalyzed by bimetallic gold–copper catalyst. *J. Catal.* **2008**, *260*, 384–386. [CrossRef]
13. Enache, D.I.; Edwards, J.K.; Landon, P.; Solsona-Espriu, B.; Carley, A.F.; Herzing, A.A.; Watanabe, M.; Kiely, C.J.; Knight, D.W.; Hutchings, G.J. Solvent-free oxidation of primary alcohols to aldehydes using Au-Pd/TiO_2 catalysts. *Science* **2006**, *311*, 362–365. [CrossRef] [PubMed]
14. Jia, A.; Lou, L.-L.; Zhang, C.; Zhang, Y.; Liu, S. Selective oxidation of benzyl alcohol to benzaldehyde with hydrogen peroxide over alkali-treated ZSM-5 zeolite catalysts. *J. Mol. Catal. A Chem.* **2009**, *306*, 123–129. [CrossRef]
15. Bansal, V.K.; Thankachan, P.P.; Prasad, R. Oxidation of benzyl alcohol and styrene using H_2O_2 catalyzed by tetraazamacrocycle complexes of Cu(II) and Ni(II) encapsulated in zeolite-Y. *Appl. Catal. A Gen.* **2010**, *381*, 8–17. [CrossRef]
16. Chen, G.; Zhou, Y.; Long, Z.; Wang, X.; Li, J.; Wang, J. Mesoporous polyoxometalate-based ionic hybrid as a triphasic catalyst for oxidation of benzyl alcohol with H_2O_2 on water. *ACS Appl. Mater. Interfaces* **2014**, *6*, 4438–4446. [CrossRef] [PubMed]
17. Della Pina, C.; Falletta, E.; Prati, L.; Rossi, M. Selective oxidation using gold. *Chem. Soc. Rev.* **2008**, *37*, 2077–2095. [CrossRef]
18. Corma, A.; García, H. Lewis Acids as Catalysts in Oxidation Reactions: From Homogeneous to Heterogeneous Systems. *Chem. Rev.* **2002**, *102*, 3837–3892. [CrossRef] [PubMed]
19. Ishii, Y.; Yamawaki, K.; Ura, T.; Yamada, H.; Yoshida, T.; Ogawa, M. Hydrogen peroxide oxidation catalyzed by heteropoly acids combined with cetylpyridinium chloride. Epoxidation of olefins and allylic alcohols, ketonization of alcohols and diols, and oxidative cleavage of 1,2-diols and olefins. *J. Org. Chem.* **1988**, *53*, 3587–3593. [CrossRef]
20. Venturello, C.; D'Aloisio, R.; Bart, J.C.J.; Ricci, M. A New peroxotungsten heteropoly anion with special oxidizing properties: Synthesis and structure of tetrahexylammonium tetra(diperoxotungsto)phosphate(3-). *J. Mol. Catal.* **1985**, *32*, 107–110. [CrossRef]
21. Kresge, C.T.; Roth, W.J. The discovery of mesoporous molecular sieves from the twenty year perspective. *Chem. Soc. Rev.* **2013**, *42*, 3663–3670. [CrossRef]

22. Alothman, A.Z. A review: Fundamental aspects of silicate mesoporous materials. *Materials* **2012**, *5*, 2874–2902. [CrossRef]
23. Linares, N.; Silvestre-Albero, A.M.; Serrano, E.; Silvestre-Albero, J.; García-Martínez, J. Mesoporous materials for clean energy technologies. *Chem. Soc. Rev.* **2014**, *43*, 7681–7717. [CrossRef]
24. Lin, H.-P.; Mou, C.-Y. Salt effect in post-synthesis hydrothermal treatment of MCM-41. *Microporous Mesoporous Mater.* **2002**, *55*, 69–80. [CrossRef]
25. Lin, H.-P.; Kao, C.-P.; Mou, C.-Y. Counterion and alcohol effect in the formation of mesoporous silica. *Microporous Mesoporous Mater.* **2001**, *48*, 135–141. [CrossRef]
26. Leontidis, E. Hofmeister anion effects on surfactant self-assembly and the formation of mesoporous solids. *Curr. Opin. Colloid Interface Sci.* **2002**, *7*, 81–91. [CrossRef]
27. Singh, S.; Patel, A. Environmentally benign oxidations of alkenes and alcohols to corresponding aldehydes over anchored phosphotungstates: Effect of supports as well as oxidants. *Catal. Lett.* **2016**, *146*, 1059–1072. [CrossRef]
28. Patel, A.; Singh, S. Undecatungstophosphate anchored to MCM-41: An ecofriendly and efficient bifunctional solid catalyst for non-solvent liquid-phase oxidation as well as esterification of benzyl alcohol. *Microporous Mesoporous Mater.* **2014**, *195*, 240–249. [CrossRef]
29. Wang, X.; Wu, G.; Li, J.; Zhao, N.; Wei, W.; Sun, Y. Selective oxidation of benzyl alcohol catalyzed by Cr(salen) complexes immobilized on MCM-41. *J. Mol. Catal. A Chem.* **2007**, *276*, 86–94. [CrossRef]
30. Hamza, A.; Srinivas, D. Selective oxidation of benzyl alcohol over copper phthalocyanine immobilized on MCM-41. *Catal. Lett.* **2009**, *128*, 434–442. [CrossRef]
31. Kumar, A.; Kumar, V.P.; Vishwanathan, V.; Chary, K.V.R. Synthesis, characterization, and reactivity of Au/MCM-41 catalysts prepared by homogeneous deposition–precipitation (HDP) method for vapor phase oxidation of benzyl alcohol. *Mater. Res. Bull.* **2015**, *61*, 105–112. [CrossRef]
32. Pal, N.; Cho, E.-B.; Kim, D. Synthesis of ordered mesoporous silica/ceria–silica composites and their high catalytic performance for solvent-free oxidation of benzyl alcohol at room temperature. *RSC Adv.* **2014**, *4*, 9213–9222. [CrossRef]
33. Cánepa, A.L.; Elías, V.R.; Vaschetti, V.M.; Sabre, E.V.; Eimer, G.A.; Casuscelli, S.G. Selective oxidation of benzyl alcohol through eco-friendly processes using mesoporous V-MCM-41, Fe-MCM-41 and Co-MCM-41 materials. *Appl. Catal. A Gen.* **2017**, *545*, 72–78. [CrossRef]
34. Dang, T.T.H.; Seeburg, D.; Radnik, J.; Kreyenschulte, C.; Atia, H.; Vu, T.T.H.; Wohlrab, S. Influence of V-sources on the catalytic performance of VMCM-41 in the selective oxidation of methane to formaldehyde. *Catal. Commun.* **2018**, *103*, 56–59. [CrossRef]
35. Corma, A.; Navarro, M.T.; Nemeth, L.; Renz, M. Sn-MCM-41—A heterogeneous selective catalyst for the Baeyer–Villiger oxidation with hydrogen peroxide. *Chem. Commun.* **2001**, 2190–2191. [CrossRef]
36. Yao, W.; Chen, Y.; Min, L.; Fang, H.; Yan, Z.; Wang, H.; Wang, J. Liquid oxidation of cyclohexane to cyclohexanol over cerium-doped MCM-41. *J. Mol. Catal. A Chem.* **2006**, *246*, 162–166. [CrossRef]
37. González Vargas, O.A.; de los Reyes Heredia, J.A.; Montesinos Castellanos, A.; Chen, L.F.; Wang, J.A. Cerium incorporating into MCM-41 mesoporous materials for CO oxidation. *Mater. Chem. Phys.* **2013**, *139*, 125–133. [CrossRef]
38. Ghesti, G.F.; de Macedo, J.L.; Parente, V.C.; Dias, J.A.; Dias, S.C. Synthesis, characterization and reactivity of Lewis acid/surfactant cerium trisdodecylsulfate catalyst for transesterification and esterification reactions. *Appl. Catal. A Gen.* **2009**, *355*, 139–147. [CrossRef]
39. Cesteros, Y.; Haller, G.L. Several factors affecting Al-MCM-41 synthesis. *Microporous Mesoporous Mater.* **2001**, *43*, 171–179. [CrossRef]
40. Ghiaci, M.; Seyedeyn-Azad, F.; Kia, R. Fast and efficient synthesis of ZSM-5 in a broad range of SiO_2/Al_2O_3 without using seeding gel. *Mater. Res. Bull.* **2004**, *39*, 1257–1264. [CrossRef]
41. Zhao, D.; Nie, C.; Zhou, Y.; Xia, S.; Huang, L.; Li, Q. Comparison of disordered mesoporous aluminosilicates with highly ordered Al-MCM-41 on stability, acidity and catalytic activity. *Catal. Today* **2001**, *68*, 11–20. [CrossRef]
42. Reddy, K.M.; Song, C. Effect of Al sources on the synthesis and acidic characteristics of mesoporous aluminosilicates of MCM-41 type. In *Studies in Surface Science and Catalysis*; Bonneviot, L., Béland, F., Danumah, C., Giasson, S., Kaliaguine, S., Eds.; Elsevier: Amsterdam, The Netherlands, 1998; Volume 117, pp. 291–299.

43. Matsumoto, A.; Chen, H.; Tsutsumi, K.; Grün, M.; Unger, K. Novel route in the synthesis of MCM-41 containing framework aluminum and its characterization. *Microporous Mesoporous Mater.* **1999**, *32*, 55–62. [CrossRef]
44. Lin, W.; Cai, Q.; Pang, W.; Yue, Y.; Zou, B. New mineralization agents for the synthesis of MCM-41. *Microporous Mesoporous Mater.* **1999**, *33*, 187–196. [CrossRef]
45. Beck, J.S.; Vartuli, J.C.; Roth, W.J.; Leonowicz, M.E.; Kresge, C.T.; Schmitt, K.D.; Chu, C.T.W.; Olson, D.H.; Sheppard, E.W.; McCullen, S.B.; et al. A new family of mesoporous molecular sieves prepared with liquid crystal templates. *J. Am. Chem. Soc.* **1992**, *114*, 10834–10843. [CrossRef]
46. Shannon, R. Revised effective ionic radii and systematic studies of interatomic distances in halides and chalcogenides. *Acta Crystallogr. Sect. A* **1976**, *32*, 751–767. [CrossRef]
47. El Haskouri, J.; Morales, J.M.; Ortiz de Zárate, D.; Fernández, L.; Latorre, J.; Guillem, C.; Beltrán, A.; Beltrán, D.; Amorós, P. Nanoparticulated silicas with bimodal porosity: Chemical control of the pore sizes. *Inorg. Chem.* **2008**, *47*, 8267–8277. [CrossRef]
48. Zhan, W.; Lu, G.; Guo, Y.; Guo, Y.; Wang, Y.; Wang, Y.; Zhang, Z.; Liu, X. Synthesis of cerium-doped MCM-48 molecular sieves and its catalytic performance for selective oxidation of cyclohexane. *J. Rare Earths* **2008**, *26*, 515–522. [CrossRef]
49. de Souza, L.K.C.; Pardauil, J.J.R.; Zamian, J.R.; da Rocha Filho, G.N.; Barrado, C.M.; Angélica, R.S.; da Costa, C.E.F. Rapid synthesis and characterization of CeMCM-41. *Powder Technol.* **2012**, *229*, 1–6. [CrossRef]
50. Bing, J.; Li, L.; Lan, B.; Liao, G.; Zeng, J.; Zhang, Q.; Li, X. Synthesis of cerium-doped MCM-41 for ozonation of p-chlorobenzoic acid in aqueous solution. *Appl. Catal. B* **2012**, *115–116*, 16–24. [CrossRef]
51. Laha, S.C.; Mukherjee, P.; Sainkar, S.R.; Kumar, R. Cerium containing MCM-41-type mesoporous materials and their acidic and redox catalytic properties. *J. Catal.* **2002**, *207*, 213–223. [CrossRef]
52. Barkam, S.; Ortiz, J.; Saraf, S.; Eliason, N.; McCormack, R.; Das, S.; Gupta, A.; Neal, C.; Petrovici, A.; Hanson, C.; et al. Modulating the catalytic activity of cerium oxide nanoparticles with the anion of the precursor salt. *J. Phys. Chem. C* **2017**, *121*, 20039–20050. [CrossRef]
53. Mihailova, B.; Valtchev, V.; Mintova, S.; Faust, A.C.; Petkov, N.; Bein, T. Interlayer stacking disorder in zeolite beta family: A Raman spectroscopic study. *Phys. Chem. Chem. Phys.* **2005**, *7*, 2756–2763. [CrossRef]
54. Zhou, L.; Li, X.; Yao, Z.; Chen, Z.; Hong, M.; Zhu, R.; Liang, Y.; Zhao, J. Transition-metal doped ceria microspheres with nanoporous structures for CO oxidation. *Sci. Rep.* **2016**, *6*, 23900. [CrossRef]
55. Khalil, K.M.S. Cerium modified MCM-41 nanocomposite materials via a nonhydrothermal direct method at room temperature. *J. Colloid Interface Sci.* **2007**, *315*, 562–568. [CrossRef]
56. Kolodziejski, W.; Corma, A.; Navarro, M.-T.; Pérez-Pariente, J. Solid-state NMR study of ordered mesoporous aluminosilicate MCM-41 synthesized on a liquid-crystal template. *Solid State Nucl. Magn. Reson.* **1993**, *2*, 253–259. [CrossRef]
57. Chao, K.J.; Wu, C.N.; Chang, H.; Lee, L.J.; Hu, S.-f. Incorporation of vanadium in mesoporous MCM-41 and microporous AFI zeolites. *J. Phys. Chem. B* **1997**, *101*, 6341–6349. [CrossRef]
58. Hill, P.J.; Shrestha, K.L.; Ishihara, S.; Ji, Q.; Ariga, K. Self-assembly: From amphiphiles to chromophores and beyond. *Molecules* **2014**, *19*, 8589–8609. [CrossRef]
59. Chen, C.-Y.; Burkett, S.L.; Li, H.-X.; Davis, M.E. Studies on mesoporous materials II. Synthesis mechanism of MCM-41. *Microporous Mater.* **1993**, *2*, 27–34. [CrossRef]
60. Dendramis, A.L.; Schwinn, E.W.; Sperline, R.P. A surface-enhanced Raman scattering study of CTAB adsorption on copper. *Surf. Sci.* **1983**, *134*, 675–688. [CrossRef]
61. Snyder, R.G.; Strauss, H.L.; Elliger, C.A. Carbon-hydrogen stretching modes and the structure of n-alkyl chains. 1. Long, disordered chains. *J. Phys. Chem.* **1982**, *86*, 5145–5150. [CrossRef]
62. Orendorff, C.J.; Ducey, M.W.; Pemberton, J.E. Quantitative correlation of Raman spectral indicators in determining conformational order in alkyl chains. *J. Phys. Chem. A* **2002**, *106*, 6991–6998. [CrossRef]
63. Jamieson Lauren, E.; Li, A.; Faulds, K.; Graham, D. Ratiometric analysis using Raman spectroscopy as a powerful predictor of structural properties of fatty acids. *R. Soc. Open Sci.* **2018**, *5*, 181483. [CrossRef]
64. Terribile, D.; Trovarelli, A.; Llorca, J.; de Leitenburg, C.; Dolcetti, G. The synthesis and characterization of mesoporous high-surface area ceria prepared using a hybrid organic/inorganic route. *J. Catal.* **1998**, *178*, 299–308. [CrossRef]

65. Hayes, S.A.; Yu, P.; O'Keefe, T.J.; O'Keefe, M.J.; Stoffer, J.O. The phase stability of cerium species in aqueous systems: I. E-pH diagram for the Ce-$HClO_4$–H_2O system. *J. Electrochem. Soc.* **2002**, *149*, C623–C630. [CrossRef]
66. Bumajdad, A.; Eastoe, J.; Mathew, A. Cerium oxide nanoparticles prepared in self-assembled systems. *Adv. Colloid Interface Sci.* **2009**, *147–148*, 56–66. [CrossRef]
67. Bouchaud, B.; Balmain, J.; Bonnet, G.; Pedraza, F. pH-distribution of cerium species in aqueous systems. *J. Rare Earths* **2012**, *30*, 559–562. [CrossRef]
68. Peppard, D.F.; Mason, G.W.; Hucher, I. Stability constants of certain lanthanide(III) and actinide(III) chloride and nitrate complexes. *J. Inorg. Nucl. Chem.* **1962**, *24*, 881–888. [CrossRef]
69. Berr, S.; Jones, R.R.M.; Johnson, J.S. Effect of counterion on the size and charge of alkyltrimethylammonium halide micelles as a function of chain length and concentration as determined by small-angle neutron scattering. *J. Phys. Chem.* **1992**, *96*, 5611–5614. [CrossRef]
70. Ye, F.; Vallhov, H.; Qin, J.; Daskalaki, E.; Sugunan, A.; Toprak, M.S.; Fornara, A.; Gabrielsson, S.; Scheynius, A.; Muhammed, M. Synthesis of high aspect ratio gold nanorods and their effects on human antigen presenting dendritic cells. *Int. J. Nanotechnol.* **2011**, *8*, 22. [CrossRef]
71. Gamboa, C.; Sepulveda, L.; Soto, R. Free energies of transfer of anions from water to cationic micelles from ionic exchange measurements. *J. Phys. Chem.* **1981**, *85*, 1429–1434. [CrossRef]
72. Xiong, G.; Li, C.; Li, H.; Xin, Q.; Feng, Z. Direct spectroscopic evidence for vanadium species in V-MCM-41 molecular sieve characterized by UV resonance Raman spectroscopy. *Chem. Commun.* **2000**, 677–678. [CrossRef]
73. Casas-Orozco, D.; Alarcón, E.; Carrero, C.A.; Venegas, J.M.; McDermott, W.; Klosterman, E.; Hermans, I.; Villa, A.-L. Influence of tin loading and pore size of Sn/MCM-41 catalysts on the synthesis of nopol. *Ind. Eng. Chem. Res.* **2017**, *56*, 6590–6598. [CrossRef]
74. Li, C. Identifying the isolated transition metal ions/oxides in molecular sieves and on oxide supports by UV resonance Raman spectroscopy. *J. Catal.* **2003**, *216*, 203–212. [CrossRef]
75. Li, L.; Chen, F.; Lu, J.-Q.; Luo, M.-F. Study of defect sites in $Ce_{1-x}M_xO_2-\delta$ (x = 0.2) solid solutions using Raman spectroscopy. *J. Phys. Chem. A* **2011**, *115*, 7972–7977. [CrossRef]
76. He, H.; Yang, P.; Li, J.; Shi, R.; Chen, L.; Zhang, A.; Zhu, Y. Controllable synthesis, characterization, and CO oxidation activity of CeO_2 nanostructures with various morphologies. *Ceram. Int.* **2016**, *42*, 7810–7818. [CrossRef]
77. Soren, S.; Bessoi, M.; Parhi, P. A rapid microwave initiated polyol synthesis of cerium oxide nanoparticle using different cerium precursors. *Ceram. Int.* **2015**, *41*, 8114–8118. [CrossRef]
78. Ghiaci, M.; Abbaspur, A.; Kia, R.; Belver, C.; Trujillano, R.; Rives, V.; Vicente, M.A. Vapor-phase alkylation of toluene by benzyl alcohol on H_3PO_4-modified MCM-41 mesoporous silicas. *Catal. Commun.* **2007**, *8*, 49–56. [CrossRef]
79. Fu, Y.; Zhan, W.; Guo, Y.; Wang, Y.; Liu, X.; Guo, Y.; Wang, Y.; Lu, G. Effect of surface functionalization of cerium-doped MCM-48 on its catalytic performance for liquid-phase free-solvent oxidation of cyclohexane with molecular oxygen. *Microporous Mesoporous Mater.* **2015**, *214*, 101–107. [CrossRef]
80. Mullins, D.R. The surface chemistry of cerium oxide. *Surf. Sci. Rep.* **2015**, *70*, 42–85. [CrossRef]
81. Zdravković, J.; Simović, B.; Golubović, A.; Poleti, D.; Veljković, I.; Šćepanović, M.; Branković, G. Comparative study of CeO_2 nanopowders obtained by the hydrothermal method from various precursors. *Ceram. Int.* **2015**, *41*, 1970–1979. [CrossRef]
82. Zhao, X.S.; Lu, G.Q.; Whittaker, A.K.; Millar, G.J.; Zhu, H.Y. Comprehensive study of surface chemistry of MCM-41 using ^{29}Si CP/MAS NMR, FTIR, pyridine-TPD, and TGA. *J. Phys. Chem. B* **1997**, *101*, 6525–6531. [CrossRef]
83. Varache, M.; Bezverkhyy, I.; Saviot, L.; Bouyer, F.; Baras, F.; Bouyer, F. Optimization of MCM-41 type silica nanoparticles for biological applications: Control of size and absence of aggregation and cell cytotoxicity. *J. Non-Cryst. Solids* **2015**, *408*, 87–97. [CrossRef]
84. Vantomme, A.; Yuan, Z.-Y.; Du, G.; Su, B.-L. Surfactant-Assisted Large-Scale Preparation of Crystalline CeO2 Nanorods. *Langmuir* **2005**, *21*, 1132–1135. [CrossRef]
85. Pan, C.; Zhang, D.; Shi, L. CTAB assisted hydrothermal synthesis, controlled conversion and CO oxidation properties of CeO_2 nanoplates, nanotubes, and nanorods. *J. Solid State Chem.* **2008**, *181*, 1298–1306. [CrossRef]

86. Galarneau, A.; Villemot, F.; Rodriguez, J.; Fajula, F.; Coasne, B. Validity of the t-plot method to assess microporosity in hierarchical micro/mesoporous materials. *Langmuir* **2014**, *30*, 13266–13274. [CrossRef]
87. Sing, K.S.W.; Everett, D.H.; Haul, R.A.W.; Moscou, L.; Pierotti, R.A.; Rouquérol, J.; Siemieniewska, T. Reporting physisorption data for gas/solid systems with special reference to the determination of surface area and porosity. *Pure Appl. Chem.* **1985**, *57*, 17. [CrossRef]
88. Anderson, M.W. Solid-state NMR as a probe of porous catalysts and catalytic processes. *Top. Catal.* **1996**, *3*, 195–220. [CrossRef]
89. Kleitz, F.; Schmidt, W.; Schüth, F. Evolution of mesoporous materials during the calcination process: Structural and chemical behavior. *Microporous Mesoporous Mater.* **2001**, *44–45*, 95–109. [CrossRef]
90. de Souza, L.K.C.; Pardauil, J.J.R.; Zamian, J.R.; da Rocha Filho, G.N.; da Costa, C.E.F. Influence of the incorporated metal on template removal from MCM-41 type mesoporous materials. *J. Therm. Anal. Calorim.* **2011**, *106*, 355–361. [CrossRef]
91. Selvaraj, M.; Park, D.W.; Ha, C.S. Well ordered two-dimensional mesoporous CeSBA-15 synthesized with improved hydrothermal stability and catalytic activity. *Microporous Mesoporous Mater.* **2011**, *138*, 94–101. [CrossRef]
92. Timofeeva, M.N.; Jhung, S.H.; Hwang, Y.K.; Kim, D.K.; Panchenko, V.N.; Melgunov, M.S.; Chesalov, Y.A.; Chang, J.S. Ce-silica mesoporous SBA-15-type materials for oxidative catalysis: Synthesis, characterization, and catalytic application. *Appl. Catal. A Gen.* **2007**, *317*, 1–10. [CrossRef]
93. Wang, Z.; Balkus, K.J. Liquid phase propylene oxidation with *tert*-butyl hydroperoxide over titanium containing wrinkled mesoporous silica. *Catal. Commun.* **2017**, *96*, 15–18. [CrossRef]
94. Grün, M.; Unger, K.K.; Matsumoto, A.; Tsutsumi, K. Novel pathways for the preparation of mesoporous MCM-41 materials: Control of porosity and morphology. *Microporous Mesoporous Mater.* **1999**, *27*, 207–216. [CrossRef]
95. Cullity, B.D. *Elements of X-ray Diffraction*, 2nd ed.; Addison-Wesley Publishing Company, Inc.: Reading, PS, USA, 1967; pp. 81–349.
96. Davis, E.A.; Mott, N.F. Conduction in non-crystalline systems V. Conductivity, optical absorption and photoconductivity in amorphous semiconductors. *Philos. Mag.* **1970**, *22*, 0903–0922. [CrossRef]
97. Igarashi, N.; Koyano, K.A.; Tanaka, Y.; Nakata, S.; Hashimoto, K.; Tatsumi, T. Investigation of the factors influencing the structural stability of mesoporous silica molecular sieves. *Microporous Mesoporous Mater.* **2003**, *59*, 43–52. [CrossRef]

catalysts

Article

Zirconia-Supported Silver Nanoparticles for the Catalytic Combustion of Pollutants Originating from Mobile Sources

Maia Montaña [1], María S. Leguizamón Aparicio [1], Marco A. Ocsachoque [1], Marisa B. Navas [1], Ivoneide de C. L. Barros [2], Enrique Rodriguez-Castellón [3], Mónica L. Casella [1] and Ileana D. Lick [1,*]

1 CINDECA (CCT La Plata- CONICET-UNLP), Departamento de Química, Facultad de Ciencias Exactas, Universidad Nacional de La Plata, Calle 47 N° 257, La Plata, Buenos Aires 1900, Argentina; maia.montana@ing.unlp.edu.ar (M.M.); mariasilvialap@quimica.unlp.edu.ar (M.S.L.A.); ocmarco@quimica.unlp.edu.ar (M.A.O.); marisanavas@quimica.unlp.edu.ar (M.B.N.); casella@quimica.unlp.edu.ar (M.L.C.)

2 Departamento de Química, Universidade Federal Rural de Pernambuco, Rua Dom Manuel de Medeiros, S/N, Dois Irmãos, Recife-PE CEP 52171-900, Brasil; iclbarros@gmail.com

3 Departamento de Química Inorgánica, Cristalografía y Mineralogía, Facultad de Ciencias, Universidad de Málaga, Campus de Teatinos, 29071 Málaga, España; castellon@uma.es

* Correspondence: ilick@quimica.unlp.edu.ar; Tel.: +54-221-421-1353

Received: 29 January 2019; Accepted: 16 March 2019; Published: 25 March 2019

Abstract: This work presents the physicochemical characterization and activity of zirconia-supported silver catalysts for the oxidation of pollutants present in diesel engine exhaust (propane, propene, naphthalene and soot). A series of silver-supported catalysts AgxZ (x = 1, 5 and 10 wt.%, Z = zirconia) were prepared, which were studied by various characterization techniques. The results show that silver is mainly found under the form of small metal nanoparticles (<10 nm) dispersed over the support. The metallic phase coexists with the AgO_x oxidic phases. Silver is introduced onto the zirconia, generating Ag–ZrO_2 catalysts with high activity for the oxidation of propene and naphthalene. These catalysts also show some activity for soot combustion. Silver species can contribute with zirconia in the catalytic redox cycle, through a synergistic effect, providing sites that facilitate the migration and availability of oxygen, which is favored by the presence of structural defects. This is a novel application of the AgO_x–Ag/ZrO_2 system in the combustion reaction of propene and naphthalene. The results are highly promising, given that the T50 values found for both model molecules are quite low.

Keywords: silver nanoparticles; zirconia; hydrocarbons; diesel soot; catalytic combustion

1. Introduction

Mobile emission sources generate a large amount of pollutants, which are emitted into the atmosphere and cause serious environmental problems, both historically and in human health. Among these pollutants are nitrogen oxides (NO_x), carbon oxides (CO_x), volatile organic compounds (VOCs), sulfur oxides (SO_x) and particulate matter, usually called soot or diesel soot [1].

Among the VOCs, different light hydrocarbons (C2–C10), known as NMHCs (non-methane hydrocarbons), are found in very low concentrations. The NMHCs that are the most difficult to eliminate are C3-C4, and polyaromatic compounds (PAHs), which present a level of high toxicity [2,3]. On the other hand, diesel soot consists of solid particles of small size, whose main component is carbon [4]. These particles are very small and can penetrate the lungs, which makes soot very harmful to health.

These pollutants can be removed by catalytic oxidation to CO_2 and water, avoiding the formation of toxic compounds. This process can be carried out in catalytic converters located in the exhaust pipe of automobiles. Moreover, to remove particulate matter, the converter must retain the particles and then regenerate itself by oxidizing this retained material. In addition, to avoid an extra energy demand, this process must function in a passive way. This implies that the oxidation reaction must occur within the operating temperature range of the exhaust pipe [5–8]. Passive filter regeneration technology is used in the commercial Continuously Regenerating Trap system (CRT, Johnson Matthey, London, UK).

A wide variety of active phases have been reported for the catalytic oxidation of contaminants present in emission sources. In this sense, different types of supports have been studied, such as: simple and compound oxides, and zeolites and clays, among others [9–12]. It is possible to differentiate between catalysts based on oxidic phases, and those based on supported metallic phases. The latter group usually contains Pt, Pd, Rh, and Au, among other noble metals. These systems are called supported platinum group metal (PGM) catalysts and present very good activity, so they can be used in several applications [13–20]. However, due to the high cost and low abundance of these metals, it is necessary to reduce the metal load used, or replace them with other less expensive phases.

In the case of VOCs, and particularly in the case of the oxidation of the remaining hydrocarbons (HC), the following oxidic systems and compounds have been reported as being active: Co_3O_4, MnO_2, CeO_2, and CuO [21–31]. The combustion reaction of particulate matter has been extensively studied in recent decades, identifying catalysts with very diverse active phases [32–41]. A meticulous study of them can be found in a review by A. Bueno Lopez et al. [32] and in a more recent review by D. Fino et al. [33]. Catalysts containing MnO_2-x, CeO_2, or alkali metals have been reported as being very active, like other noble metal free oxides systems. In addition, the metallic promotion of these oxides has been studied, to achieve better catalytic performance. In this context, the development of Ag-containing oxidic catalysts has received attention in recent years [5,42–58]. In the case of the combustion of particulate matter, these catalysts have been reported as very active [42–44,47]. There are also some studies in which different catalysts containing silver are applied in the total oxidation of remaining HC. The silver nanoparticles are of low cost, and they possess high activity and stability. Silver has been widely used as biocide, or in the development of sensors. In this sense, Ag is considered safe for the design of materials, due to its low potential hazardous effects.

In the case of polycyclic aromatic hydrocarbons (PAHs), the number of reports is much lower. In particular, the Ag/CeO_2 catalyst presents a great performance in soot oxidation reactions [55], propene and naphthalene [58]. According to the literature, the excellent results should be assigned to the ability of the Ag-containing catalysts to activate oxygen through dissociative adsorption. This capacity can be promoted synergistically with the mobility of oxygen, as provided by the support. This type of mechanism, based on the adsorption of the reactants, has been accepted as one of the feasible mechanisms in the oxidation reactions of soot and hydrocarbons [13,19,28,42,50]. Yamazaki et al. studied soot oxidation over a CeO_2–Ag catalyst with a 'rice-ball' morphology, consisting of a center composed of Ag particles, surrounded by fine CeO_2 particles. A mechanism for soot oxidation has been proposed, where active oxygen atomic species are formed on the Ag surface, from the dissociative adsorption of O_2. These species migrate to the CeO_2 interface, and form $O_n{}^{x-}$ species; then they further migrate onto soot particles, where oxidation occurs at very low temperatures [50]. A. Serve et al. studied Ag catalysts supported on yttria-stabilized zirconia (YSZ) for soot oxidation. They also proposed that Ag favors the activation of the dissociatively adsorbed oxygen, and the lattice integration of gaseous oxygen. Soot oxidation occurs through bulk O_2-YSZ [55].

On the other hand, it is proposed that oxidation reactions can follow a mechanism of the Mars-van Krevelen type. According to this mechanism, the catalytic surface can form a chemical bond with an adsorbate/reactant, for example, HC. These species can be oxidized by a surface lattice atom on the catalytic surface. When a reaction product desorbs, a vacancy is generated on the surface, which must be regenerated [59–63]. It is clear that surface species must have some redox capacity, and also the capacity to form oxygen vacancies [56]. In this context, zirconia is presented as a suitable support, due to its redox capacity [26]. In addition, when zirconia are added to a host cation in its crystalline network, the bulk and sub-surface defects are generated. This facilitates the migration of oxygen species from the bulk to surface. The presence of oxygen vacancies has been correlated with the catalytic activity in oxidation reactions; for example, the oxidation of CO and soot [8,56,58].

Various works concerning oxidation mechanisms indicate that not only the presence of oxygen vacancies is necessary, but also the presence of electrophilic species called "active oxygen O_x^-" (x = 1 or 2). Some authors even indicate that even the presence of an excessive number of surface defects can generate the deactivation of these O_x^- species, due to the generation of O_2^- species, which are nucleophilic. In a recent work, Wang et al. have studied the roles of oxygen vacancies, hydroxyls and O_x^- in different oxidation reactions, using CeO_2 and Ag/CeO_2 as catalysts [56,58]. Moreover, they have shown that the formation of O_x^- species can be promoted with the presence of Ag to obtain higher NO and soot oxidation activities. Another desirable characteristic is that the support possesses certain levels of acidity to favor the adsorption of the hydrocarbon. In this sense, there are some reports on the activity of pure zirconia for oxidation reactions [26].

Zirconia is a support that has a certain redox capacity [64]; it presents polymorphs with controllable acid–base properties and good thermal stability. Besides, in its amorphous hydrous state, it presents a high surface area that can favor the dispersion of the supported precursors, and thus avoid agglomeration during the thermal treatments [64–66].

In addition to the different reaction mechanisms that can be proposed, there are several parameters influencing the behavior of the active species in these systems. Thus, a fundamental factor to consider is the size of the metal particles (an optimum value is <10 nm) [67]. Also, the catalytic efficiency of the Ag catalysts can be related to the adsorption of some hydrocarbons on this metal [68].

In this work, it is proposed that the preparation of zirconia-supported Ag catalysts, obtained by the impregnation of $ZrO_2{\cdot}nH_2O$ (hydrogel hydrous zirconium oxide) with $AgNO_3$ solutions. The focus will be on the formation of supported Ag nanoparticles, and in the study of this catalytic system in oxidation reactions of pollutants generated in emission sources. The activity of these catalysts is analyzed in the oxidation reactions of model molecules of different hydrocarbons at very low concentrations: propane, propene, and naphthalene. It is worth mentioning that the Ag/ZrO_2 system has not yet been used for propene and naphthalene elimination. Also, their activities are analyzed in the combustion of diesel soot/particulate matter reaction. The prepared catalysts are characterized using several physicochemical techniques: N_2 adsorption–desorption, scanning electron microscopy, and energy dispersive X-ray spectroscopy (SEM-EDS), X-ray diffraction (XRD), diffuse reflectance spectroscopy (DRS-UV-vis), transmission electron microscopy (TEM), and temperature-programmed reduction (TPR). X-ray photoelectron spectroscopy (XPS) was employed to obtain information about the species at the surface level.

2. Results and Discussion

Table 1 presents the characterization results for zirconia and silver–zirconia-supported catalysts.

Table 1. Specific surface area, TPR results and crystalline phases of the studied samples.

Samples	S_{BET} (m^2g^{-1})	V_P	ZrO_2 Crystal Size (nm) *1	Theoretical H_2 Consumption *2	H_2 Consumption *3	Ag Particle Size (nm) *4
$ZrO_2{\cdot}nH_2O$	340	-	amorphous		-	
ZrO_2	50	0.09	12		-	
Ag1Z	46	0.08	14	0.0046	nd	Nd
Ag5Z	42	0.08	14	0.0232	0.0021	5.3
Ag10Z	39	0.08	15	0.0464	0.0032	6.3

*1 obtained by XRD. *2 theoretical H_2 mmol consumed per 100 mg of catalyst (Equation (1)) *3 experimental mmol H_2 consumed per 100 mg of catalyst. *4 obtained by TEM.

Textural properties were obtained by N_2 adsorption–desorption at 77 K. Figure 1 presents the adsorption–desorption isotherms of the Ag-supported catalysts. In all the cases, the obtained isotherms were type IV according to the IUPAC classification, presenting a hysteresis cycle that is associated with the process of filling of the mesopores.

The BET surface area values corresponding to the original hydrogel and the prepared catalysts are presented in Table 1. The starting hydrogel exhibits a high specific surface area value. Then, the addition of Ag and the subsequent thermal process decreases the surface area, due to the loss of the water molecules of the hydrogel, and also the increase of the metal–host–metal interactions. An increase in the Ag content decreases the specific surface area without decreasing the pore volume. The crystallization process of pure zirconia generates a system with a slightly larger area than that of the Ag-catalysts. This indicates that the thermal treatment is the predominant factor in textural changes.

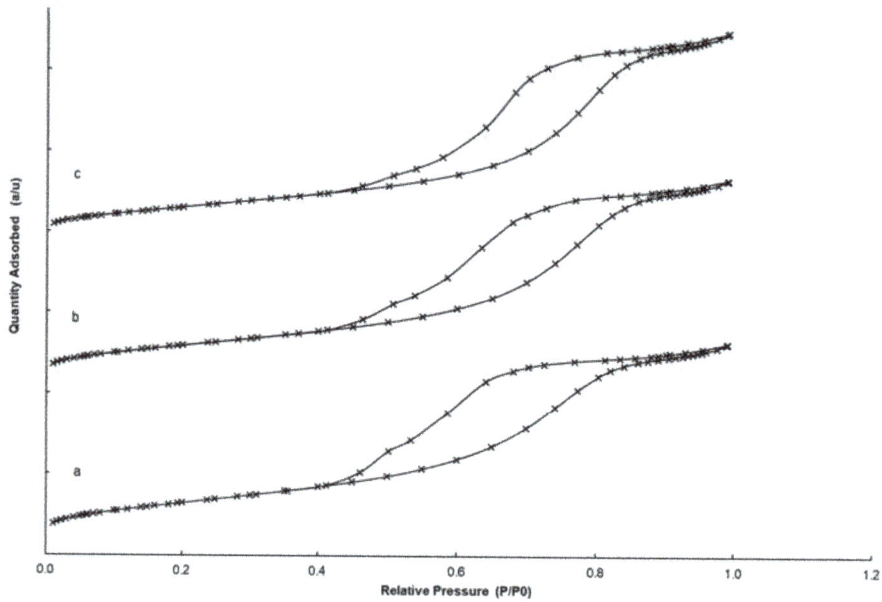

Figure 1. N_2 adsorption–desorption isotherms of the catalysts. (**a**) Ag1Z, (**b**) Ag5Z and (**c**) Ag10Z.

Figure 2 shows the XRD patterns of the support and the AgxZr catalysts. The XRD pattern of the undoped ZrO_2 support, calcined for 2 h at 600 °C presents diffraction lines that are typical of the monoclinic phase of ZrO_2, with peaks being located at 28.2, 31.5, 34.2, and 50.2° (PDF N° 03-065-1025). The diffraction lines of the meta-stable tetragonal phase, located at 30.0°, 35.0°, and 50.0° (PDF N° 01-089-7710) were identified for all AgxZ catalysts. The addition of Ag strongly influences the crystallization of amorphous hydrated zirconium oxide, and promotes the stabilization of the meta-stable tetragonal crystalline phase. This specific phase has been proposed as the active phase on certain reactions [69]. The presence of the meta-stable phase, even in the catalyst with the lowest Ag content, indicates a strong Ag—support interaction, generated in the crystallization and sintering stages.

As it has been widely reported, some ions, called "host ions" or dopants, can be incorporated into the crystal lattice of zirconia by replacing the central ion, Zr (IV). In this sense, the effective ionic radius of Ag (I) is similar to those of the most common cations used in doping zirconia, such as Y (III) and Ca (II). Furthermore, if the dopants have a lower oxidation state than the central ion in the crystalline structure, defects are generated at the bulk and surface levels, with the consequent formation of oxygen vacancies [70].

The zirconia crystallite size, obtained by Debye–Scherrer´s equation, is shown in Table 1. The results indicate that AgxZ catalysts contains nanometric zirconia crystals, with a size slightly larger than those present in pure zirconia.

On the other hand, the patterns do not exhibit diffraction lines of Ag^0 crystallites, located at 38.15°, 44.34°, and 64.51°, (PDF N° 03-065-2871), or lines of Ag_2O, located at 32.88° and 38.15° (PDF N°03-065-6811). This indicates that, if these species exist, they are present at a size that cannot be detected by XRD. It should be noted that there are no marked differences between the XRD profiles of AgxZ catalysts. Only a very slight increase in the intensity of the line at 30.0°, belonging to the support, is observed. This indicates the presence of good stabilization of the metastable tetragonal phase, and a good distribution of the supported species on the surface, since no segregated crystalline phases are observed.

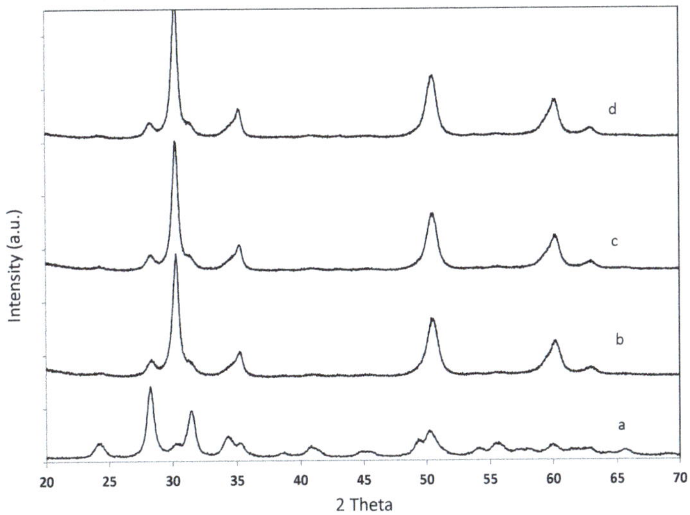

Figure 2. XRD patterns of support and catalysts. (**a**) ZrO_2, (**b**) Ag1Z, (**c**) Ag5Z, and (**d**) Ag10Z.

Figure 3 presents the SEM images and elemental maps (MAP) of the prepared catalysts. These analyses were carried out in order to determine the distribution of silver on the catalyst surface. The overlapping of zirconium and Ag distribution in the MAP images showed that the structures containing Ag are formed on the particles of the catalysts. Moreover, the Ag species are widely distributed over the Ag5Z and Ag10Z catalysts (Figure 3). Some EDS analyses were carried out for the very small areas, where the highest intensity was observed in the mapping of silver. It was found that the experimental Ag/Zr atomic ratio was greater than the nominal one.

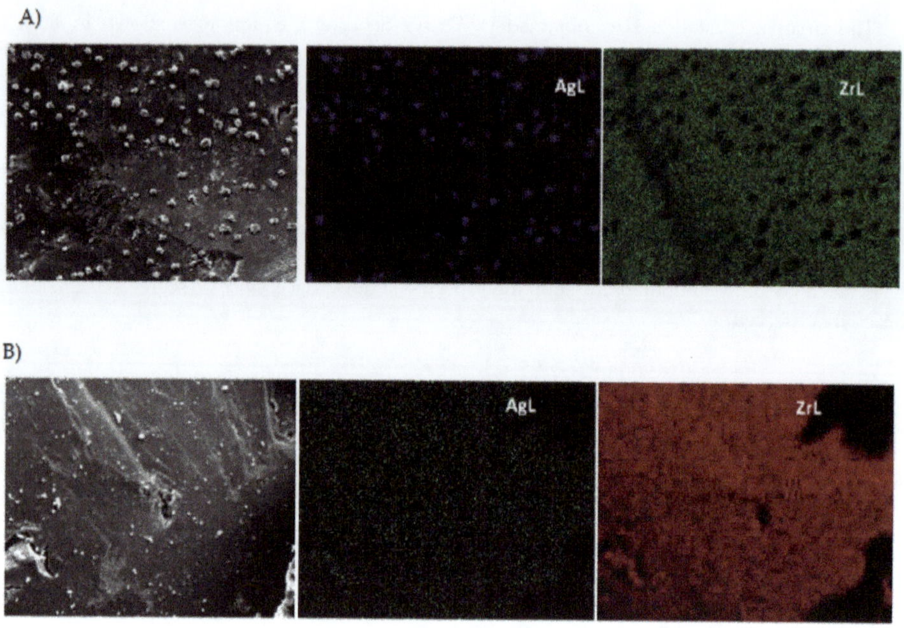

Figure 3. SEM images and elemental mapping of Ag and Zr in catalysts: (**A**) Ag5Z and (**B**) Ag10Z.

In order to analyze the presence of reducible phases in the catalysts, TPR analyses were carried out, and the results are depicted in Figure 4. The TPR diagram of the pure ZrO_2 (not shown) does not present any signal, which indicates the absence of reducible phases at the studied temperature range, although it is known that it can be partially reduced at high temperature (about 800 °C). The TPR diagrams of the catalysts containing 5 and 10 wt.% of Ag presented a signal at very low temperature (approximately 90 °C). This can be assigned to the reduction of oxidic or cationic Ag species, in low concentrations, which are well-dispersed on the support, or presenting low interactions with it. This signal also can be assigned to superficial oxygen atoms coming from its migration from zirconia [58]. The integrated area of the signal increases with the increase of Ag content. However, it is noteworthy that all the AgxZ catalysts had an extremely low level of H_2 consumption (Table 1), which is lower than the expected if the supported phase corresponds to the stoichiometric reduction of the Ag_2O oxide (Equation (1)).

$$Ag_2O + H_{2(g)} \leftrightarrow H_2O_{(g)} + 2Ag^0 \quad (1)$$

These results would indicate that silver is mainly in a reduced state. These species could have been formed during the calcination process, probably due to the reducing environment of the zirconia hydrogel-containing surface OH groups, and the low stability of the Ag oxides at high temperature. Corro et al. propose that the Ag^0 formation is associated with the low enthalpy of formation of Ag_2O at high temperature [71]. In previous studies, it has been reported that the reducibility of transition metal species increases on ZrO_2 [26,72]. It has also been proposed that the formation of oxygen vacancies affects the redox properties of the supported species. A mode of behavior that similar to that observed in this work was reported by Aneggi et al., who studied the Ag/ZrO_2 catalytic system [53]. These authors show that during the calcination process, $AgNO_3$ decomposes almost completely into metallic Ag, although they observed by HRTEM the presence of Ag_2O on the surface of the catalysts.

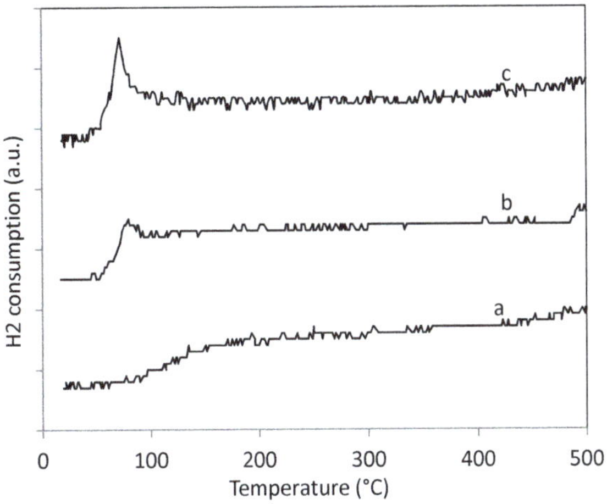

Figure 4. H_2-TPR profiles of (**a**) Ag1Z, (**b**) Ag5Z, and (**c**) Ag10Z.

In order to obtain more information about the nature of the supported Ag phases, DRS UV-Vis studies were performed. The spectra obtained for AgxZ catalysts are shown in Figure 5. An energy absorption band located between 330 nm and 390 nm can be clearly observed. In this region of the spectrum, the absorption is assigned to Ag clusters $Ag_n^{\delta+}$ (n = 2–7) or particles of metallic silver of small size (3–5 nm). As the size of the particles grows, this band can suffer a bathochromic shift, towards longer wavelengths. This is probably the reason for the slight shift of this band in the Ag10Z catalyst spectrum. The bands located at 220 and 265nm are associated with the presence of the Ag^{+1} species, and the charge transfer band of the $Ag_n^{\delta+}$ clusters respectively [67]. Although there is absorbance in the region of the plasmon resonance, 400–450 nm, these bands are not well-resolved or defined, due to the presence of energy absorption of the support, ZrO_2.

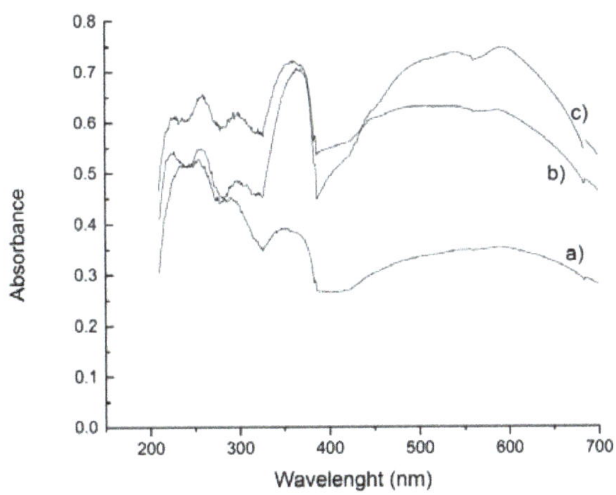

Figure 5. DRS UV-Vis signals for the prepared catalysts. (**a**) Ag1Z, (**b**) Ag5Z and (**c**) Ag10Z.

The characterization techniques performed, which provide information at a mass level, suggesting that a part of Ag is present in a metallic state, Ag^0. In order to corroborate the presence of this type of species, transmission electron microscopy (TEM) analysis were carried out (Figure 6). It is noticeable that the micrographs corresponding to Ag5Z and Ag10Z catalysts show zirconia structures with highly dispersed silver nanoparticles. In the micrographs of the Ag1Z catalyst, although some Ag particles were observed, the resolution of the technique does not allow for the visualization of a significant number of them. Ag particle size distribution was analyzed from some selected regions of the TEM micrographs; the results obtained for the mean particle size are listed in Table 1.

The histograms presented in Figure 6 show that for both Ag5Z and Ag10Z catalysts, the Ag particles had sizes of between 2 and 10 nm. The histogram of the Ag5Z catalyst exhibits a relatively narrow particle size distribution, mainly constituted of particles having diameters of between 4 and 5 nm. The increase in the silver loading of the catalysts leads to a greater contribution of the larger particles, and therefore, the average particle size increases. This increase in the silver particle size leads to a loss of metal dispersion (D%), which decreases from 23% to 17% for Ag5Z and Ag10Z, respectively.

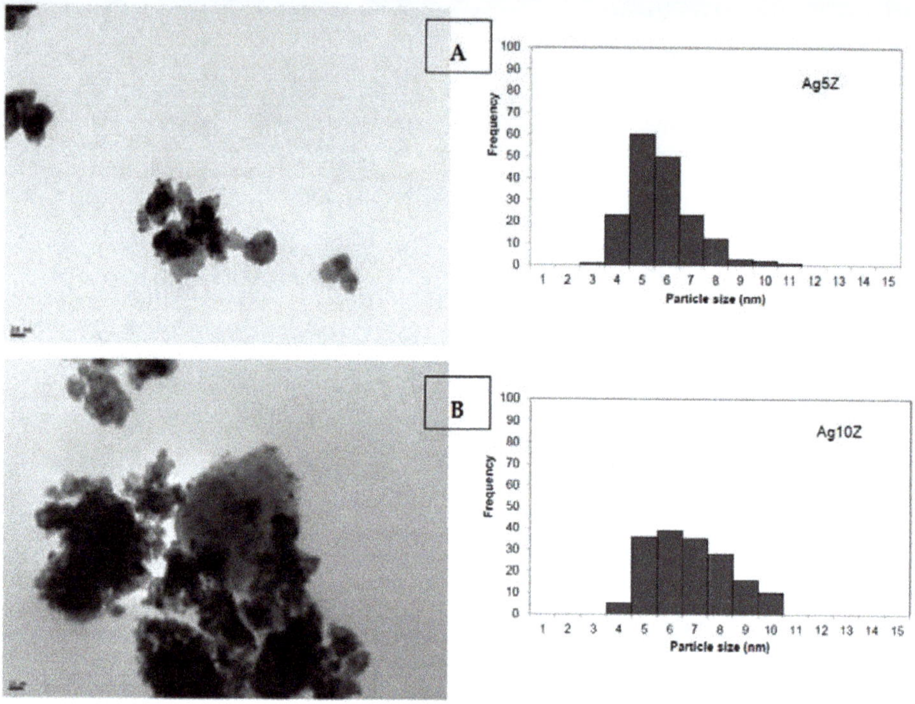

Figure 6. TEM micrographs and histograms of particle size distribution: (**A**) Ag5Z catalyst; (**B**) Ag10Z catalyst.

The TEM results agree with those obtained by H_2-TPR and UV-Vis spectroscopy. Both techniques suggested that part of the Ag was in a metallic state under the form of nano-sized particles, which cannot be observed by XRD. However, there is evidence for the existence of oxidic species in low concentration, due to the presence of signals at low temperature in the TPR analysis, and some absorption bands in UV-Vis, associated with this species.

The oxidation states of the supported species of Ag surface were studied by XPS, although several reports indicate that it is difficult to distinguish between chemical states of Ag, since the binding energies values of Ag3$d_{5/2}$ (BE) for Ag(0) and their oxides are very similar [73]. Thus, the AgMNN Auger signals were also analyzed, in order to obtain more precise information, since the chemical displacements of the Auger peaks were generally greater than the displacements of the photoelectronic peaks, therefore being more appropriate for considering the use of the modified Auger parameter (AP*) to identify the Ag oxidation states. Therefore, in this work, the XPS spectra of the catalysts in the Ag3d region and in the Auger region, AgMNN (Figure 7A,B), were studied. Table 2 presents the binding energies (BE) of the Ag3$d_{5/2}$ peak, the kinetic energy of the AgMNN Auger transition (KE), and the calculated AP* (AP* = KE + BE) for the different components of the AgMNN signals.

BE values for the Ag3$d_{5/2}$ peaks are very close to each other, for the three studied catalysts. The values are between 368.2 eV and 367.7 eV, decreasing slightly when the metallic content decreases. In the AgMNN region, the signals are wide, complex, and have several contributions (Table 2). Those whose AP* are at values of ca. 725 eV and 720 eV, can be associated with the presence of Ag in the metallic state. On the other hand, signals that are located at lower values of KE and AP* (around 717.5 eV), are associated with the presence of oxidic phases of Ag. The metallic particles are probably oxidized on their surfaces [73].

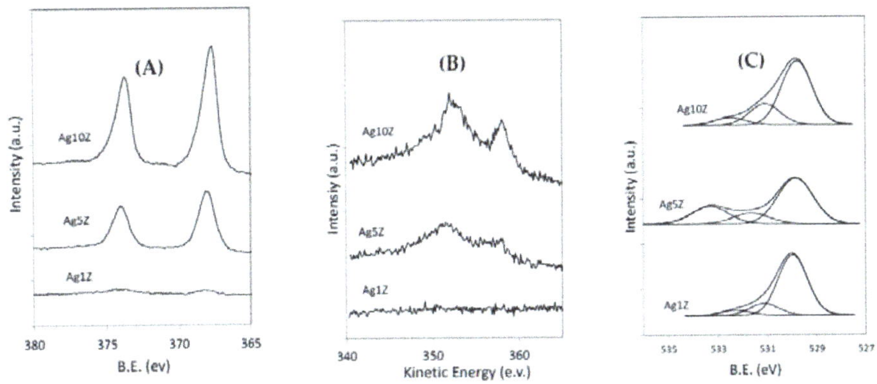

Figure 7. (**A**) Ag 3d XPS spectra, (**B**) AgMNN Auger transition spectra, and (**C**) O1s XPS spectra of the different studied catalysts.

Table 2. XPS BE (eV) values for the Ag3d, O1s, and Zr3d, Auger parameters, Ag/Zr atomic ratios, and oxygen species ratio for the studied catalysts.

	Ag1Z	Ag5Z	Ag10Z
Ag 3$d_{5/2}$	368.2	368.0	367.7
O 1s *a	529.9(77.96) 531.1(14.98) 532.3(7.01)	529.8(59.69) 531.6(15.15) 533.3(25.15)	529.8(69.2) 531.1(22.8) 532.5(8.0)
Zr 3$d_{5/2}$	182.2	181.8	182.0
KE $AgM_4N_{45}N_{45}$	-	357.5	357.9
KE $AgM_5N_{45}N_{45}$	-	352.5 349.5	352.8 349.4
AP* Ag 3$d_{5/2}$-$AgM_4N_{45}N_{45}$	-	725.5	725.6
AP* Ag 3$d_{5/2}$-$AgM_5N_{45}N_{45}$	-	720.5 717.5	720.5 717.1
Ag/Zr Atomic ratio	0.009	0.154	0.213
Theoretical Ag/Zr atomic ratio	0.011	0.060	0.127
O_β/O_α	0.28	0.67	0.45

*a between parenthesis the relative percentages of the different components of the signal are shown.

Table 2 also presents the Zr 3$d_{5/2}$ and O 1s binding energy values (in eV) from the core-level spectra signals. The spectra of the Zr 3*d* core level spectrum shows two binding energies about 182 and 184 eV, which are assigned to the doublet Zr 3$d_{5/2}$ and Zr 3$d_{3/2}$, respectively. These peaks are typical of Zr^{4+} in ZrO_2. The O 1s signal was decomposed in three components, with binding energies of about 530, 531, and 532–533 eV (Figure 7C), being related to surface oxygen species with different chemical environments. The species with binding energy between 529–530 eV, were attributed to the O_α, which is characteristic of lattice O in ZrO_2, [15,24] and those in the 531–533 eV region, were related to $O_β$, which are associated with surface oxygen vacancies, oxygen-adsorbed species, surface hydroxyl, adsorbed water molecules, or carbonate species [74]. In our case, however, the presence of carbonates was not evident in the spectra of the C1s region (not shown). The Ag5Z and Ag10Z catalysts presented a higher proportion of $O_β$ species (Table 2), which are proposed as possible active sites.

The Ag/Zr surface atomic ratios have been calculated from XPS analysis, and then compared with the theoretical composition of the bulk (Table 2). As expected, the surface presents an enrichment of Ag, due to the predominant location of Ag on the external surface of the support, both for the Ag5Z and Ag10Z catalysts.

The AgxZ catalysts, and also the pure zirconia were used in the catalytic oxidation of three model molecules of hydrocarbons (HC): propane, as a saturated linear hydrocarbon molecule; propene, as an unsaturated linear hydrocarbon molecule; and naphthalene, as a polyaromatic hydrocarbon molecule. In addition, the prepared catalysts were tested in the catalytic oxidation of particulate matter (diesel soot). Table 3 summarizes the catalytic results obtained in all of the studied reactions. The combustion reactions of hydrocarbons were carried out in fixed-bed micro-reactors. The obtained T50 values in each experiment were presented, with T50 being the temperature where 50% HC conversion is reached. The results corresponding to Tmax for the soot combustion reactions, carried out in a thermogravimetric reactor, are also presented in Table 3. Tmax is defined as being the temperature at which the derivative of the TGA curve shows a minimum. This derivative curve represents the loss of mass by combustion, which is associated with the maximum burning rate.

As can be seen, all of the studied catalysts presented a level of activity for the total oxidation of all of the selected model molecules. In addition, in all the reactions the support contributed to the activity, with its redox characteristics. The influence of Ag and its metallic load depended on the nature of the molecule to be oxidized.

Table 3. Catalytic activity of ZrO_2 and the three prepared AgxZ catalysts, for the different oxidation reactions studied.

Catalyst	T50 Propane (C_3H_8) [a]	ΔT50 (C_3H_8)	T50 Propene (C_3H_6) [a]	ΔT50 (C_3H_6)	T50 Naphthalene ($C_{10}H_8$) [b]	ΔT50 ($C_{10}H_8$)	Tmax Soot [c]	ΔTmax
Uncatalyzed reaction	600	-	630	-	430	-	650	-
ZrO_2	490	110	520	110	280	150	612	38
Ag1Z	485	115	470	160	260	170	547	97
Ag5Z	480	120	280	350	172	258	525	125
Ag10Z	450	150	210	420	210	220	425	225

[a] 1000 ppm HC, 8% O_2, m_{cat} = 100 mg, Q = 50 mL·min^{-1}; [b] 150 ppm HC, 8% O_2, m_{cat} = 100 mg, Q = 30 mL·min^{-1}; [c] m_{cat} = 30, m_{soot} = 3, loose contact, ΔT = 10 °C min^{-1}, 10% O_2, Q = 100 mL·min^{-1}.

In the case of propane combustion, the catalysts exhibited low activity, and the contribution of Ag was not significant (Table 3). Using pure zirconia, T50 was 490 °C and the addition of Ag at different concentrations slightly increased the activity. Within the AgxZ series, the most active series the propane combustion resulted was the Ag10Z catalyst. When this catalyst is employed, it is possible to reduce T50 by 150 °C, with respect to the uncatalyzed reaction (600 °C). A series of materials with greater activity than those studied here for propane combustion, are presented in the bibliography [25,27,30,31,75]. Different reports present catalytic systems where the T50 values are lower than 300 °C, even when using high-space velocities. For example, in a recent paper by X. Li et al. [30], the activities of the $CoCeO_x$ catalysts was reported and compared with other similar systems, which presented very low values of T50. The poor performance of the AgxZ catalysts can be attributed to the low capacity of the Ag species to promote the activation of propane.

In contrast, the AgxZ catalysts had very good level of activity for the oxidation of propene. The results for propene oxidation reaction, expressed as propene conversion to CO_2% vs temperature, are plotted for all the studied catalysts in Figure 8A. No other by-products or CO were observed as reaction products; CO_2 selectivity was higher than 99%. In the absence of a catalyst, T50 was 630 °C, and for pure ZrO_2, 50% propene conversion was reached at 520 °C. The addition of Ag to the system substantially increased the catalytic activity: Ag5Z and Ag10Z catalysts presented great performance, reaching 50% of CO_2 conversion at 280 °C and at 210 °C, respectively. Both catalysts exhibited high levels of conversion (> 90%) at temperatures less than or equal to 330 °C. It is evident that the catalysts are more effective for the elimination of propene than propane. The results here presented are original, since there are no reports on the activity of the Ag/zirconia system for the elimination of propone, a molecule that is classified as a VOC.

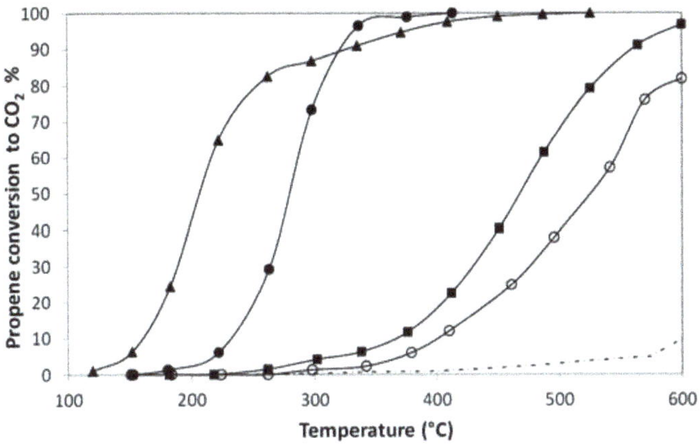

(**A**)

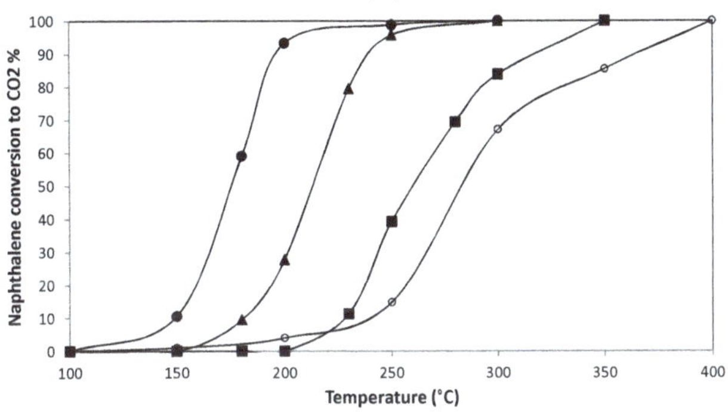

(**B**)

Figure 8. (**A**) Catalytic activity in propene oxidation: (----) without catalysts, (o) ZrO_2, ((■) Ag1Z, (●) Ag5Z, (▲) Ag10Z. (**B**) Catalytic activity in naphthalene oxidation: (o) ZrO_2, (■) Ag1Z, (●) Ag5Z, (▲) Ag10Z.

On the other hand, the prepared catalysts presented noticeable activity the naphthalene oxidation. Figure 8B shows the results obtained, expressed as naphthalene conversion to CO_2% vs temperature. In the absence of a catalyst, naphthalene is oxidized at high temperatures, reaching T50 at 430 °C. Pure zirconia presented good activity, and attained 50% conversion at 280 °C. The addition of Ag significantly promoted its activity. Thus, the Ag5Z catalyst was the most active of the series, achieving a high level of conversion at low temperatures (250 °C). Using this catalyst, T50 decreased by more than 250 °C compared to the temperature that was reached in the absence of the catalyst. When increasing the Ag content from 5 to 10%, a slight decrease in activity was observed. Probably, an increase in the size of Ag particles, and the consequent decrease in metal dispersion, generated a lower availability of active sites for the activation of naphthalene.

These results are an unprecedented contribution, taking into account that the silver–zirconia system has not been studied before, for the total oxidation of naphthalene. Besides, these results are promising, given that the T50 reached with the Ag5Z catalyst is 172 °C. This value is very low, and also comparable to previously obtained results using transition metal oxide catalysts supported on zirconia [26] and with results recently reported by M. Liu et al. [58] with the Ag/CeO_2 catalysts. Even more, as the Ag5Z catalyst was the most active, it was reused three times in the reaction cycles, keeping its activity.

Next, the activity results for diesel soot oxidation are presented. Table 3 shows the maximum oxidation rate temperature (Tmax) obtained for diesel soot combustion in the presence of air, using loose contact conditions, and carried out in a thermogravimetric reactor. Table 3 includes the differences between Tmax in a reaction with and without a catalyst (ΔTmax). The obtained values for ΔTmax indicate that both the support and the catalysts presented levels of activity. While pure zirconia exhibits poor activity, the activity of the AgxZ catalysts depends on the Ag content (Table 3, Supplementary Material). It was found that with the increase in Ag content, the Tmax decreased. In this reaction, Ag10Z catalyst was the most active of the series, reaching a ΔTmax of 225 °C. These results indicate that the presence of the Ag species generates a promoting effect.

As stated previously, although the catalysts presented activity, they depended on the Ag content and the nature of the molecule to be oxidized. The systems did not have high efficiency for propane oxidation, as it was being evident that it is not possible to activate this saturated molecule. On the other hand, they showed activity for the diesel soot oxidation, reaching combustion temperatures that were comparable with other reported catalysts containing silver, and other oxidic systems, when measurements were made in "loose contact" mode [53,76]. However, there are numerous systems in the literature that show evidence of higher activity; therefore, the presented formulations could be optimized to achieve better results. Even so the obtained value of Tmax with the Ag10Z catalyst can be considered to be acceptable, since it is in the range of temperatures at which the exhaust pipe operates.

In the oxidation of molecules containing π–π bonds, such as propene and naphthalene, the catalysts exhibited very good performance. It was achieved a significant reduction at T50, especially when using the Ag5Z and Ag10Z catalysts. As previously mentioned, these results are original, since the silver–zirconia system has not been applied to for the elimination of these unsaturated molecules. Moreover, a crystal-size effect of silver nanoparticles has been found in the catalytic performance of the system in the naphthalene combustion, which was not observed for the combustion of propene. As expected, zirconia provided some activity in all the reactions studied, and the addition of high Ag content (5 and 10 wt.%) generated an Ag–zirconia system with a greater availability of superficial oxygen species. This fact may be responsible for the acceleration of the reaction and the regeneration of the active sites, through successive redox cycles. M. Skaf et al. have demonstrated the ability for Ag species (Ag^0 and Ag^+) supported on ceria to adsorb oxygen molecules through redox cycles [77]. The Ag–zirconia synergic effect has been observed from the characterization techniques, and the formation of a metastable tetragonal phase is observed by XRD, leading to the generation of oxygen vacancies.

The coexistence of both types of species, AgO_x and Ag^0, may favor the redox capacity of the surface, with the oxidic Ag species being responsible for oxidizing the adsorbed substrate onto the catalytic surface, and hence adopting a lower oxidation state. This situation allows for an explanation of the reaction mechanism, employing Mars-van Krevelen formalism. Oxidic sites must be regenerated, and this step is favored through the capacity of the system, which contains oxygen vacancies and a high availability of adsorbed O_x^- species, as evidenced by XPS. According to this technique, the catalysts with the highest metallic loading, Ag5Z and Ag10Z, which are the most active, contain a greater proportion of oxidic species of high availability (O_β).

According to the results of TEM, DRS UV-vis, and XPS techniques, Ag is preferably found on the surface. Ag^0 nanoparticles co-exist with oxidic Ag species ($Ag^{\delta+}$ and Ag^+). According to TPR results, the presence of these oxidic phases is conditioned by the presence of reducible phases in a very low proportion. Supported Ag nanoparticles found are small (<10 nm), and well-dispersed over zirconia.

As previously mentioned, in the literature, various mechanisms that are related to this kind of systems have been proposed for oxidation reactions. Therefore, it is not possible to rule out the dissociative adsorption of molecular oxygen on Ag nanoparticles, or that a combination of both mechanisms occurs. When the activated oxygen (O_n^{x-}) is consumed during the oxidation of the substrate, it is restored by a new round of adsorption of the O_2 present in the gas phase, or by the migration of oxygen species through structural defects. These defects are present in the Ag nanoparticles and in the support, which also can chemisorb oxygen and provide mobility. Oxygen vacancies exist not only at the surface level, but also at the bulk level.

The behavior observed in naphthalene oxidation, where the most active catalyst is Ag5Z, and a later increase in the Ag content, generates a less active catalyst. Probably, this can be associated with an increase in the size of the Ag particles, with a resultant decrease of dispersion, and a lower capacity towards the adsorption of naphthalene. This point must be studied more deeply in the future, in order to correlate the catalytic results with experimental evidence. In addition, a possible geometric effect by the Ag ensembles, which cause a decrease in the capacity of naphthalene adsorption, has not yet been ruled out [68]. Probably, given the geometry of the naphthalene molecule, with two conjugated aromatic rings, its adsorption is sensitive to the structure of the catalyst surface. There is also evidence that propene combustion does not follow the same behavior, in comparison to the size of the silver crystallite, marking a difference between both molecules, which must be analyzed with greater depth in the future. All of the studied catalysts were active in the soot combustion reaction, with the Ag10Z catalyst being the most active one. Again, the ability of these catalysts to generate a synergism between the Ag species and zirconia for the activation and regeneration of oxygen is evidenced. This leads to an oxidation on the catalytic interphase, which in this case involves a gas phase and two solid phases.

3. Experimental Section

3.1. Material Preparation

Hydrous zirconium oxide, $ZrO_2 \cdot nH_2O$, was prepared by precipitation from a $ZrOCl_2 \cdot 6H_2O$ (Fluka, Buchs, Switzerland) solution with ammonium hydroxide (pH 10). The process was carried out at room temperature with constant stirring for 6 h. The pH was maintained close to 10. After filtration, the solid was washed (Cl^- negative test in the solid) and dried at 100 °C for 6 h. Portions of $ZrO_2 \cdot nH_2O$ were impregnated with an aqueous solution of $AgNO_3$, to obtain solids with a silver concentration of 1, 5, and 10 wt.% (grams of Ag per 100 g of catalyst). After drying, the samples were thermally treated at 600 °C for 2 h. The materials so obtained were named Ag1Z, Ag5Z and Ag10Z, where Z = zirconia.

3.2. Catalyst Characterization

The textural characterization of the support and catalysts was determined by using the BET method, using a Micromeritics Accusorb 2100 E apparatus (Micromeritics, Norcross, GA, USA). Samples were degassed at 100 °C prior to analysis.

Scanning electron microscopy with energy-dispersive X-ray spectroscopy (SEM-EDS) analyses were performed using a SEM Philips 505 equipment (Philips Co, Amsterdam, The Netherlands). The energy dispersive X-ray analysis of the samples was performed using an EDAX DX PRIME 10 analyzer (EDAX, New Jersey, NJ, USA) at a working potential of 15 kV.

Temperature programmed reduction (TPR) analysis was performed with a home-made equipment. In a typical run, the sample (30.0 mg) was placed in an electrically heated fixed-bed quartz micro-reactor, and heated from 50 to 800 °C, with a heating rate of 10 °C min^{-1}. In the experiments, a feed of 10% hydrogen in nitrogen (flow rate of 20 mL min^{-1}) was used. Hydrogen consumption was detected by a thermal conductivity detector.

Diffuse reflectance UV–visible spectra (DRS-UV–vis) of AgxZ catalysts were obtained with a Perkin Elmer Lambda 35 UV-VIS spectrometer (Perkin Elmer Inc., Waltham, MA, USA).

The distribution of metal particle sizes was determined by transmission electron microscopy (TEM) using a JEOL 100 CX instrument (Jeol Ltd., München, Germany), having a resolution of 6 A and an accelerating voltage of ca. 100 kV. The samples were ground and ultrasonically dispersed in distilled water. To estimate the mean particle size, the particles were considered to be spherical, and the second moment of the distribution was employed. The expression used for the calculation was:

$$d_{AV} = \frac{\sum n_i d_i^3}{\sum n_i d_i^2} \tag{2}$$

where n_i is the number of particles of d_i size. Over 200 particles were measured. The metal dispersion (D) was estimated from the Ag particle size distribution obtained from the TEM measurements by the following equation [24], assuming spherical particles:

$$D = \frac{6M_{Ag}}{\sigma\rho_{Ag}} \frac{\sum n_i d_i^2}{\sum n_i d_i^3} \tag{3}$$

where M_{Ag} and ρ_{Ag} are the molar mass (107.87 g·mol^{-1}) and density of Ag (10.5 g·cm^{-3}), respectively, and σ is the area that is occupied by 1 mol of Ag at the surface (8.75 × 10^{-16} cm^2·mol^{-1}).

XPS measurements were carried out, using a Physical Electronics spectrometer (PHI Versa Probe II Scanning XPS Microprobe, Physical Electronics, Chanhassen, MN, USA) with monochromatic X-ray Al Kα radiation (100 μm, 100 W, 20 kV, 1486.6 eV), and a dual beam charge neutralizer. The energy scale of the spectrometer was calibrated, using Cu $2p_{3/2}$, Ag $3d_{5/2}$ and Au $4f_{7/2}$ photoelectron lines at 932.7, 368.2, and 84.0 eV, respectively. Under a constant pass energy mode under 23.5 eV conditions, the Au $4f_{7/2}$ line was recorded with 0.73 eV FWHM at a binding energy (BE) of 84.0 eV. The collected XPS spectra were analyzed using PHI SmartSoft software and processed using MultiPak 9.3 package. The binding energy values were referenced to the adventitious carbon C 1s signal (284.8 eV). The recorded spectra were always fitted using Gauss–Lorentz curves. The atomic concentration percentages of the characteristic elements of the surfaces were determined, taking into account the corresponding area sensitivity factor for the differently measured spectral regions.

3.2.1. Catalytic Activity for Propane and Propene Oxidation

The catalytic activity for propane and propene oxidation was measured by using a fixed-bed quartz reactor that was electrically heated. Figure 9 shows a schematic diagram of the reaction system.

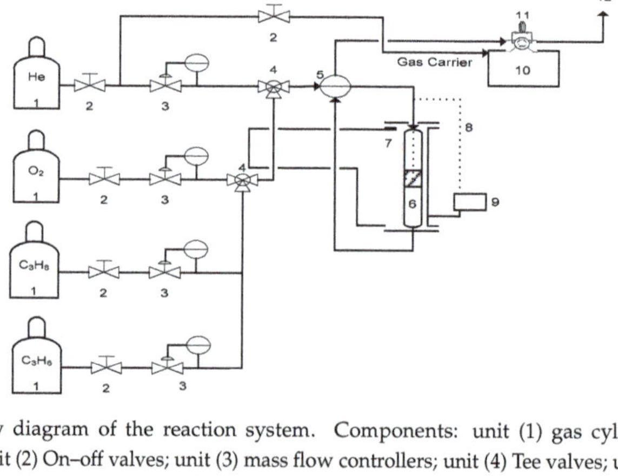

Figure 9. Flow diagram of the reaction system. Components: unit (1) gas cylinders (He, O_2, C_3H_8,C_3H_6); unit (2) On–off valves; unit (3) mass flow controllers; unit (4) Tee valves; unit (5) four-way valve; unit (6) quartz reactor; unit (7) furnace; unit (8) thermocouple type k; unit (9) thermostat; unit (10) gas chromatograph; unit (11) injection valve; unit (12) vent.

The feed employed for the propane oxidation consisted of a mixture of 1000 ppm of C_3H_8, 8% of O_2 and He to close the balance. Propene combustion was carried out using a gas mixture, having a composition of 1000 ppm of C_3H_6, 6% of O_2 and He to close the balance. For each experiment, the mass of the catalyst used was 100 mg, and the reaction temperature varied from 150 to 600 °C. A total flow rate of 50 mL·min^{-1} was used in both processes. The products were analyzed by a Shimadzu GC 2014 (Shimadzu Corporation, Kyoto, Japan) chromatograph with a thermal conductivity detector.

3.2.2. Catalytic Activity for Naphthalene Oxidation

The catalysts were tested for the oxidation of naphthalene, using an electrically heated fixed-bed quartz reactor and 100 mg of catalyst. The reaction feed consisted of 10% O_2 and 90% He, and 150 ppm naphthalene. A total flow rate of 30 mL·min^{-1} was used. Catalytic activity was measured over a temperature range from 150 to 500 °C. Data were obtained at each temperature after a certain time, in order to obtain a stable concentration of naphthalene in the gas phase. The feed flow passes through a thermostated saturator containing naphthalene. The combustion was initiated as soon as a stable vapor pressure was reached. The products were analyzed using a Shimadzu GC 2014 chromatograph with a thermal conductivity detector.

3.2.3. Catalytic Activity for Diesel Soot Oxidation

Diesel soot oxidation tests were performed in a thermogravimetric reactor (TA-50 Shimadzu, Shimadzu Corporation, Kyoto, Japan) over a range of 200 to 650 °C, with a heating rate of 10 °C·min^{-1}. The reaction mixture (8 vol.% O_2) was obtained from two feed lines, air and He respectively, individually controlled to close the balance. Printex-U was used as the model diesel soot. Before the reaction, the Printex-U was mixed with the catalyst at a 1/10 ratio, with a spatula (loose contact). The mass loss and the reaction temperature were recorded as a function of time. From the mass loss information as a function of time, the derivative curve (DTGA) was obtained, and from it, the temperature at which the combustion rate is maximum (Tmax).

4. Conclusions

This paper reported the preparation of a series of Ag-ZrO_2 catalysts having three different silver concentrations (1, 5, and 10 wt.%). The prepared catalysts were examined by using various analytical techniques (SEM-EDS, XRD, DRS, TEM, TPR, and XPS). It was observed that part of the silver was in a metallic state, in the form of nano-sized particles, and evidence was also found for the existence of oxidic species at low concentration. The catalysts were tested for the catalytic combustion of propane, propene, naphthalene, and particulate matter. Their activities depended on the Ag content and the nature of the molecule to be oxidized. Particularly, the Ag5Z and Ag10Z catalysts exhibited very good performance in the oxidation of molecules containing π–π bonds (propene and naphthalene). Particularly, these catalysts are innovative and display promising results, given that the resulting silver/zirconia system is very active for the elimination of unsaturated molecules, and can be considered a non-toxic catalyst. All of the studied catalysts were also active in the soot combustion reaction, with the Ag10Z catalyst being the most active one. These results could likely be attributed to an Ag-ZrO_2 synergic effect.

Supplementary Materials: The following are available online at http://www.mdpi.com/2073-4344/9/3/297/s1, Figure S1: TEM micrograph of Ag1Z catalyst, Figure S2: DTGA diagrams for the diesel soot catalytic results.

Author Contributions: M.L.C. and I.D.L. conceived and designed the experiments; M.S.L.A., M.B.N. and M.A.O. performed the experiments; I.d.C.L.B. and E.R.-C. contributed with the XPS spectroscopy measurements and analyses; all authors discussed the results; I.D.L. and M.L.C. wrote the manuscript. All authors read, revised, and approved the final manuscript.

Funding: This research was funded by Consejo Nacional de Investigaciones Científicas y Técnicas: PIP 0276, Agencia Nacional de Promoción Científica y Tecnológica: PICT 0737, Universidad Nacional de La Plata: X700, and Universidad Nacional de La Plata: X707.

Acknowledgments: The authors acknowledge the financial support of CONICET, ANPCyT, and UNLP, and Pablo Fetsis, Mariela Theiller, and María Laura Barbelli for characterization experiments.

Conflicts of Interest: The authors declare no conflict of interest.

References

1. Pfefferle, L.D.; Pfefferle, W.C. Catalysis in Combustion. *Catal. Rev. Sci. Eng.* **1987**, *29*, 219–267. [CrossRef]
2. Choudhary, T.V.; Banerjee, S.; Choudhary, V.R. Catalysts for combustion of methane and lower alkanes. *Appl. Catal. A* **2002**, *234*, 1–23. [CrossRef]
3. García, T.; Solsona, B.; Taylor, S.H. Naphthalene total oxidation over metal oxide catalysts. *Appl. Catal. B* **2006**, *66*, 92–99. [CrossRef]
4. Ruiz, M.L.; Lick, I.D.; Ponzi, M.; Rodríguez-Castellón, E.; Jiménez-López, A.; Ponzi, E.N. Combustion of diesel soot in NO/O_2 presence. Cesium nitrate and gold catalysts. *Appl. Catal. A* **2011**, *392*, 45–56. [CrossRef]
5. Lee, J.; Theis, J.R.; Kyriakidou, E.A. Vehicle emissions trapping materials: Successes, challenges, and the path forward. *Appl. Catal. B* **2019**, *243*, 397–414. [CrossRef]
6. Sui, L.; Yu, L. Diesel soot oxidation catalyzed by Co-Ba-K catalysts: Evaluation of the performance of the catalysts. *Chem. Eng. J.* **2008**, *142*, 327–330. [CrossRef]
7. Lick, I.D.; Carrascull, A.L.; Ponzi, M.I.; Ponzi, E.N. Zirconia-supported Cu-KNO_3 catalyst. Characterization and catalytic behavior in the catalytic combustion of soot with a NO/O_2 mixture. *Ind. Eng. Chem. Res.* **2008**, *47*, 3834–3839. [CrossRef]
8. Aranda, A.; Agouram, S.; López, J.M.; Mastral, A.M.; Sellick, D.R.; Solsona, B.; Taylor, S.H.; García, T. Oxygen defects: The key parameter controlling the activity and selectivity of mesoporous copper-doped ceria for the total oxidation of naphthalene. *Appl. Catal. B* **2012**, *127*, 77–88. [CrossRef]
9. Kamal, M.S.; Razzak, S.A.; Hossain, M.M. Catalytic oxidation of volatile organic compounds (VOCs)—A review. *Atmos. Environ.* **2016**, *140*, 117–134. [CrossRef]
10. Zhang, Z.; Jiang, Z.; Shangguan, W. Low-temperature catalysis for VOCs removal in technology and application: A state-of-the-art review. *Catal. Today* **2016**, *264*, 270–278. [CrossRef]
11. Yi, T.; Zhang, Y.; Yang, X. Combination of Pt@CeO_2/MCM-56 and CeO_2-CuO/MCM-56 to purify the exhaust emissions from diesel vehicles. *Appl. Catal. A* **2019**, *570*, 387–394. [CrossRef]
12. Banús, E.D.; Ulla, M.A.; Miró, E.E.; Milt, V.G. Structured Catalysts for Soot Combustion for Diesel Engines, In Chapter 5, in Diesel Engine—Combustion, Emissions and Condition Monitoring. *IntechOpen* **2013**, 117–142. [CrossRef]
13. Garetto, T.F.; Rincón, E.; Apesteguía, C.R. The origin of the enhanced activity of Pt/zeolites for combustion of C2–C4 alkanes. *Appl. Catal. B* **2007**, *73*, 65–72. [CrossRef]
14. Diehl, F.; Barbier, J.; Duprez, D.; Guibard, I.; Mabilon, G. Catalytic oxidation of heavy hydrocarbons over Pt/Al_2O_3. Oxidation of C_{10}^+ solid hydrocarbons representative of soluble organic fraction of Diesel soot. *Appl. Catal. A* **2015**, *504*, 37–43. [CrossRef]
15. Varela-Gandía, F.J.; Berenguer-Murcia, A.; Lozano-Castelló, D.; Cazorla-Amorós, D.; Sellick, D.R.; Taylor, S. Total oxidation of naphthalene using palladium nanoparticles supported on BETA, ZSM-5, SAPO-5 and alumina powders. *Appl. Catal. B* **2013**, *129*, 98–105. [CrossRef]
16. Demoulin, O.; Clef, B.L.; Navez, M.; Ruiz, P. Combustion of methane, ethane and propane and of mixtures of methane with ethane or propane on Pd/γ-Al_2O_3 catalysts. *Appl. Catal. A* **2008**, *344*, 1–9. [CrossRef]
17. Liotta, L.F. Catalytic oxidation of volatile organic compounds on supported noble metals. *Appl. Catal. B* **2010**, *100*, 403–412. [CrossRef]
18. Avila, M.S.; Vignatti, C.I.; Apesteguía, C.R.; Garetto, T.F. Effect of support on the deep oxidation of propane and propylene on Pt-based catalysts. *Chem. Eng. J.* **2014**, *241*, 52–59. [CrossRef]
19. Cortés-Reyes, M.; Herrera, C.; Larrubia, M.Á.; Alemany, L.J. Intrinsic reactivity analysis of soot removal in LNT-catalysts. *Appl. Catal. B* **2016**, *193*, 110–120. [CrossRef]
20. Burch, R.; Breen, J.P.; Meunier, F.C. A review of the selective reduction of NOx with hydrocarbons under lean-burn conditions with non-zeolitic oxide and platinum group metal catalysts. *Appl. Catal. B* **2002**, *39*, 283–303. [CrossRef]

21. Sanz, O.; Banús, E.D.; Goya, A.; Larumbe, H.; Delgado, J.J.; Monzón, A.; Montes, M. Stacked wire-mesh monoliths for VOCs combustion: Effect of the mesh-opening in the catalytic performance. *Catal. Today* **2017**, *296*, 76–83. [CrossRef]
22. Aranda, A.; López, J.M.; Murillo, R.; Mastral, A.M.; Dejoz, A.; Vázquez, I.; Solsona, B.; Taylor, S.H.; García, T. Total oxidation of naphthalene with high selectivity using a ceria catalyst prepared by a combustion method employing ethyleneglicol. *J. Hazard. Mater.* **2009**, *171*, 393–399. [CrossRef] [PubMed]
23. Li, W.B.; Wang, J.X.; Gong, H. Catalytic combustion of VOCs on non-noble metal catalysts. *Catal. Today* **2009**, *148*, 81–87. [CrossRef]
24. Sanchis, R.; Alonso-Domínguez, D.; Dejoz, A.; Pico, M.P.; Alvarez-Serrano, I.; García, T.; López, M.L.; Solsona, B. Eco-Friendly Cavity-Containing Iron Oxides Prepared by Mild Routes as Very Efficient Catalysts for the Total Oxidation of VOCs. *Materials* **2018**, *11*, 1387. [CrossRef]
25. Solsona, B.; Sanchis, R.; Dejoz, A.M.; García, T.; Ruiz-Rodríguez, L.; López Nieto, J.M.; Cecilia, J.A.; Rodríguez-Castellón, E. Total Oxidation of Propane Using CeO_2 and CuO-CeO_2 Catalysts Prepared Using Templates of Different Nature. *Catalysts* **2017**, *7*, 96. [CrossRef]
26. Leguizamón Aparicio, M.S.; Ocsachoque, M.A.; Gazzoli, D.; Botto, I.L.; Lick, I.D. Total Oxidation of Naphthalene with Zirconia-Supported Cobalt, Copper and Nickel catalysts. *Catalysts* **2017**, *7*, 293. [CrossRef]
27. Heynderickx, M.P.; Thybaut, J.W.; Poelman, H.; Poelman, D.; Marin, G.B. The total oxidation of propane over supported Cu and Ce oxides: A comparison of single and binary metal oxides. *J. Catal.* **2010**, *272*, 109–120. [CrossRef]
28. Morales, M.R.; Yeste, M.P.; Vidal, H.; Gatica, J.M.; Cadus, L.E. Insights on the combustion mechanism of ethanol and n-hexane in honeycomb monolithic type catalysts: Influence of the amount and nature of Mn-Cu mixed oxide. *Fuel* **2017**, *208*, 637–646. [CrossRef]
29. Mrad, R.; Cousin, R.; Poupin, C.; Aboukaïs, A.; Siffert, S. Propene oxidation and NO reduction over MgCu–Al (Fe) mixed oxides derived from hydrotalcite-like compounds. *Catal. Today* **2015**, *257*, 98–103. [CrossRef]
30. Li, X.; Li, X.; Zeng, X.; Zhu, T. Correlation between the physicochemical properties and catalytic performances of micro/mesoporous CoCeOx mixed oxides for propane combustion. *Appl. Catal. A* **2019**, *572*, 61–70. [CrossRef]
31. Solsona, B.; Davies, T.E.; García, T.; Vázquez, I.; Dejoz, A.; Taylor, S.H. Total oxidation of propane using nanocrystalline cobalt oxide and supported cobalt oxide catalysts. *Appl. Catal. B* **2008**, *84*, 176–184. [CrossRef]
32. Hernández-Giménez, A.M.; Castelló, D.L.; Bueno-López, A. Diesel soot combustion catalysts: Review of active phases. *Chem. Pap.* **2014**, *68*, 1154–1168. [CrossRef]
33. Fino, D.; Bensaid, S.; Piumetti, M.; Russo, N. A review on the catalytic combustion of soot in diesel particulate filters for automotive applications: From powder catalysts to structured reactors. *Appl. Catal. A* **2016**, *509*, 75–96. [CrossRef]
34. Comelli, N.A.; Ruiz, M.L.; Leguizamón Aparicio, M.S.; Merino, N.A.; Cecilia, J.A.; Rodríguez-Castellón, E.; Lick, I.D.; Ponzi, M.I. Influence of the synthetic conditions on the composition, morphology of CuMgAl hydrotalcites and their use as catalytic precursor in diesel soot combustion reactions. *Appl. Clay Sci.* **2018**, *157*, 148–157. [CrossRef]
35. Banús, E.B.; Sanz, O.; Milt, V.G.; Miró, E.E.; Montes, M. Development of a stacked wire-mesh structure for diesel soot combustion. *Chem. Eng. J.* **2014**, *246*, 353–365. [CrossRef]
36. Alcalde-Santiago, V.; Davó-Quiñonero, A.; Lozano-Castelló, D.; Bueno-López, A. On the soot combustion mechanism using 3DOM ceria catalysts. *Appl. Catal. B* **2018**, *234*, 187–197. [CrossRef]
37. Bueno-López, A.; Lozano-Castelló, D.; McCue, A.J.; Anderson, J.A. NOx storage and reduction over copper-based catalysts. part 3: Simultaneous NOx and soot removal. *Appl. Catal. B* **2016**, *198*, 266–275.
38. Legutko, P.; Pęza, J.; Villar Rossi, A.; Marzec, M.; Jakubek, T.; Kozieł, M.; Adamski, A. Elucidation of unexpectedly weak Catalytic effect of doping with cobalt of the cryptomelane and birnessite systems Active in Soot Combustion. *Top. Catal.* **2019**. [CrossRef]
39. Atribak, I.; Bueno-Lopez, A.; García-García, A.; Navarro, P.; Frías, D. Catalytic activity for soot combustion of birnessite and cryptomelane. *Appl. Catal. B* **2010**, *93*, 267–273. [CrossRef]
40. Li, Q.; Wang, X.; Chen, H.; Xin, Y.; Tian, G.; Lu, C.; Zhang, Z.; Zheng, L. K-supported catalysts for diesel soot combustion: Making a balance between activity and stability. *Catal. Today* **2016**, *264*, 171–179. [CrossRef]
41. Ruiz, M.L.; Lick, I.D.; Ponzi, M.I.; Ponzi, E.N. Catalysts of alkaline nitrates supported on oxides for the diesel soot combustion. Deactivation by hydro-treatment and CO_2. *Catal. Commun.* **2013**, *34*, 45–51. [CrossRef]

42. Nossova, L.; Caravaggio, G.; Couillard, M.; Ntais, S. Effect of preparation method on the performance of silver-zirconia catalysts for soot oxidation in diesel engine exhaust. *Appl. Catal. B* **2018**, *225*, 538–549. [CrossRef]
43. Pecchi, G.; Dinamarca, R.; Campos, C.M.; Garcia, X.; Jimenez, R.G.; Fierro, J.L. Soot Oxidation on Silver-Substituted $LaMn_{0.9}Co_{0.1}O_3$ Perovskites. *Ind. Eng. Chem. Res.* **2014**, *53*, 10090–10096. [CrossRef]
44. Haneda, M.; Towata, A. Catalytic performance of supported Ag nano-particles prepared by liquid phase chemical reduction for soot oxidation. *Catal. Today* **2015**, *242*, 351–356. [CrossRef]
45. Lee, C.; Jeon, Y.; Kim, T.; Tou, A.; Park, J.-I.; Einaga, H.; Shul, Y.-G. Ag-loaded cerium-zirconium solid solution oxide nano-fibrous webs and their catalytic activity for soot and CO oxidation. *Fuel* **2018**, *212*, 395–404. [CrossRef]
46. Luo, M.-F.; Yuan, X.-X.; Zheng, X.-M. Catalyst characterization and activity of Ag–Mn, Ag–Co and Ag–Ce composite oxides for oxidation of volatile organic compounds. *Appl. Catal. A* **1998**, *175*, 121–129. [CrossRef]
47. Yamazaki, K.; Sakakibara, Y.; Dong, F.; Shinjoh, H. The remote oxidation of soot separated by ash deposits via silver–ceria composite catalysts. *Appl. Catal. A* **2014**, *476*, 113–120. [CrossRef]
48. Gao, Y.; Duan, A.; Liu, S.; Wu, X.; Liu, W.; Weng, D. Study of Ag/$CexNd_{1-x}O_2$ nanocubes as soot oxidation catalysts for gasoline particulate filters: Balancing catalyst activity and stability by Nd doping. *Appl. Catal. B* **2017**, *203*, 116–126. [CrossRef]
49. Shimizu, H.; Kawachi, H.; Satsuma, A. Study of active sites and mechanism for soot oxidation by silver-loaded ceria catalyst. *Appl. Catal. B* **2010**, *96*, 169–175. [CrossRef]
50. Yamazaki, K.; Tomoyuki Kayama, T.; Dong, F.; Hirofumi Shinjoh, H. A mechanistic study on soot oxidation over CeO_2–Ag catalyst with 'rice-ball' morphology. *J. Catal.* **2011**, *282*, 289–298. [CrossRef]
51. Yamaura, H.; Takahashi, H.; Fukuoka, M.; Nishibori, M.; Yamaguchi, S.; Yahiro, H. PM oxidation over Ag-loaded perovskite-type oxide catalyst prepared by thermal decomposition of heteronuclear cyano-complex precursor. *Catal. Today* **2018**. [CrossRef]
52. Castoldi, L.; Aneggi, E.; Matarrese, R.; Bonzi, R.; Llorca, J.; Trovarelli, A.; Lietti, L. Silver-based catalytic materials for the simultaneous removal of soot and NOx. *Catal. Today* **2015**, *258*, 405–415. [CrossRef]
53. Aneggi, E.; Llorca, J.; de Leitenburg, C.; Dolcetti, G.; Trovarelli, A. Soot combustion over silver-supported catalysts. *Appl. Catal. B* **2009**, *91*, 489–498. [CrossRef]
54. Deng, X.; Li, M.; Zhang, J.; Hu, X.; Zheng, J.; Zhang, N.; Chen, B.H. Constructing nano-structure on silver/ceria-zirconia towards highly active and stable catalyst for soot oxidation. *Chem. Eng. J.* **2017**, *313*, 544–555. [CrossRef]
55. Serve, A.; Boreave, A.; Cartoxia, B.; Pajot, K.; Vernoux, P. Synergy between Ag nanoparticles and yttria-stabilized zirconia for soot oxidation. *Appl. Catal B* **2019**, *242*, 140–149. [CrossRef]
56. Wang, H.; Luo, S.; Zhang, M.; Liu, W.; Wu, X.; Liu, S. Roles of oxygen vacancy and $Ox-$ in oxidation reactions over CeO_2 and Ag/CeO_2 nanorod model catalysts. *J. Catal.* **2018**, *368*, 365–378. [CrossRef]
57. Makkee, M.; Jelles, S.J.; Moulijn, J.A. Cerium and Platinum Based Diesel Fuel Additives in the Diesel Soot Abatement Technology. In *Catalysis by Ceria and Related Materials*, 1st ed.; Trovarelli, A., Ed.; Imperial College Press: London, UK, 2002; Volume 2, pp. 391–406.
58. Liu, M.; Wu, X.; Shuang Liu, S.; Gao, Y.; Chen, Z.; Ma, Y.; Ran, R.; Weng, D. Study of Ag/CeO_2 catalysts for naphthalene oxidation: Balancing the oxygen availability and oxygen regeneration capacity. *Appl. Catal. B* **2017**, *219*, 231–240. [CrossRef]
59. Mars, P.; van Krevelen, D.W. Oxidations carried out by means of vanadium oxide catalysts. *Chem. Eng. Sci.* **1954**, *3*, 41–59. [CrossRef]
60. Vannice, A. An analysis of the Mars–van Krevelen rate expression. *Catal. Today* **2007**, *123*, 18–22. [CrossRef]
61. Bueno-López, A.; Krishna, K.; Makkee, M.; Moulijn, J.A. Active oxygen from CeO_2 and its role in catalyzed soot oxidation. *Catal. Lett.* **2005**, *99*, 203–205. [CrossRef]
62. Hosseini, M.; Barakat, T.; Cousin, R.; Aboukais, A.; Su, B.-L.; De Weireld, G. Catalytic performance of core–shell and alloy Pd–Au nanoparticles for total oxidation of VOC: The effect of metal deposition. *Appl. Catal. B* **2012**, *111*, 218–224. [CrossRef]
63. Guillén-Hurtado, N.; García-García, A.; Bueno-López, A. Isotopic study of ceria catalyzed soot oxidation in the presence of NOx. *J. Catal.* **2013**, *299*, 181–187. [CrossRef]
64. Kauppi, E.I.; Honkala, K.; Krause, A.O.I.; Kanervo, J.M.; Lefferts, L. ZrO_2 Acting as a Redox Catalyst. *Top. Catal.* **2016**, *59*, 823–832. [CrossRef]

65. Wyrwalski, F.; Lamonier, J.-F.; Siffert, S.; Aboukaïs, A. Additional effects of cobalt precursor and zirconia support modifications for the design of efficient VOC oxidation catalysts. *Appl. Catal B* **2007**, *70*, 393–399. [CrossRef]
66. Campa, M.C.; Ferraris, G.; Gazzoli, D.; Pettiti, I.; Pietrogiacomi, D. Rhodium supported on tetragonal or monoclinic ZrO_2 as catalyst for the partial oxidation of methane. *Appl. Catal B* **2013**, *142*, 423–431. [CrossRef]
67. Kolobova, E.; Pestryakov, A.; Mamontov, G.; Kotolevich, Y.; Bogdanchikova, N.; Farias, M.; Vosmerikov, A.; Vosmerikova, L.; Cortes Corberán, V. Low-temperature CO oxidation on Ag/ZSM-5 catalysts: Influence of Si/Al ratio and redox pretreatments on formation of silver active sites. *Fuel* **2017**, *188*, 121–131. [CrossRef]
68. Rekha, T.N.; Umadevi, M.; Rajkumar, B.J.M. Structural and spectroscopic study of adsorption of naphthalene on silver. *J. Mol. Struct.* **2015**, *1079*, 155–162. [CrossRef]
69. Xie, Y.; Tang, Y. Spontaneous monolayer dispersion of oxides and salts onto surfaces of supports: Applications to heterogeneous catalysis. *Adv. Catal.* **1990**, *37*, 1–43.
70. Valigi, M.; Gazzoli, D.; Dragone, R.; Gherardi, M.; Minelli, G. Nickel oxide-zirconium oxide: Ni^{2+} incorporation and its influence on the phase transition and sintering of zirconia. *J. Mater. Chem.* **1995**, *5*, 183–189. [CrossRef]
71. Corro, G.; Vidal, E.; Cebada, S.; Pal, U.; Buñuelos, F.; Vargas, D.; Guilleminot, E. Electronic state of silver in Ag/SiO_2 and Ag/ZnO catalysts and its effect on diesel particulate matter oxidation: An XPS study. *Appl. Catal. B* **2017**, *216*, 1–10. [CrossRef]
72. Centi, G.; Cerrato, G.; D'Angelo, S.; Finardi, U.; Giamello, E.; Morterra, C.; Perathoner, S. Catalytic behavior and nature of active sites in copper-on-zirconia catalysts for the decomposition of N_2O. *Catal. Today* **1996**, *27*, 265–270. [CrossRef]
73. Ferraria, A.M.; Carapeto, A.P.; do Rego, A.M. X-ray photoelectron spectroscopy: Silver salts revisited. *Vacuum* **2012**, *86*, 1988–1991. [CrossRef]
74. Galtayries, A.; Sporken, R.; Riga, J.; Blanchard, G.; Caudano, R. XPS comparative study of ceria/zirconia mixed oxides: Powders and thin film characterization. *J. Electron. Spectrosc. Relat. Phenom.* **1998**, *88*, 951–956. [CrossRef]
75. Okal, J.; Zawadzki, M. Combustion of propane over novel zinc aluminate-supported ruthenium catalysts. *Appl. Catal. B* **2011**, *105*, 182–190. [CrossRef]
76. Lee, C.; Shul, Y.-G.; Einaga, H. Silver and manganese oxide catalysts supported on mesoporous ZrO_2 nanofiber mats for catalytic removal of benzene and diesel soot. *Catal. Today* **2017**, *281*, 460–466. [CrossRef]
77. Skaf, M.; Hany, S.; Aouad, S.; Labaki, M.; Abi-Aad, E.; Aboukaïs, A. Adsorption of probe molecules to investigate by EPR the redox properties of silver loaded on ceria. *Chem. Phys.* **2019**, *517*, 131–137. [CrossRef]

catalysts

Article

Synthesis, Performance and Emission Quality Assessment of Ecodiesel from Castor Oil in Diesel/Biofuel/Alcohol Triple Blends in a Diesel Engine

Beatriz Hurtado [1], Alejandro Posadillo [2], Diego Luna [1,*], Felipa M. Bautista [1], Jose M. Hidalgo [3], Carlos Luna [1], Juan Calero [1], Antonio A. Romero [1] and Rafael Estevez [1]

[1] Departamento de Química Orgánica, Universidad de Córdoba, Campus de Rabanales, Ed. Marie Curie, 14014 Córdoba, Spain; q02huceb@uco.es (B.H.); qo1baruf@uco.es (F.M.B.); qo2luduc@uco.es (C.L.); p72camaj@gmail.com (J.C.); qo1rorea@uco.es (A.A.R.); rafa_20_15@hotmail.com (R.E.)

[2] Seneca Green Catalyst S.L., Campus de Rabanales, 14014 Córdoba, Spain; seneca@uco.es

[3] Institute of Physical Chemistry, Polish Academy of Sciences, ul. Kasprzaka 44/52, 01-224 Warsaw, Poland; jose.hidalgo@unicre.cz

* Correspondence: diego.luna@uco.es; Tel.: +34-957212065

Received: 13 November 2018; Accepted: 20 December 2018; Published: 3 January 2019

Abstract: This research aims to promote the use of second-generation biofuels based mainly on Castor oil, which is not adequate for food use, and Sunflower oil as a standard reference for recycled oils. They have been applied in the production of Ecodiesel, a biofuel that integrates glycerol as monoglyceride, employing sodium methoxide as homogeneous catalyst and ethanol as solvent, but operating in milder conditions than in the synthesis of conventional biodiesel in order to obtain a kinetic control of the selective transesterification. The behavior of biofuels has been evaluated in a conventional diesel engine, operating as an electricity generator. The contamination degree was also evaluated from the opacity values of the generated smokes. The different biofuels here studied have practically no differences in the behavior with respect to the power generated, although a small increase in the fuel consumption was obtained in some cases. However, with the biofuels employed, a significant reduction, up to 40%, in the emission of pollutants is obtained, mainly with the blend diesel/castor oil/alcohol. Besides, it is found that pure Castor oil can be employed directly as biofuel in triple blends diesel/biofuel/alcohol, exhibiting results that are very close to those obtained using biodiesel or Ecodiesel.

Keywords: castor oil; biofuel; selective transesterification; ecodiesel; biodiesel; diesel engine; electricity generator; smoke opacity; Bacharach opacity

1. Introduction

Nowadays, regardless of the introduction of vehicles that incorporate electric or hydrogen engines, the gradual incorporation of biofuels as substitute of fossil fuels is still mandatory [1]. The use of biofuels palliate the depletion of fossil fuels, minimize the negative impact of greenhouse gases, producing less amount of carbon monoxide, sulfur dioxide and unburned hydrocarbons than fossil fuel [2,3], and also allow continued use of the fleet of cars currently existing, estimated at more than a billion, without modifying the compression ignition (C.I.) engines [4]. Furthermore, biofuels and, specifically the biodiesel, can be easily integrated into the logistic of the global transportation system [5,6]. The gradual replacement of fossil fuels by others of renewable nature involves the introduction of blends diesel/biofuel. In this sense, the objectives pursued by the EU are estimated at 20% of biofuel in the blend in 2020 and 30% in 2030. Despite these objectives are apparently not difficult to achieve, the enormous amount of glycerol produced during the synthesis of biodiesel makes that other approaches can be considered in order to accomplish the fossil replacement to a higher extent. In this respect, a viable option could be the use of unprocessed vegetable oils in double blends with conventional diesel. This is theoretically possible because all the relevant physicochemical properties, for its use as fuels in conventional diesel engines, are comparable to conventional diesel, with the exception of the viscosity, much higher in oils than in diesel [7]. Another approach is related to the use of alcohols in triple blends with diesel and different biofuels. According to EN 14214, the presence of alcohols in fuel and biofuels does not constitute any inconvenience. In fact, according to recent research, the presence of ethanol and other short-chain alcohols has a favorable effect on the emissions of the biofuels [8–11], so it is advisable the addition of certain quantities of ethanol to the diesel No. 2 standard. These mixtures constitute the so-called E diesel, oxidiesel or oxygenated diesel, which apart from reducing the emissions of the C.I. engines, improves the flow properties (viscosity) and the essential parameters that limit the application of diesel when operating at low temperatures [12], like the "cloud point" (CP), "pour point" (PP), cold filter plugging point temperature (CFPP), or point of occlusion of the cold filter (POFF), viscosity, and emission levels of the motors, without any significant negative effect in most of the parameters that define the quality of biodiesel [13–16]. Thus, the use of alcohols in triple mixtures diesel/biofuel/alcohol would allow to replace larger amounts of fossil diesel than those achieved using double diesel/biofuel blends. In fact, the utility of the triple diesel/biodiesel/ethanol blends has been patented under the name of Diesterol [17,18]. These blends reduce emissions, viscosity and flash point, together to a slight reduction of the engine power [13–18].

Considering the biofuel synthesis and taking into account the available technology, the conventional biodiesel production described by the standard EN 14214, present as the main drawback the glycerol generated as byproduct, which is a 10% by weight of the total of biodiesel produced [19]. For instance, the Lurgi's biodiesel fabrication technology is based on two successive transesterification reactions of TG with methanol to form FAME and glycerol in the presence of an alkaline catalyst [20]. At the end of the reaction, the mixture is neutralized by adding hydrochloric acid. A subsequent counter-current washing step removes by-product components and gives a "ready for use" biodiesel after final drying step. The washing step of biodiesel to eliminate glycerol residues, that must be less than 0.02%, provides an additional complication. On one hand, it requires processing with high energy costs. On the other hand, it requires a high consumption of water. A possible solution to this problem is the production of a new type of biodiesel that integrates glycerol in the form of a soluble derivative. Thus, the production of glycerol is avoided, and, at the same time, the atomic performance of the process is increased, since all the reactive raw materials are transformed into a biofuel [21,22]. In this sense, our Research Group has accomplished the transesterification of triglycerides with ethanol to produce monoglycerides (MGs) as soluble derivatives of glycerol employing different lipases as catalysts. Hence, through the partial transesterification of one mole of triglyceride (TG) with ethanol, two moles of ethyl esters (FAEE) and one mole of monoglyceride (MG) are generated, obtaining a biofuel called Ecodiesel, Figure 1 [23–29].

However, the high cost of lipases compels to search for a more cost-effective process for producing Ecodiesel, such as the employ of supported KF or CaO as heterogeneous catalysts [28,29]. Thus, the Ecodiesel, constituted by the 2/1 mixture of FAME/MG can be obtained under kinetic control of the chemical process, using a catalyst less basic than the alkali metals usually employed as homogeneous catalysts in the biodiesel production. However, further attempts to reduce the economic cost of the process should be investigated.

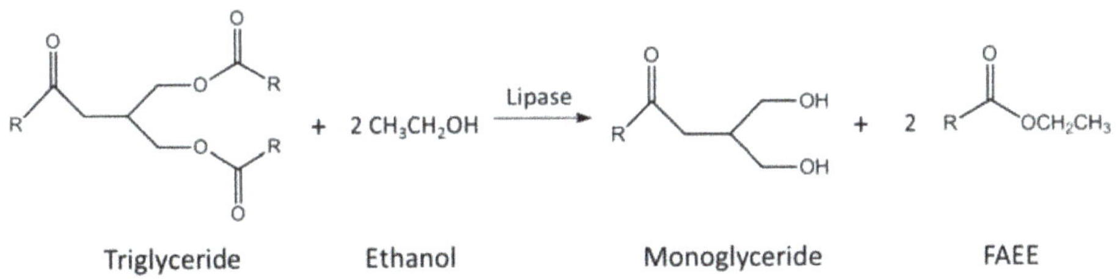

Figure 1. Selective transesterification of a triglyceride molecule through the application of enzymatic catalysis to produce Ecodiesel, a biofuel similar to biodiesel, constituted by two molecules of ethyl esters of fatty acids and a monoglyceride molecule.

In addition to the abovementioned drawback of glycerol generation during the biodiesel production, another important issue related to the biodiesel production is the ethical conflicts surrounding the production of biodiesel from edible oils, instead of using these edible oils exclusively for feeding purposes. To overcome this, the production of the so-called second-generation biofuels, which are obtained from non-edible oils and also from recycled waste oils, using procedures with minimum waste generation and high atom efficiency, are being taken into account [30,31]. Among the non-edible oils, Castor oil (*Ricinus communis* L.) should be a promising option because it has a large international market, since it is widely used in chemical and pharmaceutical industry [32,33] and it can be grown on marginal lands and in semi-arid climates [34]. However, it is not considered a good raw material to produce biofuels [35] because of the high kinematic viscosity that it exhibits (241.5 cSt). Likewise, the kinematic viscosity of the biodiesel obtained by its transesterification with methanol is also too high for being employed in combustion engines (15 cSt). In fact, very recent studies [36–39] have determined that Castor oil biodiesel can be only used in a 20–30% in mixtures with fossil diesel, i.e., BCO20 to BCO30. In this respect, some researches, regarding the use of triple diesel/biodiesel/alcohol blends in order to increase the amount of biodiesel of castor oil in these blends, have been initiated [13].

Another possibility, which is barely studied at present, is that Castor oil can be employed in its pure form blending with fossil diesel [40–42]. Castor oil exhibits, in fact, advantageous rheological properties for being employed directly as biofuel, such as high solubility in very diverse organic compounds, compatibility with polar liquids of low viscosity, such as alcohols; high values of Cetane Index (81.1) and Flammability Point (229 °C); low cloud point −10 to −18 °C, (−23 °C in FAME), and crystallization (pour point) −30 °C (−45 °C in FAME), which allow its use in motor oil high performance, as a lubricant and in hydraulic braking systems [34,43].

Hence, in this research, two different aspects have been addressed. On one hand, to further reduce the economic cost of the Ecodiesel production, a basic homogeneous catalytic process at the experimental conditions that are soft enough to achieve the transesterification of the esters of primary alcohols, positions 1 and 3 of glycerol, without affecting the carbon 2, a secondary alcohol, which is less reactive, has been investigated. To do so, a commercial sodium methoxide is employed as catalyst. Furthermore, in order to avoid the glycerol production that surely occurs employing methanol and an homogeneous catalyst, the possibility of using ethanol as solvent in a methanol/ethanol mixture to attenuate the higher reactivity of methanol has been studied.

On the other hand, to solve the problem of the high viscosity of Castor oil, this research deals with the possibility of increasing the amount of renewable material in biofuel blends, here so-called diesel additive, through the application of diesel/biofuel/alcohol triple blends. In this sense, the most suitable double and triple blends, according to their viscosity values, are applied as a biofuel in a conventional diesel engine, operating as an electric generator. Besides, the quality of the emissions obtained by using these biofuels was also evaluated from the opacity values of the generated smokes.

2. Results

2.1. Synthesis of Ecodiesel by Kinetic Control of the Conventional Procedure of Biodiesel Production

As aforementioned, to obtain a selective alcoholysis of triglycerides, the use of milder conditions than those usually employed in conventional biodiesel production (FAME) is required. As can be seen in Table 1, the biodiesel production from two different oils, sunflower oil (SO) and castor oil (CO), reached a total conversion and selectivities higher than 90% to FAME + MG at 60 °C, 60 min of reaction time and employing 4 mL of sodium methoxide per 100 mL of oil. For its part, in order to substitute the lipases by a homogeneous catalyst in the Ecodiesel synthesis, the use of ethanol to reduce the glycerol production has been accomplished, due to ethanol is less reactive than methanol and can avoid the break of the ester bond of secondary alcohol). Table 2 shows the results obtained in the selective ethanolysis of Sunflower oil (SO) at room temperature (25 °C) and 30 min of reaction time, using different proportions of ethanol (EtOH) and methanol (MeOH).

Table 1. Results obtained in the transesterification of different oils. Reaction conditions: 100 mL of oil, 20 mL of methanol and 4 mL of sodium methoxide, 60 °C and 60 min of reaction time. In all the cases, the conversion is 100%. Sel. = FAME + MG. In the methanol fraction, 13 mL in this case, 2–4 g of glycerol and 3.5–6.5 g of MG were obtained.

Vegetable Oil	Oil Viscosity (cSt)	Selectivity (%)	FAME (%)	MG (%)	DG (%)	Biodiesel Viscosity (cSt)
Sunflower	37.37	91.9	85.3	6.6	8.1	4
Castor	227.0	100	90.0	10.0	0.0	11

Table 2. Results obtained in the transesterification of 500 mL of Sunflower oil. Reaction conditions 25 °C, 30 min of reaction time and 300 rpm. The neutralization is carried out with of H_3PO_4. No phase separation was observed.

N°	EtOH (mL)	MeOH (mL)	NaOMe (mL)	Visc. (cSt)	Sel. (%)	FAE (%)	MG (%)	DG (%)
1	25	35	10	6.60	89.1	64.5	24.6	10.9
2	25	25	10	10.95	86.9	67.6	19.3	13.1
3	20	20	7	13.18	80.7	64.3	15.4	19.3
4	30	30	5	9.43	77.1	63.4	13.7	22.9
5	25	25	5	13.15	80.0	64.9	15.1	20.0
6	30	25	5	11.77	77.8	58.5	19.3	22.2

As can be seen in Table 2, non-appreciable differences in the viscosity of the mixtures have been obtained. Furthermore, the selectivities of the reactions are between 77% and 90%. Thus, we have selected the reaction conditions more favourable from an economic point of view (Entry 5), since the least amount of catalyst and alcohols are employed. This proportion can be easily transformed to a ratio if we consider the reaction of 100 mL of Sunflower oil. In this case, the reaction mixture would be oil/EtOH/MeOH/NaOMe ratio of 100/5/5/1.

Once the mixture of reactants was fixed, different reaction parameters such as reaction temperature (Table S1) and reaction time were also studied. Table 3 compiles the conversion values and selectivity to different products in the Ecodiesel synthesis from SO (EcoSO) and CO (EcoCO), at different reaction times, 30 °C and with a mixture of methanol/ethanol. It is remarkable that, operating at the same experimental conditions as those employed in the Ecodiesel production over different enzymatic extracts [23–27] but with an oil/EtOH/MeOH/NaOMe ratio of 100/5/5/1, it is possible to obtain biofuels that can be mixed with fossil diesel for being employed in diesel C.I. engines, by reducing the starting viscosity of the oil in practically 1/3 (Tables 1 and 3). Besides, an atomic efficiency of practically 100% is obtained. According to complementary studies, the biofuels showed in Tables 2 and 3 can be used in mixtures with diesel fossil, up to 20% of Biofuel in the total mixture for Ecodiesel of sunflower oil (Table S2), and up to 15% if the Biodiesel is obtained from castor oil [41]. In these blends, the viscosity values drop to the limits established in EN 14214 standards.

Table 3. Results obtained in the transesterification of castor oil. Reaction conditions: 100 mL of either Castor oil or Sunflower oil (*), 5mL of ethanol, 5 mL of methanol and 1.0 mL of sodium methoxide, at 30 °C and 300 rpm. The neutralization is carried out with 0.1 mL of H_3PO_4. No phase separation was observed.

Time (min)	Conv. (%)	Sel. (%)	FAME (%)	MG (%)	DG (%)	TG (%)	Ecodiesel Visc. (cSt)
15	67.3	44.0	34.1	9.9	23.3	32.7	93
30	71.1	42.7	30.4	12.3	28.4	28.9	106
45	73.3	52.0	40.9	11.1	21.3	26.7	64
60	78.0	50.8	40.3	10.5	27.2	22.0	76
15 *	58.9	23.2	19.6	3.6	35.7	41.1	12
30 *	86.7	38.9	32.4	6.5	47.8	13.3	12
45 *	100	84.3	70.8	13.5	15.7	0.0	11
60 *	100	66.6	54.4	12.2	33.4	0.0	11

2.2. Characterization of Biofuel Mixtures, with Fossil Diesel and Alcohols, to Allow Their Use in Conventional Compression Ignition Engines, without Any Modification

2.2.1. Double Blends: Diesel/Biofuels

The mixtures of oils with diesel exhibit very different viscosity values, depending on the percentages of oil in the diesel, and also depending on the oil employed. However, the rheological properties of these mixtures are hardly affected by the nature of the oils, mainly at low oil concentrations, so it is possible to obtain mixtures with a 10% v/v of oil in diesel, suitable of being use in conventional diesel engines, conforming to EN 14214. Thus, it is very remarkable that, despite the high viscosity of castor oil, it behaves analogously to sunflower oil when it is blending with diesel. In fact, considering BSO10 blends, the viscosity reached a 3.6 cSt while a BCO10 blend reaches a viscosity value of 4.5 cSt.

Taking into account the use of different mixtures of diesel with Ecodiesel, with lower viscosity than the starting oils, higher values of fossil fuel replacement can be achieved. Thus, with sunflower oil, employed in this research as a reference for waste cooking oils (second generation biofuel), it is possible to obtain Ecodiesel (EcoSO) under the experimental conditions as in Table 3, with viscosities in the range of 12 to 15 cSt, whereas in a blend of diesel with a 30% of EcoSO, a viscosity value of 5.1 cSt is obtained, suitable to be employed directly in conventional diesel engines. In addition, we must not lose the perspective of how the biofuels is considered to be employed, i.e., in different mixtures, 20% in 2020, and 30% in 2030. To do so, the process described in Tables 2 and 3 seems to be the most appropriate to produce the biofuels in the most easy (and economical) way, although obviously they can be obtained with higher quality, even with a similar quality to biodiesel. However, higher concentrations of catalyst and alcohols must be used for this purpose, implying then a higher cost.

Regardless of the procedure employed for the Ecodiesel synthesis, it is important to establish what percentages of mixture with diesel can be employed in every case. Thus, for the Ecodiesel obtained with the experimental conditions indicated in Table 3 (viscosity = 64 cSt), the maximum mixing level for being employed in C.I. engine is 15%, attaining a viscosity value of 5.8 cSt. However, if a lower viscosity Ecodiesel is employed, either EcoCO or EcoSO, obtained by using higher concentrations of catalyst and alcohol, they can be employed even up to 30% blending with diesel, Figure 2. These results show that EcoCO exhibits similar rheological properties as BCO when they are blended with diesel [41], up to 25% v/v.

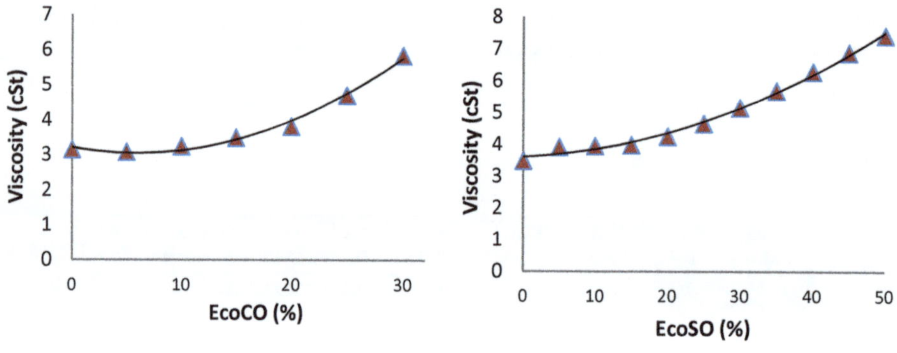

Figure 2. Kinematic viscosity values (cSt), obtained at 40 °C, of mixtures composed for (**a**) increasing amounts of castor oil ecodiesel (EcoCO) in fosil diesel, and (**b**) increasing amounts of sunflower oil ecodiesel (EcoSO) in fosil diessel.

2.2.2. Triple Blends: Diesel/Biofuels/Short Chain Alcohols

In Figure 3, the viscosity values of different triple blends Diesel/EcoSO/alcohol, are collected. Two different alcohols, ethanol and 2-propanol have been employed. First of all, it should be highlighted that the blends were prepared adding increasing amounts of a mixture of Ecodiesel/alcohol 4:1 (v/v). This 4:1 proportion has the maximum amount of ethanol for the blends to be stable, taking into account that the subsequent adding to diesel fuel will be further dilute the mixture (triple blends). Thus, these Ecodiesel/alcohol mixtures present a viscosity of 7.75 cSt with ethanol y 7.32 cSt with 2-propanol, so they cannot be applied as biofuels because of their high viscosity values (the diesel engine works at viscosities between 3–5 cSt, according to EN 14214). However, by adding different amount of diesel to these double mixtures, a sufficient reduction of the viscosity values was achieved, in the range of 3–5 cSt, which can be perfectly employed in a C.I. engine. Taking into account the viscosity results shown in Figure 3a, with a blend diesel/EcoSO/ethanol 60/30/10, it is possible to incorporate a 40% of renewable compounds as diesel additive, which is almost the double than that achieved with the blends diesel/Ecodiesel, where a maximum of 25% of EcoSO in diesel was obtained (Figure 2b). Regarding the data obtained with 2-propanol, Figure 3a, with only a 40% of diesel in the blend, we are able to obtain a biofuel with a suitable viscosity for being employed in C.I. engines, replacing a 5% more of diesel than using ethanol.

Then, the high capability shown by alcohols to reduce the viscosity of their blends with EcoSO can be also employed to optimize the blends of these alcohols with oils. This can be especially useful in the case of castor oil, which according to the data collected in Figure 4, has a greater mixing capability with alcohols than EcoCO itself. This behavior can be explained by the high content of hydroxyl groups of Castor oil, exceptional among fatty acids, which confer it a high capability for being blended with alcohols, including ethanol, in any proportion. The appropriate viscosity values, lower than 5.0 cSt, are reached when the concentration of alcohols in the blends are higher than 60%. However, in EcoCO/alcohol blends, Figure 4b, the expected decrease in viscosity was not observed, taking into account the much lower viscosity of the Ecodiesel, compared to the starting Castor oil. Therefore, there is no advantages in using EcoCO instead of pure Castor oil.

Furthermore, as can be seen in Figure 4, any blend of CO either with ethanol or isopropanol, in which the alcohol content was above 65% exhibits the appropriate viscosity for being use in conventional diesel engines. This is especially interesting in the case of ethanol, due to its renewable character, which means that these mixtures of castor oil with ethanol, 35/65, are 100% renewable biofuels. However, it is foreseeable that mixtures of CO with alcohols would not have an adequate behavior in terms of the power reached and the high consumption of biofuel, due to the low calorific power of the alcohols.

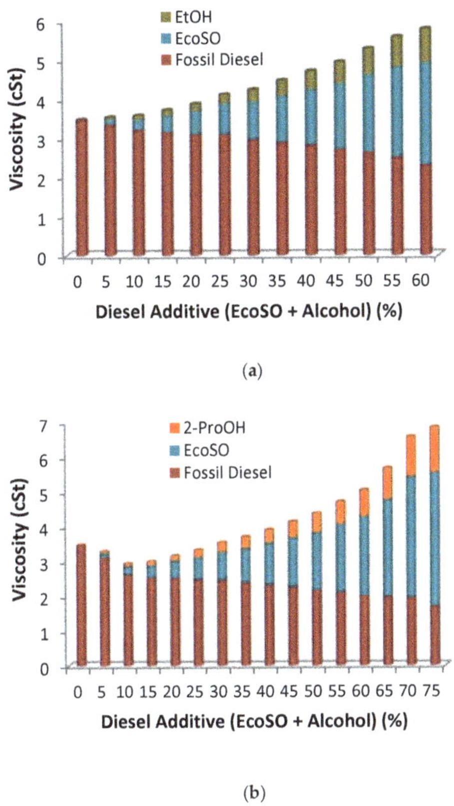

Figure 3. Viscosity values of the different triple mixtures obtained blending conventional Fossil Diesel, Ecodiesel from sunflower oil (EcoSO) and either ethanol (**a**) or 2-Propanol (**b**). In all the cases, the Ecodiesel/alcohol ratio is constant, with 25% alcohol.

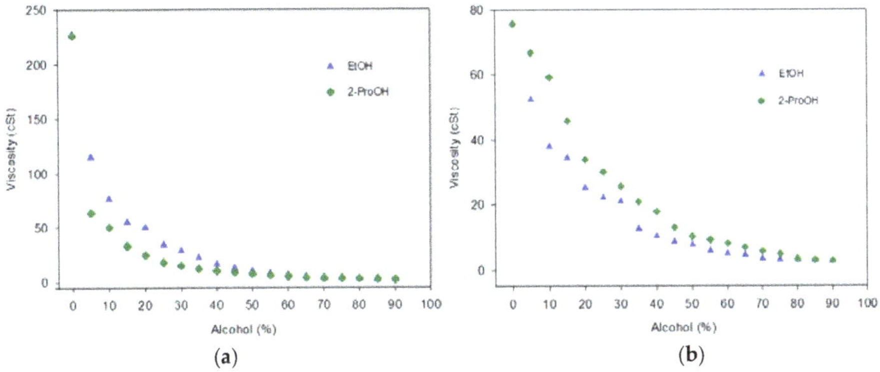

Figure 4. Viscosity values of the different double mixtures obtained blending ethanol (1.3 cSt) or isopropanol (1.63 cSt) with castor oil (227.0) cSt (**a**) and with the Ecodiesel from castor oil (EcoCO) (**b**).

This behavior makes possible the assumption that triple blends diesel/EcoCO/alcohol and diesel/CO/alcohol can be employed as biofuels. In this respect, the triple mixtures diesel/EcoCO/ethanol and diesel/EcoCO/2-propanol have been prepared by blending fossil diesel with increasing amounts of a 1:1 blend of EcoCO/alcohol, and the viscosity values obtained for those blends are shown in Figure 5.

According to the results shown in Figure 5a, it is possible to incorporate a 40% diesel additive, e.g., 60/20/20 in the blend diesel/EcoCO/ethanol, so it is possible to achieve a higher substitution of diesel fuel than that achieved with the double blends diesel/Ecodiesel. Furthermore, no differences in the viscosity values obtained for the blends using either ethanol or 2-propanol were observed (Figure 5).

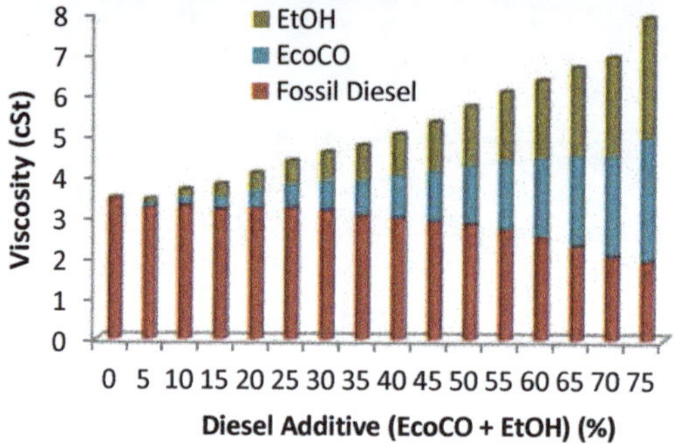

(**a**)

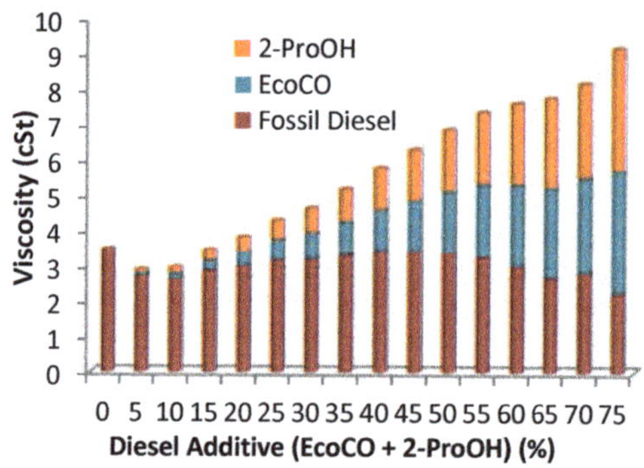

(**b**)

Figure 5. Viscosity values of the different triple mixtures obtained blending Ecodiesel from castor oil (EcoCO), a conventional diesel and either ethanol (**a**) or isopropanol (**b**). In all cases, the Ecodiesel/alcohol ratio is 1:1.

However, when Castor oil is directly employed in these triple blends, diesel/CO/alcohol, a different behavior depending on the alcohol employed is observed. As can be seen in Figure 6, for blends diesel/Castor oil/alcohol, the replacement of fossil diesel by diesel additive in amounts higher than 20% is not possible, e.g., 80/10/10. This fact is due to the low solubility of ethanol and diesel. However, if 2-propanol is employed, a 40% of diesel fuel can be replace by diesel additives, as can be seen for the blend 60/20/20, diesel/Castor oil/2-propanol.

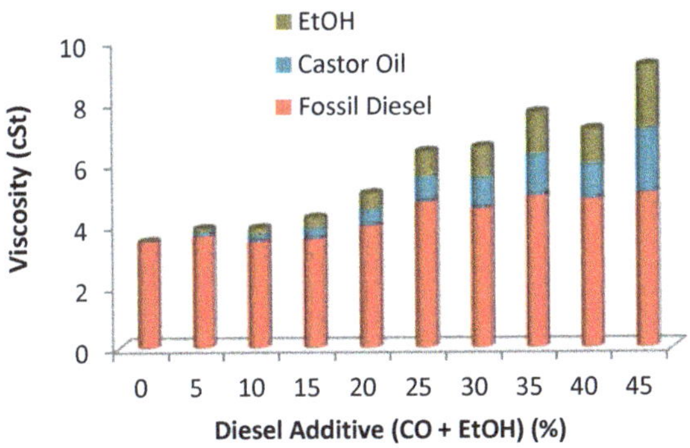

(a)

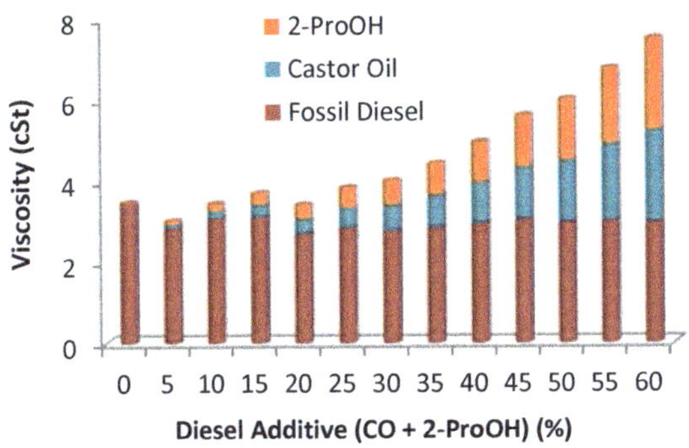

(b)

Figure 6. Viscosity values of the different triple mixtures obtained blending Castor Oil (CO), fossil diesel and either ethanol (**a**) or isopropanol (**b**). In all cases, the Ecodiesel/alcohol ratio is constant, 1:1.

These observations are really important, since Castor oil can be employed directly in the blends mixing with 2-propanol up to 40% of the total blend. Thus, there is no need to carry out the transesterification of this oil in order to get a biofuel from this raw material.

2.3. *Evaluation of Different Biofuels from Their Behavior in a Conventional Internal Combustion Engine*

2.3.1. Double Blends Diesel/Biofuels

At this point, it must be indicated that the physico-chemical properties of Ecodiesel (corrosion, calorific value, density, cetane index, viscosity and several properties of biofuels at low temperatures) to be employed as biofuel, are similar to those exhibited by Biodiesel. Therefore, only a very short number of rheological properties are relevant enough for being taken into account, considering its application in motor tests. These properties, i.e., viscosity, pour point (PP) and cloud point (CP), are collected in Table 4 for the different blends here studied. In principle, all the blends shown in Table 4 can be employed as biofuels in conventional diesel engines without any modification. However, the presence of MG in the Ecodiesel slightly increases the viscosity values of the mixtures, being this fact more noticeable in blends with more than 40% of Ecodiesel. In this sense, it is not advisable the use of double blends in which Ecodiesel is present in a percentage higher than 50%.

Once the reological properties were obtained, all the blends collected in Table 4 were tested in an internal combustion engine. For a better comparison, one more experiment at the same conditions but operating with a conventional diesel, was also carried out.

Table 4. Rheological properties of diesel/biofuel blends, either with EcoSO, or conventional Biodiesel from sunflower oil (BSO).

Blends	Biofuel in Blend (%)	Viscosity (cSt)	Pour Point (°C)	Cloud Point (°C)
BSO10	10	3.17	−15	−10
BSO20	20	3.30	−10	−12
BSO30	30	3.50	−10	−12
BSO40	40	3.83	−9	−11
BSO50	50	4.16	−9	−8
EcoSO10	10	3.93	−11	−10
EcoSO20	20	4.42	−10	−9
EcoSO30	30	5.15	−9	−8
EcoSO40	40	6.86	−9	−7
EcoSO50	50	7.13	−8	−6

The results of power generation and opacity (contamination parameter) obtained with all the double blends, diesel/Biodiesel, are shown in Figure 7. First of all, it can be seen that at 4 kW of Power demand, the highest Power Generation is achieved by the engine, regardless of the biofuel employed. In addition, all the blends exhibited similar values of Power generation, which was also analogous to that exhibited by the fossil diesel. However, considering the opacity parameter, it is highlighted that the lowest opacity value was obtained with the BSO50 blend, which presents also the highest content of Biodiesel.

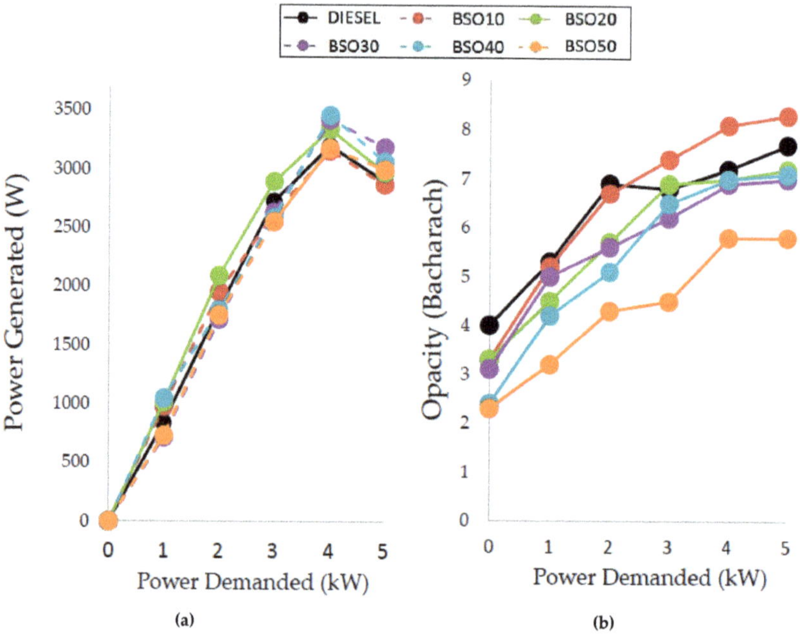

Figure 7. (**a**) Power generated (in Watts), based on the power demanded (in kWatts) and (**b**) Opacity values generated in the smokes (in Bacharach units) as a function of the power demanded of different double mixtures of Fossil Diesel with Biodiesel of Sunflower Oil (BSO).

The double blends diesel/EcoSO also exhibited a better power generated values than the fossil diesel at high power demands, although their behavior at power demands of 1, 2 and 3 kWatts is pretty similar, even with the lowest amount of EcoSO in the blend, Figure 8a. Furthermore, the values of opacity are better than that obtained with fossil diesel and, in general, better than those obtained with the diesel/biodiesel blends, Figure 8b.

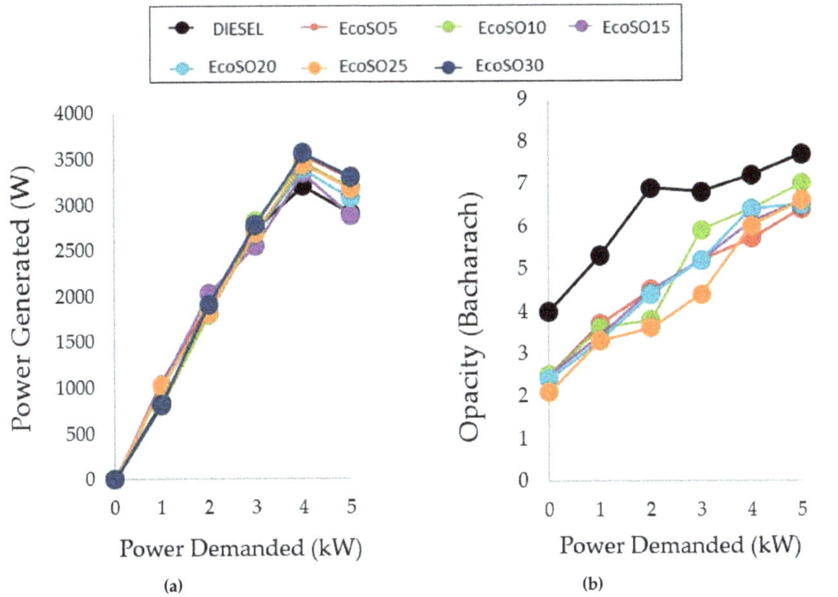

Figure 8. Results obtained by using different mixtures of Ecodiesel from sunflower oil with diesel fossil: (**a**) Power values generated (in Watts), based on the power demanded (in kWatts). (**b**) Opacity values generated in the smokes (in Bacharach units) as a function of the power demanded.

2.3.2. Triple Blends: Diesel/Biofuels/Short Chain Alcohols

The behavior of the more characteristics triple blends diesel/Ecodiesel/alcohol has been investigated to estimate the influence of the percentages of Diesel Additive (Ecodiesel + alcohol) in the rheological properties abovementioned, i.e., viscosity and pour point and cloud point temperatures. Thus, the results obtained for the most characeristic blends are compiled in Table 5.

Table 5. Rheological properties of triple blends diesel/biofuel/alcohol, in the proportions indicated as percentages.

Entry	Diesel/Biofuel/Alcohol (%)	Viscosity (cSt)	Pour Point (°C)	Cloud Point (°C)
1	Diesel (100)	3.47	−19	−4
2	Diesel/EcoSO/EtOH (50/40/10)	4.72	−9	−7
3	Diesel/EcoSO/2-ProOH (35/50/15)	5.20	−12	−9
4	Diesel/EcoCO/2-ProOH (50/25/25)	5.60	−13	−11
5	Diesel/CO/EtOH (50/25/25)	5.93	−7	−6
6	Diesel/CO/2-ProOH (50/25/25)	5.81	−8	−5

The same as with double blends, the triple ones were tested in a diesel combustion engine and the results of power generated, opacity and consumption are shown in Figures 9 and 10. On one hand, the power generated is similar to that obtained with fossil diesel, mainly up to 3 kW of power demanded. However, two different behaviors have been observed from 4 kW, and it is related to the ethanol taking part of the mixture. As it was explained before, ethanol has a low calorific power. In fact, the higher the amount of ethanol in the blend, the lower power generated at 5 kW (entry 5, Figure 9).

On the other hand, it is noteworthy the results related to the opacity for all the triple blends here studied. These opacity values are quite lower than that obtained with fossil diesel, indicating the great importance of these blends to reduce the emissions of pollutants. Among all of these blends, the behavior of the Castor oil is surprising, no matter with which alcohol is blended. In fact, in addition to the lowest values of opacity, the lowest consumption in L/h were achieved at high power demanded, even better than that obtained with fossil diesel, Figure 10.

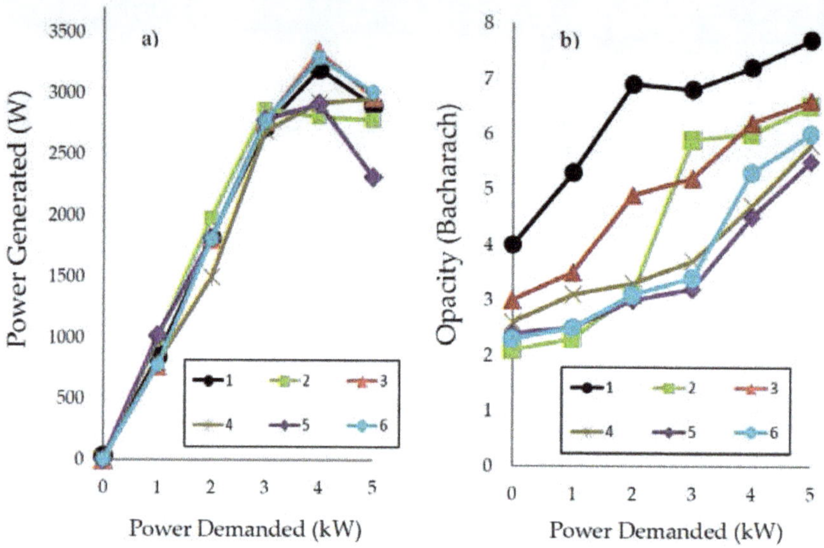

Figure 9. (**a**) Power generated values (in Watts), based on the power demanded (kWatts) and (**b**) Opacity values generated in the smokes (in Bacharach units) as a function of the power demanded of the blends of Table 5.

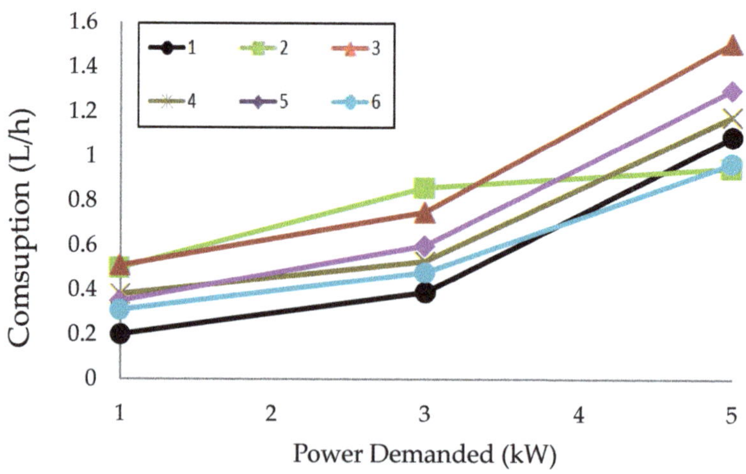

Figure 10. Consumption values of the blends in Table 5 as function of the power demanded of the engine.

3. Discussion

3.1. Synthesis of Ecodiesel by Kinetic Control of the Conventional Procedure of Biodiesel Production

From the results here obtained, we can conclude that among all the available procedures to reduce the viscosity of the oils by chemical reaction, the selective alcoholysis exhibits the greater atomic efficiency (100%), taking full advantage of the triglyceride molecule as biofuel [21,22]. As can be observed from Tables 1–3, by this procedure, the viscosity of the Ecodiesel is slightly higher than that exhibited by the biodiesel. This fact can be explained because the total alcoholysis generates three molecules of FAME. But, in exchange, glycerol generation, in a 10% by weight of the total biodiesel produced, is avoided.

Another important aspect from the economical point of view is that the synthesis of biodiesel needs, in general, more energetic reaction conditions. In fact, complementary studies about the Ecodiesel production [44], Table S2, have also shown that the reaction temperatures in the range of 20 to 60 °C, produce very little influence on the biofuel obtained, since the viscosity in this range is reduced by less than 10%. This fact implies that the activation energy in the alcoholysis of the secondary carbon of glycerol (position 2) is higher than corresponding to the activation of the primary alcohols (positions 1 and 3 of glycerol). This, among other things, would justify that both processes take place through a different reaction mechanism. From a practical point of view, this implies that the synthesis of Ecodiesel, regardless of other parameters involved, it should always be performed in the range 20–40 °C, usually at room temperature. Furthermore, as can be seen in Table 2, the use of EtOH together with MeOH allow the 1,3-selective transesterification over sodium methoxide, attaining an Ecodiesel similar to that obtained with lipases, improving the process from an economic point of view.

In addition, another important aspect in the evaluation of the homogeneous catalytic synthesis of Ecodiesel here reported is related to its application to waste cooking oils. This process can be applied to waste oils subjected to a very simple cleaning process, since after the synthesis reaction of the Ecodiesel it is necessary to carry out a very simple filtering operation, to eliminate the salts of alkaline phosphates produced by the reaction with phosphoric acid. In the filtering process, the solid residues usually mixed with the wasted oil are also easily eliminated. In other words, biofuels can be "cleaned" in the same synthesis process of Ecodiesel. Therefore, with pure oils, the only waste is practically reduced to the Na_3PO_4 salts produced in the last neutralization step. Furthermore, the presence of ethanol as a co-reactant or solvent of methanol, works very positively because it improves the viscosity, and prevents that methanol becomes separate of the biofuel.

Everything raised above leads us to conclude that the production of Ecodiesel in mild conditions, employing homogeneous basic catalysts improves by far the previous studies which implies the use of lipases [23–26] or heterogeneous basic catalysts [28,29], not only from a technical point of view but also form the economic one, above all if we consider that this Ecodiesel can be also employed as Biofuel in different double and triple blends.

3.2. Evaluation of Different Biofuels from Their Behavior in a Conventional Internal Combustion Engine

From the oils here studied as raw materials in the Ecodiesel production, sunflower oil and castor oil, both Ecodiesel produced, EcoSO and EcoCO, exhibited similar rheological properties as those described in the literature for the Biodiesel from Castor oil [41], as they can be mixed in a 25% with fossil diesel, Figure 2, to run in conventional diesel engines. However, it is possible to achieve similar, or even higher, replacement of fossil diesel using pure castor oil in triple mixtures with alcohols. The peculiar structure of Castor oil allows a high solubility with both, fossil diesel and alcohols, including ethanol (Figure 4). Thus, triple mixtures can be obtained with Castor oil (Figure 6) with a high level of fossil diesel replacement, with no need of transforming the starting oil into Ecodiesel (Figures 4 and 5).

Taking into account the two different alcohols proposed in this research, and, according to the results shown in Figures 5 and 6, if ethanol is employed in the triple mixtures, the EcoCO favors a higher substitution of fossil diesel than the use of pure Castor oil, 60/20/20 and 80/10/10, respectively.

However, if isopropanol is employed, similar results are obtained with EcoCO and pure CO. Anyway, a 40% of fossil diesel substitution was always achieved.

These double and triple blends have been evaluated according to the viscosity, as this parameter gives information about which ones can be considered for its use in a diesel engine, as the limits of viscosity required by conventional diesel engines are in the range 3–7 cSt, although European regulations (EN 14214) establish the interval 3–5 cSt, Figures 3–6.

Regarding the behavior of the blends at low temperatures, Table 5, conventional biodiesel and ecodiesel, in double and triple mixtures, solidify at temperatures around −10 °C and are completely frozen at temperatures around −12 °C. In contrast, the corresponding temperatures for diesel fossil are around −16 to −18 °C, for pour point and cloud point, respectively. Thus, for the values here obtained, it can be observed that in blends whose percentage of biofuel is higher, the temperatures of pour point and cloud point are higher, solidifying before. Independently on the slightly worst temperatures of pour and cloud point than those exhibited by diesel, the blends studied in the present work gave acceptable results for environmental conditions where the temperatures are not reduced below −10 °C. In addition, the use of any alcohol as an additive would influences positively these parameters, since its addition promotes a better performance against low temperatures.

The blends, which meet the requirements for being employed as fuel in diesel engines, were tested at different values of power demanded. As can be seen in Figures 7 and 8, independently of the biofuel tested, a stabilization of the generated power values takes place above 4 kW of demanded power. Likewise, the values of power generated are slightly higher when blends with Ecodiesel were employed, especially in the proportions 20% and 30%, Figures 7a and 8a.

About the pollution produced during the combustion of the different samples, the results show that the higher the power demanded, the higher the emission of pollutants, as it was expected. In fact, an increase in the power demand from a power equivalent to the engine working at idling speed to the highest demand of 4 or 5 kW, the soot formation doubles. However, it is very remarkable that for blends diesel/biodiesel with biodiesel content higher than 10% (BS20, BS30, BS40 and BS50) and blends diesel/Ecodiesel with Ecodiesel content higher than 5%, the opacity values obtained were always lower than that obtained with fossil diesel, Figures 7b and 8b.

Regarding the triple blends diesel/biofuel/alcohol, the power generated values and the opacity as a function of the power demanded are shown in Figure 9. On one hand, it should be noted that all the samples exhibited similar values of power generated than diesel fuel (Figure 9a), although those in which ethanol is employed as alcohol showed the lower ones, especially the blend diesel/CO/EtOH (50/25/25). However, if 2-propanol is employed, independently on the biofuel, EcoSO, EcoCO or CO, very good results of power generated were obtained.

On the other hand, it must be remarked the surprisingly low opacity values obtained with all the triple blends tested, more than 10% lower than the opacity obtained with diesel, Figure 9b.

According to the engine fuel consumption, the results in Figure 10 show that EcoSO in triple blends with isopropanol exhibit higher consumption than diesel, whereas blended with ethanol exhibit lower consumption than diesel. For its part, the blends with EcoCO and CO behave the opposite, i.e., a lower consumption was obtained using 2-propanol. Thus, considering the results previously exposed, the triple blend diesel/CO/2-proOH (50/25/25) exhibit the best performance, with higher power generation, lower emission of pollutants and also lower fuel consumption, indicating its feasibility for being employed in diesel engines.

4. Materials and Methods

4.1. Evaluation of Sodium Methoxide as Homogeneous Catalyst in the Selective Alcoholysis to Obtain Ecodiesel

The selective alcoholysis of the Sunflower oil (food quality) and Castor oil (Panreac, Castellar Del Valles, Spain) were carried out over commercial sodium methoxide (Panreac, Sodium methylate solution 30% in methanol PS) as homogeneous catalyst. The reactions were performed in a 250 mL flask immersed in a temperature-controlled water bath, at atmospheric pressure using methanol as alcohol. Different reaction conditions have been studied, such as temperature, 20–60 °C, and reaction time, Figure 11.

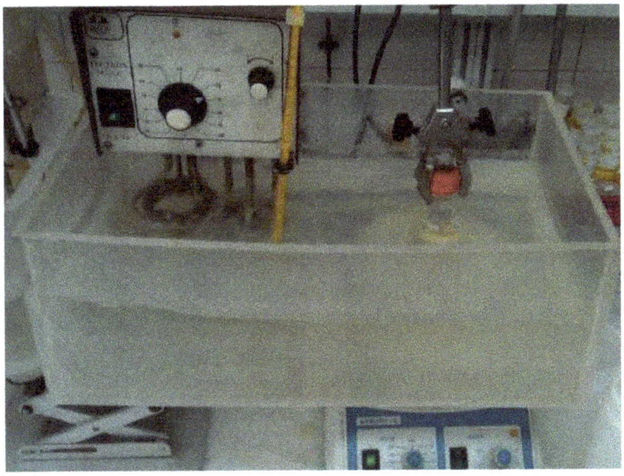

Figure 11. Experimental dispositive for the transesterification reaction.

In general, 120 mL of oil (0.1 mol) were mixed with variable proportions of methanol and sodium methoxide. The changes are easily visualized as a ratio, for example, 100/10/1 indicates 100 mL of oil, 10 mL of methanol and 1 mL of sodium methoxide. Variable amounts of ethanol are sometimes also incorporated in some experiments, which act as a solvent, given its lower activity with respect to methanol. It has been taken as standard measure that 0.1 mol = 91.0 g of oil. The weigh was performed on a Mettler AJ50 precision balance (precision of ± 0.01 g), depositing the sample directly into a 250 mL two-neck round bottom flask.

4.2. Determination of the Content of FAMEs, FAEEs and Glycerides (MG, DG and TG) in the Reaction Products by Gas Chromatography

The determination of the content of methyl esters, ethyl esters and different glycerides in the biofuel samples has been carried out by a chromatographic method developed in previous researches [23]. Thus, an HP 5890 Series II gas chromatograph with a HT5 (25 m × 0.32 mm I.D × 0.1 μm, SGE, Supelco) Aldrich Chemie capillary column and equipped with a flame ionization detector (FID) was employed. Cetane (n-hexadecane) is used as an internal standard

This method basically consists of a modification and integration of two official methods, UNE EN ISO 14103 (esters) and UNE EN ISO 14105 (glycerides), to quantify the content of glycerol, ethyl esters and glycerides (mono, di and triglycerides), respectively. The ethyl esters of palmitic acid, stearic acid, oleic acid, linoleic acid, and linolenic acid were commercially obtained from AccuStandard (New Haven, CT, USA), and hexadecane (cetane) was obtained from Sigma-Aldrich (St. Louis, MO, USA).

Considering that castor oil or sunflower oil are constituted by a mixture of fatty acids (mainly linoleic, oleic, palmitic, and stearic acids, in sunflower oil, or ricinoleic in castor oil) in variable proportions, the results obtained are expressed as the relative amounts of the corresponding methyl esters (FAME, fatty acid methyl esters), monoglycerides (MG), and diglycerides (DG) that are integrated in the chromatogram. The number of diglycerides (DG) and triglycerides (TG) that have not reacted is calculated from the difference to the internal standard (cetane). Thus, the Conversion includes the total amount of triglyceride transformed (FAEE + MG + DG) in the methanolysis process, and Selectivity refers to the relative amount of FAEE + MG obtained.

4.3. Determination of Kinematic Viscosity of Biofuels

The kinematic viscosity has been measured in an Ostwald-Cannon-Fenske capillary viscometer (Proton Routine Viscometer 33200, size 150), determining the time required for a certain volume of liquid to pass between two marked points on the instrument, placed in an upright position. From the flow time (t), expressed in seconds, we obtain the kinematic viscosity expressed in centistokes, $\upsilon = C \cdot t$. Where C is the calibration constant of the measurement system in mm^2/s^2, which is specified by the manufacturer (0.040350 mm^2/s^2 at 40 °C, in this case). All measures have been carried out in duplicate and are presented as the average of both, proving that there is no greater variation of 0.35% between measures, as required by the standard ASTM (American Society for Testing and Materials) D2270-79, Method for calculating viscosity index from kinematic viscosity at 40 and 100 °C.

4.4. Determination of the Pour Point and Cloud Point of Biofuels

Cloud point and the Pour point are determined by introducing the different double or triple samples, of different composition, in a digitally controlled temperature refrigerator for twenty-four hours; after this time the loss of transparency of the solutions is evaluated. The appearance of turbidity in the samples is indicative that the cloud point temperature has been reached (cloud point). After a progressive decrease in temperature, the samples are kept under observation until they stop flowing (pour point).

4.5. Assessment of Energy Performance and Air Pollutant Emissions in a Diesel Engine Electric Generator, Fueled with Different Blends of Biofuels

The mechanical and environmental characterization of a compression ignition diesel engine has been carried out, working at a rate of 3000 rpm coupled to an AYERBE electric generator, 5KVA, 230v type AY4000MN, for the generation of electricity, operating at a crankshaft constant rotation rate and under different degrees of demand for electrical power. This is achieved by connecting heating plates of 1000 watts each one (Figure 12a). This diesel engine will operate at a constant rate of rotation of the crankshaft and torque, so that the different values of electrical power obtained will be an exact consequence of the mechanical power obtained after the combustion of the corresponding biofuel. Different tests are obtained by providing to the engine double and triple mixtures of different biofuels in different percentages. The electrical power generated can be easily determined from the product of the potential difference (or voltage) and the electric current intensity (or amperage), equation (1), both obtained by means of a voltmeter-ammeter [45,46].

$$\text{Electrical Power Generated (Watts)} = \text{voltage (Volts)} \times \text{amperage (Amps)} \quad (1)$$

The consumption of the diesel engine with the different biofuels employed was calculated estimating the speed of consumption of the engine, with a given fuel, when operating under a determined demand of electric power.

On the other hand, the contamination degree is evaluated regarding the opacity of the smoke generated in the combustion process. This is obtained by using an opacimeter—TESTO 308 opacity meter—under the operating conditions studied (Figure 12b). All the results obtained with the biofuels evaluated were compared with the corresponding measurements obtained with conventional diesel. The opacimeter, is a device designed to estimate the amount of soot emitted by diesel engines. Unlike gasoline engines, where the amount of carbon monoxide (CO) and hydrocarbons (HC) is measured to assess the quality of combustion and toxic emissions, only the amount of coal is analyzed in diesel engines (soot). These are tiny particles in suspension which can not be treated as a gas, that is, they can not be quantified through the gas analyzer. The opacimeter is basically composed of three components: measuring chamber, analyzer and a portable terminal. The outlet of the exhaust pipe of the engine is connected, through a pipe and a hose, with the measuring chamber and partial samples of the exhaust gases are taken. It is called partial since only part of the gases enter the machine and the rest is lost in the atmosphere. The gases enter inside a tube and through a sensor the intensity of the light (turbidity) is measured, to then calculate the density of the particles. The tube has a source of halogen light at one end and a receiver at the other so that when there is no gas inside the tube, the light intensity is not affected. The result of such measurements is reflected visually on a filter paper. Besides, this value can be expressed as a percentage of Opacity (being 100% totally cloudy and 0% totally clear) or as an equivalent number called the k value (Opacity Bacharach) the scale runs from white (0 Bacharach unit) to black (9 Bacharach units), as established by ASTM D 2156-94, Standard Test Method for Smoke Density in Flue Gases from Burning Distillate Fuels [47]. It must be clarified that all the measured were repeated at least three times, attaining an experimental error lower than 5%.

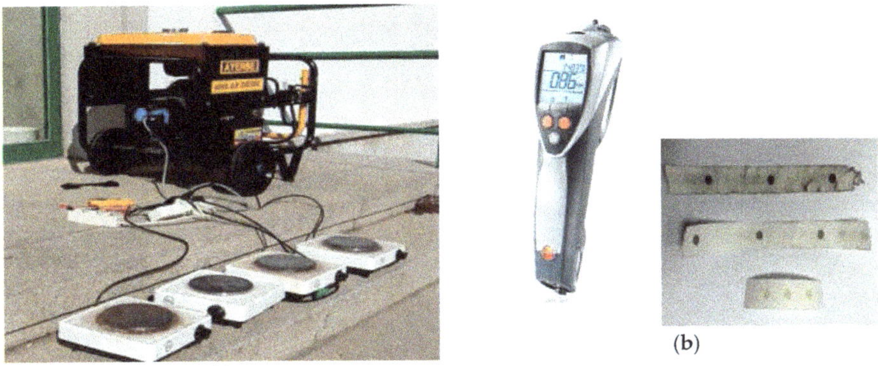

(**b**)

(**a**)

Figure 12. (**a**) Electrogenerator AYERBE, 5KVA, 230v tipo AY4000MN, heating plates of 1000 watts of power each and voltmeter-ammeter devise (yellow colour, on the floor); (**b**) TESTO 308 opacity meter, which operates as established by ASTM D 2156-94, Standard Test Method for Smoke Density in Flue Gases from Burning Distillate Fuels.

5. Conclusions

In this research, some Biofuels have been synthesized employing a commercial basic homogeneous catalyst, sodium methoxide, at milder conditions than those employed to obtain conventional biodiesel, to favor a selective 1,3 methanolysis of triglycerides, achieving a 100% atomic efficiency in the synthesis of Ecodiesel. In this respect, sunflower oil and castor oil have been evaluated. Besides, the behavior of diesel/biofuel blends, as well as the behavior of several triple blends, diesel/biofuel/alcohol have been also studied. Furthermore, the differences between Ecodiesel and conventional biodiesel, as well as the alcohol employed in the blends, in what proportions must be added together with fossil diesel, to obtain the rheological properties to operate in conventional diesel engines without making any modifications have been also determined. The main conclusions obtained in this research can be summarized as follows:

It has been demonstrated, for the first time, that the use of EtOH as solvent in the 1,3-selective transesterification of triglycerides improve the diffusion between methanol and TG, allowing its reaction at milder conditions, since an increase in the temperature of the reaction does not improve the Ecodiesel production.

Ecodiesel from both, sunflower and castor oil, has been successfully obtained at the reaction conditions: Oil/ethanol/methanol/catalyst ratio 100/5/5/1, employing sodium methoxide as homogeneous catalyst at 30 °C and at a stirring speed of 300 rpm and 15 min of reaction time.

This research has shown that the chemical route here proposed for the synthesis of Ecodiesel is able to reduce the production costs to a great extent, in comparison with enzymatic routes and heterogeneous catalysed routes.

About the rheological properties of the double and triple blends here studied, it has been verified that the viscosity, Pour Point and Cloud Point values of the different samples allow their use as biofuels in conventional diesel engines.

The results obtained using a compression ignition diesel engine show that 4 kW of power demand gives the greatest engine power generation, independently on the fuel employed.

With the blends diesel/biodiesel, a 50% of fossil diesel can be replaced by a biofuel obtained from a renewable source, whereas in the blends diesel/Ecodiesel, the amount of fossil diesel that can be replaced is lower, 30%. This fact is due to the presence of MG in the Ecodiesel, which increase the viscosity above 6 cSt.

The use of triple blends diesel/biofuel/alcohol allows us to increase the diesel replacement, even employing directly Ecodiesel or Castor oil. Thus, it is possible to use pure Castor oil in the proportions 50/25/25, diesel/CO/2-propanol, in conventional diesel engines, with a performance and level of consumption like fossil diesel, but with an appreciably lower emission of pollutants. In the case of the Ecodiesel of sunflower oil, (and used cooking oils, for use as second generation biofuels) it is possible to obtain triple mixtures with any ethanol in a proportion 50/40/10 diesel/EcoSO/ethanol which also exhibits very good results in terms of consumption and emission of pollutants.

Supplementary Materials: The following are available online at http://www.mdpi.com/2073-4344/9/1/40/s1, Figure S1: Chromatograms obtained in the sunflower oil alcoholysis (a). The initial sunflower oil is in black (a). Commercial diesel fuel chromatogram (b), Table S1: Viscosity values of different blends diesel/Ecodiesel from Castor oil (EcoCO), Table S2: Viscosity and Selectivity values of Ecodiesel of Sunflower oil (EcoSO) and reaction products obtained at the same reaction conditions as in Table 2, after 60 min. The Conversion was also 100%. Sel. = FAE + MG, being FAE = FAME + FAEE.

Author Contributions: This research article is part of the doctoral thesis of B.H., and A.P., directed by D.L., J.M.H and F.M.B., who in general conceived and designed the experiments. C.L., J.C., A.A.R. and R.E. made substantive intellectual contributions to this study, making substantial contributions to conception and design of it, as well as to the acquisition, analysis and interpretation of data. Furthermore, D.L. and R.E wrote the paper. All the authors have been also involved in drafting and revising the manuscript, so that everyone has given final approval of the current version to be published in Catalysts Journal.

Funding: This research received no external funding.

Acknowledgments: This research is supported by the MEIC funds (Project ENE 2016-81013-R), Junta de Andalucía and FEDER (P11-TEP-7723), that cover the costs to publish in open access.

Conflicts of Interest: The authors declare no conflict of interest.

Abbreviations

FAME	fatty acids methyl esters, components of conventional biodiesel
DMC	dimethyl carbonate
FAEE	Fatty Acid Ethyl Ester
FAE	esters of fatty acids
MG	monoglycerides or monoacylglycerols
DG	diacylglycerols
TG	triacylglycerols or triglycerides
GC	gas chromatograph
CO	Castor oil
SO	Sunflower oil
BCO	Biodiesel obtained from Castor oil
BSO	Biodiesel obtained from Sunflower oil
C.I.	Compression Ignition engines
CP	Cloud Point
PP	Pour point
CFPP	Cold filter plugging point temperature
POFF	Point of occlusion of the cold filter

References

1. Arutyunov, V.S.; Lisichkin, G.V. Energy resources of the 21st century: Problems and forecasts. Can renewable energy sources replace fossil fuels? *Russ. Chem. Rev.* **2017**, *86*, 777–804. [CrossRef]
2. Nigam, P.S.; Singh, A. Production of liquid biofuels from renewable resources. *Prog. Energy Comb. Sci.* **2011**, *37*, 52–68. [CrossRef]
3. Cheng, F.; Li, X. Preparation and Application of Biochar-Based Catalysts for Biofuel Production. *Catalysts* **2018**, *8*, 346. [CrossRef]
4. Chu, S.; Majumdar, A. Opportunities and challenges for a sustainable energy future. *Nature* **2012**, *488*, 294–303. [CrossRef] [PubMed]
5. Ridjan, I.; Mathiesen, B.V.; Connolly, D.; Dui, N. The feasibility of synthetic fuels in renewable energy systems. *Energy* **2013**, *57*, 76–84. [CrossRef]
6. Kumar, A.A.; Sharma, S. Potential non-edible oil resources as biodiesel feedstock: An Indian perspective. *Renew. Sustain. Energy Rev.* **2011**, *15*, 1791–1800. [CrossRef]
7. Rakopoulos, D.C.; Rakopoulos, C.D.; Giakoumis, E.G.; Dimaratos, A.M.; Founti, M.A. Comparative environmental behaviour of bus engine operating on blends of diesel fuel with four straight vegetable oils of Greek origin: Sunflower, cottonseed, corn and olive. *Fuel* **2011**, *90*, 3439–3446. [CrossRef]
8. Lapuerta, M.; Armas, O.; Herreros, J.M. Emissions from a diesel–bioethanol blend in an automotive diesel engine. *Fuel* **2008**, *87*, 25–31. [CrossRef]
9. Karabektas, M.; Hosoz, M. Performance and emission characteristics of a diesel engine using isobutanol–diesel fuel blends. *Renew. Energy* **2009**, *34*, 1554–1559. [CrossRef]
10. Atmanlı, A.; Ileri, E.; Yüksel, B. Experimental investigation of engine performance and exhaust emissions of a diesel engine fueled with diesel–n-butanol–vegetable oil blends. *Energy Convers. Manag.* **2014**, *81*, 312–321. [CrossRef]
11. Campos-Fernández, J.; Arnal, J.M.; Gómez, J.; Dorado, M.P. A comparison of performance of higher alcohols/diesel fuel blends in a diesel engine. *Appl. Energy* **2012**, *95*, 267–275. [CrossRef]
12. Navaneethakrishnan, P.; Vasudevan, D. Experimental Study on Performance and Exhaust Emission Characteristics of a C.I. Engine Fueled with Tri Compound Oxygenated Diesel Fuel Blends. *Indian J. Sci. Technol.* **2015**, *8*, 96–102. [CrossRef]
13. Yilmaz, N. Comparative analysis of biodiesel-ethanol-diesel and biodiesel-methanol-diesel blends in a diesel engine. *Energy* **2012**, *40*, 210–213. [CrossRef]
14. Yilmaz, N.; Vigil, F.M.; Donaldson, A.B.; Darabseh, T. Investigation of CI engine emissions in biodiesel–ethanol–diesel blends as a function of ethanol concentration. *Fuel* **2014**, *115*, 790–793. [CrossRef]
15. Shahir, S.A.; Masjuki, H.H.; Kalam, M.A.; Imran, A.; Ashraful, A.M. Performance and emission assessment of diesel–biodiesel–ethanol/bioethanol blend as a fuel in diesel engines: A review. *Renew. Sustain. Energy Rev.* **2015**, *48*, 62–78. [CrossRef]
16. Aydın, F.; Ogüt, H. Effects of using ethanol-biodiesel-diesel fuel in single cylinder diesel engine to engine performance and emissions. *Renew. Energy* **2017**, *103*, 688–694. [CrossRef]
17. Ghobadian, B.; Yusaf, T.; Najafi, G.; Khatamifar, M. Diesterol: An environment-friendly IC engine fuel. *Renew. Energy* **2009**, *34*, 335–342. [CrossRef]
18. Noorollahi, Y.; Azadbakht, M.; Ghobadian, B.; Noorollahi, Y. The effect of different diesterol (diesel–biodiesel–ethanol) blends on small air-cooled diesel engine performance and its exhaust gases. *Energy* **2018**, *142*, 196–200. [CrossRef]
19. Quispe, C.A.G.; Coronado, C.R.; Carvalho, J.A., Jr. Glycerol: Production, consumption, prices, characterization and new trends in combustion. *Renew. Sustain. Energy Rev.* **2013**, *27*, 475–493. [CrossRef]
20. Oh, P.P.; Lau, H.L.N.; Chen, J.; Chong, M.F.; Choo, Y.M. A review on conventional technologies and emerging process intensification (PI) methods for biodiesel production. *Renew. Sustain. Energy Rev.* **2012**, *16*, 5131–5145. [CrossRef]
21. Calero, J.; Luna, D.; Sancho, E.D.; Luna, C.; Bautista, F.M.; Romero, A.A.; Posadillo, A.; Berbel, J.; Verdugo-Escamilla, C. An overview on glycerol-free processes for the production of renewable liquid biofuels, applicable in diesel engines. *Renew. Sustain. Energy Rev.* **2015**, *42*, 1437–1452. [CrossRef]

22. Calero, J.; Luna, D.; Sancho, E.D.; Luna, C.; Posadillo, A.; Bautista, F.M.; Romero, A.A.; Berbel, J.; Verdugo, C. Technological challenges for the production of biodiesel in arid lands. *J. Arid Environ.* **2014**, *102*, 127–138. [CrossRef]
23. Verdugo, C.; Luque, R.; Luna, D.; Hidalgo, J.M.; Posadillo, A.; Sancho, E.D.; Rodriguez, S.; Ferreira-Dias, S.; Bautista, F.; Romero, A.A. A comprehensive study of reaction parameters in the enzymatic production of novel biofuels integrating glycerol into their composition. *Bioresour. Technol.* **2010**, *101*, 6657–6662. [CrossRef] [PubMed]
24. Luna, C.; Sancho, E.; Luna, D.; Caballero, V.; Calero, J.; Posadillo, A.; Verdugo, C.; Bautista, F.M.; Romero, A.A. Biofuel that keeps glycerol as monoglyceride by 1,3-selective ethanolysis with pig pancreatic lipase covalently immobilized on AlPO4 support. *Energies* **2013**, *6*, 3879–3900. [CrossRef]
25. Luna, C.; Verdugo, C.; Sancho, E.D.; Luna, D.; Calero, J.; Posadillo, A.; Bautista, F.M.; Romero, A.A. Production of a biodiesel-like biofuel without glycerol generation, by using Novozym 435, an immobilized Candida antarctica lipase. *Bioresour. Bioprocess.* **2014**, *1*, 11. [CrossRef]
26. Luna, C.; Verdugo, C.; Sancho, E.D.; Luna, D.; Calero, J.; Posadillo, A.; Bautista, F.M.; Romero, A.A. Biocatalytic behaviour of immobilized Rhizopus oryzae lipase in the 1,3-selective ethanolysis of sunflower oil to obtain a biofuel similar to biodiesel. *Molecules* **2014**, *19*, 11419–11439. [CrossRef] [PubMed]
27. Luna, C.; Luna, D.; Bautista, F.M.; Estevez, R.; Calero, J.; Posadillo, A.; Romero, A.A.; Sancho, E.D. Application of Enzymatic Extracts from a CALB Standard Strain as Biocatalyst within the Context of Conventional Biodiesel Production Optimization. *Molecules* **2017**, *22*, 2025. [CrossRef]
28. Calero, J.; Cumplido, G.; Luna, D.; Sancho, E.D.; Luna, C.; Posadillo, A.; Bautista, F.M.; Romero, A.A.; Verdugo-Escamilla, C. Production of a Biofuel that Keeps the Glycerol as a Monoglyceride by Using Supported KF as Heterogeneous Catalyst. *Energies* **2014**, *7*, 3764–3780. [CrossRef]
29. Calero, J.; Luna, D.; Sancho, E.D.; Luna, C.; Bautista, F.M.; Romero, A.A.; Posadillo, A.; Verdugo, C. develoment of a new biodiesel that integrates glycerol, by using CaO as heterogeneous catalyst, in the partial methanolysis of sunflower oil. *Fuel* **2014**, *7*, 94–102. [CrossRef]
30. Dias, J.M.; Araújo, J.M.; Costa, J.F.; Alvim-Ferraz, M.C.M.; Almeida, M.F. Biodiesel production from raw castor oil. *Energy* **2013**, *53*, 58–69. [CrossRef]
31. Kim, K.H.; Lee, O.K.; Lee, E.Y. Nano-Immobilized Biocatalysts for Biodiesel Production from Renewable and Sustainable Resources. A review. *Catalysts* **2018**, *8*, 68. [CrossRef]
32. Bello, E.I.; Makanju, A. Production, Characterization and Evaluation of Castor oil Biodiesel as Alternative Fuel for Diesel Engines. *J. Emerg. Trends Eng. Appl. Sci.* **2011**, *2*, 525–530.
33. Scholz, V.; Silva, J.N. Prospects and risks of the use of castor oil as a fuel. A review. *Biomass Bioenergy* **2008**, *32*, 95–100. [CrossRef]
34. Patel, V.R.; Dumancas, G.G.; Viswanath, L.C.K.; Maples, R.; Subong, B.J.J. Castor Oil: Properties, Uses, and Optimization of Processing Parameters in Commercial Production. *Lipid Insights* **2016**, *9*, 1–12. [CrossRef]
35. Knothe, G. "Designer" Biodiesel: Optimizing Fatty Ester Composition to Improve Fuel Properties. *Energy Fuels* **2008**, *22*, 1358–1364. [CrossRef]
36. Berman, P.; Nizri, S.; Wiesman, Z. Castor oil biodiesel and its blends as alternative fuel. *Biomass Bioenergy* **2011**, *35*, 2861–2866. [CrossRef]
37. Panwar, N.L.; Shrirame, H.Y.; Rathore, N.S.; Jindal, S.; Kurchania, A.K. Performance evaluation of a diesel engine fueled with methyl ester of castor seed oil. *Appl. Therm. Eng.* **2010**, *30*, 245–249. [CrossRef]
38. Valente, O.S.; Silva, M.J.; Pasa, V.M.D.; Belchior, C.R.P.; Sodré, J.R. Fuel consumption and emissions from a diesel power generator fuelled with castor oil and soybean biodiesel. *Fuel* **2010**, *89*, 3637–3642. [CrossRef]
39. Nabi, M.N.; Akhter, M.S.; Shahadat, M.M.Z. Improvement of engine emissions with conventional diesel fuel and diesel–biodiesel blends. *Bioresour. Technol.* **2006**, *97*, 372–378. [CrossRef]
40. Atabani, A.E.; Silitonga, A.S.; Ong, H.C.; Mahlia, T.M.I.; Masjuki, H.H. Non-edible vegetable oils: A critical evaluation of oil extraction, fatty acid compositions, biodiesel production, characteristics, engine performance and emissions production. *Renew. Sustain. Energy Rev.* **2013**, *18*, 211–245. [CrossRef]
41. Gokdogan, O.; Eryilmaz, T.; Yesilyurt, M.K. Thermophysical properties of castor oil (*Ricinus communis* L.) biodıesel and its blends. *CT&F* **2015**, *6*, 95–128.
42. Akanaw, T.T.; Moges, H.G.; Babu, R.; Bisrat, D. Castor Seed from Melkasa Agricultural Research Centre, East Showa, Ethiopia and it's biodiesel performance in Four Stroke Diesel Engine. *Int. J. Renew. Energy Dev.* **2014**, *3*, 3–99. [CrossRef]

43. Mobarak, H.M.; Mohamad, E.N.; Masjuki, H.H.; Kalam, M.A.; Mahmud, K.A.H.A.; Habibullah, M.; Ashraful, A.M. The prospects of biolubricants as alternatives in automotive applications. *Renew. Sustain. Energy Rev.* **2014**, *33*, 34–43. [CrossRef]
44. Posadillo, A. ptimization of Experimental Conditions for the Production of Second Generation Biofuels that Integrate Glycerol. Universidad de Córdoba, 2016. Available online: https://helvia.uco.es/handle/10396/13237 (accessed on 1 October 2015).
45. Lee, W.J.; Liu, Y.-C.; Mwangi, F.K.; Chen, W.H.; Lin, S.-L.; Fukushima, Y.; Liao, C.-N.; Wang, L.C. Assessment of energy performance and air pollutant emissions in a diesel engine generator fueled with water-containing ethanol–biodiesel–diesel blend of fuels. *Energy* **2011**, *36*, 5591–5599. [CrossRef]
46. Chaves, L.I.; de Souza, S.N.M.; Rosa, H.A.; Bariccatti, R.A.; Nogueira, C.E.C.; Secco, D.; Wazilewski, W.T.; Avaci, A.B.; Brenneisen, P.J.; da Silva, M.J. Comparative performance of a diesel cycle engine using blends of diesel and biodiesel. *Afr. J. Biotechnol.* **2012**, *11*, 14578–14585. [CrossRef]
47. Blanco, M.; Coello, J.; Maspoch, S.; Puigdomènech, A.; Peralta, X.; González, J.M.; Torres, J. Correlating Bacharach opacity in fuel oil exhaust. prediction of the operating parameters that reduce it. *Oil Gas Sci. Technol. IFP* **2000**, *55*, 533–541.

catalysts

MDPI

Article

Biodiesel Production (FAEEs) by Heterogeneous Combi-Lipase Biocatalysts Using Wet Extracted Lipids from Microalgae

Alejandra Sánchez-Bayo [1], Victoria Morales [2], Rosalía Rodríguez [1], Gemma Vicente [1] and Luis Fernando Bautista [2,*]

1 Department of Chemical, Energy and Mechanical Technology, ESCET, Universidad Rey Juan Carlos, Móstoles, 28933 Madrid, Spain; alejandra.sanchezbayo@urjc.es (A.S.-B.); rosalia.rodriguez@urjc.es (R.R.); gemma.vicente@urjc.es (G.V.)

2 Department of Chemical and Environmental Technology, ESCET, Universidad Rey Juan Carlos, Móstoles, 28933 Madrid, Spain; victoria.morales@urjc.es

* Correspondence: fernando.bautista@urjc.es; Tel.: +34-914-888-501

Received: 31 January 2019; Accepted: 18 March 2019; Published: 25 March 2019

Abstract: The production of fatty acids ethyl esters (FAEEs) to be used as biodiesel from oleaginous microalgae shows great opportunities as an attractive source for the production of renewable fuels without competing with human food. To ensure the economic viability and environmental sustainability of the microbial biomass as a raw material, the integration of its production and transformation into the biorefinery concept is required. In the present work, lipids from wet *Isochrysis galbana* microalga were extracted with ethyl acetate with and without drying the microalgal biomass (dry and wet extraction method, respectively). Then, FAEEs were produced by lipase-catalyzed transesterification and esterification of the extracted lipids with ethanol using lipase B from *Candida antarctica* (CALB) and *Pseudomonas cepacia* (PC) lipase supported on SBA-15 mesoporous silica functionalized with amino groups. The conversion to FAEEs with CALB (97 and 85.5 mol% for dry and wet extraction, respectively) and PC (91 and 87 mol%) biocatalysts reached higher values than those obtained with commercial Novozym 435 (75 and 69.5 mol%). Due to the heterogeneous nature of the composition of microalgae lipids, mixtures with different CALB:PC biocatalyst ratio were used to improve conversion of wet-extracted lipids. The results showed that a 25:75 combi-lipase produced a significantly higher conversion to FAEEs (97.2 mol%) than those produced by each biocatalyst independently from wet-extracted lipids and similar ones than those obtained by each lipase from the dry extraction method. Therefore, that optimized combi-lipase biocatalyst, along with achieving the highest conversion to FAEEs, would allow improving viability of a biorefinery since biodiesel production could be performed without the energy-intensive step of biomass drying.

Keywords: FAEEs; biodiesel; mixed biocatalysts; lipases; microalgae

1. Introduction

Nowadays, we are facing a major energy crisis not only caused by the decline of fossil reserves but also by the problems caused by their use. Therefore, there is a need to look for new sources of cleaner, safer and renewable energy. Currently, the research is focuses in obtaining advanced biofuels to ensure the economic viability and environmental sustainability of biomass as a raw material [1]. Oleaginous microalga are promising species that can accumulate large amount of lipids (>20 wt% of their biomass). They constitute an attractive source for the production of biofuels such as biodiesel without competing with human food [2].

Biodiesel can be generated from the lipids in microalgae by the esterification of their free fatty acids (FFA) and the transesterification of their saponifiable lipids with methanol or ethanol into their corresponding methyl or ethyl esters (FAMEs or FAEEs, respectively) in the presence of a catalyst. Due to its high content of free fatty acids in the lipids extracted from microalgae, the most common way to carry out the biodiesel process is by using acid catalysts. This generates certain disadvantages not only in the process, such as high energy consumption, corrosion of materials and the difficulty of transesterifying triglycerides, but also in the post-reaction treatments, such as the recovery of the catalyst [3]. In recent years, new catalysts based on enzymes (biocatalysts) are being developed to avoid these problems [4], achieving better selectivity and specificity. Immobilized lipases can be an alternative because they are capable of carrying out the transesterification and esterification reactions at lower temperatures energy cost, facilitating the recovery of the catalyst and the purification of glycerol [5]. However, the main problem of biocatalysts is the presence of water in the reaction medium that can cause the hydrolysis of esters. In addition, the use of certain solvents such as methanol can lead to inactivation of the catalyst [6,7].

Due to the high cost of enzymes, the use of heterogeneous biocatalysts is required for the economic viability of the whole process since they can be reused. In addition, the immobilization of enzymes in solid carriers can increase thermal and chemical stability and protects enzyme molecules from denaturation. There are different techniques of enzymatic immobilization: binding to the support, confinement or encapsulation and cross-linking [8]. The most common technique is to bind the enzyme onto the support by covalent or ionic attachment or by physical adsorption [9]. To carry out this immobilization, materials such as mesoporous silicas (SBA-15, SBA-16, MCM-41, FDU-12) [4,10–13] or carbon nanotubes [14,15] have been used. The above materials have large pores or cavities where the enzymes can be housed. However, it is usually necessary to modify the surface of support to achieve a better anchorage by functionalization with amine, chloride, sulfur or phenol groups [10,13,16–19].

Because of the high specificity of lipases for different substrates, catalytic activity of lipases depends on the source organism producing the enzyme. Most of biodiesel production from microalgal oil has been done with lipases from *Pseudomonas cepacia* [15,20], *P. fluorescens* [15,21], *Thermomyces lanuginosus* [15,22], *Candida rugosa* [15,23] and *C. antarctica* [4,15]. The last one is frequently used in the form of the commercial catalyst Novozym 435 [6,15,24,25]. Although most of studies reported the use of single lipases, some authors showed that the yield of FAEE could increase when some lipases are used in combination [5,20]. Therefore, the concept of combi-lipase biocatalysts have been recently developed to improve the biodiesel production processes [26–28]. However, these studies reported the lipase-catalyzed synthesis of biodiesel using vegetable oils such as palm, soybean or spent coffee grounds, but they did not use the more complex microalgal oil as raw material for biodiesel production.

In this work, biodiesel (FAEEs) were produced using enzyme catalysis from lipids extracted with ethyl acetate from both wet and dry biomass from microalgae. For this purpose, two enzymes from different origins were assessed, namely lipase B from the fungus *Candida antarctica* (CalB) and lipase from the bacterium *Pseudomonas cepacia* (PC) which were supported on amino-functionalized SBA-15. The results were compared to those obtained with the commercial Novozym 435® catalyst. In addition, both enzymes were combined in different proportions to evaluate the synergistic activity and selectivity towards biodiesel production.

2. Results and Discussions

2.1. Characterization of the Synthesized Enzymatic Catalyst

This section contains the most relevant results of the synthesis of the mesoporous silica material SBA-15 as well as the modifications thereof when incorporating amino groups, glutaraldehyde and the enzymes.

The textural properties are shown in Table 1. The pore surface and diameter were reduced slightly after functionalization of the surface with amino groups and further glutaraldehyde linkage as spacer-arm to favor the subsequent binding of the enzyme. Before functionalization, pore size of SBA-15 was 59.6 Å, and the surface area was the largest, 847.3 m^2/g, and both parameters decreased to

56.3 Å and 841.8 m^2/g, respectively, as chain length increased after amine group and glutaraldehyde introduction [29]. However, pore volume and surface area were drastically reduced by ~70% with respect to the original SBA-15 after the introduction of enzymes.

Table 1. Textural properties and enzyme fixed with synthesized mesoporous materials.

Material	S_{BET} (m^2/g) [1]	D_P (Å) [2]	V_P (cm^3/g) [3]	Protein/Material (mg/g)
SBA-15	847.3	59.6	1.307	-
SBA-15-NH_2	843.2	58.5	1.233	-
SBA-15-NH_2G	841.8	56.3	1.186	-
SBA-15-NH_2G-PC	275.4	55.0	0.378	35
SBA-15-NH_2G-CalB	262.3	56.2	0.393	33

[1] BET surface area, [2] BJH pore diameter, [3] Pore volume at $P/P_0 = 0.95$.

The amount of enzyme loaded in each biocatalyst was determined by mass balance measuring the enzyme concentration in the supernatant solution using the Bradford's assay. The immobilization values were 35 and 33 mg/g for SBA-15-NH_2G-PC and SBA-15-NH_2G-CalB, respectively, similar to those obtained by Bautista et al. [4] (36.1 mg/g) and slightly lower than that obtained by Serra et al. [30] (44 mg/g), both using CalB as enzyme. The slightly higher incorporation of enzyme in the case of PC may be due to the fact that its size is somewhat smaller (3 × 3.2 × 6.6 nm) [31] and can enter into the pores more easily than CalB (3 × 4 × 6 nm) [32].

Regarding the structure, XRD (Figure 1) shows three reflection peaks at $2\theta = 0.97°$, 1.6° and 1.9° corresponding to *p6mm* mesoporous hexagonal symmetry (planes (100), (110) and (200), respectively) [33]. Based on the results obtained, the immobilization of enzymes covalently bound to the material did not modify its structure [10].

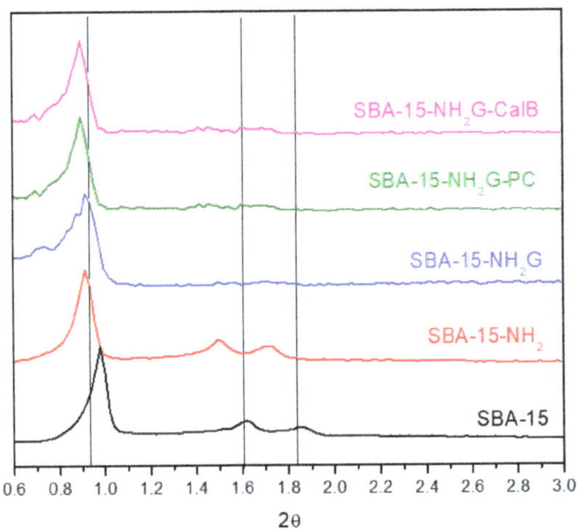

Figure 1. X-ray diffractograms of the mesoporous materials.

Figure 2 shows TEM images obtained from SBA-15 (a) and modified material with anchored lipases from PC (b) and CalB (c). It is clearly noted that both samples show the arrays of long-range mesopore channels, which are similar to the image of SBA-15 reported in the literature [19]. When the SBA-15 contains enzymes partially occupying the interior of its pores, a morphology showing the different degrees of sharpness of the channels, similar to that described by Mohammadi et al. [34], can be observed. However, as explained by Abdullah et al. [11], the fact that the modification of the surface of the SBA-15 material was made by grafting a posteriori prevents major structural changes from taking place and only intervening in the formation of new links with Si-O-Si that are on the surface.

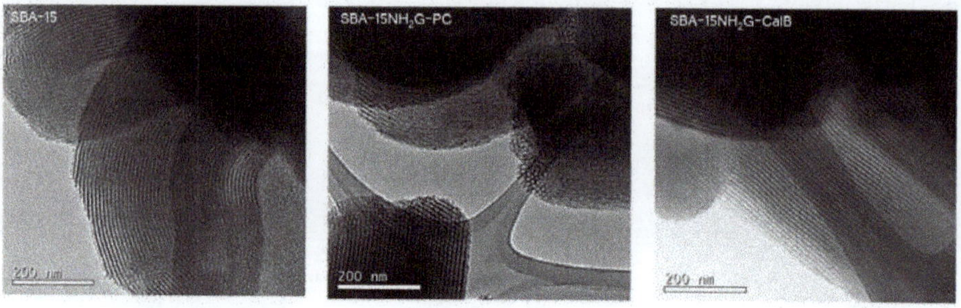

(a) **(b)** **(c)**

Figure 2. TEM micrographs: (**a**) SBA-15, (**b**) SBA-15-NH_2G-PC, (**c**) SBA-15-NH_2G-CalB.

The ^{29}Si MAS-NMR spectrum (Figure 3) of pure silica shows one wide signal within the range −90 to −110 ppm that correspond to the $Si(OSi)_4$ (Q^4), $HOSi(OSi)_3$ (Q^3) and $(HO)_2Si(OSi)_2$ (Q^2) sites of the silica framework [35]. Around a displacement −60 to −78 ppm, a broad signal can be seen that corresponds to two signals created by the Si-C links corresponding to $RSi(OSi)_3$ (T^3) and $RSi(HO)(OSi)_2$ (T^2) [36]. These last two signals are due to the incorporation of the organic part ($CH_3CH_2CH_2NH_2$) by linking on the surface of the material. The peak areas were calculated after Gaussian deconvolution of the spectra considering the associated species and the respective intensities. The results of the integrations (Table 2) show that the greatest contribution to the areas is given by the peak corresponding to a Si atom linked to four Si-O structures, i.e., corresponding to the Q^4 signal. Overlapped with this peak, two signals of smaller area corresponding to the formation of one or two Si-OH bonds, signals Q^3 and Q^2 respectively. When the amino group was incorporated by grafting on the SBA-15 surface by using 3-aminopropyltriethoxysilane (APTES), the material SBA-15-NH_2 was formed, showing a significant increase in T signals (from 0% to ~10% for both T^2 and T^3) and a concomitant decrease in Q signals, which evidenced the formation of Si-C bonds. However, the addition of glutaraldehyde (SBA-15-NH_2G) and CalB or PC lipases (SBA-15-NH_2G-CalB and SBA-15-NH_2G-PC, respectively) did not modified the peak contribution of T and Q signals significantly, proving that these incorporations are according to the immobilization scheme shown in Figure 4, where the silica framework was not affected thereafter. These results confirm that the structure is not modified when both organic modifiers and lipases are introduced [37].

Table 2. Relative peak area in the ^{29}Si NMR spectra.

Material	Q^4 (%)	Q^3 (%)	Q^2 (%)	T^3 (%)	T^2 (%)
SBA-15	83.8	13.4	2.77	-	-
SBA-15-NH_2	66.0	10.2	3.42	10.32	10.23
SBA-15-NH_2G	67.3	9.27	3.21	11.26	9.87
SBA-15-NH_2G-PC	68.5	7.9	2.8	10.76	9.96
SBA-15-NH_2G-CalB	63.9	7.8	2.49	11.87	12.21

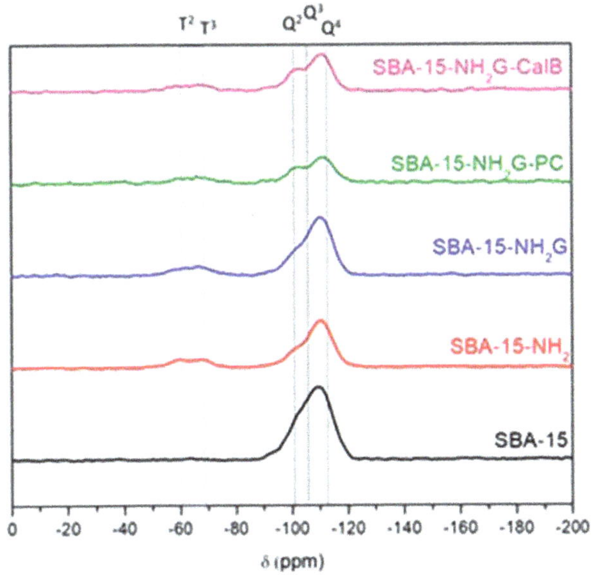

Figure 3. Solid state ^{29}Si MAS-NMR spectra of the different materials.

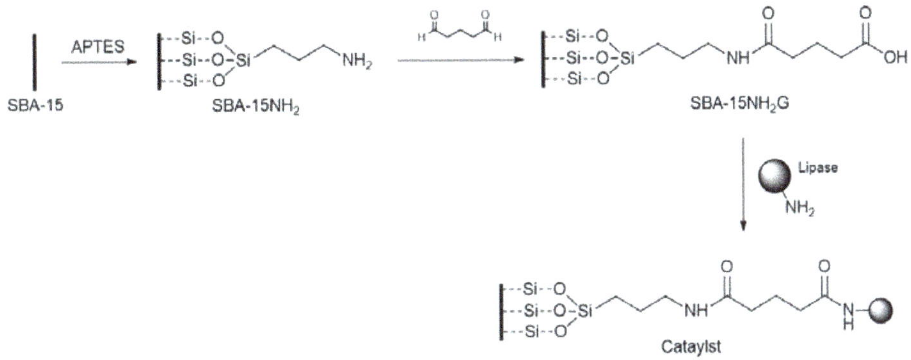

Figure 4. Synthesis of the enzymatic catalyst supported on SBA-15.

The elemental analysis shows how the composition of hydrogen, carbon, nitrogen and oxygen varies as new functional groups, amino and aldehyde are introduced into the structure, as well as with the incorporation of enzymes. The results shown in Table 3 are coherent since the initial mesoporous structure contains only silicon and hydrogen. The presence of nitrogen, carbon and oxygen are due to the incorporation of the different functional groups in the SBA-15. The introduction of the amino group from APTES increases the nitrogen concentration to ~2.8%, but the incorporation of glutaraldehyde decreases the nitrogen content to 2.2%, due to the addition of carbon, hydrogen and oxygen into the structure. Similar results were reported previously for different functionalization of SBA-15 materials [18]. Conversely, an increase in all the elements after lipase immobilization was observed in SBA-15NH$_2$G-PC and SBA-15NH$_2$G-CalB because the incorporation of the lipase molecules added N, C and H, as expected.

Table 3. Elemental analysis of synthesized mesoporous materials.

Material	N (%)	C (%)	H (%)	S (%)
SBA-15	0.00 ± 0.00	0.00 ± 0.00	0.19 ± 0.01	n.d.
SBA-15NH$_2$	2.77 ± 0.01	8.01 ± 0.04	1.97 ± 0.02	n.d.
SBA-15NH$_2$G	2.2 ± 0.1	15.1 ± 0.5	2.23 ± 0.09	n.d.
SBA-15NH$_2$G-PC	2.59 ± 0.02	16.8 ± 0.2	2.71 ± 0.04	n.d.
SBA-15NH$_2$G-CalB	2.49 ± 0.01	21.07 ± 0.02	3.28 ± 0.02	0.01 ± 0.01

n.d: not detected (lower than the detection limit).

After water removal at T < 100 °C, thermogravimetric analysis shows a weight loss between 200 and 300 °C that corresponds to glutaraldehyde followed by the loss of the amino group at temperatures between 300 and 600 °C (Figure 5). The results of this analysis show an organic loss, corresponding to propylamine ($CH_3CH_2CH_2NH_2$), of 11.8 wt% in all the materials. The nitrogen content present in the material SBA-15-NH$_2$ according to this test (2.8 wt%) corroborates the results of elemental analysis (Table 3).

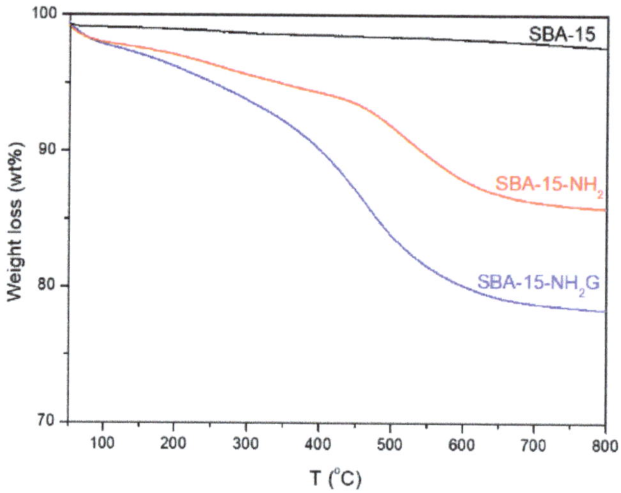

Figure 5. Thermogravimetric analysis of SBA-15 and functionalized materials.

A weight loss of 7.48% is observed in the material SBA-15-NH$_2$G, which implies an incorporation of glutaraldehyde of 37 wt% with respect to the amount of available amino groups. This indicates that part of amino groups incorporated into the material did not form bonds with glutaraldehyde. In addition, this test allowed to corroborate the results of nitrogen content (2.35 wt%) with that obtained by elemental analysis (2.23 wt%).

2.2. Production of FAEEs Using Single Lipase Biocatalysts

The activity of both synthesized biocatalysts (SBA-15-NH$_2$G-CalB and SBA-15-NH$_2$G-PC) towards FAEEs production from wet and dry extracted lipids from *I. galbana* was evaluated and compared with the commercial Novozym 435. The results were assessed based on the molar conversion of saponifiable lipids measured by ^{1}H NMR (Table 4). As it can be seen, the lipid extraction method had a large impact on lipase-catalyzed conversion to FAEEs. When the microalgal oil, used as feedstock, was extracted by the wet method, the results showed a reduction in FAEEs production from 97 to 85.5 mol% and from 91 to 87 mol% using CalB and PC lipase-based biocatalysts, respectively. The commercial Novozym 435 also showed a negative effect when lipids extracted by the wet method were used. Microalgal oils from both the wet and dry extraction methods did not contain water. Therefore, the decrease in conversion may be caused by the extraction of water-soluble lipase-inhibitor compounds during the wet route that are not present in the oil extracted from dry biomass, because lipase activity was negatively affected by polar compounds causing inhibition or denaturation [38].

Table 4. Conversion to FAEEs using single lipases.

Lipid Extraction Method	Catalyst	Conversion (mol%)
Dry	CalB	97 ± 1
	PC	91 ± 2
	N435	75 ± 2
Wet	CalB	85.5 ± 0.8
	PC	87 ± 2
	N435	69.5 ± 0.6

The highest conversion to FAEEs (97 mol%) using dry extracted lipids with ethyl acetate from *I. galbana* microalgae was achieved using SBA-15-NH_2G-CalB as biocatalyst. For the wet extraction route, both CalB and PC lipase-based biocatalysts showed similar activity (85.5 and 87 mol%, respectively) so that SBA-15-NH_2G-CalB seems to be more sensitive to the presence of water or other polar compounds likely extracted during the wet process. However, in all cases, the synthesized biocatalysts SBA-15-NH_2G-CalB and SBA-15-NH_2G-PC were significantly more active towards FAEEs production than the commercial Novozym 435, proving the better performance of both biocatalysts synthesized in the present work.

2.3. *Production of FAEEs Using Combi-Lipase Biocatalysts*

In recent years, the search for new biocatalysts calls for the combination of different enzymes of different specificity to produce a higher production yield of FAEEs [26,27]. Consequently, a biocatalyst formed by the combination in different proportions of CalB and PC supported on SBA-15 modified with amino groups and glutaraldehyde is used in order to evaluate the FAEEs production in the same conditions used previously, using the lipids extracted by wet route with ethyl acetate as raw material.

Table 5 shows the synergistic effect of different combi-lipases on the conversion to FAEEs, along with those values achieves with the corresponding single-lipase biocatalysts It is observed that the combi-lipase 25:75 produced the highest conversion to FAEEs (97.2 mol%) which is more than 10% higher than those produced by the biocatalysts formed by CalB or PC alone. This is a remarkable result because it demonstrates that it is possible to design a catalyst containing an optimized mixture of different lipases that maximizes the conversion to FAEEs.

Table 5. Conversion to FAEEs using combi-lipases.

Calb:PC Lipase Ratio	Conversion (mol%)
0:100	87 ± 2
25:75	97.2 ± 0.5
50:50	81 ± 2
75:25	89.7 ± 0.3
100:0	85.0 ± 0.8

The fatty acid profile of FAEEs produced by the optimized 25:75 SBA-15-NH_2G-CalB:SBA-15-NH_2G-PC is shown in Table 6. The major saturated fatty acids are myristic (C14:0) and palmitic (C16:0), while palmitoleic (C16:1) and oleic (C18:1) acids represented the main monounsaturated and linolenic (C18:3) the most abundant polyunsaturated fatty acid. It is important to highlight the value of linolenic acid (20.2%), a concentration about 10% higher than that regulated by EN 14,214 standard (Table 7). This would require further actions after biodiesel production to meet the required specifications, such as mixing with biodiesel from other lipid sources whose linolenic acid content is lower than that of the microalgae used in this work.

Table 6. Fatty acid profile of FAEEs obtained with combi-lipase 25:75 SBA-15-NH_2G-CalB:SBA-15-NH_2G-PC.

Fatty Acid		Composition (%)
Myristic	C14:0	9.0
Palmitic	C16:0	11.3
Palmitoleic	C16:1	3.6
Stearic	C18:0	0.6
Oleic	C18:1	14.5
Linoleic	C18:2	5.1
Linolenic	C18:3	20.2
Arachidic	C20:0	0.5
Behenic	C22:0	18.2
Erucic	C22:1	0.2
Lignoceric	C24:0	16.8
Saturated		56.3
Monounsaturated		18.3
Polyunsaturated		25.4

Table 7. Biodiesel properties obtained by the combi-lipase 25:75 SBA-15-NH_2G-CalB:SBA-15-NH_2G-PC.

Property	EN 14214	Combi-Lipase 25:75
Iodine value (g I_2/100 g)	<120	72
Group I metals (Na, K) (mg/kg)	<5	179.1
Group II metals (Ca, Mg) (mg/kg)	<5	205.1
Phosphorus (mg/kg)	<10.0	250.6
Sulfur (mg/kg)	<10	0
Monoglycerides (%)	<0.7	0.6
Diglycerides (%)	<0.2	<0.1
Triglycerides (%)	<0.2	<0.1
Linolenic esters (%)	<12	20.2

Other key properties of the biodiesel produced by the combi-lipase biocatalyst 25:75 SBA-15-NH_2G-CalB:SBA-15-NH_2G-PC using the lipids obtained after wet extraction with ethyl acetate are shown in Table 7. Iodine value as well as mono, di and triglyceride content and sulfur content fulfill the European standard EN 14214. However, metals and phosphorus content are higher than those regulated by the above standard, which would require a further purification stage.

2.4. *Study of the Reuse of the Heterogeneous Enzymatic Catalyst*

As previously mentioned, one of the key features of heterogeneous catalysts for their economic viability is their possible recovery and reuse in order to reduce the costs of the biodiesel production process [39]. Therefore, after each reaction, the catalyst was washed with 3 mL of ethanol and dried in order to remove possible impurities that could affect the conversion [40]. The combi-lipase 25:75 SBA-15-NH_2G-CalB:SBA-15-NH_2G-PC was reused for 10 consecutive cycles under the same operating conditions (Figure 6).

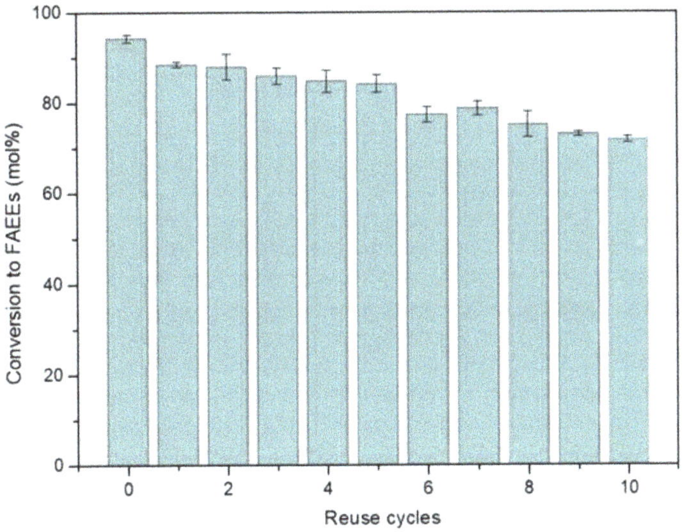

Figure 6. Reuse of combi-lipase 25:75 SBA-15-NH2G-CalB:SBA-15-NH2G-PC.

The results showed that the conversion was above 80 mol% during the first five cycles, decreasing by 10% with respect to the first reaction. Subsequently, the conversion gradually decreased, reaching a conversion of around 70 mol% after the tenth reuse These results are comparable although better than others reported in the literature for the lipase from *P. cepacia* where after the first use, 10% of the activity was lost and in the fourth cycle the activity was reduced by 53% [41]. This loss of activity can be caused by a possible deterioration of the biocatalyst by enzyme poisoning [42].

3. Materials and Methods

3.1. Microalga

The microalga used in this work was *Isochrysis galbana* and it was supplied by AlgaEnergy S.A. (Madrid, Spain). This strain has been selected due to its high lipid content (33.5 wt%), which makes it favorable for the production of FAEEs.

3.2. Lipid Extraction

The extraction of lipids from the microalga, in the dry biomass procedure, was carried out with a dry microalga:solvent ratio of 1:20, using ethyl acetate as solvent. The samples were stirred vigorously with the help of a vortex (Finepcr, Gunpo-si, South Korea) for 5 min at 11,000 rpm at room temperature to obtain the lipids contained in the microalgae. In the case of the extraction of lipids from wet biomass, the same procedure was carried out starting from a suspension of microalga in water with a concentration of 50 g/L. Lipid yields using ethyl acetate were 17.9 $\pm$ 0.7 wt% and 17.6 $\pm$ 0.1 wt% from dry and wet microalgal biomass, respectively.

3.3. Synthesis of Lipase-Based Biocatalysts

The production of FAEEs was carried out using lipase B from *Candida antarctica* (CalB) and lipase from *Pseudomonas cepacia* (PC), both from Sigma-Aldrich (St. Louis, MO, USA), supported on SBA-15 modified with amino groups and glutaraldehyde following protocols previously described [43–45] (Figure 4).

In a typical synthesis for SBA-15, 8 g of block copolymer surfactant Pluronic 123 (Sigma-Aldrich. St. Louis, MO, USA) were dissolved at room temperature under stirring in 250 mL of 1.9 M HCl

(Scharlab. Barcelona, Spain). The solution was heated up to 40 °C and 17.8 g of TEOS (tetraethyl orthosilicate, Sigma-Aldrich) were added to the solution. The resultant mixture was then stirred at that temperature for 20 h and hydrothermally aged at 110 °C for further 24 h. The template was removed by calcination at 550 °C for 5 h at a heating rate of 1.8 °C/min [44].

The protocol for covalent immobilization of lipases (Figure 4) was adapted from Wang's procedure [36]. Thus, 2 g of calcined mesoporous material SBA-15 were immersed into a solution of 1 g of APTES (3-aminopropyltriethoxysilane, Sigma-Aldrich) in 30 mL of anhydrous toluene (Sigma-Aldrich). The mixture was refluxed at 110 °C for 24 h under inert nitrogen atmosphere. Then, the suspension was filtered and washed three times with anhydrous toluene. The solid was placed in glass vials and vacuum dried for 24 h at 110 °C, yielding the amine-functionalized support named SBA-15-NH_2. Then, 1 g of SBA-15-NH_2 was blended with 1 mL of 25 vol% aqueous glutaraldehyde (Sigma-Aldrich) and 9 mL of 0.1 M phosphate buffer (Sigma-Aldrich. St. Louis, MO, USA) for 2 h. The solid material (SBA-15-NH_2G) was filtered and washed with phosphate buffer.

The functionalized materials were used as carriers for lipase immobilization. 100 mg of amino-functionalized and glutaraldehyde linked mesoporous material was blended with 1 mL of a 2.5 mg/mL stock solution of lipase in 4 mL of 0.1 M phosphate buffer at pH = 7.0 and the mixture was shaken at 200 rpm and 25 °C for 3 h. Then, the material was filtered and washed three times with 5 mL phosphate buffer. The final biocatalysts produced were named as SBA-15-NH_2G-CalB and SBA-15-NH_2G-PC, containing immobilized CalB and PC lipase, respectively.

3.4. Characterization of Biocatalysts

The evaluation of the textural properties of SBA-15 mesoporous materials was performed by N_2 adsorption/desorption isotherms in a Tristar model 3000 equipment (Micromeritics, Norcross, GA, USA). Prior to the analysis, a degasification stage is carried out based on the controlled heating of the samples up to 250 °C and 120 °C for inorganic and organically functionalized samples, respectively, and the use of a nitrogen flow in a SmartPrep type degasser. The specific surface area was calculated by the Brunauer-Emmett-Teller (BET) method, the pore diameter was measured through the Barret-Joyner-Halenda technique (BJH) and pore volume was determined by using the Harkins-Jura technique.

X-ray diffraction experiments were performed in a X'Pert MPD (Philips, Amsterdam, The Netherlands) using monochromatic Cu Kα radiation, with a wavelength of 1.54 Å. The step size was 0.02° with an accumulation time per step of 5 s for a sweep of angles 2θ of 5–50°.

Transmission electron microscopy (TEM) analysis was performed with a Tecnai 20 microscope (Philips, Amsterdam, The Netherlands) with a resolution of 0.27 nm and ±70° inclination of the sample, equipped with an EDX detector.

^{29}Si NMR spectroscopy allow to determine the relationship between condensed and non-condensed silicon species, obtaining structural information about the mesoporous siliceous material [37]. Solid-state ^{29}Si MAS-NMR experiments were performed on a Bruker Avance III/HD 400 MHz spectrometer fitted with a 9.4 T magnetic field. These nuclei resonate at 79.41 MHz. An H/X 7.5 mm MAS probe and ZrO2 rotors spinning at 6 kHz were used. In addition, to allow an accurate quantification of silicon groups' ^{29}Si MAS-NMR spectra using one pulse with spinal-64 ^{1}H decoupling were obtained. The ^{29}Si experiments were performed for 500 scans, $\pi/2$ pulse of 4.0 µs, and 60 s of repetition time. ^{29}Si chemical shifts were externally referenced to tetramethylsilane.

The content of hydrogen, carbon, nitrogen, sulfur and oxygen was measured in a Vario EL III element analyzer (Elementar Analysensysteme GmbH, Langenselbold, Germany) using sulphanilic acid as standard.

To evaluate the thermal stability of the SBA-15 materials, thermogravimetric analysis (TGA) were carried out under inert atmosphere (nitrogen) in an 1100 TGA/DSC1 model (Mettler Toledo. Columbus, OH, USA). The thermal analysis ranged from 40 °C to 800 °C with a ramp of 10 °C/min.

Finally, the amount of enzyme immobilized was calculated by mass balance measuring the enzyme concentration in the supernatant solution using the Bradford's protein assay (Bio-Rad Laboratories, Hercules, CA, USA) with bovine serum albumin (BSA) as standard [46]. All the analyses were performed in triplicate.

3.5. FAEEs Production

FAEE production experiments were carried out in 12 mL glass pressure tubes (Sigma-Aldrich). To carry out the reaction, 0.1 g of extracted lipids were added to the reaction tube together with 15 mg of the synthesized biocatalyst [4]. The reaction mixture was maintained at 50 °C and 300 rpm during 24 h. After the reaction time, the mixture was cooled to room temperature (25 °C) and vacuum filtered to recover the catalyst. The liquid phase containing FAEE was washed twice with 10 mL of water, then with 10 mL of hexane:diethylether (80:20) and, finally, with 10 mL of water again. Then, the solvent was removed in a rotary evaporator (Heidolph Instruments, Schwabach, Germany) and the purified FAEEs were obtained. To carry out a comparative study, the commercial biocatalysts Novozym 435® (N345) (Novozymes A/S. Bagsværd, Denmark) was used following the same protocol. Conversion of saponifiable lipids into FAEEs was measured by ^{1}H-NMR analyses (Figure 7) performed in a Varian Mercury Plus 400 unit (Varian. Palo Alto, CA, USA), following the procedure reported in the literature [47], according to Equation (1)

$$\text{FAEE conversion (mol\%)} = 100 \cdot \frac{A_{\delta 1} - A_{\delta 2}}{A_{\delta 3}} \tag{1}$$

where $A_{\delta 1}$ is the area of the peaks corresponding to glyceryl methylenic hydrogens of the oil together with the hydrogen of the ethoxy group of the ethyl ester appearing within the range 4.10–4.17 ppm, $A_{\delta 2}$ is the area of glyceryl methylenic hydrogens of oil integrated from 4.28 to 4.32 ppm and $A_{\delta 3}$ corresponds to the area of the α-acyl methylenic hydrogens in the oil, in the ethyl esters and in the free fatty acids, between 2.2 and 2.4 ppm.

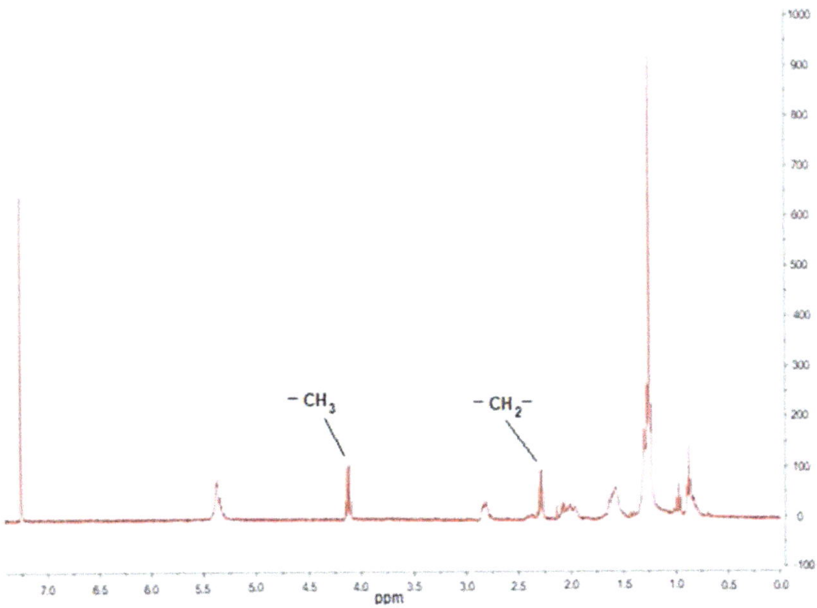

Figure 7. Typical ^{1}H-NMR spectrum of FAEE.

FAEE production reactions were also carried out using combi-lipase biocatalysts with the following SBA15-NH_2G-CalB:SBA15-NH_2G-PC ratios: 25:75, 50:50 and 75:25 wt%. Finally, successive reactions were carried out under the same conditions to assess the reuse capacity.

3.6. FAEE Characterization

The characteristics of the FAEE to be used as biodiesel must be according to EN 14,214 standard, in which the protocol for determining the iodine value of the biodiesel obtained is included. In addition, the content of mono-, di- and triglycerides was determined by thin layer chromatography (TLC) following the method described by Vicente et al. [2], where the sample was diluted with hexane. The TLC plate was developed, stained with iodine and digitized. The software Un-Scan-It Gel 6.1 (Silk Scientific Inc., Orem, UT, USA) was used for the quantification of each lipid fraction using the corresponding standards. The non-saponifiable extracted matter was determined by a gravimetric procedure described elsewhere [48].

Fatty acid profile of FAEE was performed by gas chromatography in a CP-3900 gas chromatograph (Varian) fitted with flame ionization detector and a TRB-FFAP capillary column (60 m length, 0.32 mm I.D., 0.25 μm film thickness. Teknokroma, Barcelona, Spain) following EN 14,103 standard.

The sulfur content was measured by elemental analysis in a Vario EL III elemental analyzer and metals were determined by inductively coupled plasma atomic emission spectroscopy in a Vista AX CCD simultaneous ICP-AES equipment coupled to a spectrophotometer (Varian. Palo Alto, CA, USA) following EN 14,538 (for Na, K, Ca and Mg) and EN 14,107 (for P) standards, respectively.

4. Conclusions

Lipids from *Isochrysis galbana* extracted with ethyl acetate using both dry and wet biomass were used to produce biodiesel (FAEEs) with heterogeneous lipase-based biocatalysts. Lipase B from *Candida antarctica* (CalB) and lipase from *Pseudomonas cepacia* (PC) were covalently immobilized on amino-functionalized SBA-15. Their catalytic activity was compared to that of the commercial catalyst Novozym 435. The conversion to FAEEs using the synthesized biocatalysts SBA-15-NH_2G-CalB (97 and 85.5 mol% from dry and wet extraction, respectively) and SBA-15-NH_2G-PC (91 and 87 mol%) resulted in higher conversions for both dry and wet ones than those obtained with the commercial catalyst Novozym 435 (75 and 69.5 mol%, respectively). Due to the heterogeneous nature of the lipid composition of the microalgae, mixtures with different proportions of CALB:PC biocatalysts were used to improve the conversion of wet extracted lipids. The results showed that a combi-lipase biocatalyst 25:75 SBA-15-NH_2G-CalB:SBA-15-NH_2G-PC produced a conversion of FAEE (97.2 mol%) significantly higher than those from wet extracted oil by biocatalysts containing a single lipase and similar to the maximum conversion using oils from wet extraction. Therefore, the combi-lipase presented can help to improve the viability of a microalgal biorefinery since the biomass-drying step for biodiesel production can be avoided.

Author Contributions: A.S.-B. performed the experimental work (synthesis of biocatalysts, catalytic experiments and sample characterization) and made a first draft of the manuscript. V.M. and R.R. analyzed and discussed the results of biocatalysts characterization and V.M. collaborated in the writing of the first draft of the manuscript. L.F.B. and G.V. devised the experimental work, analyzed the catalytic tests and wrote the final version of the manuscript. The listed authors have contributed substantially to this work.

Funding: The authors acknowledge the support from INSPIRA1 project (S2013/ABI-2783) financed by Comunidad de Madrid and co-financed by FEDER "A way of making Europe"; and BIOHIDROALGA project (ENE2017-83696-R) financed by Ministerio de Ciencia, Innovación y Universidades and co-financed by FEDER "A way of making Europe".

Conflicts of Interest: The authors declare no conflict of interest.

References

1. Demirbas, A. Political, Economic and Environmental Impacts of Biofuels: A Review. *Appl. Energy* **2009**, *86*, S108–S117. [CrossRef]
2. Vicente, G.; Bautista, L.F.; Rodríguez, R.; Gutiérrez, F.J.; Sádaba, I.; Ruiz-Vázquez, R.M.; Torres-Martínez, S.; Garre, V. Biodiesel Production from Biomass of an Oleaginous Fungus. *Biochem. Eng. J.* **2009**, *48*, 22–27. [CrossRef]
3. Carrero, A.; Vicente, G.; Rodríguez, R.; del Peso, G.L.; Santos, C. Synthesis of Fatty Acids Methyl Esters (FAMEs) from *Nannochloropsis gaditana* Microalga Using Heterogeneous Acid Catalysts. *Biochem. Eng. J.* **2015**, *97*, 119–124. [CrossRef]
4. Bautista, L.F.; Vicente, G.; Mendoza, Á.; González, S.; Morales, V. Enzymatic Production of Biodiesel from *Nannochloropsis gaditana* Microalgae Using Immobilized Lipases in Mesoporous Materials. *Energy Fuels* **2015**, *29*, 4981–4989. [CrossRef]
5. Guldhe, A.; Singh, B.; Mutanda, T.; Permaul, K.; Bux, F. Advances in Synthesis of Biodiesel via Enzyme Catalysis: Novel and Sustainable Approaches. *Renew. Sustain. Energy Rev.* **2015**, *41*, 1447–1464. [CrossRef]
6. Navarro López, E.; Robles Medina, A.; Esteban Cerdán, L.; González Moreno, P.A.; Macías Sánchez, M.D.; Molina Grima, E. Fatty Acid Methyl Ester Production from Wet Microalgal Biomass by Lipase-Catalyzed Direct Transesterification. *Biomass Bioenergy* **2016**, *93*, 6–12. [CrossRef]
7. Park, J.-Y.; Park, M.S.; Lee, Y.-C. Advances in Direct Transesterification of Algal Oils from Wet Biomass. *Bioresour. Technol.* **2015**, *184*, 267–275. [CrossRef] [PubMed]
8. Dalla-Vecchia, R.; Nascimento, M.D.G.; Soldi, V. Aplicações Sintéticas de Lipases Imobilizadas Em Polímeros. *Quim. Nova* **2004**, *27*, 623–630. [CrossRef]
9. Hanefeld, U.; Gardossi, L.; Magner, E. Understanding Enzyme Immobilisation. *Chem. Soc. Rev.* **2009**, *38*, 453–468. [CrossRef]
10. Rios, N.S.; Pinheiro, M.P.; dos Santos, J.C.S.; de S. Fonseca, T.; Lima, L.D.; de Mattos, M.C.; Freire, D.M.G.; da Silva, I.J.; Rodríguez-Aguado, E.; Gonçalves, L.R.B. Strategies of Covalent Immobilization of a Recombinant *Candida antarctica* Lipase B on Pore-Expanded SBA-15 and Its Application in the Kinetic Resolution of (R,S)-Phenylethyl Acetate. *J. Mol. Catal. B Enzym.* **2016**, *133*, 246–258. [CrossRef]
11. Abdullah, A.Z.Z.; Sulaiman, N.S.S.; Kamaruddin, A.H.H. Biocatalytic Esterification of Citronellol with Lauric Acid by Immobilized Lipase on Aminopropyl-Grafted Mesoporous SBA-15. *Biochem. Eng. J.* **2009**, *44*, 263–270. [CrossRef]
12. Canilho, N.; Jacoby, J.; Pasc, A.; Carteret, C.; Dupire, F.; Stébé, M.J.; Blin, J.L. Isocyanate-Mediated Covalent Immobilization of *Mucor miehei* Lipase onto SBA-15 for Transesterification Reaction. *Colloids Surf. B Biointerfaces* **2013**, *112*, 139–145. [CrossRef]
13. Zhou, Z.; Piepenbreier, F.; Marthala, V.R.R.; Karbacher, K.; Hartmann, M. Immobilization of Lipase in Cage-Type Mesoporous Organosilicas via Covalent Bonding and Crosslinking. *Catal. Today* **2015**, *243*, 173–183. [CrossRef]
14. Zniszczoł, A.; Herman, A.P.; Szymańska, K.; Mrowiec-Białoń, J.; Walczak, K.Z.; Jarzebski, A.; Boncel, S. Covalently Immobilized Lipase on Aminoalkyl-, Carboxy- and Hydroxy-Multi-Wall Carbon Nanotubes in the Enantioselective Synthesis of Solketal Esters. *Enzym. Microb. Technol.* **2016**, *87–88*, 61–69. [CrossRef]
15. Dhawane, S.H.; Kumar, T.; Halder, G. Recent Advancement and Prospective of Heterogeneous Carbonaceous Catalysts in Chemical and Enzymatic Transformation of Biodiesel. *Energy Convers. Manag.* **2018**, *167*, 176–202. [CrossRef]
16. Rios, N.S.; Pinheiro, M.P.; Lima, M.L.B.; Freire, D.M.G.; da Silva, I.J.; Rodríguez-Castellón, E.; de Sant'Ana, H.B.; Macedo, A.C.; Gonçalves, L.R.B. Pore-Expanded SBA-15 for the Immobilization of a Recombinant *Candida antarctica* Lipase B: Application in Esterification and Hydrolysis as Model Reactions. *Chem. Eng. Res. Des.* **2018**, *129*, 12–24. [CrossRef]
17. Ashjari, M.; Mohammadi, M.; Badri, R. Selective Concentration of Eicosapentaenoic Acid and Docosahexaenoic Acid from Fish Oil with Immobilized/Stabilized Preparations of *Rhizopus oryzae* Lipase. *J. Mol. Catal. B Enzym.* **2015**, *122*, 147–155. [CrossRef]
18. Lee, J.S.; Yim, J.H.; Jeon, J.K.; Ko, Y.S. Polymerization of Olefins with Single-Site Catalyst Anchored on Amine-Functionalized Surface of SBA-15. *Catal. Today* **2012**, *185*, 175–182. [CrossRef]

19. Yiu, H.H.P.; Wright, P.A.; Botting, N.P. Enzyme Immobilisation Using SBA-15 Mesoporous Molecular Sieves with Functionalised Surfaces. *J. Mol. Catal. Enzym.* **2001**, *15*, 81–92. [CrossRef]
20. Tongboriboon, K.; Cheirsilp, B.; H-Kittikun, A. Mixed Lipases for Efficient Enzymatic Synthesis of Biodiesel from Used Palm Oil and Ethanol in a Solvent-Free System. *J. Mol. Catal. B Enzym.* **2010**, *67*, 52–59. [CrossRef]
21. Guldhe, A.; Singh, B.; Rawat, I.; Permaul, K.; Bux, F. Biocatalytic Conversion of Lipids from Microalgae *Scenedesmus obliquus* to Biodiesel Using *Pseudomonas fluorescens* Lipase. *Fuel* **2015**, *147*, 117–124. [CrossRef]
22. He, Y.; Wu, T.; Wang, X.; Chen, B.; Chen, F. Cost-Effective Biodiesel Production from Wet Microalgal Biomass by a Novel Two-Step Enzymatic Process. *Bioresour. Technol.* **2018**, *268*, 583–591. [CrossRef]
23. Guldhe, A.; Singh, P.; Renuka, N.; Bux, F. Biodiesel Synthesis from Wastewater Grown Microalgal Feedstock Using Enzymatic Conversion: A Greener Approach. *Fuel* **2019**, *237*, 1112–1118. [CrossRef]
24. Bauer, G.; Lima, S.; Chenevard, J.; Sugnaux, M.; Fischer, F. Biodiesel via in Situ Wet Microalgae Biotransformation: Zwitter-Type Ionic Liquid Supported Extraction and Transesterification. *ACS Sustain. Chem. Eng.* **2017**, *5*, 1391–1937. [CrossRef]
25. Castillo López, B.; Esteban Cerdán, L.; Robles Medina, A.; Navarro López, E.; Martín Valverde, L.; Hita Peña, E.; González Moreno, P.A.; Molina Grima, E. Production of Biodiesel from Vegetable Oil and Microalgae by Fatty Acid Extraction and Enzymatic Esterification. *J. Biosci. Bioeng.* **2015**, *119*, 706–711. [CrossRef]
26. Banerjee, A.; Singh, V.; Solanki, K.; Mukherjee, J.; Gupta, M. Combi-Protein Coated Microcrystals of Lipases for Production of Biodiesel from Oil from Spent Coffee Grounds. *Sustain. Chem. Process.* **2013**, *1*, 1–14. [CrossRef]
27. Alves, J.S.; Vieira, N.S.; Cunha, A.S.; Silva, A.M.; Záchia Ayub, M.A.; Fernandez-Lafuente, R.; Rodrigues, R.C. Combi-Lipase for Heterogeneous Substrates: A New Approach for Hydrolysis of Soybean Oil Using Mixtures of Biocatalysts. *RSC Adv.* **2014**, *4*, 6863–6868. [CrossRef]
28. Bajaj, A.; Lohan, P.; Jha, P.N.; Mehrotra, R. Biodiesel Production through Lipase Catalyzed Transesterification: An Overview. *J. Mol. Catal. B Enzym.* **2010**, *62*, 9–14. [CrossRef]
29. Lee, H.W.; Cho, H.J.; Yim, J.H.; Kim, J.M.; Jeon, J.K.; Sohn, J.M.; Yoo, K.S.; Kim, S.S.; Park, Y.K. Removal of Cu(II)-Ion over Amine-Functionalized Mesoporous Silica Materials. *J. Ind. Eng. Chem.* **2011**, *17*, 504–509. [CrossRef]
30. Serra, E.; Mayoral, Á.; Sakamoto, Y.; Blanco, R.M.; Díaz, I. Immobilization of Lipase in Ordered Mesoporous Materials: Effect of Textural and Structural Parameters. *Microporous Mesoporous Mater.* **2008**, *114*, 201–213. [CrossRef]
31. Sun, Q.; Fu, C.W.; Aguila, B.; Perman, J.; Wang, S.; Huang, H.Y.; Xiao, F.S.; Ma, S. Pore Environment Control and Enhanced Performance of Enzymes Infiltrated in Covalent Organic Frameworks. *J. Am. Chem. Soc.* **2018**, *140*, 984–992. [CrossRef] [PubMed]
32. Rabbani, G.; Ahmad, E.; Khan, M.V.; Ashraf, M.T.; Bhat, R.; Khan, R.H. Impact of Structural Stability of Cold Adapted *Candida antarctica* Lipase B (CaLB): In Relation to PH, Chemical and Thermal Denaturation. *RSC Adv.* **2015**, *5*, 20115–20131. [CrossRef]
33. Gómez-Orozco, S.Y.; Huirache-Acuña, R.; Pawelec, B.; Fierro, J.L.G.; Rivera-Muñoz, E.M.; Lara-Romero, J.; Alonso-Nuñez, G. Characterizations and HDS Performances of Sulfided NiMoW Catalysts Supported on Mesoporous Titania-Modified SBA-15. *Catal. Today* **2018**, *305*, 152–161. [CrossRef]
34. Mohammadi, M.; Habibi, Z.; Gandomkar, S.; Yousefi, M. A Novel Approach for Bioconjugation of *Rhizomucor miehei* Lipase (RML) onto Amine-Functionalized Supports; Application for Enantioselective Resolution of Rac-Ibuprofen. *Int. J. Biol. Macromol.* **2018**, *67*, 139–146. [CrossRef] [PubMed]
35. Zhao, X.S.; Lu, G.Q.; Whittaker, A.K.; Millar, G.J.; Zhu, H.Y. Comprehensive Study of Surface Chemistry of MCM-41 Using 29Si CP/MAS NMR, FTIR, Pyridine-TPD, and TGA. *J. Phys. Chem. B* **1997**, *33*, 6525–6531. [CrossRef]
36. Wang, C.; Zhou, G.; Li, Y.-J.; Lu, N.; Song, H.; Zhang, L. Biocatalytic Esterification of Caprylic Acid with Caprylic Alcohol by Immobilized Lipase on Amino-Functionalized Mesoporous Silica. *Colloids Surfaces A Physicochem. Eng. Asp.* **2012**, *406*, 75–83. [CrossRef]
37. Simonutti, R.; Comotti, A.; Bracco, S.; Sozzani, P. Surfactant Organization in MCM-41 Mesoporous Materials As Studied by ^{13}C and ^{29}Si Solid-State NMR. *Chem. Mater.* **2001**, *13*, 771–777. [CrossRef]
38. Guncheva, M.; Yancheva, D.; Ossowicz, P.; Janus, E. Structural Basis for the Inactivation of Candida Rugosa Lipase in the Presence of Amino Acid Ionic Liquids. *Bulg. Chem. Commun.* **2017**, *49*, 132–136.

39. Amini, Z.; Ilham, Z.; Ong, H.C.; Mazaheri, H.; Chen, W.-H. State of the Art and Prospective of Lipase-Catalyzed Transesterification Reaction for Biodiesel Production. *Energy Convers. Manag.* **2017**, *141*, 339–353. [CrossRef]
40. Galgali, A.; Gawas, S.D.; Rathod, V.K. Ultrasound Assisted Synthesis of Citronellol Laurate by Using Novozym 435. *Catal. Today* **2018**, *309*, 133–139. [CrossRef]
41. Piligaev, A.V.; Sorokina, K.N.; Samoylova, Y.V.; Parmon, V.N. Lipid Production by Microalga *Micractinium* Sp. IC-76 in a Flat Panel Photobioreactor and Its Transesterification with Cross-Linked Enzyme Aggregates of *Burkholderia Cepacia* Lipase. *Energy Convers. Manag.* **2018**, *156*, 1–9. [CrossRef]
42. Navarro López, E.; Robles Medina, A.; González Moreno, P.A.; Esteban Cerdán, L.; Martín Valverde, L.; Molina Grima, E. Biodiesel Production from *Nannochloropsis Gaditana* Lipids through Transesterification Catalyzed by *Rhizopus Oryzae* Lipase. *Bioresour. Technol.* **2016**, *203*, 236–244. [CrossRef] [PubMed]
43. Lai, J.Q.; Hu, Z.L.; Wang, P.W.; Yang, Z. Enzymatic Production of Microalgal Biodiesel in Ionic Liquid [BMIm][PF_6]. *Fuel* **2012**, *95*, 329–333. [CrossRef]
44. Zhao, D.; Feng, J.; Huo, Q.; Melosh, N.; Glenn, H.; Chmelka, B.F.; Stucky, G.D.; Zhao, D.; Feng, J.; Huo, Q.; et al. Triblock Copolymer Syntheses of Mesoporous Silica with Periodic 50 to 300 Angstrom. *Science* **1998**, *279*, 548–552. [CrossRef]
45. Salis, A.; Pinna, M.; Monduzzi, M.; Solinas, V. Biodiesel Production from Triolein and Short Chain Alcohols through Biocatalysis. *J. Biotechnol.* **2005**, *119*, 291–299. [CrossRef] [PubMed]
46. Bradford, M.M. A Rapid and Sensitive Method for the Quantitation of Microgram Quantities of Protein Utilizing the Principle of Protein-Dye Binding. *Anal. Biochem.* **1976**, *72*, 248–254. [CrossRef]
47. Jaiswal, S.K.; Tejo Prakash, N.; Prakash, R. ^{1}H NMR Based Quantification of Ethyl Ester in Biodiesel: A Comparative Study of Product-Dependent Derivations. *Anal. Chem. Lett.* **2016**, *6*, 518–525. [CrossRef]
48. Panreac Quimica. *Métodos Analíticos En Alimentaria*; Montplet & Esteban: Barcelona, Spain, 1999.

catalysts

MDPI

Article

Use of Co/Fe-Mixed Oxides as Heterogeneous Catalysts in Obtaining Biodiesel

Norma Gutiérrez-Ortega [1,*], Esthela Ramos-Ramírez [2,*], Alma Serafín-Muñoz [1], Adrián Zamorategui-Molina [1] and Jesús Monjaraz-Vallejo [2]

1 Environmental Engineering Laboratory, Department of Civil Engineering, Engineering Division, University of Guanajuato, 36000 Guanajuato, México; sermuah@ugto.mx (A.S.-M.); zamorategui@ugto.mx (A.Z.-M.)

2 Advanced Materials and Processes Laboratory, Department of Chemistry, Division of Natural and Exact Sciences, University of Guanajuato, 36000 Guanajuato, México; jdj.monjarazvallejo@ugto.mx

* Correspondence: normagut@ugto.mx (N.G.-O.); ramosre@ugto.mx (E.R.-R.); Tel.: +52-473-732-00-06 (ext. 2227) (N.G.-O.); +52-473-732-00-06 (ext. 1457) (E.R.-R.)

Received: 1 February 2019; Accepted: 26 April 2019; Published: 29 April 2019

Abstract: Catalyst-type mixed metal oxides with different compositions and Co/Fe ratios were obtained from layered double hydroxides to be used as heterogeneous catalysts in the production of biodiesel. The effect of the Co/Fe ratio on the precursors of the catalysts was analyzed, considering their thermal, textural and structural properties. The physicochemical properties of the catalysts were determined by thermogravimetric analysis (differential scanning calorimetry and thermogravimetric), X-ray diffraction, Fourier-transform infrared spectroscopy, Scanning Electron Microscopy-Energy Dispersive X-ray spectroscopy and N_2-physisorption. The conversion to biodiesel using the different catalysts obtained was determined by diffuse reflectance infrared Fourier-transform spectroscopy and ^{1}H-Nuclear magnetic resonance spectroscopy, allowing us to correlate the effect of the catalyst composition with the catalytic capacity. The conditions for obtaining biodiesel were optimized by selecting the catalyst and varying the percentage of catalyst, the methanol/oil ratio and the reaction time. The catalysts reached yields of conversion to biodiesel of up to 96% in 20 min of reaction using only 2% catalyst. The catalyst that showed the best catalytic activity contains a mixture of predominant crystalline and amorphous phases of $CoFe_2O_4$ and Na_xCoO_2. The results suggest that cobalt is a determinant in the activity of the catalyst when forming active sites in the crystalline network of mixed oxides for the transesterification of triglycerides, with high conversion capacity and selectivity to biodiesel.

Keywords: Cobalt ferrite; layered double hydroxides; ethylesters; biofuels; hydrotalcite; transesterification

1. Introduction

Currently, one of the main challenges facing humanity is to reduce the use of petroleum fuels and increase the production of fuels from low carbon sources [1]. A viable alternative is biofuels, which have the advantages of being easily extractable from biomass, biodegradable, non-toxic and environmentally friendly [2,3]. Biodiesel is a liquid fuel derived from triglycerides of animal, vegetable and even microbial origin [4,5]. Biodiesel can be obtained by a great variety of techniques such as direct mixtures, microemulsions, pyrolysis and transesterification [6]. The transesterification technique is the most used because it is easy to carry out; however, the type of catalyst used is important to ensure that the reaction has direct effects on the efficiency of the process and the quality of the products. The catalysts used can be homogeneous, heterogeneous or enzymatic (Figure 1) [7–9].

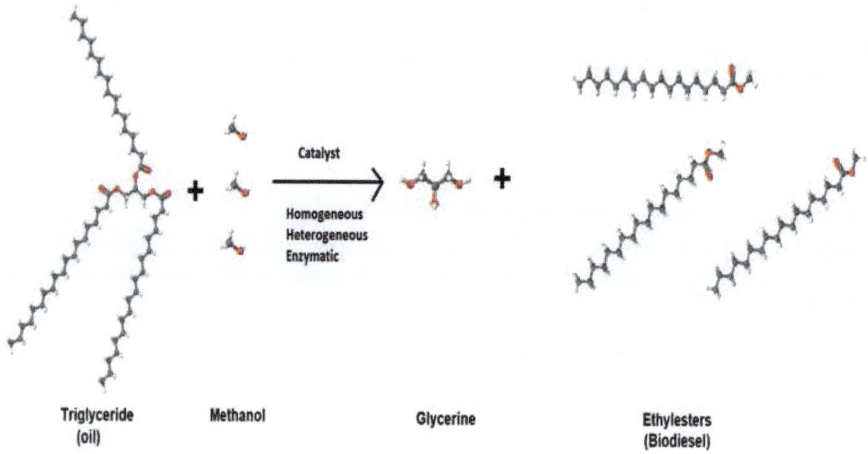

Figure 1. Transesterification reaction for biodiesel production.

Each of the catalytic processes used to obtain biodiesel presents advantages and disadvantages. For example, enzymatic transesterification is considered the most effective and environmentally sustainable method to produce biodiesel; however, its reaction speed is very low, costs are extremely high, and scaling is complicated [10]. Basic homogenous transesterification is easy to operate on an industrial level, has a high catalytic activity and the reaction times are short [11], but has disadvantages, such as that the presence of free fatty acids and water produces soap affecting the entire production process, and that this method generates large volumes of wastewater and therefore has a high environmental impact. On the other hand, heterogeneous catalysts have gained interest in the production of biodiesel, because they are neither consumed nor dissolved in the reaction mixture, which facilitates the separation of the product and the reuse of the catalyst. The main disadvantage of this approach is that it is still in the pilot escape phase and problems of diffusion, immiscibility and mass transfer efficiency are still not resolved at the operational level, which affects the speed of the reaction [12–14].

Recently, the heterogeneous catalysts that have presented greater effectiveness and potential to be used in the production of biodiesel are simple and mixed metal oxides, zeolites, anion exchange resins and carbon-based catalysts, among others [15–20]. Layered double hydroxides (LDH) are a family of natural and synthetic compounds that are characterized by a laminar structure-type hydrotalcite (HT). The sheets are octahedral networks containing metals with divalent and trivalent positive charges that generate a positive residual charge which is neutralized in the presence of anions and water molecules in the interlaminar space. Their molecular formula is $[M^{2+}{}_{(1-x)}M^{3+}{}_{x}(OH)_2][A^{n-}]_{x/n}{\cdot}yH_2O$ [21,22]. Layered double hydroxides can be prepared by different methods such as coprecipitation, hydrolysis with urea, sol–gel, hydrothermal treatment, combustion and mechanical synthesis, among others, with the most common and economic method being coprecipitation [23–25]. Studies have reported that the synthesis method, the conditions of synthesis, the type of cations and structural anions, the metal precursors and the metal ratio M^{2+}/M^{3+} are determining factors on the properties and applications of LDH [26–29]. An important element to improve the catalytic activity of LDH is the thermal treatment to form mixed spinel-type oxides, specifically in the production of biodiesel by heterogeneous catalysis [30,31]. The most relevant studies have shown good results of conversion to biodiesel, but these reactions were found to require a long synthesis time, high temperatures or critical conditions, high percentages of catalyst and large amounts of methanol, among other factors. Table 1 shows examples of investigations in which catalysts of the LDH type have been used, which mainly consist of metals such as Mg, Al, Zn and Fe.

Table 1. Obtaining biodiesel using different layered double hydroxides (LDH)-type catalysts.

Catalyst	Synthesis Conditions	Biodiesel Conversion (%)	References
Mg/Al = 3 HT [1]	Soybean oil:methanol 1:12, 5 wt% of catalyst, at 180 °C, 700 rpm, 180 min, with microwave irradiation	96	Coral, 2019 [30]
Mg/Fe = 5 HT calcined at 500 °C	Rapeseed oil:methanol 1:9, 5% catalyst, 400 rpm at 120 °C for 8 h in Batch autoclave reactor	68	Hajek, 2018 [31]
Comercial Zn/Al = 3.3 HT-KF	Jatropha oil 1:30, 10% catalyst, reflux for 3 h and 30 min in Batch microwave reactor without reporting temperature	50	Fatimah, 2018 [32]
LDH at 300 °C	Oil:alcohol, 50 mg cat:10 mmol oil, 1:30, 65 °C, 3 h	30	Tajuddi, 2018 [33]
Mg/Al = 7 HT at 500 °C	Oil:alcohol 1:48, 6% catalyst, 60 °C, 1 atm, 24 h	92	Navajas, 2018 [34]
Mg/Al = 3 HT at 500 °C	Soybean oil:methanol 1:90, 1000 rpm, 150 °C, 0.4 g cat:0.0045 mol oil, 180 min, autoclave pressure	97	Dhawan, 2018 [35]
Mg/Al = 2 HT nanoparticles at 500 °C	Jatropha oil 1:40, 5%, 3.3–5.5 MPa, 200 °C, 4 h	54	Wang, 2015 [36]
Mg/Al = 3 HT	Soybean oil:methanol 1:14, N_2, 1380 kPa, 115 °C, 4% catalyst, 2 h	95	Prado, 2014 [37]
Mg/Al = 3 HT-10% Fe at 480 °C	Seed oil:methanol 1:6, 1% catalyst, 80 °C, 100 min	96	Macala, 2008 [38]
MgCoAl-LDH at 500 °C	Canola oil: 1:16 methanol, 2% catalyst, 100 °C, 25 atm and 300 min	95	Li, 2009 [39]

[1] HT: hydrotalcite or layered double hydroxide

The above shows a panorama highlighting the need to continue looking for new materials that achieve better efficiencies in the conversion to biodiesel in the shortest possible time and using the smoothest and most controllable reaction conditions.

The modification that has been given to the LDH containing Fe, both in the laminar structure and impregnated, have shown efficient conversion results, but with long reaction times or with systems that require pressure [31,38]. On the other hand, cobalt is a catalytically active element that has been used in many catalytic processes, including transesterification. The use of catalysts with Co content to obtain biodiesel has given interesting results, with the case of the use of materials of type Co(II)-organic compounds in microwave reactors that have reached 80% conversion in 12 h [40]. Other studies have used cobalt supported, for example, in zeolite ZSM5 with in reaction times of 8 h to 95 °C and 80% conversion [41]. Additionally, a study reports the direct esterification of fatty acids to form biodiesel by using a mixture of oxides with Co, Sn and Al with 1:32 oil/methanol ratio, 2% catalyst and 60 °C, reaching 65% conversion in 4 h [42].

The present work focuses on the obtaining of a new catalysts obtained by thermal activation of layered double hydroxides Co/Fe, to be used in the esterification and transesterification reactions to produce biodiesel. The use of Co and Fe as active metals in crystalline networks of the simple and mixed oxides type is proposed, considering mainly that the cobalt is a transition element that can contribute to the formation of active sites for the simultaneous esterification and transesterification of the cooking oil to biodiesel. The effect of the variation of the Co/Fe ratio of the LDH precursor on the physicochemical and catalytic properties of its calcination products at 700 °C for catalytic activation is studied. Complementarily, the best operating conditions for obtaining biodiesel by heterogeneous catalysis were determined, the foregoing to generate a viable alternative for the industrial production of biodiesel under simple operating conditions.

2. Results

2.1. Catalyst Characterization

This section shows the results of the characterization of the catalytic precursors Layered Double Hydroxides (LDH) identified as LDH1, LDH2, LDH3 and LDH4 corresponding to the ratio Co/Fe = 1, 2, 3 and 4 respectively, as well as the products of calcination of LDH identified as MO-LDH1, MO-LDH2, MO-LDH3 and MO-LDH4.

2.1.1. Thermal Evolution of Layered Double Hydroxides (LDH) to Co/Fe-Mixed Oxides (MO-LDH)

Figure 2a shows the thermogravimetric analysis (TGA) of the LDH, corresponding to the different precursors of the mixed Co/Fe oxides. As can be seen, the LDH2 precursor presents a first major weight loss associated with the elimination of water around 200 °C corresponding to 12% of its original weight, while the LDH3 and LDH4 precursors present a weight loss of only 3.5% in that same temperature range. The second weight loss occurs between 600 and 700 °C which for the solid LDH1 amounts to a loss of 22%, LDH2 of 33%, LDH3 of 34.5% and LDH4 of 46.5%, the latter being the highest total loss reached before losing 50% of its mass at 1100 °C. Thermograms by differential scanning calorimetry (DSC) are shown in Figure 2b, where it can be seen that the different Co/Fe precursors synthesized present similar thermal evolutions. The main signals correspond to the dehydration of the molecules housed in the interlaminar at 300 °C and the decarbonation and dehydroxylation at 680 °C associated with the weight loss described in the TGA, which forms the crystalline and amorphous phases of the simple and mixed oxides. In the case of the LDH3 and LDH4 solids, an additional reaction can be seen at 998 °C associated with the sintering of the spinel of the cobalt ferrite ($CoFe_2O_4$).

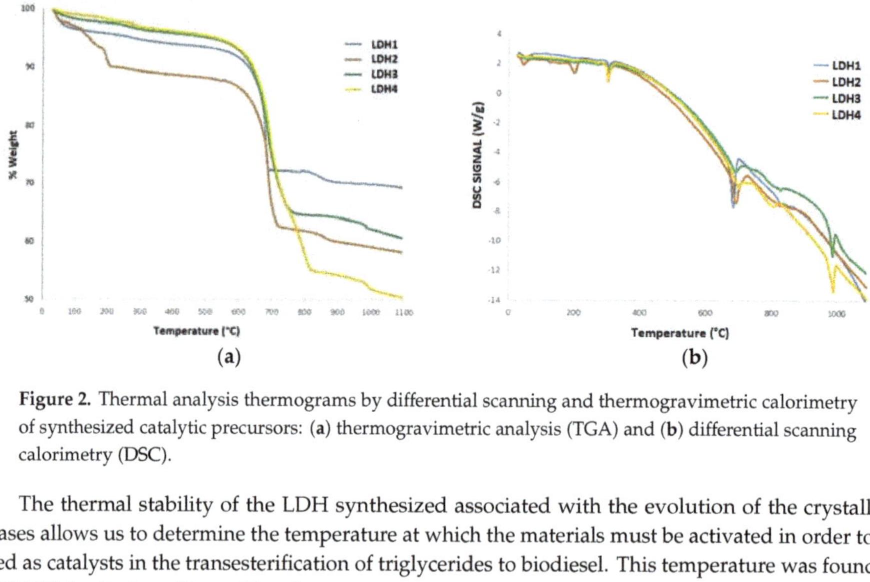

(a) (b)

Figure 2. Thermal analysis thermograms by differential scanning and thermogravimetric calorimetry of synthesized catalytic precursors: (**a**) thermogravimetric analysis (TGA) and (**b**) differential scanning calorimetry (DSC).

The thermal stability of the LDH synthesized associated with the evolution of the crystalline phases allows us to determine the temperature at which the materials must be activated in order to be used as catalysts in the transesterification of triglycerides to biodiesel. This temperature was found to be 700 °C for the transition of the phases.

2.1.2. Modification of Functional Groups by Thermal Activation of the Catalyst

Figure 3 shows the FTIR spectra of the LDHs prior to activation (Figure 3a) compared to the spectra of the catalysts obtained by calcination at 700 °C (Figure 3b). As can be seen, the thermal decomposition profiles are correlated to the activation temperature of the catalysts. A broad band of low intensity close to 3300 cm^{-1} can be observed in all spectra of Figure 3b, which is associated with hydroxyl groups, where in the case of the catalysts MO-LDH1 and MO-LDH2 it is a double band associated with the interaction of hydroxyl and carbonates that could be reabsorbed on the surface given the precursor reconstruction effect. This signal attributed to the hydroxyl groups is of low intensity for the catalysts MO-LDH3 and MO-LDH4. With respect to the presence of carbonate groups, these signals are observed between 1200 and 1600 cm^{-1}. In the case of MO-LDH1 the band belonging to carbonates is not observed, for MO-LDH2 most of the carbonates are interlaminar, in the MO-LDH3 the carbonates are free, that is, they are adsorbed on the surface and the MO-LDH4 carbonates are not observed. For the case of the modifications in the interlaminar bonds of the metals when evolving from hydroxides to oxides, these signals can be identified in the region between 400 and 1000 cm^{-1}. For the catalyst MO-LDH3 and MO-LDH4, more intense bands can be observed that belong to the Fe–O, Co–O and Fe–O–Co bonds, these signals are characteristic of an infrared spectrum of the cobalt ferrite. which are less intense for the catalysts MO-LDH1 and MO-LDH2 given the over positioning some signals of simple and mixed oxides.

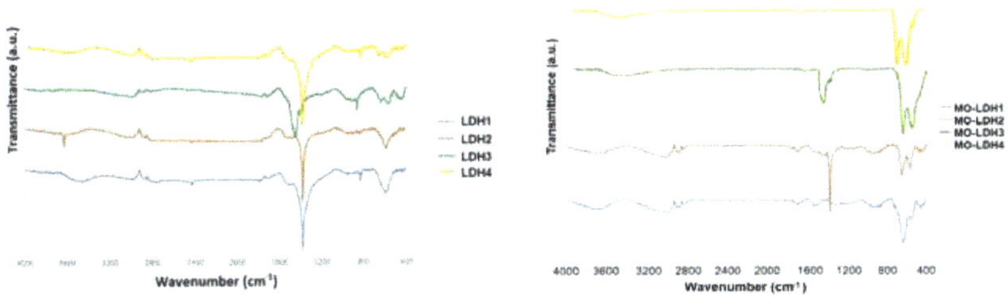

(**a**) (**b**)

Figure 3. Spectra of infrared spectroscopy of the catalysts: (**a**) LDH synthesized and (**b**) LDH calcined at 700 °C.

2.1.3. Identification of Crystalline Phases Present in the Catalysts

Figure 4 shows the X-ray diffraction patterns of the catalysts obtained by calcination at 700 °C. The solid MO-LDH1 does not present the crystalline phase of $CoFe_2O_4$ (cobalt ferrite) but many noise signals associated with amorphous phases. This confirms our observations of the infrared spectra, showing that when cobalt ferrite is present it is still amorphous, in addition to the presence of small crystalline phases secreted by Fe_3O_4, CoO and CoO_2. The solid MO-LDH2 can be observed in the presence of the crystalline phase of $CoFe_2O_4$ accompanied by segregated phases of Fe_3O_4, CoO_2 and CoO. In the case of solid MO-LDH3, a better crystallinity of $CoFe_2O_4$ is observed, as evidenced in the FTIR spectra, in addition to the presence of a segregated phase of CoO_2 and $CoNa_xO_2$. In the case of MO-LDH4, the $CoFe_2O_4$ phase is observed, combined with $CoNaxO_2$, CoO_2 and CoO [43,44].

For the solids MO-LDH3 and MO-LDH4, the presence of sodium in the crystalline networks is due to the fact that the amount of iron is less than cobalt, which allows cobalt to interact with sodium at the moment of synthesis in basic medium, favoring the integration of sodium in the crystalline network and modifying the properties of the catalysts.

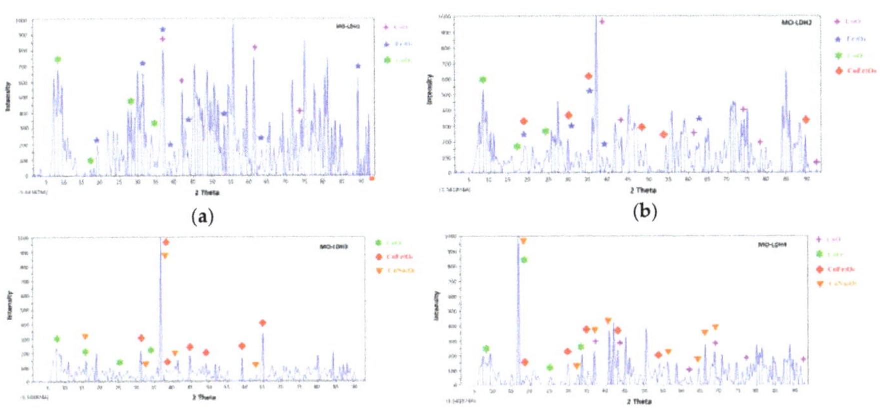

Figure 4. X-ray diffraction patterns of the catalysts obtained at 700 ° C: (**a**) MO-LDH1, (**b**) MO-LDH2, (**c**) MO-LDH3 and (**d**) MO-LDH4.

Given the variation in the composition and segregation of the crystalline phases in the different catalysts obtained from the calcination products at 700 °C of the layered double hydroxides with different Co/Fe molar ratio, the effect of the presence may be analyzed of the different simple and mixed oxides of Co and Fe with respect to the catalytic activity.

2.1.4. Determination of Textural Properties of the Catalysts

Figure 5 shows the physisorption isotherms of N_2 for the catalysts obtained by calcination at 700 °C. The adsorption isotherms for all solids belong to type III, corresponding to physical adsorption in multilayers by a free surface, which is characteristic of macroporous solids (pore size greater than 50 nm). However, a variation in the curl of hysteresis was observed, especially for the solid MO-LDH1. The reason why the solid MO-LDH1 presents a broad curl is associated with an incomplete thermal evolution so that it does not yet collapse the structure of the oxides. The rest of the catalysts show narrow curls, which reflect a collapse of the interlaminar spaces associated with a decrease in the specific area.

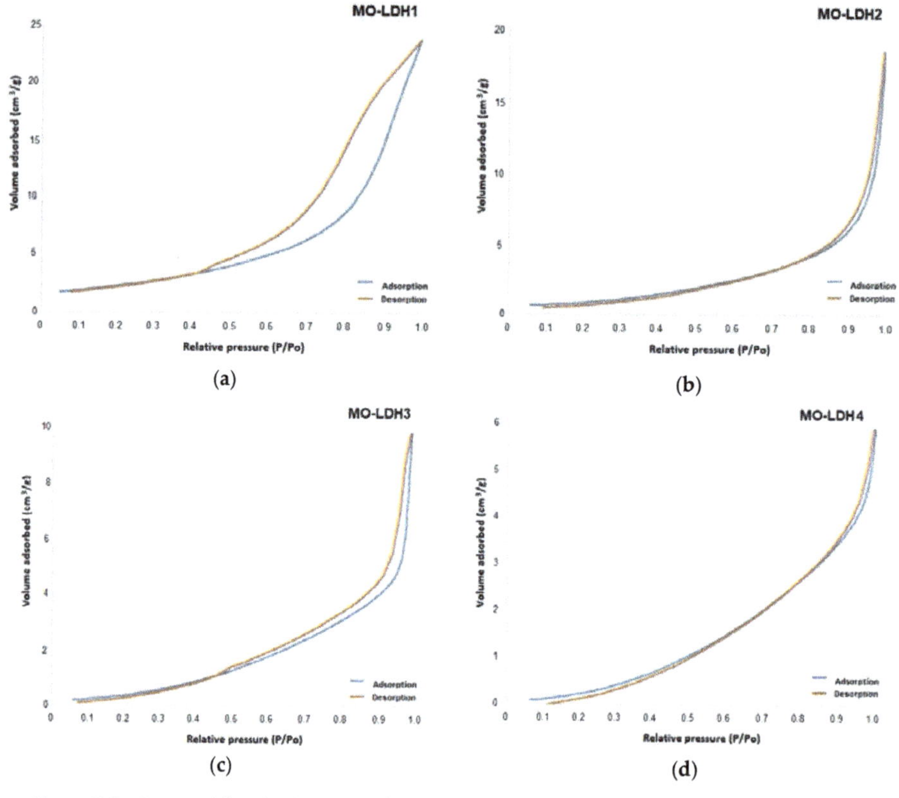

Figure 5. Isotherms of the physisorption of N_2 of the catalysts obtained at 700 °C: (**a**) MO-LDH1, (**b**) MO-LDH2, (**c**) MO-LDH3 and (**d**) MO-LDH4.

Table 2 shows the results of the textural characterization of mixed oxides of different molar ratios: Brunauer, Emmett and Teller (BET)-area, diameter and pore volume.

Table 2. Textural properties of the catalysts obtained at 700 °C.

Catalyst	BET-Area (m^2/g)	Pore Diameter (nm)	Pore Volume (cm^3/g)
MO-LDH1	8.37	83	0.031
MO-LDH2	3.18	38	0.009
MO-LDH3	2.92	54	0.007
MO-LDH4	1.48	61	0.006

As can be seen, the catalysts show a trend of the value of the specific area in correlation with the metal molar ratio Co/Fe and with the thermal decomposition profile of the catalysts. For all cases the areas are less than 10 m^2/g, which suggests the degree of collapse of the laminar structure associated with thermal evolution to simple and mixed metal oxides. In relation to the pore diameter, only the MO-LDH2 catalyst has mesoporous size, and the rest exhibit a larger size corresponding to macropores.

2.1.5. Surface Analysis of the Particles of the Catalysts

Figure 6 shows the photographs obtained by scanning electron microscopy of the catalysts. As can be observed, the solid MO-LDH1 and MO-LDH2 show an amorphous texture with the presence of few hexagonal crystalline structures, which is because the crystallization by thermal evolution has not concluded for all the crystalline phases of the mixed and simple oxides. For the solids MO-LDH3 and MO-LDH4, the presence of hexagonal particles in cumulus and in the form of thin, defined sheets increases markedly. The hexagonal sheets form due to the large amount of mixed oxide Co/Fe present in the different catalysts, mainly in the MO-LDH3 catalyst. In the case of the MO-LDH4 catalyst, very large hexagonal prisms resulting from the ordering of the catalyst sheets can be clearly observed.

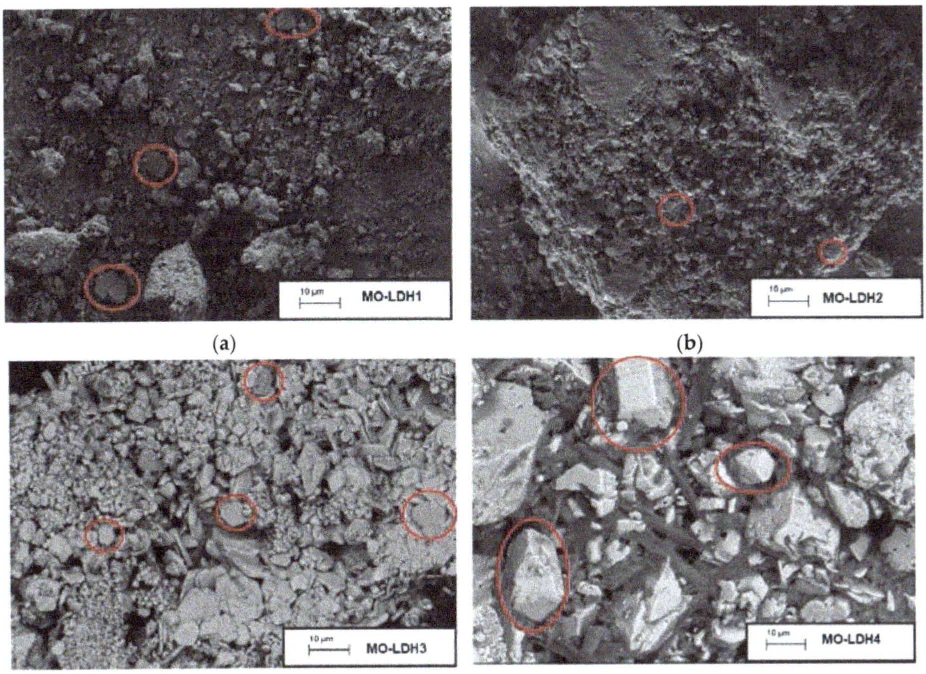

(a) (b) (c) (d)

Figure 6. Scanning microscopy images of the catalysts obtained at 700 °C: (**a**) MO-LDH1, (**b**) MO-LDH2, (**c**) MO-LDH3 and (**d**) MO-LDH4.

2.1.6. Elemental Composition on the Surface of the Particles

Finally, in Figure 7 is shown an elemental analysis performed on the catalysts obtained by calcination at 700 °C. As can be seen, the general composition of the mixed oxides is similar for all the catalysts, which was to be expected because they were synthesized with the same methodology. The variation between the catalysts lies in the variation in the metallic Co/Fe ratio in the synthesis of the catalytic precursors, as well as in the proportion of the simple and mixed metal oxides present in the catalyst. The elements present in the catalysts are: cobalt, iron, sodium, oxygen and carbon, where by correlation with the diffractograms are in the form of metal oxides for the case of cobalt, iron and sodium; whereas the carbon by confirmation of the FTIR spectra corresponds to the CO_2 physisorbed on the surface of the catalysts.

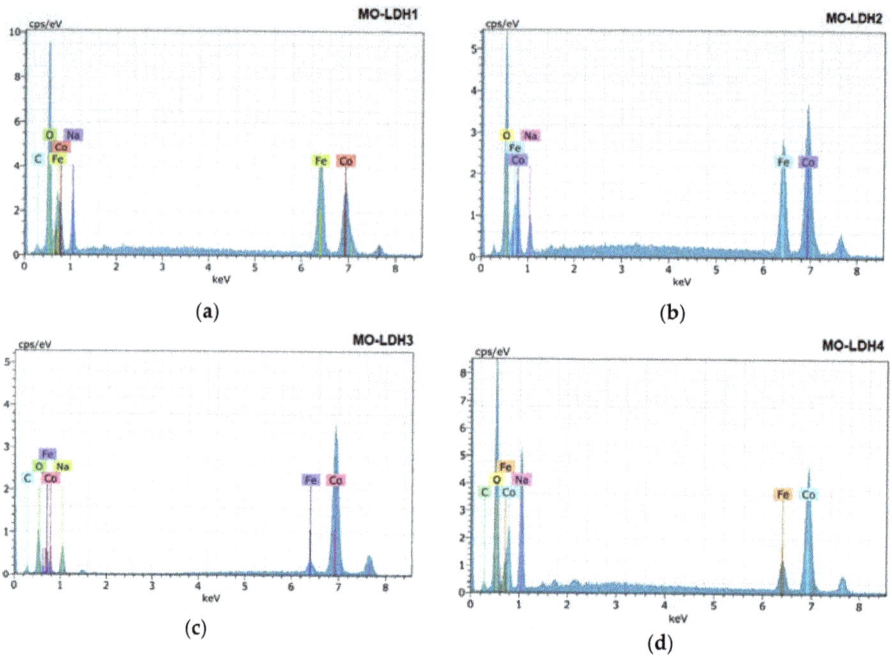

Figure 7. Elemental analysis of the catalysts obtained at 700 °C: (**a**) MO-LDH1, (**b**) MO-LDH2, (**c**) MO-LDH3 and (**d**) MO-LDH4.

Table 3 shows the results of the determination of the Co/Fe molar ratio in the catalytic precursors compared with that of the catalysts.

Table 3. Comparison of the variation of the Co/Fe molar ratio on the surface of the precursor particles and their corresponding mixed oxides.

Sample	Molar Ratio	Sample	Molar Ratio
LDH1	1.04	MO-LDH1	0.85
LDH2	1.85	MO-LDH2	1.94
LDH3	2.45	MO-LDH3	3.93
LDH4	4.10	MO-LDH4	5.63

As can be seen, for the case of catalytic precursors, the theoretical molar ratio of Co/Fe is very close to the values of the molar ratio calculated from the average quantification on the surface of the particles. After the heat treatment, it can be seen that for the case of the MO-LDH1 and MO-LDH2 solids the Co/Fe ratio varies very little, which can be attributed to the homogeneity of the dispersion of metals in the particles. These solids were also homogeneous after thermal evolution. In the case of the MO-LDH3 and MO-LDH4 catalysts, a decrease in the Co/Fe molar ratio is observed, which can be attributed to the fact that the particles are more crystalline with defined edges, which concentrate the iron in the crystalline network, thus decreasing the amount of iron on the surface.

2.2. Evaluation of Biodiesel Quality

The physical and rapid qualitative parameter that was used to determine whether transesterification was performed for the conversion of triglycerides to biodiesel was the kinematic viscosity, starting from the fact that the oil prior to transesterification had a viscosity of 32 cp, while biodiesel with a high percentage of conversion presents values below 6 cp according to ASTMD6751, implying that a decrease in viscosity is associated with the transesterification of the fatty acids present in the oil [45].

Once it was confirmed that the viscosity had been reduced to values close to the norm, we proceeded to the quantification and confirmation of the conversion by diffuse reflectance infrared Fourier-transform spectroscopy (DRIFTS) and ^{1}H-Nuclear magnetic resonance (NMR) spectroscopy [46], as well as the verification of the international parameters of density, corrosion of the copper foil and turbidity. These parameters can be used to verify the feasibility of the products' application as biofuels in internal combustion engines.

Table 4 shows the best conversion percentages obtained in the shortest reaction times tested for obtaining biodiesel using a 3% catalyst, a molar oil/methanol ratio of 1:12, a reaction temperature of 65 °C and constant agitation at 1400 rpm.

Table 4. Quality parameters of biodiesel obtained using the different catalysts.

Catalyst	Time (min)	Viscosity (cp)	Density (kg/m^3)	Acidity Index (mg$_{KOH}$/g)	Corrosion of Copper Foil	Turbidity (°C)	DRIFTS Conversion (%)	^{1}H-NMR Conversion (%)
MO-LDH1	60	6.92 ± 0.3	885 ± 6	0.18	1a	<1	70.4 ± 3.4	63.3
MO-LDH2	60	6.55 ± 0.1	883 ± 5	0.20	1a	<1	80.1 ± 3.4	82.0
MO-LDH3	30	3.71 ± 0.2	881 ± 2	0.19	1b	<1	93.2 ± 3.3	92.7
MO-LDH4	30	3.66 ± 0.1	875 ± 5	0.20	1a	<1	96.2 ± 3.4	96.0

It can be seen that the four catalysts achieved a conversion of triglycerides to biodiesel by lowering the viscosity of the 32 cp oil to values lower than 7 cp. In the case of the MO-LDH1 and MO-LDH2 catalysts, at a reaction time of 30 min, the viscosities were 14.24 and 15.27 cp, respectively, which indicated that the conversion had not yet finished. When the reaction time was extended to 60 min, a decrease in viscosity values close to 7 cp was observed, associated with conversion values greater than 70%. On the other hand, for the case of the MO-LDH3 and MO-LDH4 catalysts, at 30 min they had already reached viscosities below 4 cp and achieved conversions greater than 90%.

Figure 8 shows the ^{1}H-NMR spectra which were used to calculate the % conversion to biodiesel from the M signal corresponding to the methyl group of the biodiesel methylesters that is formed as the reaction is carried out. Associated with the M signal is the G zone corresponding to the monoglycerides, diglycerides and triglycerides of the oil, which disappear in the conversion to methylesters.

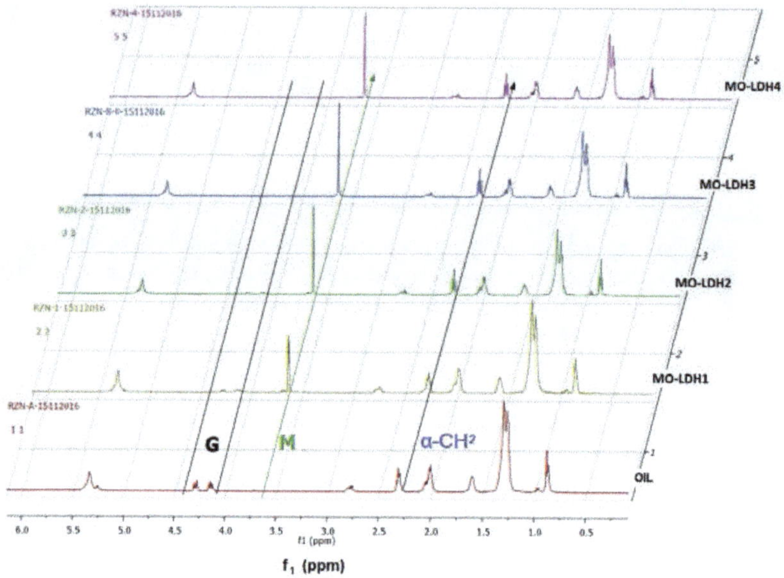

Figure 8. Nuclear magnetic resonance spectra of ^1H in the transesterification reaction for biodiesel production as a function of reaction time.

With this analysis it can be verified that at 60 min the MO-LDH1 and MO-LDH2 catalysts have not yet been able to complete the transesterification reaction of all triglycerides in the oil, but there is a high percentage of conversion greater than 60%. It can be observed that in the case of the MO-LDH1 catalyst, the activity reached at 60 min was 63.3%, which can be attributed to the fact that it consists of particles of amorphous material with a small amount of simple CoO, CoO_2 and Fe_3O_4, which provide a greater specific area than the rest of the catalysts, but with fewer active sites for catalytic conversion. The MO-LDH2 catalyst has a better conversion capacity, reaching 82% in 60 min, because the catalyst contains the mixed oxide $CoFe_2O_4$ in addition to the segregated phases of Fe_3O_4, CoO_2 and CoO. For the case of the MO-LDH3 and MO-LDH4 catalysts, a high conversion to biodiesel with values of 92.7% and 96%, respectively, can be observed. In the NMR spectra it can be observed that there are no signals in the G zone, which implies that they have already completed the transesterification reaction for the formation of the methyl esters when a high intensity of the corresponding M signal is observed. This is associated with better conversion capacity of the MO-LDH3 and MO-LDH4 catalysts, attributed to the presence of crystalline particles of the mixed oxides $CoFe_2O_4$ and $CoNa_xO_2$.

With respect to the complementary quality parameters shown in Table 4, related to the acidity index, corrosion of the sheet and turbidity of the biodiesels obtained with the MO-LDH3 and MO-LDH4 catalysts, it can be seen that these parameters are within the standard ranges. Thus, it is confirmed that the biodiesel obtained with these catalysts can be used in an internal combustion engine.

As the MO-LDH4 catalyst exhibits a better conversion capacity, reaching 96% and presenting a viscosity decrease to 3.66 cp, this catalyst was used to correlate and improve the reaction conditions for obtaining biodiesel with respect to the decrease of the reaction time, the amount of catalyst and the aomount of methanol used in the synthesis.

In relation to the reaction time, a study of the conversion capacity of the catalyst as a function of time was carried out, from 5 to 30 min, while maintaining the reaction conditions of 3% catalyst, a molar ratio of oil/methanol of 1:12, a reaction temperature of 65 °C and constant stirring at 1400 rpm. Table 5 shows the results of the biodiesel conversion values quantified by ^1H-NMR obtained with the catalyst MO-LDH4 at times below 30 min, also indicating the viscosity values as a quality parameter of the biodiesel.

Table 5. Percentage of conversion of the biodiesel obtained with the MO-LDH4 catalyst at different reaction times.

Time (min)	Viscosity (cp)	^{1}H-NMR Conversion (%)
5	4.14	81.33
10	3.67	89.33
15	3.26	90.00
20	3.43	94.00
25	2.88	91.33
30	3.22	96.00

It can be seen that the MO-LDH4 catalyst has the capacity to convert the oil to biodiesel in only 5 min with viscosity values that comply with the norm, reaching a conversion of 81%. These results exceed the capacity found for the MO-LDH2 catalyst, which required 60 min to reach the same conversion capacity, thus confirming the fast selectivity and conversion capacity of the catalyst as well as the important role of mixed oxides in the process. After 5 min, the biodiesel complies with the viscosity quality parameter. With respect to the variation of the conversion% as a function of time, it can be seen that from 5 to 30 min the values of the conversion increase gradually from 81.3% to 96%. The activity of the MO-LDH4 catalyst, attributed to the presence of particles of the mixed oxides $CoFe_2O_4$ and $CoNa_xO_2$ and of the simple oxides CoO_2 and CoO, is favored by the crystalline structure of the catalyst particles, as well as by the macroporosity that allows access to active sites. An advantage of the catalyst is that the formation of catalytically active sites is favored by the formation of mixed crystalline structures.

The combination of the different particles containing Co species with different oxidation states in both the simple oxides and in the mixed with Fe and Na, as a catalytic composition, allows the formation of different acidic and basic sites for the esterification of the free fatty acids present in the oil, as well as for the transesterification of the fatty acids from triglycerides to biodiesel. The process that could be developing in the catalyst is proposed based on the recent kinetic model of transesterification in two sites of the Eley–Rideal type, where the mechanism has three important steps, the first step is the adsorption of both methanol and oil on the surface of the catalyst, in the second step a tetrahedral intermediate is produced from the alkoxide group which attacks the positively polarized carbon of the triacylglyceride in both the liquid phase and the catalyst surface, and the third step involves the production of a fatty acid ester and the release of a diacylglycerides. this procedure is repeated with the following two fatty acids bound to the glycerol [47].

Table 6 shows the viscosity values obtained by decreasing the amount of methanol used in the synthesis while preserving the synthesis conditions of 3% catalyst, a reaction temperature of 65 °C, constant agitation at 1400 rpm and a reaction time of 30 min. It can be seen that the efficiency of the catalyst is not significantly affected by decreasing the oil/methanol ratio from 1:12 to 1:9 and to 1:6. This represents an operational advantage since an excess of methanol as a reagent is not required to displace the reaction to products, as in most procedures reported by heterogeneous catalysis [8,22]. In this case, the conversion capacity is even slightly improved by decreasing the volume of methanol in the reaction.

Table 6. Percentage of conversion of the biodiesel obtained with the MO-LDH4 catalyst at different oil/methanol ratios.

Molar ratio	Viscosity (cp)	^{1}H-NMR Conversion (%)
6	3.43	96
9	3.51	96
12	3.57	95

The effect of the decrease of the amount of the catalyst used in the transesterification reaction was tested, while maintaining the conditions of a reaction temperature of 65 °C, an oil/ethanol molar ratio of 1:12, constant agitation at 1400 rpm and a reaction time of 30 min. Table 7 shows that the efficiency of the process is not significantly affected by decreasing the amount of the catalyst from 3 to 2%, allowing a reduction of up to one third of the catalyst.

Table 7. Percentage of conversion of the biodiesel obtained with the MO-LDH4 catalyst at different oil/methanol ratios.

% Catalyst	Viscosity (cp)	^{1}H-NMR Conversion (%)
1	5.27	90
2	3.52	95
3	3.57	96

Finally, to verify the best synthesis conditions considering the simultaneous decrease of the catalyst, methanol and reaction time, biodiesel was obtained with 2% catalyst, a oil/methanol molar ratio of 1:6 and a reaction time of 20 min. Thus, we achieved a biodiesel with quality parameters that comply with the ASTM standards, as reported in Table 8.

Table 8. Quality parameters of the biodiesel obtained using the MO-LDH4 catalyst under optimal synthesis conditions.

Parameters	Limits ASTM	Biodiesel Obtained
Density (kg/m^3)	860–900	884
Viscosity (cp)	1.9–6.0	3.89
Corrosion of copper foil	1 class	1 class
Acidity index (mg$_{KOH}$/g)	0.5	0.2
Turbity	not established	<1
% of biodiesel conversion	not established	90.24%
Methylester composition by Gas Chromatography-Flame Ionization Detector (GC-FID)	not established	C18:1 ^{9C}, C18:2 ^{9C,12C}, C18:3 ^{9C,12C,15C}

It can be seen that the biodiesel obtained by Gas Chromatography-Flame Ionization Detector (CG-FID) is mainly composed of the methyl esters of oleic acid (C18: 1 cis-9), linoleic acid (C18: 2 cis-9,12) and linolenic acid (C18: 3 cis-9,12,15), and in a lower proportion of palmitic acid (C16: 0) and stearic acid (C18: 0), which were identified according to the chromatograms shown in Figure 9. These results confirm the high selectivity of MO-LDH4 towards unsaturated fatty acids.

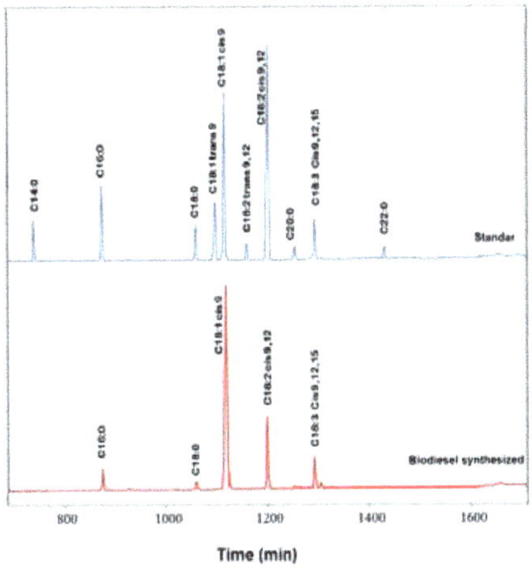

Figure 9. Chromatograms of biodiesel obtained with 2% MO-LDH4 catalyst compared to the Supelco standard for Fatty Acid Methyl Esters (FAME).

The results of the analysis of the biodiesel composition obtained confirm the formation of the methyl esters of the respective fatty acids that formed the oil used as raw material in the transesterification reaction. The biodiesel obtained complies with the quality standards set by the ASTM international standards. The quality of the biodiesel was improved by the selectivity and high conversion capacity of the catalyst.

3. Materials and Methods

3.1. Synthesis of the Catalyst

The catalysts were synthesized by the procedure registered in the patent process MX2015009362 of the Mexican Institute of Industrial Property [48], as described below. A series of Co/Fe layered double hydroxides samples was prepared by the coprecipitation method with a molar ratio $Co^{2+}/Fe^{3+} = 1$, 2, 3 and 4 using a stoichiometric amount of cobalt nitrate hexahydrate and iron nitrate nonahydrate. The aqueous solutions containing the metal salts were coprecipitated in a basic medium with an alkaline solution of sodium hydroxide (2M) and sodium bicarbonate (1M) at a constant pH of 11.5. After complete precipitation of the mixture, the precipitates were washed with water at room temperature until obtaining a pH of 9. Subsequently the precipitates were dried in an oven at 70 °C for 24 h to obtain the LDH as catalytic precursors. The materials were identified as LDH1, LDH2, LDH3 and LDH4 corresponding to the Co/Fe ratio = 1, 2, 3 and 4 respectively. The catalytically active phases were obtained by calcination of the precursors at 700 °C for 4 h. The calcination products of the layered doble hydroxides were identified as corresponding MO-LDH1, MO-LDH2, MO-LDH3 and MO-LDH4.

3.2. Physicochemical Characterization of the Catalyst

The catalysts were characterized by (a) differential scanning calorimetry (DSC) and thermogravimetric analysis (TGA) with a Universal (Irapuato, Mexico) V4.5A TA Instruments: SDT Q600 V20.5 Build 15; (b) X-ray powder diffraction (XRD) in Inel Equinox (Guanajuato, Mexico) with an X-ray tube coupled with a copper anode using monochromatic CuKα radiation; (c) Fourier transform infrared spectroscopy (FTIR) in a Bruker Tensor (Guanajuato, Mexico) 27 with OPUS software version 6.5; (d) textural properties in an ASAP (Guanajuato, Mexico) 2010 and (e) scanning electron microscopy (SEM) using an X Carl Zeiss Evo (Guanajuato, Mexico) HD-15 microscope with an integrated X-ray dispersive energy microanalysis system.

3.3. Production of Biodiesel

The production of biodiesel was developed by the procedure registered in the patent process MX2015009362 of the Mexican Institute of Industrial Property [48], as described below. To obtain biodiesel, a batch reactor was used with the operating conditions in a reflux system at 65 °C, constant agitation at 1400 rpm and a pressure atmosphere. The alcohol used in the synthesis was anhydrous methanol (Golden Bell, Guadalajara, Mexico) and the source of triglycerides was commercial cooking oil. The variables used in the synthesis of biodiesel were the oil/methanol molar ratio (1:6, 1:9 and 1:12), the percentage of oil/catalyst (1, 2 and 3% w/w) and the reaction time (30 and 60 min). Initially, 3 g of calcined LDH was added to 50 mL methanol in a 500 mL three-necked glass flask equipped with a reflux condenser. The mixture was vigorously mixed under atmospheric and heated until the temperature reached 50 °C. Then, 100 mL of cooking oil, was added to the reactor and the temperature was increased to 65 °C. The mixture was stirred at 1400 rpm during the reaction time (30 min). At the end of the reaction, methanol was recovered by evaporation, catalyst was recovered by vacuum filtration, and subsequently, glycerol, and biodiesel were separated by decantation.

3.4. Evaluation of the Quality of Biodiesel

The quality parameters of the biodiesel were evaluated in accordance with the ASTM international standards to verify that the quality indicated in the ASTM D6751 standard was met. The viscosity was determined in accordance with ASTM D445. The density was determined according to ASTM D121718. The ASTM D66419 standard was used for the acidity index. The corrosion of copper foil was evaluated according to ASTM D13020. Turbidity was measured according to ASTM 250021. The quantification of the conversion of biodiesel by FTIR with diffuse reflectivity (DRIFTS) was conducted as established in ASTM D7371, as well as by proton nuclear magnetic resonance (^{1}H-MNR). The determination of the methyl esters contained in the biodiesel was carried out in accordance with the ASTM D6584 standard.

4. Conclusions

Catalytic precursors of the layered double hydroxides type with different Co/Fe ratios were synthesized in a basic medium, which allowed us to obtain catalysts that showed a correlation between their physicochemical properties and their catalytic capacity in the conversion of triglycerides to biodiesel. The best conversion capacity was presented by the MO-LDH4 catalyst with a 96% conversion to biodiesel by transesterification using a molar alcohol/oil ratio of 6:1, with a reaction time of 20 min and the catalyst percentage of 2%, mainly formed by a mixture of crystalline particles in the form of stacked sheets. The high capacity and selectivity for the conversion of triglycerides to biodiesel shown by the MO-LDH4 catalyst is attributed to the fact that it is a composite catalyst that contains a combination of the simple oxides CoO_2 and CoO and the mixed oxides $CoFe_2O_4$ and $CoNa_xO_2$ where the Co with different oxidation states in combination with O, Fe and Na, favors the presence of acidic and basic sites for the transesterification esterification of the oil to biodiesel.

5. Patents

MX2015009362 Esthela Ramos Ramírez, Norma Leticia Gutiérrez-Ortega, José de Jesús Monjaraz Vallejo, "Obtaining Cobalt Ferrite for its application as a catalyst in the optimization of esterification and transesterification reactions of fatty acids for the biodiesel production process" ("Obtención de Ferrita de Cobalto para su aplicación como catalizador en la optimización de las reacciones de esterificación y transesterificación de ácidos grasos para el proceso de producción de biodiesel" titule in spanish), request patent filed with the Mexican Institute of Industrial Property (IMPI Spanish acronym) Publication date: January 20, 2017.

Author Contributions: Conceptualization, G.-O.N. and R.-R.E.; methodology, G.-O.N. and M.-V.J.; validation, G.-O.N. and R.-R.E.; formal analysis, S.-M.A.; investigation, Z.-M.A.; resources, S.-M.A.; data curation, Z.-M.A.;

writing—original draft preparation, G.-O.N.; writing—review and editing, R.-R.E.; project administration, R.-R.E.; funding acquisition, G.-O.N.

Funding: This research received no external funding. Funding was received from the resources of the university of Guanajuato.

Acknowledgments: This work was supported by the Professors Improvement Program (PROMEP Spanish acronym) of the Public Education Secretary (SEP Spanish acronym), and the Direction of Research and Postgraduate Course Support (DAIP Spanish acronym) from the University of Guanajuato. The authors thank the Directorate of Research Support and the Postgraduate Program of the University of Guanajuato for funding and CONACYT for the scholarship received for the completion of postgraduate studies. In addition, Dr. Ramón Zarraga and Dr. Ricardo Navarro from the University of Guanajuato are thanked for the support with the NMR and SEM-EDX techniques, respectively.

Conflicts of Interest: The authors declare no conflict of interest.

References

1. Oliveira, F.C.; Coelho, S.T. History, evolution, and environmental impact of biodiesel in Brazil: A review. *Renew. Sust. Energ. Rev.* **2017**, *75*, 168–179. [CrossRef]
2. García-Olivares, A.; Ballabrera-Poy, J. Energy and mineral peaks, and a future steady state economy. *Technol. Forecast. Soc. Change* **2015**, *95*, 587–598. [CrossRef]
3. Oh, Y.K. Recent developments and key barriers to advanced biofuels: A short review. *Bioresour. Technol.* **2018**. [CrossRef]
4. Semwal, S.; Arora, A.K.; Badoni, R.P.; Tuli, D.K. Biodiesel production using heterogeneous catalysts. *Bioresour. Technol.* **2011**, *102*, 2151–2161. [CrossRef] [PubMed]
5. Natarajan, Y.; Nabera, A.; Salike, S.; Tamilkkuricil, V.D.; Pandian, S.; Karuppan, M.; Appusamy, A. An overview on the process intensification of microchannel reactors for biodiesel production. *Chem. Eng. Process.* **2019**, *136*, 163–176. [CrossRef]
6. Ge, J.C.; Yoon, S.K.; Choi, N.J. Using Canola Oil Biodiesel as an Alternative Fuel in Diesel Engines: A Review. *Appl. Sci.* **2017**, *7*, 881. [CrossRef]
7. Tariq, M.; Al, S.; Khalid, N. Activity of homogeneous and heterogeneous catalysts, spectroscopic and chromatographic characterization of biodiesel: A review. *Renew. Sust. Energ. Rev.* **2012**, *16*, 6303–6316. [CrossRef]
8. Shan, R.; Lu, L.; Shi, Y.; Yuan, H.; Shi, J. Catalysts from renewable resources for biodiesel production. *J. Energ. Conver. Manage.* **2018**, *178*, 277–289. [CrossRef]
9. Chang, C.C.; Teng, S.; Yuan, M.H.; Ji, D.R.; Chang, C.Y.; Chen, Y.H.; Shie, J.L.; Ho, C.; Tian, S.Y.; Andrade-Tacca, C.A.; et al. Esterification of Jatropha Oil with Isopropanol via Ultrasonic Irradiation. *Energies* **2018**, *11*, 1456. [CrossRef]
10. Christopher, L.P.; Kumar, H.; Zambare, V.P. Enzymatic biodiesel: Challenges and opportunities. *Appl. Energ.* **2014**, *119*, 497–520. [CrossRef]
11. Reyero, I.; Arzamendi, G.; Zabala, S.; Gandía, L.M. Fuel Kinetics of the NaOH-catalyzed transesterification of sunflower oil with ethanol to produce biodiesel. *Process. Technol.* **2015**, *129*, 147–155. [CrossRef]
12. Abdullah, S.H.Y.S.; Hanapi, N.H.M.; Azid, A.; Umar, R.; Juahir, H.; Khatoon, H.; Endut, A. A review of biomass-derived heterogeneous catalyst for a sustainable biodiesel production. *Renew. Sust. Energ. Rev.* **2017**, *70*, 1040–1051. [CrossRef]
13. Yan, S.; Maggio, C.D.; Mohan, S.; Kim, M.; Salley, S.O.; Simon, K.Y. Advancements in Heterogeneous Catalysis for Biodiesel Synthesis. *Top. Catal.* **2010**, *53*, 721–736. [CrossRef]
14. Dimian, A.C.; Srokol, Z.W.; Mittelmeijer-Hazeleger, M.C.; Rothenberg, G. Interrelation of Chemistry and Process Design in Biodiesel Manufacturing by Heterogeneous Catalysis. *Top. Catal.* **2010**, *53*, 1197–1201. [CrossRef]
15. Chouhan, A.P.S.; Sarma, A.K. Modern heterogeneous catalysts for biodiesel production: A comprehensive review. *Renew. Sust. Energ. Rev.* **2011**, *15*, 4378–4399. [CrossRef]
16. Lee, H.V.; Juan, J.C.; Hin, T.Y.Y.; Ong, H.C. Environment-Friendly Heterogeneous Alkaline-Based Mixed Metal Oxide Catalysts for Biodiesel Production. *Energies* **2016**, *9*, 611. [CrossRef]

17. Chang, K.L.; Lin, Y.C.; Jhang, S.R.; Cheng, W.L.; Chen, S.C.; Mao, S.Y. Rapid Jatropha-Castor Biodiesel Production with Microwave Heating and a Heterogeneous Base Catalyst Nano-Ca$(OH)_2$/Fe_3O_4. *Catalysts* **2017**, *7*, 203. [CrossRef]
18. Du, L.; Ding, S.; Li, Z.; Lv, E.; Lu, J.; Ding, J. Transesterification of castor oil to biodiesel using NaY zeolite-supported La_2O_3 catalysts. *Energ. Convers. Manage.* **2018**, *173*, 728–734. [CrossRef]
19. Jaya, N.; Selvan, B.K.; Vennison, S.J. Synthesis of biodiesel from pongamia oil using heterogeneous ion-exchange resin catalyst. *Ecotox. Environ. Safe.* **2015**, *121*, 3–9. [CrossRef]
20. Tanga, Z.E.; Lima, S.; Panga, Y.L.; Ongb, H.C.; Lee, K.T. Synthesis of biomass as heterogeneous catalyst for application in biodiesel production: State of the art and fundamental review. *Renew. Sust. Energ. Rev.* **2018**, *92*, 235–253. [CrossRef]
21. Cavani, F.; Trifiro, F.; Vaccari, A. Hydrotalcite–type anionic clays: Preparation, properties and applications. *Catal. Today* **1992**, *11*, 173–301. [CrossRef]
22. Mills, S.J.; Christy, A.G.; Génin, J.M.R.; Kameda, T.; Colombo, F. Nomenclature of the hydrotalcite supergroup: natural layered double hydroxides. *Mineral. Mag.* **2012**, *76*, 1289–1336. [CrossRef]
23. Othman, M.R.; Helwani, Z.; Martunus, W.J.N. Synthetic hydrotalcites from different routes and their application as catalysts and gas adsorbents: a review. *Appl. Organometal Chem.* **2009**, *23*, 335–346. [CrossRef]
24. Jitianu, M.; Balasoiu, M.; Zaharescu, M.; Jitianu, A.; Ivanov, A. Comparative Study of Sol-Gel and Coprecipitated Ni-Al Hydrotalcites. *J. Sol.-Gel. Sci. Techn.* **2000**, *19*, 453–457. [CrossRef]
25. Qu, J.; Sha, L.; Wu, C.; Zhang, Q. Applications of Mechanochemically Prepared Layered Double Hydroxides as Adsorbents and Catalysts: A Mini-Review. *Nanomaterials* **2019**, *9*, 80. [CrossRef]
26. Miyata, S. Physico-chemical properties of synthetic hydrotalcites in relation to composition. *Clays. Clay. Miner.* **1980**, *28*, 50–56. [CrossRef]
27. Roelofs, J.C.A.A.; Bokhoven, J.A.; Dillen, A.J.; Geus, J.W.; Jong, K.P. The Thermal Decomposition of Mg±Al Hydrotalcites: Effects of Interlayer Anions and Characteristics of the Final Structure. *Chem. Eur. J.* **2002**, *8*, 5571–5579. [CrossRef]
28. Iqbal, M.A.; Fedel, M. Effect of Synthesis Conditions on the Controlled Growth of MgAl–LDH Corrosion Resistance Film: Structure and Corrosion Resistance Properties. *Coatings* **2019**, *9*, 30. [CrossRef]
29. Yan, A.L.; Wang, X.C.; Cheng, J.P. Research Progress of NiMn Layered Double Hydroxides for Supercapacitors: A Review. *Nanomaterials* **2018**, *8*, 747. [CrossRef]
30. Coral, N.; Brasil, H.; Rodrigues, E.; Costa, C.E.F.; Rumjanek, V. Microwave-modified hydrotalcites for the transesterification of soybean oil. *Sust. Chem. Pharm.* **2019**, *11*, 49–53. [CrossRef]
31. Hájek, M.; Tomášová, A.; Kocík, J.; Podzemná, V. Statistical evaluation of the mutual relations of properties of Mg/Fe hydrotalcites and mixed oxides as transesterification catalysts. *Appl. Clay. Sci.* **2018**, *154*, 28–35. [CrossRef]
32. Fatimah, I.; Rubiyanto, D.; Nugraha, J. Preparation, characterization, and modelling activity of potassium fluoride modified hydrotalcite for microwave assisted biodiesel conversion. *Sust. Chem. Pharm.* **2018**, *8*, 63–70. [CrossRef]
33. Tajuddin, N.A.; Manayil, J.C.; Isaacs, M.A.; Parlett, C.M.A.; Lee, A.F.; Wilson, K. Alkali-Free Zn–Al Layered Double Hydroxide Catalysts for Triglyceride Transesterification. *Catalysts* **2018**, *8*, 667. [CrossRef]
34. Navajas, A.; Campo, I.; Moral, A.; Echave, J.; Sanz, O.; Montes, M.; Odriozola, J.A.; Arzamendi, G.; Gandía, L.M. Outstanding performance of rehydrated Mg-Al hydrotalcites as heterogeneous methanolysis catalysts for the synthesis of biodiesel. *Fuel* **2018**, *211*, 173–181. [CrossRef]
35. Dhawan, M.S.; Yadav, G.D. Insight into a catalytic process for simultaneous production of biodiesel and glycerol carbonate from triglycerides. *Catal. Today* **2018**, *309*, 161–171. [CrossRef]
36. Wang, Y.T.; Fang, Z.; Zhang, F.; Xue, B.J. One-step production of biodiesel from oils with high acid value by activated Mg–Al hydrotalcite nanoparticles. *Bioresour. Technol.* **2015**, *193*, 84–89. [CrossRef] [PubMed]
37. Prado, R.G.; Almeida, G.D.; Carvalho, M.M.O.; Galvao, L.M.; Bejan, C.C.C.; Costa, L.M.; Pinto, F.G.; Tronto, J.; Pasa, V.M.D. Multivariate Method for Transesterification Reaction of Soybean Oil Using Calcined Mg–Al Layered Double Hydroxide as Catalyst. *Catal. Lett.* **2014**, *144*, 1062–1073. [CrossRef]
38. Macala, G.S.; Robertson, A.W.; Johnson, C.L.; Day, Z.B.; Lewis, R.S.; White, M.G.; Iretskii, A.V.; Ford, P.C. Transesterification Catalysts from Iron Doped Hydrotalcite-like Precursors: Solid Bases for Biodiesel Production. *Catal. Lett.* **2008**, *122*, 205–209. [CrossRef]

39. Li, E.; Ping Xu, Z.; Rudolph, V. MgCoAl–LDH derived heterogeneous catalysts for the ethanol transesterification of canola oil to biodiesel. *Appl. Catal. B* **2009**, *88*, 42–49. [CrossRef]
40. Peña-Rodríguez, R.; Márquez-López, E.; Guerrero, A.; Chiñas, L.E. Hydrothermal synthesis of cobalt (II) 3D metal-organic framework acid catalyst applied in the transesterification process of vegetable oil. *Mater. Lett.* **2018**, *217*, 117–119. [CrossRef]
41. Jamilah; Krisnandia, Y.K.; Sihombing, R. Synthesis and characterization of mesoporous Co/ZSM5 catalyst and activity study on transesterification reaction. *AIP Conf. Proc.* **2016**, *1729*, 020042. [CrossRef]
42. Nuithitikul, K.; Prasitturattanachai, W.; Hasinc, W. Comparison in catalytics activities of sulfated Cobalt-Tin and sulfated aluminum-Tin mixed oxiddes for esterification of free fatty acis to produce methyl esteres. *Energy Procedia* **2017**, *138*, 75–80. [CrossRef]
43. Nasrollahzadeh, M.; Bagherzadeh, M.; Karimi, H. Preparation, characterization and catalytic activity of $CoFe_2O_4$ nanoparticles as a magnetically recoverable catalyst for selective oxidation of benzyl alcohol to benzaldehyde and reduction of organic dyes. *J. Colloid. Interface Sci.* **2016**, *465*, 271–278. [CrossRef] [PubMed]
44. Soares, J.M.; Conceição, O.L.A.; Machado, F.L.A.; Prakash, A.; Radha, S.; Nigam, A.K. Magnetic properties of the nanocomposite $CoFe_2O_4$/FeCo-FeO at a high H/T regime. *J. Magn. Magn. Mater.* **2015**, *374*, 192–196. [CrossRef]
45. Knothe, G.; Steidley, K.R. Kinematic viscosity of biodiesel fuel components and related compounds. Influence of compound structure and comparison to petrodiesel fuel components. *Fuel* **2005**, *84*, 1059–1065. [CrossRef]
46. Kollar, S.R.M.; Novotny, E.H.; Nascimento, C.J.; Suarez, P.A.Z. Nuclear Magnetic Resonance (1.40 T) and Mid Infrared (FTIR-ATR) Associated with Chemometrics as Analytical Methods for the Analysis of Methyl Ester Yield Obtained by Esterification Reaction. *J. Braz. Chem. Soc.* **2017**, *28*, 1917–1925. [CrossRef]
47. Olutoye, M.A.; Hameed, B.H. ScienceDirectKinetics and deactivation of a dual-site heterogeneous oxide catalystduring the transesterification of crude jatropha oil with methanol. *J. Taibah. Univ. Sci.* **2016**, *10*, 685–699. [CrossRef]
48. World Intellectual Property Organization. Patent Scope. Available online: https://patentscope.wipo.int/search/es/detail.jsf?docId=MX194961824&redirectedID=true (accessed on 31 January 2019).

catalysts

MDPI

Article

Improved Etherification of Glycerol with *Tert*-Butyl Alcohol by the Addition of Dibutyl Ether as Solvent

Carmen M. Dominguez *, Arturo Romero and Aurora Santos

Chemical Engineering and Materials, Faculty of Chemical Sciences, Universidad Complutense de Madrid, Ciudad Universitaria S/N, 28040 Madrid, Spain; aromeros@ucm.es (A.R.); aursan@ucm.es (A.S.)
* Correspondence: carmdomi@ucm.es; Tel.: +34-913-944-171

Received: 26 March 2019; Accepted: 17 April 2019; Published: 23 April 2019

Abstract: The etherification of glycerol with *tert*-butyl alcohol in the presence of acid catalysts gives rise to the production of ethers (monoethers, diethers and triethers) of high added-value, which can be used as oxygenated additives in fuels. This reaction is limited by the thermodynamic equilibrium, which can be modified by the addition of solvents that selectively solubilize the products of interest along with *tert*-butyl alcohol, leading to the progress of the reaction. In this work, it has been demonstrated that the addition of dibutyl ether allows shifting the reaction equilibrium, increasing the production of diethers. From the study of the main operating conditions, it was determined that an increase in the concentration of the solvent has a positive effect on the selectivity towards the production of diethers, the concentration of the catalyst (a commercial ion exchange resin, Amberlyst 15, named A-15) and the reaction temperature were also determining variables. Working with concentrations of *tert*-butyl alcohol above the stoichiometric one did not report great advantages. The optimal operating conditions to maximize the conversion of glycerol and the selectivity towards diethers were: 70 °C, 20% catalyst (referred to the total starting mass of the system), the stoichiometric ratio of glycerol:*tert*-butyl alcohol (G:TB = 1:3) and 1:2 molar ratio of dibutyl ether:*tert*-butyl alcohol. A study of three consecutive reaction cycles showed the high stability of the catalyst, obtaining identical results.

Keywords: etherification; glycerol; *tert*-butyl alcohol; dibutyl ether; A-15; catalyst stability

1. Introduction

The renewable and biodegradable character of biodiesel has made it an interesting alternative to the use of conventional fuels [1–3]. Its use as combustible for diesel engines has greatly grown in recent decades [4] and is expected to grow further in the next years [5,6]. Biodiesel is a mixture of methyl esters of fatty acids [7], synthesized from animal or vegetable fats, which are non-toxic, biodegradable and renewable resources [8], obtaining glycerol (G) (about 1 kg of glycerol is generated for every 9 kg of biodiesel) as the main by-product of the process [4,9]. As a direct consequence of the increase in biodiesel production, the price of glycerol has considerably dropped [6]. Therefore, finding solutions for the valorization of this by-product to added value chemicals could improve the economy of the process [10]. Thus, the production of biodiesel on a large scale will be aided if adequate technologies capable of converting glycerol into added value chemicals are developed [11]. There are several industrial processes that use glycerol as raw material (oligomerization/polymerization, pyrolysis and gasification, selective oxidation, steam reforming, selective transesterification, etherification to fuel-oxygenates, etc.) [12,13]. Another alternative is the production of glycerol carbonate (an important glycerol derivative commonly used as a solvent in cosmetics, personal care items and medicine) from the catalytic oxidative carbonylation of the parent compound [14]. Among the existing options, an interesting alternative for the excess of glycerol generated is its transformation into oxygenated additives for liquid fuels [15], which can enhance the combustion efficiency in internal combustion engines and reduce the emission of pollutants, particularly, *tert*-butyl ethers of glycerol [7]. One possibility for obtaining these oxygenated compounds is the conversion of glycerol into ethers from its etherification with isobutene (IB) [16–22]. The use of gaseous isobutene involves the typical drawbacks of a complex three-phase system (mass transfer phenomena, security issues, etc.) [7] and other alternatives have been studied. Thus, it has been shown that the use of *tert*-butyl alcohol (TB), which is a by-product of polypropylene production, instead of isobutene, could be preferable since (i) The reagents are in the same phase, (ii) prevents the oligomerization of isobutene [9,23] and (iii) allows to overcome the technological problems arising from the need to use solvents (i.e., dioxane, dimethyl sulfoxide) able to dissolve glycerol [7,24].

The etherification of glycerol with *tert*-butyl alcohol proceeds according to a consecutive path giving rise to a mixture of five *tert*-butyl ethers, namely monoethers (M1, M2), diethers (D1, D2) and, eventually, triether (T) (Figure 1). These compounds, mainly diethers and triethers, are considered excellent additives for diesel fuels (diesel, biodiesel and their mixtures) [9,19,22]. The solubility of monoethers in diesel fuel is quite low. Therefore, in order to avoid an additional separation step, the etherification of glycerol should address the generation of higher glycerol ethers (di- and triethers) [16]. On the other hand, it has been demonstrated that the production of triether is usually highly limited due to steric hindrance [7,9]. Thus, the selectivity of glycerol towards the production of di-substituted ethers should be maximized. Otherwise, it should to be noted the undesired (and inevitable) reaction of dehydration of TB to isobutene (IB) (secondary reaction) and the generation of water in each stage of the reaction, which negatively affects de etherification equilibrium (Figure 1) and competes with the reagents (glycerol and *tert*-butyl alcohol) for the active adsorption sites of the catalyst [7,25].

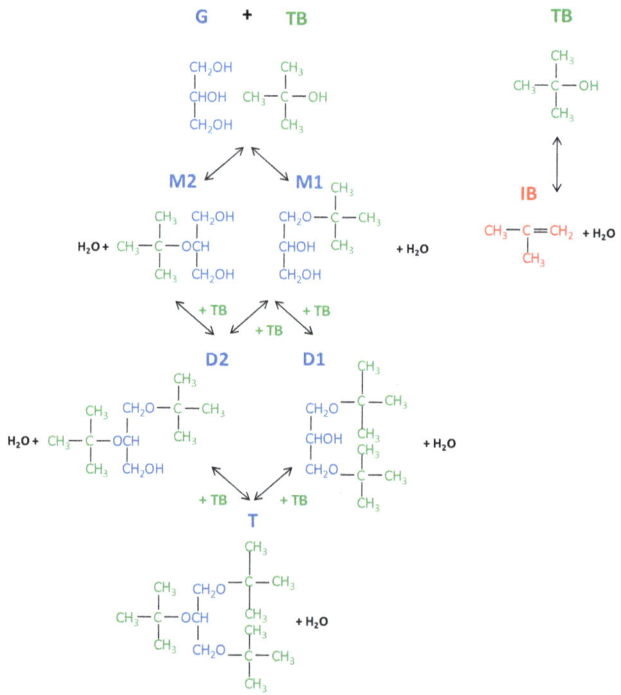

Figure 1. Reaction scheme of glycerol (G) etherification with *tert*-butyl alcohol (TB) and secondary reaction to isobutene (IB) from TB dehydration. Adapted from Pico et al., 2012 [9].

Interesting results were obtained when water was removed from the reaction medium, leading to an improvement in ethers production [7,25]. The dehydration of the reaction system (etherification of glycerol with *tert*-butyl alcohol with a commercial ion exchange resin as catalyst) by using zeolites as water adsorbents after 6 h reaction time resulted in a net increase in the yield of diethers from 28.5% to 41.5% [7]. Ozbay et al. (2012) [25] also found a significant enhancement of glycerol conversion and diether yield by the in situ removal of water generated during the sequential etherification reactions (the selectivity towards diether increased from 20 to 33% when a zeolite:catalyst ratio of 4:1 was used). Another possibility (very little studied so far) could be adding a solvent (not miscible with water or with low water solubility) to the reaction medium able to selectively dissolve TB along with the generated ethers, allowing in this way the progress of the reaction. Roze et al. (2013) found that performing the reaction in toluene, which works as a water removing agent, influenced the etherification reaction significantly increasing the conversion of glycerol and the yield of glycerol ethers [26]. Two other solvents, 1,4-dioxane and ethanol, were tested, but lead to worse results [26].

This reaction requires the use of acid catalysts, such as sulfuric acid, ion exchange resins, zeolites, mesostructured silicas, etc. [9,26]. Amongst the heterogeneous catalysts (more advantageous from an environmental point of view due to their easy separation at the end of the process and its possible reuse), the most convenient ones have turned out to be ion exchange resins [27]. Among them, the commercial Amberlite 15 resin (A-15) has given interesting results in the etherification of glycerol with *tert*-butyl alcohol [7,9,23] due to its high acidity and good stability. This catalyst exhibited higher activity than other acid systems such as Nafion® on amorphous silica and a home-made mesoestructurated silica supported acid catalyst because of the wider pore diameter of the A-15, which allowed easier accessibility of the reagent molecules [7]. Pico et al. (2013) [23] found that this catalyst also showed better performance in the etherification of glycerol with *tert*-butyl alcohol (higher glycerol conversion and selectivity towards diethers) than other commercial ion-exchange resins (Ambertite 200 and Amberlite IRC-50) because of its higher acidity and better textural properties. However, the maximum selectivity of glycerol towards the products of interest (diethers) in the presence of A-15 catalyst was relatively low (28.5% [7] and 21% [23]). The explanation to this fact was that the thermodynamic equilibrium was reached. Accordingly, further research to modify this equilibrium and increase the yield to di- and triethers is required.

The objective of the present work is to shift the equilibrium of the glycerol etherification reaction with TB towards the production of diethers. To do this, dibutyl ether (DBE), non-miscible with water (DBE solubility in water = 0.113 g L^{-1}), has been introduced as a solvent into the reaction system to selectively solubilize the ethers, partially solubilize TB and minimize the presence of water in this phase. Dibutyl ether was selected among other organic solvents due to its interesting properties: Non-polar nature, low solubility in water, low reactivity, low toxicity, high flash point, etc.

In this way, working with a biphasic system, the progress of the reaction from monoethers to diethers could be favored. A-15 heterogeneous catalyst, a commercial ion-exchange resin, accepted as the most suitable catalyst for this reaction, has been used in this study. The influence of the main operating variables (DBE concentration, catalyst concentration, reaction temperature and TB concentration) has been evaluated with the aim of finding the optimal conditions to maximize the yield of the reaction and the selectivity towards the products of interest. Finally, the stability of the catalyst in three consecutive reaction cycles has been demonstrated.

2. Results and Discussion

The results obtained have been discussed and evaluated according to the conversion of reagents, glycerol (G) and *tert*-butyl alcohol (TB) (Equations (1) and (2)) and the selectivity towards the generated ethers (M, D and T) at different reaction times, calculated from the following expressions:

$$X_G(\%) = \left(1 - \frac{[G]_{t=x}}{[G]_{t=0}}\right) \cdot 100 \quad (1)$$

$$X_{TB}(\%) = \left(1 - \frac{[TB]_{t=x}}{[TB]_{t=0}}\right)\cdot 100 \quad (2)$$

$$S_M(\%) = \frac{[M]_{t=x}}{[M]_{t=x} + [D]_{t=x} + [T]_{t=x}}\cdot 100 \quad (3)$$

$$S_D(\%) = \frac{[D]_{t=x}}{[M]_{t=x} + [D]_{t=x} + [T]_{t=x}}\cdot 100 \quad (4)$$

$$S_T(\%) = \frac{[T]_{t=x}}{[M]_{t=x} + [D]_{t=x} + [T]_{t=x}}\cdot 100 \quad (5)$$

The terms S_M and S_D include the selectivity towards the two monoethers (M1 + M2) and diethers (D1 + D2), respectively (see Figure 1). The sum of selectivities (S_M + S_D + ST) is equal to 100%. The selectivity towards the triether (T) was in all the experiments below 1% (probably due to steric hindrance [7,9]) so it has not been considered in the discussion of results.

In addition to the parameters discussed above, the selectivity towards the production isobutene (IB), an undesired product of the reaction (Figure 1), and calculated from Equation (6) at 360 min of reaction time, has also been taken into account:

$$S_{IB}(\%) = \frac{[IB]_{t=360}}{[M]_{t=360} + 2\cdot[D]_{t=360} + 3\cdot[T]_{t=360} + [IB]_{t=360}}\cdot 100 \quad (6)$$

2.1. Preliminary Results: Solubility Study

The aim of introducing DBE as a solvent in the etherification reaction between TB and G is to improve the production of diethers (compounds with higher added value than monoethers). For that purpose, it is necessary that the first products of the reaction (M1 and M2), as well as a fraction of TB (to continue the etherification reaction), are selectively solubilized in the organic phase (DBE), keeping most of the water generated in the other phase. In this way, the reaction equilibrium could be modified.

The distribution of the different compounds selected to carry out this essay (see Section 4.3.1. Solubility test) in the two phases (named as hydrophilic and hydrophobic phases), determined by gas-chromatographic and Karl Fischer analyses, is collected in Table 1. As can be seen, 100% of G remained in the hydrophilic phase, while most of TB (77%) and M1 (61%) were solubilized in the hydrophobic phase. Thus, the presence of DBE led to a different compounds-distribution, obtaining M1 (reaction product) and TB (reagent necessary for the progress of the reaction) preferentially in the hydrophobic phase. In view of these results, most of the water (84%) was removed from the reaction phase, an improvement in the selectivity towards the products of interest can be expected.

Table 1. Distribution of the compounds involved in the etherification of G in the hydrophilic and hydrophobic phases.

Compound	Hydrophilic Phase (%)	Hydrophobic Phase (%)
TB	23	77
DBE	2	98
M1	39	61
G	100	0
water	84	16

TB (*tert*-butanol), DBE (dibutyl ether), M1 (monoether, 1-terbutoxipropano-2,3-diol), G (Glycerol).

2.2. Role of DBE as Solvent in the Etherification of Glycerol

To verify this hypothesis (modification of the reaction equilibrium and positive effect of DBE on the etherification reaction of G with TB), the selectivity of G towards the reaction products (M and D) obtained in the presence of DBE (experiment R5, Table 3, Materials and Methods section) was compared, for the same glycerol conversion value (X_G = 82%), with that obtained in the absence of DBE (data obtained elsewhere at similar operating conditions [9]). The results are depicted in Figure 2. As can be seen, the introduction of DBE into the reaction system led to a substantial increase in the selectivity towards the diethers (S_D increased from 21% to 40% when DBE was added to the system), confirming the positive role of this solvent.

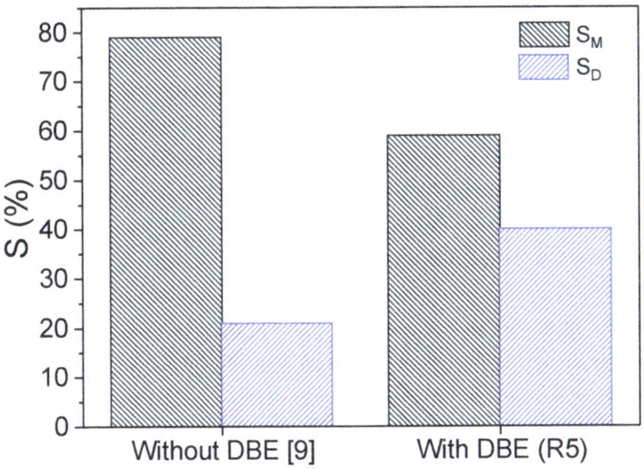

Figure 2. Effect of DBE on the selectivity towards the reaction products (M and D).

On the other hand, the equilibrium constants of reactions giving rise to the generation of M and D ($K_{eq,M}$ and $K_{eq,D}$, respectively) in the absence and the presence of DBE have been also compared. The expression of the equilibrium constants without were obtained from Pico et al. (2012) [9] (Equations (7) and (8)), whereas the corresponding equilibrium constants in the presence of DBE were experimentally obtained by using the Equations (9) and (10) (R1, t = 360 min):

$$K_{eq,\,M} = \exp\left(26.37 - \frac{8639.46}{T}\right) \quad (7)$$

$$K_{eq,D} = \exp\left(21.51 - \frac{8083.91}{T}\right) \quad (8)$$

$$K_{eq,M(DBE)} = \frac{[M]\cdot[H2O]}{[TB]\cdot[G]} \quad (9)$$

$$K_{eq,D(DBE)} = \frac{[D]\cdot[H2O]}{[TB]\cdot[M]} \quad (10)$$

In Equations (9) and (10), the concentration of reactants and products have been calculated considering a pseudo-homogeneous phase, being expressed all of them in mmol kg^{-1}. The values obtained for the equilibrium constants are summarized in Table 2.

Table 2. Equilibrium constants in the absence and presence of DBE as solvent.

Equilibrium Constant	without DBE	with DBE
$K_{eq,M}$	3.26	3.42
$K_{eq,D}$	0.13	1.16

$K_{eq,M}$ was slightly modified by the presence of DBE, while $K_{eq,D}$ greatly increased. Thus, it has been confirmed that the addition of DBE as a solvent to the reaction system is beneficial since it shifts the reaction equilibrium towards the formation of diethers.

2.3. *Operation Condition Study*

Once the positive effect of DBE on the etherification of glycerol with TB was confirmed, a study of the main operating variables was carried out, including the concentration of the solvent and the catalyst, the operating temperature and the concentration of TB.

2.3.1. Effect of DBE Concentration

The effect of DBE concentration was evaluated by working at 70 °C, 4 bar, the stoichiometric reagents molar ratio (G:TB = 1:3) and a catalyst concentration of 20% (referred to the initial mass of the system) using two molar ratios of DBE:TB, 1:2 and 1:5. The results of reagents conversion (X_G, X_{TB}) and products selectivity (S_M, S_D) with reaction time are depicted in Figure 3a,b, respectively.

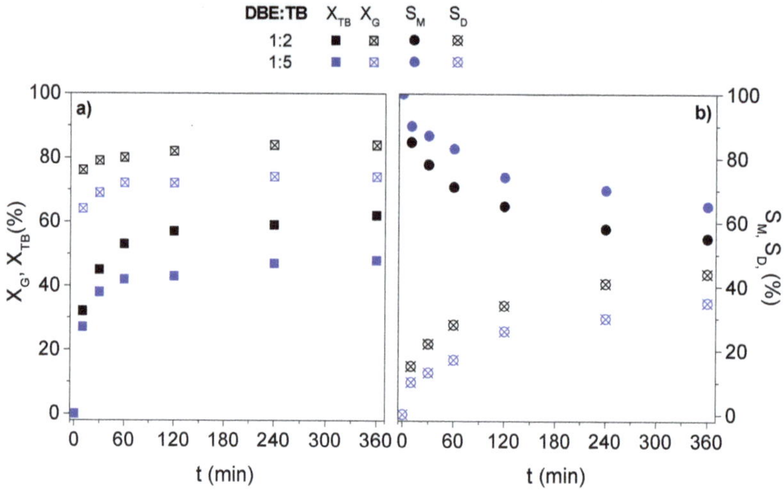

Figure 3. Influence of DBE concentration on G and TB conversion (**a**) and M and D selectivity (**b**) with reaction time. Operating conditions: 70 °C, 4 bar, G:TB = 1:3, C_{cat} = 20% (runs R1 and R2 of Table 3).

The reagents conversion (G and TB) was very fast in the early stages of the reaction (t < 60 min) (Figure 3a). From this reaction time, the conversion of glycerol hardly changed until the end of the experiment ($X_G \approx 85\%$, t = 360 min) and that of TB progressively increased. G conversion was always higher than TB conversion since the stoichiometric dose of these reagents were used and G only participates in the first step of the reaction (Figure 1).

As the etherification of G with TB is a series reaction, it is considered that the selectivity towards M is maximum at zero reaction time. Once produced, M1 and M2 react with TB giving rise to the generation of D1 and D2 (Figure 1). Thus, as the reaction proceeds, the selectivity towards M decreases, while the selectivity towards D increases (Figure 3b).

An increase in DBE concentration led to higher conversions of both reagents (Figure 3a) and therefore, greater production of the interest compounds. Thus, the conversion of G and TB at 360 min increased from 74 and 48% to 84 and 62%, respectively, when using a DBE:TB ratio = 1:5 and 1:2. As can be seen in Figure 3b, the selectivity towards the diethers was also favored when increasing the solvent concentration (from 35 to 44%, at 360 min reaction time). Therefore, a higher proportion of DBE in the reaction medium (with respect to the reagents) increased the production of diethers (greater conversion of reagents and selectivity towards these compounds).

The selectivity towards the production of IB at 360 min reaction time slightly increased with the concentration of DBE (from 35 to 39 when using a ratio DBE:TB of 1:5 and 1:2, respectively), as can be seen in Figure 4a.

Finally, it must be pointed out that the addition of DBE to the reaction system is not a major problem because TB and DBE can be easily recycled from the hydrophobic phase (by distillation) and being reused in a new cycle. Besides, this leads to the recovery of purified ethers from the reaction media. Thus, it was decided to work with the highest concentration tested of this solvent in the following experiments (DBE:TB = 1:2).

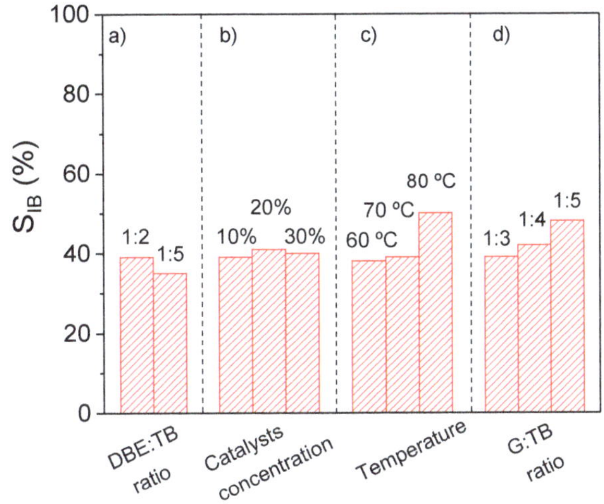

Figure 4. Effect of different operating variables on IB selectivity. Operating conditions: 70 °C, G:TB = 1:3, C_{cat} = 20% (**a**), 70 °C, G:TB = 1:3, DBE:TB = 1:2 (**b**), G:TB = 1:3, DBE:TB = 1:2, Ccat = 20% (**c**) 70 °C, DBE:TB = 1:2, C_{cat} = 20% (**d**).

2.3.2. Effect of Catalyst Concentration

In previous works, it was found that the optimal concentration of A-15 in the etherification of G with TB was 7.5% [9]. However, considering that the reaction equilibrium has been modified due to the addition of DBE, this parameter may have been affected and it must be studied again.

Thus, the influence of A-15 concentration working with three catalyst loads (10, 20 and 30%) and maintaining constant the other operating variables (70 °C, 4 bar, G:TB = 1:3, DBE:TB = 1:2) has been evaluated. A control experiment without catalyst was also performed, obtaining negligible conversions of the reagents during 360 min (data not shown), which is in agreement with the results previously published in the absence of co-solvent [7]. As can be seen in Figure 5a, the conversion of G greatly increased by increasing the concentration of A-15 from 10 to 20%, but this increase was negligible for higher concentrations. Surprisingly, the conversion of TB when using a catalyst concentration of 20% (R1) was higher than the corresponding to a catalyst load of 30% (R4). A catalyst content in the reaction medium of 30% could be excessive, resulting probably in the formation of aggregates, which reduces the active area of the catalyst and therefore, part of the catalytic surface could remain inaccessible to the reagents.

Furthermore, the selectivity towards the diethers greatly improved (from 26 to 44% at 360 min reaction time) by increasing the concentration of the catalyst from 10 to 20%, but no improvement was obtained when operating with higher catalyst concentrations (Figure 5b).

On the other hand, the concentration of catalyst seems to have no significant effect on the selectivity towards isobutene (Figure 4b). Consequently, it has been estimated that the optimal catalyst concentration in this system is 20%.

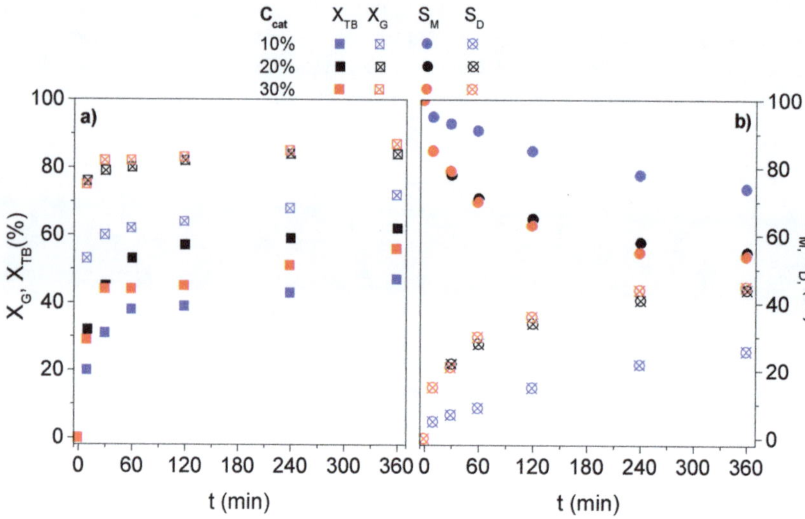

Figure 5. Influence of catalyst concentration on G and TB conversion (**a**) and M and D selectivity (**b**) with reaction time. Operating conditions: 70 °C, 4 bar, G:TB = 1:3, DBE:TB = 1:2 (R3, R1 and R4).

2.3.3. Effect of Operating Temperature

The equilibrium of the reaction is determined by the operating temperature, so this variable plays a key role. Thus, knowing the effect of this parameter on the etherification of G with TB in the presence of DBE is necessary. The etherification reaction was carried out at three different temperatures (60, 70 and 80 °C), always below the maximum temperature at which the catalyst can be subjected to (120 °C, property provided by the supplier) and maintaining invariable the other operating conditions (4 bar, G:TB = 1:3, DBE:TB = 1:2, C_{cat} = 20%). Moreover, it has been previously observed that when working at higher temperatures (≥ 90 °C), glycerol was not totally converted due to the occurrence of de-etherification reactions [7,19].

As expected, an increase in the operating temperature led to higher reaction rates and consequently, higher conversions of G and TB at the equilibrium, due to the endothermic nature of the reactions (this effect was especially notable in the case of TB, Figure 6a). These results are in agreement with those published when no solvent was added to the reaction medium. Similarly, as the reaction temperature increased, the selectivity towards the formation of diethers also increased (Figure 6b). Once the reaction equilibrium was reached (360 min) at 80 °C (R6), S_D was higher than S_M (Figure 6b). The introduction of DBE in the reaction system led to a new reaction equilibrium that allowed operating at higher temperatures. It should be noted that at this temperature (80 °C), in the absence of DBE, the selectivity towards the diethers was only 23% [9].

Otherwise, the secondary reaction of IB production was also favored by an increase in temperature (Figure 4c), and the yield of the reaction towards the products of interest decreased. For this reason, and in order to reach a compromise between the conversion of reagents, the selectivity towards the products of interest and the selectivity towards IB, 70 °C has been selected as the most convenient reaction temperature.

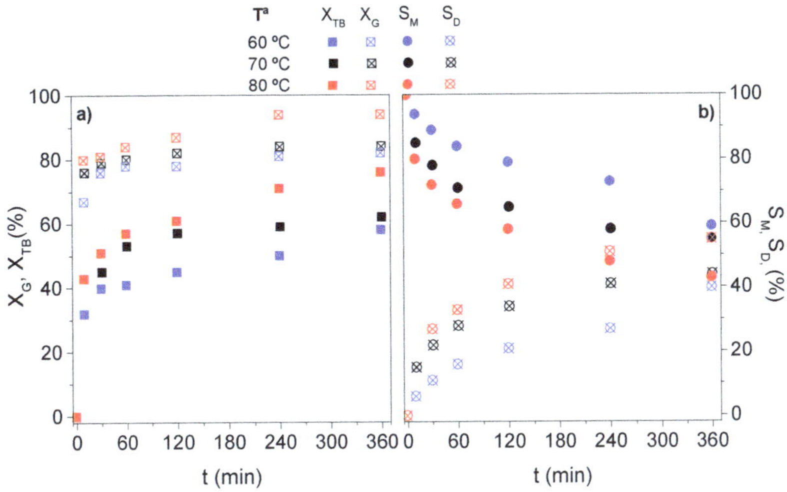

Figure 6. Influence of reaction temperature on G and TB conversion (**a**) and M and D selectivity (**b**) with reaction time. Operating conditions: 4 bar, G:TB = 1:3, DBE:TB = 1:2, C_{cat} = 20% (R5, R1 and R6).

2.3.4. Effect of *Tert*-Butyl Alcohol (TB) Concentration

Three G:TB molar ratios have been tested, one corresponding to the stoichiometric ratio between both reagents for a total conversion to triether (G:TB = 1:3, R1) and two others in which TB was in excess (G:TB = 1:4 and 1:5, R7 and R8, respectively), maintaining constant the other operating conditions (70 °C, 4 bar, DBE:TB = 1:2, Ccat = 20%). The results have been depicted in Figure 7a,b. The conversion of G was hardly affected by the concentration of TB (≈ 85% at 360 min, regardless of the G:TB ratio, Figure 7a). Consequently, the extension of the reaction did not change when working with an excess of TB. As expected, the conversion of TB decreased when increasing the concentration of this reagent.

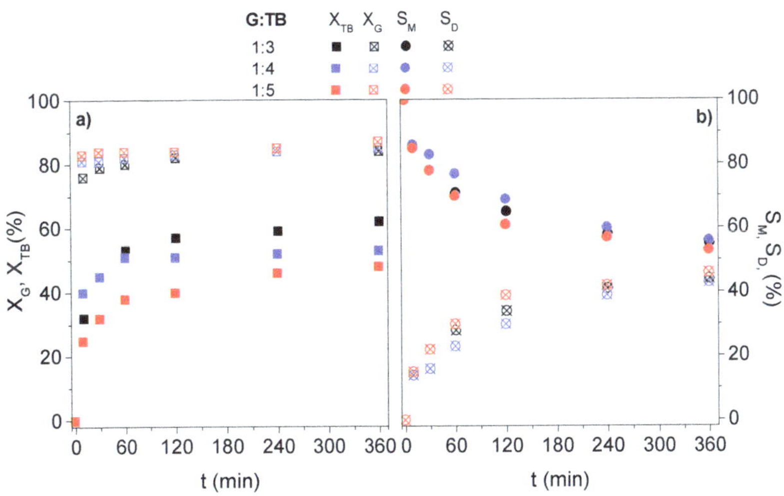

Figure 7. Influence of reaction temperature on G and TB conversion (**a**) and M and D selectivity (**b**) with reaction time. Operating conditions: 70 °C, 4 bar, DBE:TB = 1:2, C_{cat} = 20% (R1, R7 and R8).

Regarding the selectivity towards mono and diethers, it has been shown that a modification in the G:TB ratio did not affect the reaction in terms of products distribution and therefore, no significant changes were observed in the selectivity towards M and D (Figure 7b). These results are in agreement with those obtained by Frustreri et al. (2009) when working with a similar reaction system in the absence of DBE [7]. Finally, the generation of non-desired reaction products, IB (Figure 4d) and water (data do not shown) increased when increasing TB concentration. The selectivity towards IB increased from 39 to 48% when the ratio G:TB was 1:3 and 1:5, respectively (Figure 4d). Thus, it was determined that working with an excess of TB is not necessary in this system, thus reducing the costs of the process.

2.4. Catalyst Stability

The stability of the catalyst is a key factor for the economy of the process, especially taking into account the high concentration of catalyst used in the present study (20%, referred to the total starting mass of the system).

Once the reaction was stopped (360 min), the catalyst was filtered (by using a vacuum pump), washed with distilled water and stored in methanol [22,23]. Before undergoing a new catalytic cycle, the catalyst was separated from methanol and dried for 12 h at 60 °C. The stability study was carried out under the experimental conditions of R1 (70 °C, 4 bar, DBE:TB = 1:3, DBE:TB = 1:2 and C_{cat} = 20%), considered the most appropriate conditions from the previous results (2.3. Operation Condition Study). As can be seen in Figure 8a,b, the evolution of the conversion of both reagents (G and TB) and the selectivity towards the products of interest (M and D) with reaction time were constant during the three consecutive reaction cycles. Moreover, the selectivity towards secondary reactions (IB production) was not modified (data not shown). Thus, A-15 showed excellent stability under the selected operating conditions.

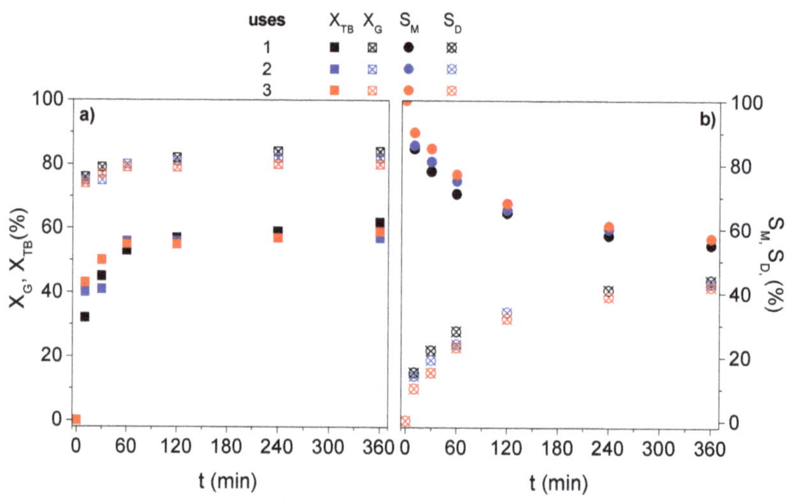

Figure 8. Influence of reaction temperature on G and TB conversion (**a**) and M and D selectivity (**b**) with reaction time. Operating conditions: 70 °C, 4 bar, G:TB = 1:3, DBE:TB = 1:2, C_{cat}=20% (R1, R9 and R10).

3. Conclusions

It has been demonstrated that the addition of dibutyl ether (DBE) as solvent in the etherification of glycerol (G) with *tert*-butyl alcohol (TB) in the presence of an acid solid catalyst (ion exchange commercial resin Amberlyst, A-15) had a positive effect since this compound allowed shifting the reaction equilibrium towards the formation of diethers (D).

The study of the main operating conditions revealed that an increase in the concentration of DBE improved the conversion of reagents (G and TB) and the selectivity towards the products of interest (D) without modifying the production of isobutene (IB), an undesired by-product. The concentration of catalyst played an important role, increasing the production and selectivity to diethers. It has been determined that the optimal concentration of A-15 was 20%. On the other hand, an increase in the operating temperature (from 60 to 80 °C) greatly increased the selectivity towards diethers. However, it is necessary to reach a compromise value (T^a = 70 °C) to minimize the undesired production of IB. Finally, the molar ratio G:TB had negligible influence on the performance of the reaction. Therefore, it has been concluded that working with the stoichiometric ratio (G:TB = 1:3) is the most convenient option.

The reuse of the catalyst in three consecutive reaction cycles showed its high stability, achieving identical results in terms of reagents conversion and selectivity towards the products of interest.

4. Materials and Methods

4.1. Reagents

Anhydrous glycerol (G, $C_3H_8O_3$, purity ≥ 99.8%), supplied by Fluka (Bucharest, Romania), and anhydrous *tert*-butyl alcohol (TB, $C_4H_{10}O$, purity ≥ 99.5%), provided by Sigma-Aldrich (Darmstadt, Germany), were employed as reagents. Dibutyl ether (DBE, $C_8H_{18}O$, purity ≥ 99.3%), purchased by Sigma-Aldrich was used as solvent to improve the performance of the etherification reaction. Pyridine reagent grade (C_5H_5N, purity ≥ 99.5%), supplied by Scharlau (Madrid, Spain), was used as solvent for the preparation of the samples before analysis. 1-terbutoxipropano-2,3-diol (M1) was used for the calibration of the gas chromatograph (GC). 1-Pentanol ($C_5H_{12}O$, purity ≥ 99%) supplied by Sigma-Aldrich (Darmstadt, Germany), was employed as internal standard compound (ISTD) in the GC analysis. Hydranal-Composite 5 and dry Hydranal-Methanol, both from Sigma-Aldrich (Darmstadt, Germany), were used as reactive compound and solvent, respectively, for the water analysis. Methanol (CH_4O, purity ≥ 99.8%, Sigma-Aldrich, -Darmstadt, Germany) was used to wash the catalyst between successive reactions.

4.2. Catalyst

Commercial acid ion-exchange resin Amberlyst 15 (A-15) was supplied by Fluka (Bucharest, Romania) and used as received, without further treatments. This catalyst was selected by Pico et al. (2012) as the most interesting material among three types of commercial ion exchange resins (i.e., Amberlyst 15, Amberlite 200 and Amberlite IRC-50) tested in the etherification of glycerol [9]. The A-15 resin showed the highest activity (glycerol conversion) and selectivity towards the desired product (diethers, D).

The BET surface area, average pore volume, external surface area, and average pore diameter are $S_{BET} = 44\ m^2\ g^{-1}$, $V_p = 0.338\ cm^3\ g^{-1}$, $A_t = 44\ m^2\ g^{-1}$, and $d_p = 50 \times 10^{-8}$ m, respectively. The exchange capacity of the catalysts studied (NaCl 0.5 mol L^{-1} and NaOH 0.1 mol L^{-1}) is 4.05 eq Kg^{-1}. Further information and details about A-15 characterization can be found elsewhere [9].

4.3. Etherification Experiments

4.3.1. Solubility Test

Firstly, a solubility test of the compounds involved in the reaction (G, TB), as well as the only ether commercially available (M1), was carried out in a water:DBE system (1:1). The aim of this preliminary experiment is to study the distribution of reagents and products between both phases aqueous (hydrophilic) and organic (hydrophobic). This test was carried out at atmospheric pressure and ambient temperature, so the quantitative results could not be directly extrapolated to those obtained under the reaction conditions, but they will be representative to know the distribution of these compounds between both phases.

The reagents were added to a glass reactor and stirred vigorously with a vortex-type stirrer during 10 min. After that, the aqueous and organic phases were separated with the aid of a decanter and immediately analyzed.

4.3.2. Etherification Reactions

Etherification reactions were carried out batch-wise in an autoclave reactor (BüchiGlasUster. Uster Switzerland, 50 mL) with magnetic stirring at constant temperature and pressure (N_2, 4 bar). The stirring speed selected (1000 rpm) guarantee the absence of external mass-transfer resistances [9]. At these conditions, internal mass-transfer limitations were discarded in a previous work [9]. The reactor was held at constant temperature by immersion in a hot water bath. The reactor temperature was continuously measured and controlled with a thermocouple immersed in the liquid reaction medium.

In a typical run, the catalyst was charged into the vessel of the reactor with the appropriate volume of the solvent (DBE). Then, the reactor was stoppered, heated and pressurized under nitrogen atmosphere to the desired conditions. When the selected temperature was reached, the reagents (G and TB) were injected and the stirring started (the total mass of the system was fixed at 40 g). This time was considered the starting point of the reaction. During reaction time, liquid samples (emulsified) were periodically withdrawn from the reactor and immediately injected in a vial (submerged in crushed ice) containing a known volume of cold pyridine. The diluted samples were subsequently analyzed. Gaseous isobutylene (IB), continuously generated from TB dehydration (Figure 1), was recovered with a cold trap filled with pyridine (located just after the reactor to liquefy the sample) during the whole reaction time. The concentration of IB was measured at the end of the run to check the mass balance and calculate the selectivity of TB towards this non-desired compound. After 360 min, the reaction was finally stopped by cooling the reactor.

Table 3 summarizes the operating conditions selected for each experiment. Etherification runs were performed with two solvent concentrations (DBE:TB molar ratios used were 1:2 and 1:5), three different catalyst loading (from 10 to 30 %, referred to the total starting mass, G + TB + DBE), three temperatures within the range 60–80 °C, and three different concentrations of TB (the G:TB molar ratios used were 1:3, 1:4 and 1:5). The molar ratio G:TB = 1:3 corresponds to the stoichiometric ratio. Once the optimum reaction conditions have been selected, the stability of the catalyst was studied during three consecutive reaction cycles.

Table 3. Operating conditions of the etherification runs.

Run	DBE:TB	C_{cat} (%)	T^a (°C)	G:TB	Catalyst Use
R1	1:2	20	70	1:3	1st use
R2	1:5	20	70	1:3	1st use
R3	1:2	10	70	1:3	1st use
R4	1:2	30	70	1:3	1st use
R5	1:2	20	60	1:3	1st use
R6	1:2	20	80	1:3	1st use
R7	1:2	20	70	1:4	1st use
R8	1:2	20	70	1:5	1st use
R9	1:2	20	70	1:3	2nd use
R10	1:2	20	70	1:3	3rd use

4.3.3. Analytical Methods

The progress of the reaction was followed by analyzing liquid samples at different reaction times.

The reactor composition (TB, G, IB and the different ethers generated, M1, M2, D1, D2 and T) was analyzed by means of an Agilent 6850 gas chromatograph (Santa Clara, CA, USA) fitted with a flame ionization detector (GC-FID). An HP Innowax chromatography column (30 m length

× 0.32 mm internal diameter) was used as stationary phase. The chromatographic conditions selected were the following: Initial oven temperature 40 °C, final oven temperature 220 °C, program rate 20 °C min^{-1}. 1-Pentanol was used as internal standard, ISTD, to minimize the error in the quantification of the different compounds. Commercial 1-*tert*-butoxypropane-2,3-diol (M1) was employed to obtain the corresponding response factor for this compound, which was extrapolated to the non-commercial products such as 2-*tert*-butoxypropane-1,3-diol (M2), 1,3-di-*tert*-butoxypropan-2-ol (D1), 1,2-di-*tert*-butoxypropan-3-ol (D2), and 1,2,3- tri-*tert*-butoxypropane (T).

Water concentration (generated in the different steps of the reaction) was measured at the end of the experiment (360 min) with a quantitative Karl Fischer analyzer (Titromatic 1S Crison, Barcelona, Spain).

Author Contributions: A.S. and A.R. made substantive intellectual contributions to this study, especially with the conceptualization and the methodology used. C.M.D. conceived and performed the experiments and wrote the paper. All the authors have been also involved in drafting and revising the manuscript, so that everyone has given final approval of the current version to be published in Catalysts Journal.

Funding: This research received no external funding.

Acknowledgments: Carmen M. Domínguez acknowledges the Spanish MICINN for the "Juan de la Cierva" post-doctoral grant (FJCI-2016-28462).

Conflicts of Interest: The authors declare no conflict of interest.

References

1. Demirbas, A. Progress and recent trends in biofuels. *Prog. Energy Combust. Sci.* **2007**, *33*, 1–18. [CrossRef]
2. Sani, Y.M.; Daud, W.M.A.W.; Aziz, A.R.A. Activity of solid acid catalysts for biodiesel production: A critical review. *Appl. Catal. A* **2014**, *470*, 140–161. [CrossRef]
3. Knothe, G.; Krahl, J.; Van Gerpen, J. *The Biodiesel Handbook*; Elsevier: Amsterdam, The Netherlands, 2015.
4. Meher, L.; Sagar, D.V.; Naik, S. Technical aspects of biodiesel production by transesterification—A review. *Renew. Sustain. Energy Rev.* **2006**, *10*, 248–268. [CrossRef]
5. Melero, J.A.; Vicente, G.; Paniagua, M.; Morales, G.; Muñoz, P. Etherification of biodiesel-derived glycerol with ethanol for fuel formulation over sulfonic modified catalysts. *Bioresour. Technol.* **2012**, *103*, 142–151. [CrossRef]
6. Jaworski, M.A.; Vega, S.R.; Siri, G.J.; Casella, M.L.; Salvador, A.R.; López, A.S. Glycerol etherification with benzyl alcohol over sulfated zirconia catalysts. *Appl. Catal. A* **2015**, *505*, 36–43. [CrossRef]
7. Frusteri, F.; Arena, F.; Bonura, G.; Cannilla, C.; Spadaro, L.; Di Blasi, O. Catalytic etherification of glycerol by tert-butyl alcohol to produce oxygenated additives for diesel fuel. *Appl. Catal. A* **2009**, *367*, 77–83. [CrossRef]
8. Bozbas, K. Biodiesel as an alternative motor fuel: Production and policies in the European Union. *Renew. Sustain. Energy Rev.* **2008**, *12*, 542–552. [CrossRef]
9. Pico, M.P.; Romero, A.; Rodríguez, S.; Santos, A. Etherification of glycerol by tert-butyl alcohol. *Ind. Eng. Chem. Res.* **2012**, *51*, 9500–9509. [CrossRef]
10. Pico, M.P.; Rodríguez, S.; Santos, A.; Romero, A. Etherification of glycerol with benzyl alcohol. *Ind. Eng. Chem. Res.* **2013**, *52*, 14545–14555. [CrossRef]
11. Marchetti, J.; Miguel, V.; Errazu, A. Possible methods for biodiesel production. *Renew. Sustain. Energy Rev.* **2007**, *11*, 1300–1311. [CrossRef]
12. Silva, P.H.R.; Gonçalves, V.L.C.; Mota, C.J.A. Glycerol acetals as anti-freezing additives for biodiesel. *Bioresour. Technol.* **2010**, *101*, 6225–6229. [CrossRef]
13. Sivaiah, M.V.; Robles-Manuel, S.; Valange, S.; Barrault, J. Recent developments in acid and base-catalyzed etherification of glycerol to polyglycerols. *Catal. Today* **2012**, *198*, 305–313. [CrossRef]
14. Gabriele, B.; Mancuso, R.; Salerno, G.; Veltri, L.; Costa, M.; Dibenedetto, A. A general and expedient synthesis of 5-and 6-membered cyclic carbonates by palladium-catalyzed oxidative carbonylation of 1, 2-and 1, 3-diols. *ChemSusChem* **2011**, *4*, 1778–1786. [CrossRef]
15. Karinen, R.; Krause, A. New biocomponents from glycerol. *Appl. Catal. A* **2006**, *306*, 128–133. [CrossRef]
16. Klepáčová, K.; Mravec, D.; Kaszonyi, A.; Bajus, M. Etherification of glycerol and ethylene glycol by isobutylene. *Appl. Catal. A* **2007**, *328*, 1–13. [CrossRef]

17. Klepáčová, K.; Mravec, D.; Bajus, M. Tert-butylation of glycerol catalysed by ion-exchange resins. *Appl. Catal. A.* **2005**, *294*, 141–147. [CrossRef]
18. Izquierdo, J.F.; Montiel, M.; Pales, I.; Outon, P.R.; Galan, M.; Jutglar, L.; Villarrubia, M.; Izquierdo, M.; Hermo, M.P.; Ariza, X. Fuel additives from glycerol etherification with light olefins: State of the art. *Renew. Sustain. Energy Rev.* **2012**, *16*, 6717–6724. [CrossRef]
19. Melero, J.A.; Vicente, G.; Morales, G.; Paniagua, M.; Moreno, J.M.; Roldán, R.; Ezquerro, A.; Pérez, C. Acid-catalyzed etherification of bio-glycerol and isobutylene over sulfonic mesostructured silicas. *Appl. Catal. A* **2008**, *346*, 44–51. [CrossRef]
20. Zhao, W.; Yang, B.; Yi, C.; Lei, Z.; Xu, J. Etherification of glycerol with isobutylene to produce oxygenate additive using sulfonated peanut shell catalyst. *Ind. Eng. Chem. Res.* **2010**, *49*, 12399–12404. [CrossRef]
21. Lee, H.J.; Seung, D.; Filimonov, I.N.; Kim, H. Etherification of glycerol by isobutylene. Effects of the density of acidic sites in ion-exchange resin on the distribution of products. *Korean J. Chem. Eng.* **2011**, *28*, 756–762. [CrossRef]
22. Lee, H.J.; Seung, D.; Jung, K.S.; Kim, H.; Filimonov, I.N. Etherification of glycerol by isobutylene: Tuning the product composition. *Appl. Catal. A* **2010**, *390*, 235–244. [CrossRef]
23. Pico, M.P.; Rosas, J.M.; Rodríguez, S.; Santos, A.; Romero, A. Glycerol etherification over acid ion exchange resins: Effect of catalyst concentration and reusability. *J. Chem. Technol. Biotechnol.* **2013**, *88*, 2027–2038. [CrossRef]
24. Chang, J.; Chen, D. Optimization on the etherification of glycerol with tert-butyl alcohol. *J. Taiwan Inst. Chem. Eng.* **2011**, *42*, 760–767. [CrossRef]
25. Ozbay, N.; Oktar, N.; Dogu, G.; Dogu, T. Effects of sorption enhancement and isobutene formation on etherification of glycerol with tert-butyl alcohol in a flow reactor. *Ind. Eng. Chem. Res.* **2012**, *26*, 8788–8795. [CrossRef]
26. Roze, M.; Kampars, V.; Teivena, K.; Kampare, R.; Liepiņš, E. Catalytic etherification of glycerol with alcohols. *Mat. Sci. Appl. Chem.* **2013**, *28*, 67–72. [CrossRef]
27. Melero, J.A.; Iglesias, J.; Morales, G. Heterogeneous acid catalysts for biodiesel production: Current status and future challenges. *Green Chem.* **2009**, *11*, 1285–1308. [CrossRef]

catalysts

MDPI

Article

Effect of Metal Loading in Unpromoted and Promoted CoMo/Al_2O_3–TiO_2 Catalysts for the Hydrodeoxygenation of Phenol

J. Andrés Tavizón-Pozos [1,2,*], Carlos E. Santolalla-Vargas [3], Omar U. Valdés-Martínez [1] and José Antonio de los Reyes Heredia [1,*]

1 Departamento de Ingeniería de Procesos e Hidráulica, Universidad Autónoma Metropolitana-Iztapalapa, Av. San Rafael Atlixco No. 86, Col. Leyes de Reforma 1ª Secc., Iztapalapa, C.P. 09310 Ciudad de México, Mexico; omurvama@gmail.com

2 Cátedras CONACyT-Área Académica de Química, Universidad Autónoma del Estado de 5 Hidalgo, Carr. Pachuca-Tulancingo Km. 4.5, C.P. 42184 Pachuca, Hidalgo, Mexico

3 Departamento de Biociencias e Ingeniería, Centro Interdisciplinario de Investigaciones y Estudios sobre Medio Ambiente y Desarrollo, Instituto Politécnico Nacional, C.P. 07340 Ciudad de Mexico, Mexico; csantolallav@ipn.mx

* Correspondence: jarh@xanum.uam.mx (J.A.d.l.R.H.); jesus.tavizon@conacyt.mx (J.A.T.-P.)

Received: 29 May 2019; Accepted: 14 June 2019; Published: 19 June 2019

Abstract: This paper reports the effects of changes in the supported active phase concentration over titania containing mixed oxides catalysts for hydrodeoxygenation (HDO). Mo and CoMo supported on sol–gel Al_2O_3–TiO_2 (Al/Ti = 2) were synthetized and tested for the HDO of phenol in a batch reactor at 5.5 MPa, 593 K, and 100 ppm S. Characterization results showed that the increase in Mo loading led to an increase in the amount of oxide Mo species with octahedral coordination (Mo^{Oh}), which produced more active sites and augmented the catalytic activity. The study of the change of Co concentration allowed prototypes of the oxide species and their relationship with the CoMo/AT2 activity to be described. Catalysts were tested at four different Co/(Co + Mo) ratios. The results presented a correlation between the available fraction of Co^{Oh} and the catalytic performance. At low Co^{Oh} fractions (Co/(Co + Mo) = 0.1), Co could not promote all MoS_2 slabs and metallic sites from this latter phase performed the reaction. Also, at high Co/(Co + Mo) ratios (0.3 and 0.4), there was a loss of Co species. The Co/(Co + Mo) = 0.2 ratio presented an optimum amount of available Co^{Oh} and catalytic activity since the XPS results indicated a higher concentration of the CoMoS phase than at a higher ratio.

Keywords: hydrodeoxygenation; phenol; Al_2O_3-TiO_2; CoMo; CoMoS; biofuels; MoS_2

1. Introduction

Studies on the transformation of fast pyrolysis oils from lignocellulosic biomass as an alternative to produce clean and renewable transportation fuels have been increasing for the last years [1–3]. Bio-oil, rich in oxygenated compounds (30%–40%), needs to be upgraded to enhance its heating value, chemical and thermal stability, and miscibility with fossil fuels [4]. To achieve this, catalytic hydrodeoxygenation (HDO) can be used to eliminate oxygen and to hydrogenate instaurations for hydrocarbons chains. HDO can proceed in a wide range of temperatures (200–400 °C) and pressure (1–7 MPa) [5,6]. Also, as some oxygenated molecules present in pyrolysis oils are soluble in water, aqueous phase reactions can be carried out in the presence of a catalyst [7,8]. However, the aqueous hydrodeoxygenation may only be effective for some fractions of the bio-oil. For the more refractory molecules, this process could take advantage of the current technology of typical hydrotreatment processes used for petroleum feeds since upgraded bio-oil can be blended with them and be transformed in the same step [5]. Several works have been focused on the improvement of their catalytic properties, for instance selectivity, activity, and active phase dispersion [2,9–12]. These studies have used resulting lignin representative probe molecules, for example, guaiacol, eugenol, furans, cresol, catechol, anisole, and phenol, to understand catalytic performance and functionalities [7,13–15]. Particularly, phenol has been used as a probe molecule since its reactions could expose information about the reactivity of the C_{AR}–OH bond and hydrogenation capacity of the catalyst. Additionally, it is formed as a remnant of more complicated oxygen containing molecules [10,14,16,17]. HDO of phenol can proceed by two pathways: Hydrogenolysis or direct dehydrogenation (DDO) and hydrogenation (HYD) [10,17]. The DDO route consists in the cleavage of the CAR–OH bond (414 kJ/mol) to form benzene. For its part, the HYD route proceeds by hydrogenation of the π bonds of the aromatic ring to generate an oxygenated intermediate (O–I, cyclohexanol, and cyclohexanone). Therefore, the C–OH bond splits to produce cyclohexene and subsequently cyclohexane [10,17].

Typical hydrotreatment $CoMo/Al_2O_3$ sulfided catalysts have been used in HDO with probe molecules with promising results [10,13,18,19]. It is widely accepted that the active sites are located at the edges of the formed sulfided slabs. The Mo-edge has been attributed the role of hydrogenation due to its metallic character, whereas the active sites for hydrogenolysis path (C–O scission) have been identified as coordinatively unsaturated sites (CUSs) of sulfur at the S-edge on the MoS_2 phase [20,21]. These sulfided vacancies possess an electrophilic character, where oxygen from phenol can adsorb and SH groups provide hydrogen to carry out the C–O bond cleavage [5,22]. When Co is added as a promoter atom, hydrotreatment activity increases significantly by the formation of the CoMoS II phase as proposed by Topsoe et al. [21]. In this highly active phase, it has been proposed that Co has a preference to be located at the S-edge on the MoS_2 phase, which may promote the hydrogenolysis pathway [22,23]. Nonetheless, in HDO reactions, there is still debate concerning the way that the oxygenated functional groups of the molecules react on the different sulfide phases. On this regard, to achieve improvements of the catalysts by increasing the concentration of the CoMoS phase, the synergic effect of the Co concentration in the active sites for these reactions needs to be understood [24–26].

To develop highly active CoMoS catalysts supported on alumina, metal–support interaction (MSI) between oxide precursors for both metal sulfides is an important factor to tune for the supported active phase structure, dispersion, reducibility, and promotion [20–22]. One possible way is to modify Al-based mixed oxides, such as Al_2O_3–TiO_2 (AT). This mixed oxide has been proposed as an alternative support to take advantage of the combined properties of alumina and titania [27–33]. It had displayed notorious improvements in the textural and physicochemical properties in comparison to alumina or titania supports [30–34]. Particularly, CoMoS supported over sol–gel synthetized Al_2O_3–TiO_2 catalysts with an Al/Ti = 2 ratio (called "AT2") showed that the interactions between the metal oxide phases and the support enhance the formation of the sulfided active phases compared to alumina, leading to higher activities [32,35–37]. In a previous work of the HDO of phenol, it was shown that CoMo/AT2 was twice as active and was more selective to the DDO route than $CoMo/Al_2O_3$ catalyst [34]. These results were attributed to a decrease in the MSI due to the presence of titania in the support. The differences in activity and selectivity were attributed to a higher fraction of supported MoO_x species with octahedral coordination (Mo^{Oh}) on AT2 than on alumina. These Mo^{Oh} species are easier to reduce than MoO_x with tetrahedral coordination (Mo^{Th}), which were more abundant on alumina [34]. As it has been reported, variations on metal loading may be induced in the formation of oxide species and therefore in the active phase formation [26]. Additionally, the Co/(Co + Mo) ratio must be optimized in order to avoid promoter loss into the support and segregation to complete the understanding of the synergic effect of the promoter on the MoS_2 in CoMo/AT2 catalysts. However, there is no information about the effects of metal loadings on the formation of different species over AT supports, and their impact on the activity and selectivity of HDO reactions. Therefore, the objective of this work was gain further insight on the Co/(Co + Mo) ratio and its effect on C–O bond cleavage of CoMoS/AT catalyst for the HDO process using phenol as the model molecule.

2. Results and Discussion

2.1. Effect of Mo Loading

2.1.1. Catalytic Activity

The catalytic hydrodeoxygenation (HDO) experiments results, for non-promoted Mo catalysts are shown in Figure 1.

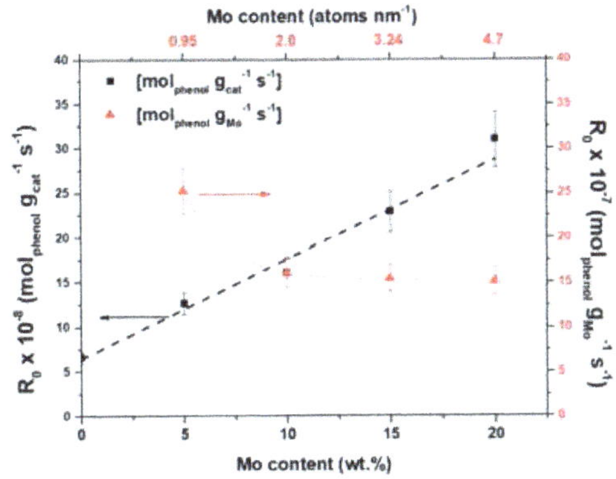

Figure 1. Normalized initial reaction rate of HDO of phenol for unpromoted Mo/AT2 catalysts at different Mo loadings at 5.5 MPa, 593 K, and 100 ppm S, (■) by gram of catalyst, (▲) by gram of Mo.

These results showed that the AT2 support was active and the initial reaction rate increased linearly with the Mo loading. Specifically, 5 wt.% Mo exhibited twice the activity of the support, while the activity for the 20 wt.% Mo catalyst was also five times higher. When the initial reaction rates were quantified by gram of supported Mo, a decrease was observed when the metal loading increased. However, at loadings higher than 10 wt.% Mo, activity remained constant. This may indicate that at this Mo loading, a monolayer was achieved. This result seems to be adequate to the system, since the Mo/Al_2O_3 monolayer coverage was near 10 wt.% Mo and 6.6 wt.% for Mo/TiO_2 [38–40]. Selectivity data at 20% of phenol conversion are presented in Figure 2.

It was observed that the AT2 support alone mainly generated incomplete hydrogenation products, i.e., oxygenated intermediates (O–I, cyclohexanone, and cyclohexanol) and cyclohexene. Therefore, AT2 support presented a higher hydrogenation functionality than hydrogenolysis. However, it is possible that there were not enough sites to achieve cyclohexene hydrogenation. When Mo was added, even at low loadings, the product yields changed to an increase in benzene and cyclohexene production compared with the AT2 support. Also, as all Mo catalyst yields did not present significant changes, it is possible that the active sites' nature was the same.

It has been suggested that S-edge sites are responsible for the hydrogenolysis pathway, while Mo-edge sites are responsible for hydrogenation reactions [5]. On this basis, it is possible to suggest that S-edge sites predominate in all Mo/AT2 catalysts since the hydrogenation of cyclohexene to cyclohexane was limited. This resulted in the capability to cleave the C_{AR}–OH and the C–OH (C = O) bonds from the O–I to produce benzene and cyclohexene, respectively. However, the O–I yield decreased with the Mo loading. Therefore, it is possible to propose that both hydrogenolysis processes may occur in the same type of sites and their abundance increased with the Mo loading.

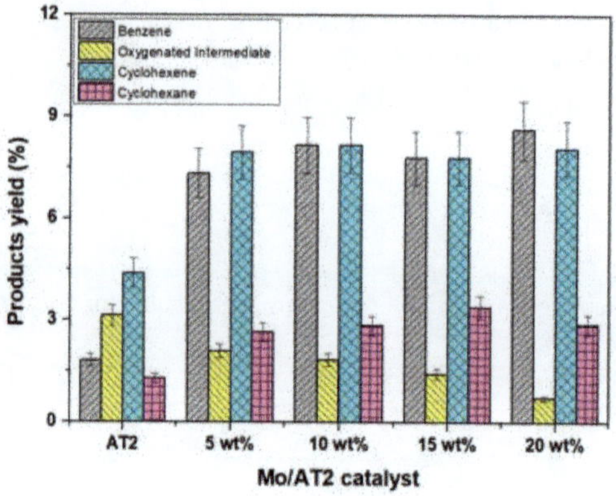

Figure 2. Products yields of HDO of phenol for unpromoted Mo/AT2 catalysts at different Mo loadings at 5.5 MPa, 593 K, and 100 ppm S.

2.1.2. Diffuse Reflectance Spectroscopy UV-Vis

Figure 3 gives the DRS UV-vis spectra for the calcined MoO_x catalysts supported on AT2. For all studied samples, DRS UV-vis spectra showed a single signal located from 200 to 600 nm. For the AT2 sample, a simple band was visible between 200 and 350 nm and corresponds to the metal–ligand charge transfer (MLCT) for $O^{2-} \rightarrow Ti^{4+}$ [41].

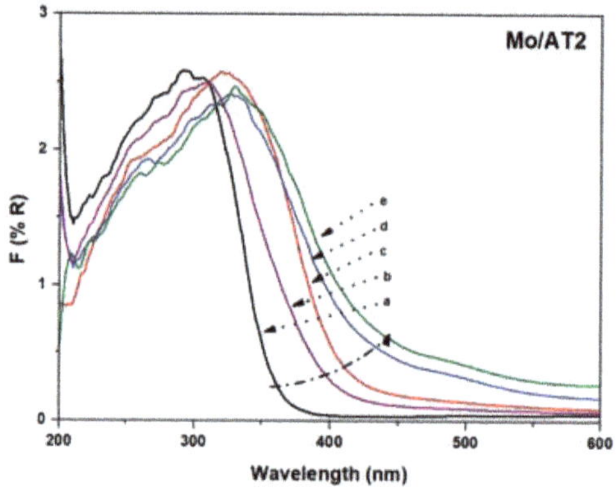

Figure 3. Diffuse reflectance UV-vis spectra of the unpromoted Mo/AT2 catalysts calcined at 673 K with different Mo loadings: **a**) AT2, (**b**) 5 wt.%, (**c**) 10 wt.%, (**d**) 15 wt.%, and (**e**) 20 wt.%.

For the supported Mo catalysts, it could be considered that the band between 200 and 400 nm included the signals of Mo and titania [36]. The MLCT band for $O^{2-} \rightarrow Mo^{6+}$ was located between 200 and 400 nm. In this band, Mo with tetrahedral coordination (Mo^{Th}) (MoO_4^{2-}, $Mo_2O_7^{2-}$) were located in the 200–300 nm range. Besides, Mo with octahedral coordination (Mo^{Oh}) from heptamolybdates and octamolybdates were included between 300 and 400 nm [41–43]. Even when a clear band assignation is complex, in a comparison between AT2 support with the different catalysts, it is possible to observe a shift in the reflectance bands to longer wavelengths with increments of the Mo loading. This could indicate a higher concentration of Mo^{Oh} compared with Mo^{Th} at high Mo loadings.

2.1.3. Laser Raman Spectroscopy

Figure 4 displays the laser Raman spectra for 10 and 15 wt.% Mo calcined catalysts.

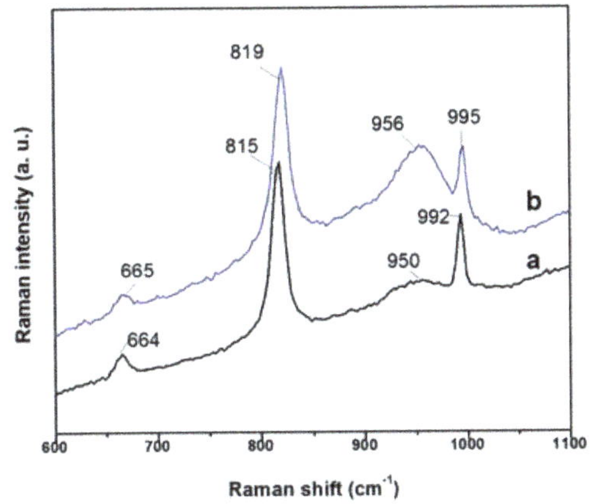

Figure 4. Laser Raman spectra of Mo/AT2 catalyst calcined at 673 K with different Mo loadings: (**a**) 10 wt.%, (**b**) 15 wt.%.

For the 10 wt.% catalyst, four peaks were displayed at 664, 815, 950, and 992 cm^{-1}. These signals presented a slight shift to a higher stretching frequency for 15 wt.% Mo due to the increment of metal content. The band located at 664 cm^{-1} was attributed to symmetric Mo-O-Mo deformations. Besides, 815 and 992 cm^{-1} frequencies were related to antisymmetric Mo-O-Mo stretching for bulk-like MoO_3 crystallites [26,44]. The band at 950 cm^{-1} was assigned to terminal Mo = O symmetric and asymmetric stretching bonds for coexisting polymolybdates ($Mo_7O_{24}{}^{6-}$ and $Mo_8O_{36}{}^{4-}$, Mo^{Oh}). The intensity of this signal increased for the 15 wt.% Mo loading, suggesting that more octahedral species could be present compared with the 10 wt.% Mo sample, in agreement with UV-vis results. This signal involved heptamolybdates and octamolybdates species, which were normally found around 920, 945, and 965 cm^{-1} [45,46]. Hence, this could indicate that MoO_3 structures are formed before the total coverage of the monolayer.

2.1.4. Temperature Programed Reduction

Unpromoted oxide Mo catalysts and AT2 support TPR profiles are shown in Figure 5. The AT2 TPR profile (Figure 5a) presented two reduction peaks attributed to the reduction of surface Ti species at 900 K and bulk Ti species ($Ti^{4+} \rightarrow Ti^{3+}$) at 1050 K [31,47].

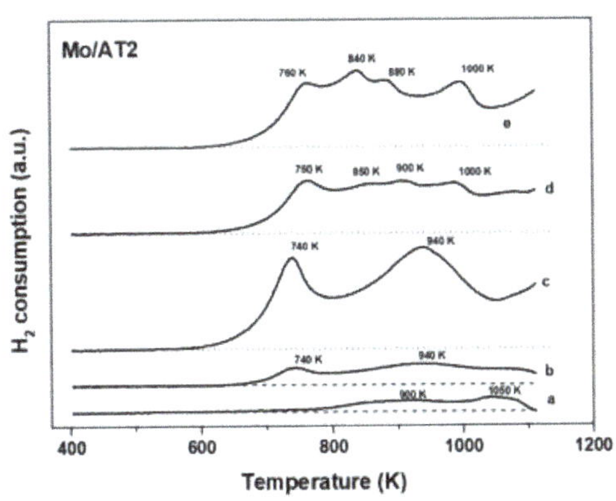

Figure 5. TPR profiles for unpromoted Mo/AT2 catalysts calcined at 673 K at different Mo loadings. (**a**) AT2, (**b**) 5 wt.%, (**c**) 10 wt.%, (**d**) 15 wt.%, and (**e**) 20 wt.%.

However, according to Platanitis et al. [47], only approximately 24% of anatase could be reduced on pure TiO_2. Hence, it is expected that less than this percentage of titania could be reduced in the mixed oxide support. In the Mo containing samples, all catalysts showed reduction peaks corresponding to $Mo^{6+} \rightarrow Mo^{4+}$ and $Mo^{4+} \rightarrow Mo^0$ at low temperatures (400–800 K) and high temperatures (>800 K), respectively [48–50]. Particularly, for the 5 wt.% Mo catalysts, a peak centered at 740 K corresponded to an easily reducible Mo species, possibly in octahedral coordination [50]. The signal located at 940 K was attributed to Mo strongly interacting with the support. Furthermore, the 10 wt.% Mo catalysts also showed two signals at 740 and 940 K. However, it is possible that the 940 K peak could not only be assigned for the Mo^{Th} species reduction, but also to a contribution from the MoO_3 species, as LRS results showed. The TPR profile of 15 and 20 wt.% Mo catalysts presented four signals at 780, 850, 880, and 1000 K. The peaks located between 780 and 900 K could be attributed to the reduction of Mo^{Oh} and a mixture of Mo^{Oh} and Mo^{Th}, respectively. The high temperature peaks (880 and 1000 K) could be caused by the presence of Mo^{Th} and bulk MoO_3 [47,49,50]. The absence of an 850 K peak at 5 and 10 wt.% Mo would indicate that at low Mo loadings, there may not be a notorious mixture of Mo^{Oh} and Mo^{Th} and more Mo^{Oh} was formed when Mo loading was augmented.

From the impregnation of Mo, the ammonium heptamolybdate solution for the 5 wt.% Mo catalyst's pH value was 5.2 and decreased to 4.5 for the 20 wt.% Mo catalyst. Consequently, ionic polymolybdates complexes, including $[H_3MoO_{24}]^{3-}$, $[H_2Mo_7O_{24}]^{4-}$, $[Mo_7O_{24}]^{6-}$ $[Mo_8O_{26}]^{4-}$, and $[HMo_7O_{24}]^{5-}$ species, were predominant in the solution as has been mentioned by the literature [46,51]. Additionally, since the AT2 isoelectric point value was 7.6, acid terminal OH groups were predominant on the surface at the impregnation pH. Hence, anionic polymolybdates species could anchor to the support by electrostatic interactions [52]. On this basis, at low metal loadings (5 wt.% Mo), the anionic Mo species could be well dispersed on the surface. As the metal loading increased, the anionic species were closer to each other and conglomerated, generating large Mo species that would have a weaker metal–support interaction than those present at low metal loadings. Characterization analysis showed that Mo^{Oh} oxide species increased with the metal loading. However, as Mo^{Th} species are not clearly visible in the DRS results, the Raman and TPR profiles confirmed the presence of a fraction of them and MoO_3 crystallites. These results may indicate that Mo dispersion decreased at high Mo loadings.

The presence of MoO_3 at the 10 wt.% Mo catalyst, and the catalytic results shown in Figure 1, indicated that the monolayer coverage would be complete at 20 wt.% Mo and that MoO_3 can be formed before it [42,53]. This behavior seems to be quite similar to the Mo/Al_2O_3 catalyst, which due to the parallel configuration of the hydroxyl groups of the support, leads to a rearrangement of the Mo species during calcination. In contrast, Mo/TiO_2 tends to form MoO_3 after total coverage of the monolayer, caused by the homogeneity of its hydroxyl groups [26]. Therefore, supported polymolybdates on AT2 may present a configuration more alike to alumina than titania. However, due to the presence of the latter, the generation of MoO_3 was delayed. On this basis, it can be suggested that this could occur just before the total formation of the monolayer.

Catalytic evaluation indicated that the AT2 support was active and presented a selectivity to partial hydrogenation. In contrast, when Mo was supported, even at low metal loadings, the selectivity changed to cyclohexane and benzene production. Nevertheless, since all unpromoted catalysts' selectivity did not present significant changes, it is possible to suggest that the generation of products was due to a high number of active sites with the same nature. Consequently, the non-covered parts of the support could contribute slightly to the production of the O–I and cyclohexene, whereas the MoS_2 phase contributed to the generation of DDO and HYD sites. Considering that Mo^{Oh} were easier to reduce than Mo^{Th}, more MoS_2 would be present at high Mo loadings (>10 wt.% Mo). In this sense, Mo^{Th} species may not change the selectivity; however, some of them would not be completely sulfided, i.e., only the number of active sites changed. This is deduced by the fact that even in the presence of CS_2, benzene was still produced. Due to the presence of a sulfiding agent (CS_2), the production of benzene was limited by the competition of CS_2 for the electrophilic sites located at the S-edge [5,10,34,54]. Then, as Mo^{Oh} increased, more S-edge sites were formed, and the resistance to inhibition was improved.

2.2. *Effect of Co Loading*

2.2.1. Catalytic Activity

The initial reaction rates for Co/AT2 and AT2 are presented in Table 1. The Co/AT2 catalyst was 1.3 times more active than the AT2 support. This means that Co containing sample presented active sites that improved the catalytic activity. Nevertheless, this activity was 0.3 times lower than that for the 5 wt.% Mo/AT2 sample (see Figure 1). The selectivity showed that cyclohexane was the main product on Co/AT2 in contrast with cyclohexene on AT2.

Table 1. Initial reaction rate of the HDO of phenol at 5.5 MPa and 593 K and product yields at 15% of the conversion for AT2 support and Co/AT2 sulfided catalyst.

Catalyst	R_0 ($\times 10^8$mol·(g_{cat}·s)$^{-1}$)	Product Yields at 15% of Phenol Conversion			
		Benzene (%)	O-I (%)	Cyclohexene (%)	Cyclohexane (%)
AT2	6.6 ± 0.6	3	4	6	2
Co/AT2	9.1 ± 0.9	3	1	2	9

The O–I yield decreased when Co was present on the catalyst, while benzene production remained the same. On this basis, the Co sulfide phase did not provide enough sites to cleave the C_{AR}–OH bond (468 kJ/mol). However, the high production of cyclohexane indicated that O–I could transform (339 kJ/mol) into cyclohexene to hydrogenate in a further step [17]. Nevertheless, these active sites presented a more hydrogenating character than hydrogenolysis, leading to cyclohexane. Furthermore, since Co did not cover the entire surface of the support, the AT2 support may have had a role. Since the AT2 support showed selectivity to O–I and cyclohexene, it is possible that these products were generated by the support in Co/AT2. In this sense, the AT2 support could have provided Bronsted acid sites, whereas the Co_9S_8 phase provided metallic sites, which have a hydrogenating character.

The synergic effect of Co for the HDO of phenol activity is presented in Figure 6.

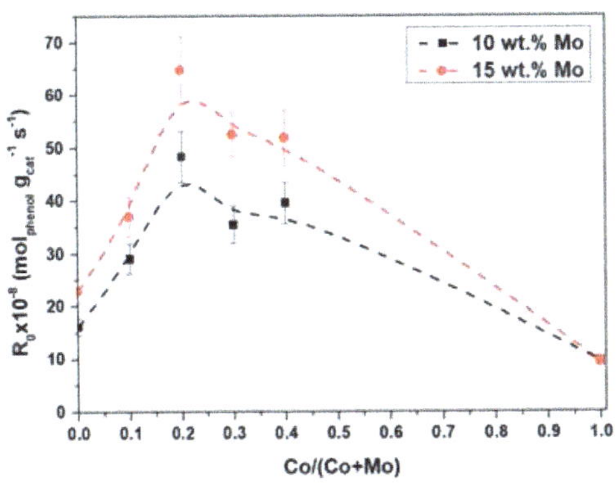

Figure 6. Initial reaction rates of HDO of phenol at different Co/(Co + Mo) ratios in the promoted CoMo/AT2 catalyst (■) at 10 wt.% Mo and (●) 15 wt.% Mo.

For both series, the dependence of the activity by the concentration of Co on the catalyst presented parallel volcano type curves with a maximum at Co/(Co + Mo) = 0.2. Since the concentration of Mo was higher at the promoted 15 wt.% Mo catalysts than the 10 wt.% Mo catalysts, the initial reaction rate increased 1.3 times. Table 2 presents a synergic factor to compare the activity of the promoted with the unpromoted catalysts, at different Co/(Co + Mo) ratios.

Table 2. Synergic effect in the HDO of phenol with CoMo/AT2 at different Co concentrations.

Co/(Co + Mo)	Synergic Factor
0.1	1.6
0.2	2.8
0.3	2.3
0.4	2.2

Following the volcano like curve, the Co/(Co + Mo) = 0.1 ratio presented a lower promoting factor (1.6) compared with the other samples. At an atomic ratio of 0.2, a maximum synergic factor was found (2.8) and this decreased by 20% when Co was loaded at higher atomic ratios. This indicated that Co loading played an important role in the interaction with MoS_2 slabs and subsequently the promotion of active sites. The selectivity changed for the different atomic ratios as illustrated in Table 3.

Table 3. Direct deoxygenation and hydrogenation route ratio of HDO of phenol at 20% of the phenol conversion for the CoMo/AT2 catalyst.

Co/(Co + Mo)	DDO/HYD
0.1	1.75 ± 0.1
0.2	1.60 ± 0.1
0.3	1.36 ± 0.1
0.4	1.27 ± 0.1

In all catalysts, phenol was the main product, following the direct deoxygenation (DDO) route by direct incision of the C_{AR}–OH bond. However, hydrogenated products (HYD), including O-I, cyclohexene, and cyclohexane, increased with the Co loading. As Co/AT2 showed, Co_9S_8 functionalities promoted the HYD route, thus the increment of the Co concentration led to the formation of this sulfide. At low Co loadings (Co/(Co + Mo) = 0.1), a fraction of DDO sites were promoted by this element at the edges of the MoS_2 phase [55]. At the Co/(Co + Mo) = 0.2 ratio, a high concentration of the CoMoS phase could be formed due to this ratio presenting the higher activity. Since not all MoS_2 would be promoted, the presence of Co_9S_8 could increase, leading to an increase in HYD selectivity. In this sense, at the Co/(Co + Mo) = 0.3 and 0.4 ratios, Co not only promoted the DDO route at the edges of the CoMoS phase but also contributed to the HYD route due to the presence of the Co_9S_8 segregated phase.

2.2.2. Diffuse Reflectance Spectroscopy UV-Vis

The Co/AT2 calcined catalyst spectra are presented in Figure 7.

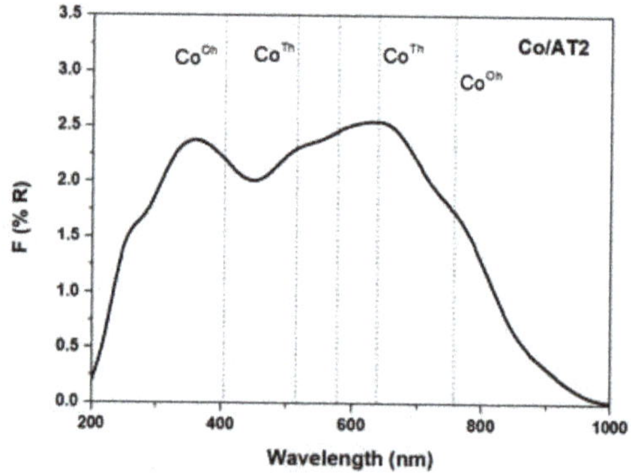

Figure 7. DR UV-vis spectra of the Co/AT2 calcined at 673 K.

It is possible to observe a shoulder centered at 260 nm corresponding to the MLCT $O^{2-} \rightarrow TiO^{4+}$ bands. Then, the Co transition bands were in a wide signal between 400 and 1000 nm. The first signal at 400 nm was attributed to the Co^{2+} species with octahedral coordination (Co^{Oh}) [56]. However, like the Mo results, this band was overlaid with the titania MLCT and the analysis was difficult. The wide signal involves a triplet centered at 500, 580, and 630 nm, corresponding to d–d transitions of the Co^{2+} species with tetrahedral coordination (Co^{Th}) and strongly interacting with alumina ($CoAl_2O_4$) [57]. Finally, a band near 750 nm was assigned to Co^{3+} and Co^{2+} species with octahedral coordination in Co_3O_4 [57]. On this basis, Co^{Th} species were more abundant than Co^{Oh} species since the respective were more intense. This was due to the absence of Mo and the low Co loading (2 wt.%) on the support. Co could be well dispersed on the support in small particles, which interacted strongly with the alumina present in the support. Nonetheless, as titania was highly dispersed in the alumina matrix [58], it could avoid the migration of Co into the support. This could promote the formation of Co^{Oh}, which is a Co_9S_8 precursor. The promoted supported catalysts' DR spectra are presented in Figure 8, showing two main bands, one located at 200–400 nm, and the other at 500 – 800 nm.

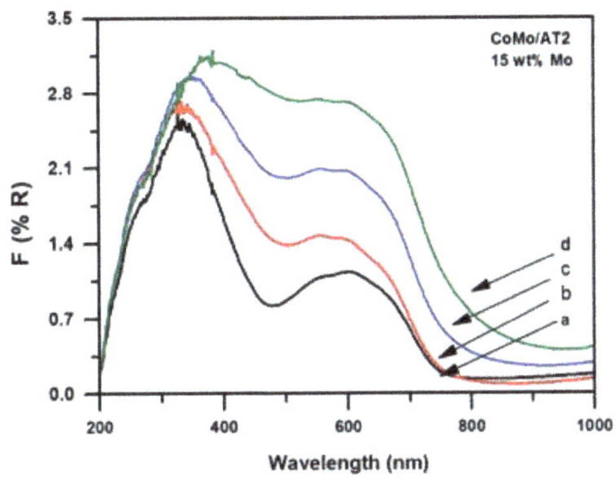

Figure 8. DR UV-vis spectra of the CoMo supported catalysts calcined at 673 K at different Co/(Co + Mo) ratios: (**a**) 0.1, (**b**) 0.2, (**c**) 0.3, and (**d**) 0.4.

Considering the previous unpromoted Mo and Co supported catalyst spectra, the fist signal (200–400 nm) involved the titania, and Mo and Co bands. However, the second signal only involved the Co species. In comparison to Figure 7, it is possible to observe an increase in the intensity of these bands with the Co/(Co + Mo) ratio. Also, a shift to near infrared was detected as the Co loading was increased. Additionally, the shoulder located at 400 nm (Co^{Oh}) increased its intensity at high atomic ratios. This may indicate that more Co with octahedral coordination was present at Co/(Co + Mo) = 0.4 compared with the other catalysts. Nevertheless, the Co^{Th} transition bands increased as well. Since Co may find its migration into the titania support difficult, these Co species could be part of $CoMoO_4$ in which Co was in octahedral coordination [59,60]. To achieve a proper comparison between the Co^{Oh} and Co^{Th} species, Gaussian deconvolution (not showed) was carried out considering the representative signals reported in the literature [56,57,61]. Since $CoMoO_4$ is considered a bad precursor for the CoMoS phase [62], the Co^{Oh} present in this oxide complex was considered as Co^{Th}. It is possible to calculate the ratio between Co^{Oh} and Co^{Th} with the area under each peak and using the $F(R\infty)^{Oh}/(F(R\infty)^{Oh} + F(R\infty)^{Th})$ equation [63]. The results are shown in Figure 9.

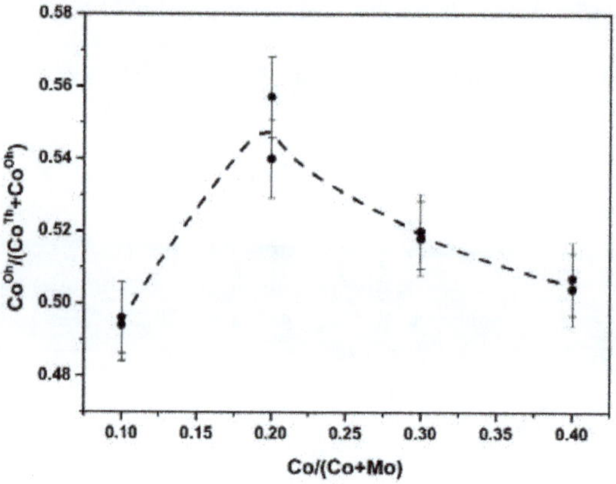

Figure 9. Correlation between Co^{Oh} and Co loading of the promoted CoMo catalyst with 10 wt.% and 15 wt.% Mo calcined at 673 K.

The correlation between Co^{Oh} and Co loading presented a volcano type curve with a maximum at the Co/(Co + Mo) = 0.2 ratio. At low Co/(Co + Mo) ratios, the concentration of the promoting atom was inadequate to interact with Mo. Hence, it resulted in small particles that interacted strongly with the support, leading to an incomplete promotion. In contrast, at high Co/(Co + Mo) ratios, the promoting atom was in excess. Thus, it resulted in the formation of $CoMoO_4$ and highly dispersed particles, which did not interact with Mo but with the support. In this sense, at the maximum Co^{Oh} ratio, there was enough Co^{Oh} to interact properly with the MoO_X species, leading to the formation of the CoMoS phase. The amount of Co^{Oh} and Co^{Th} obtained from DRS UV-vis analysis was correlated with the activity, as presented in Figure 10.

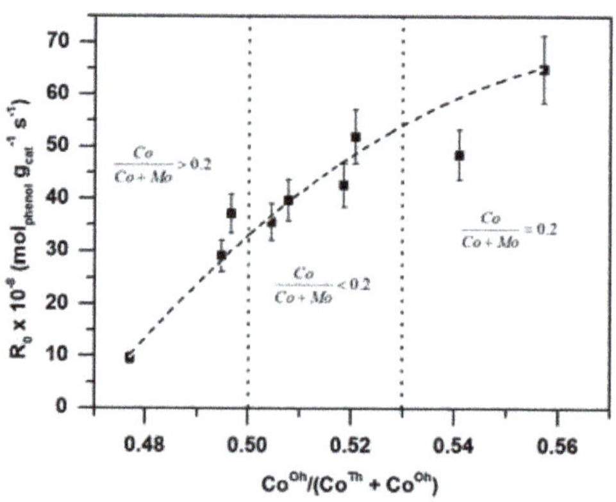

Figure 10. Correlation between the initial reaction rate of the HDO of phenol and octahedral Co of 10 wt.% and 15 wt.% Mo supported catalysts.

This correlation showed that as Co^{Oh} was increased, the initial reaction rate also increased. This is due to the proper promotion of Mo with Co^{Oh} to form the CoMoS phase. At low Co loadings, more Co^{Th} species were present on the support and their sulfidation was less than the Co^{Oh} species, as has been reported before [34,64]. At the optimum Co/(Co + Mo) ratio, the amount of Co^{Oh} promoted the proper formation of the CoMoS phase and a fraction of Co could be Co^{Th}. When Co was in excess, a fraction of it promoted the CoMoS phase, but also a fraction interacted strongly with Mo, giving place to $CoMoO_4$. However, the Co_3O_4 phase was also be present and could be transformed into Co_9S_8.

2.2.3. Temperature Programmed Reduction

In Figure 11, the Co/AT2 calcined catalyst's TPR profile is presented. It is possible to observe four signals centered at 650, 780, 1030, and 1098 K.

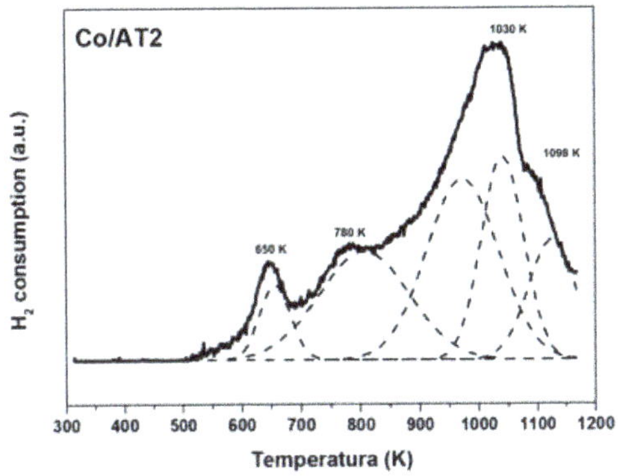

Figure 11. TPR profile of Co/AT2 catalyst calcined at 673 K.

By Gaussian deconvolution, an extra peak was found in the 1030 K peak. The first signal (650 K) was assigned to the reduction of Co_3O_4 crystals to CoO, i.e., the Co^{Oh}, $Co^{3+} \rightarrow Co^{2+}$ [65,66]. The signal at 780 K corresponded to the reduction of superficial well dispersed Co^{3+} species. Also, the deconvolution resulting peak at 923 K could be attributable to $Co^{2+} \rightarrow Co^0$, whereas, the second resulting peak at 1040 K may be referred to the $CoAl_2O_4$ species [50]. However, these peaks may have contributed to the initial titania ions' reduction to Ti^{3+}. Finally, the last peak at 1090 K was the result of the $Ti^{4+} \rightarrow Ti^{3+}$ reduction. Note that the Co^{Th} species, with strong interaction with the support, were more abundant than Co^{Oh}. This result confirms the DR UV-vis results since the Co metal–support interaction is strong enough to generate Co^{Th} species that are difficult to reduce. Therefore, their sulfidation was incomplete and less active phase was formed. As was previously seen in the literature [67], Co metal–support interactions decrease when Mo is present on the surface of the support. According to these results, the Co/AT2 catalyst presented a higher amount of Co^{Th} than Co^{Oh} due to the interaction with alumina. However, both species could be sulfided and form Co_9S_8, which had metallic sites with selectivity to the HYD route. The benzene production was inhibited by competition for the hydrogenolysis active sites by the presence of H_2S in the reactor. Additionally, the amount of these sites would be less than the hydrogenation sites. Since Co did not totally cover the support, the latter could contribute to the production of the O–I and cyclohexene.

The CoMo/AT2 catalyst's TPR profiles are presented in Figure 12. In general, all TPR profiles were similar to each other and showed three main peaks at 750, 950, and 1080 K.

The Gaussian deconvolution of these TPR profiles developed two additional peaks at 640 K and 820 K. The peak at 640 K corresponded to Co^{Oh}, $Co^{3+} \rightarrow Co^{2+}$ as shown in Figure 11 as well. The second peak, at 750 K, was assigned to the $Mo^{6+} \rightarrow Mo^{4+}$ reduction [51]. The third peak, at 820 K, could be assigned to a mixture of $Co^{2+} \rightarrow Co^0$ and $Mo^{4+} \rightarrow Mo^0$ [68]. The fourth peak near 940 K was caused by Mo strongly interacting with the support and involved $CoMoO_4$ and $CoAl_4O_3$ species [63]. Finally, the last high temperature peak belonged to the support.

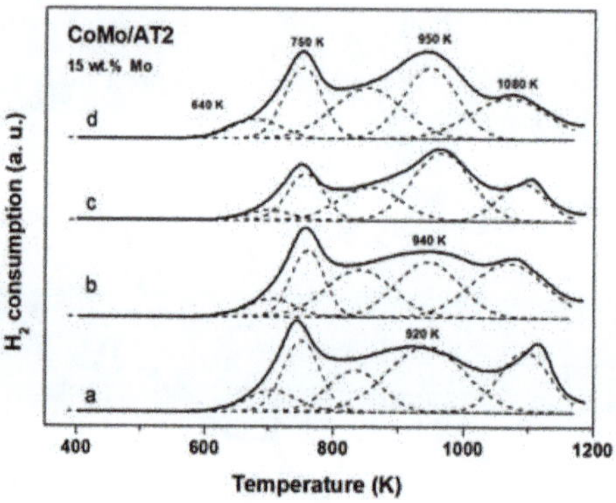

Figure 12. TPR profiles of CoMo/AT2 calcined catalyst at 15 wt.% Mo with different Co/(Co + Mo) ratios: (**a**) 0.1, (**b**) 0.2, (**c**) 0.3, and (**d**) 0.4.

In general, the reduction peaks did not present significant shifts to low temperatures with the increase in the Co loading. The Co/(Co + Mo) = 0.1 ratio catalyst showed a slight shift of the second main peak to low temperatures due to the low Co concentration, i.e., the contribution by the Mo reduction was predominant. By its part, the Co/(Co + Mo) = 0.2 catalyst consumed more H_2 in the first peak than in the second one, meaning that more octahedral Co and Mo species were present in this catalyst. This could lead to a better promotion of the active phase and therefore an improvement of the activity. At Co/(Co + Mo) = 0.4, the peak located at 640 K was more intense than in the other samples. This signal could be the result of the high concentration of segregated Co_3O_4. Additionally, since more segregated Co was present on these catalysts, Co^{Th} was present in a minor fraction.

2.2.4. X-Ray Photoelectron Spectroscopy

Figures 13 and 14 exhibit the XPS spectra for Mo3d and Co2p core levels for sulfided CoMo/AT2 catalysts at the Co/(Co + Mo) = 0.2 ratio with 15 wt.% Mo.

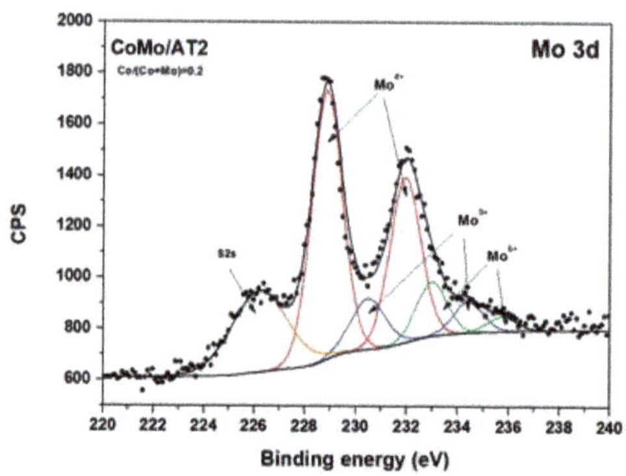

Figure 13. XPS deconvolution of Mo3d core level for sulfided 15 wt.% Mo CoMo/AT2 catalysts at the Co/(Co + Mo) = 0.2 ratio.

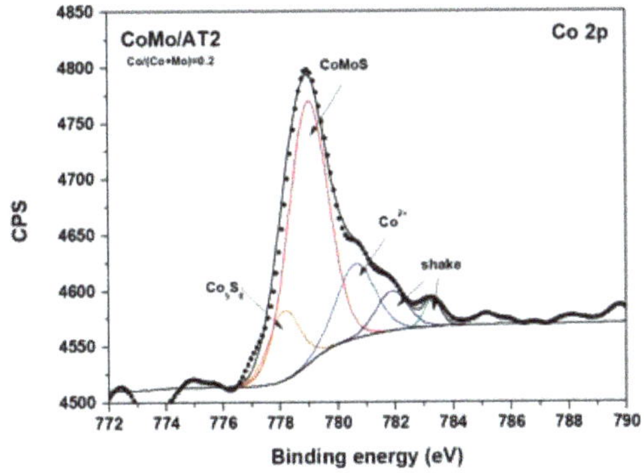

Figure 14. XPS deconvolution of Co 2p core level for sulfided 15 wt.% Mo CoMo/AT2 catalysts at the Co/(Co + Mo) = 0.2 ratio.

The Mo3d spectra displayed a doublet for the two spin orbit components, $3d_{5/2}$ and $3d_{3/2}$, located at 228.76 and 231.8 eV, respectively. Additionally, the S2s core levels' band was detected at 226.3 eV. The XPS for the Mo3d5/2 decomposition showed peaks attributed to Mo in sulfide (Mo^{4+}), oxysulfide (Mo^{5+}), and oxidic (Mo^{6+}) species. These peaks were located at 228.79 eV, 230.4 eV, and 232.80 eV, respectively [38,69,70].

For the $Co2p_{3/2}$ core levels displayed in Figure 14, the Co_9S_8, CoMoS, and Co^{2+} oxide species were found by decomposition of the main signal. The binding energies for Co species were identified at 778.16 eV for Co_9S_8, at 778.76 eV for CoMoS, and at 780.7 eV for oxidic Co^{2+} [38,69,71]. The CoMo/AT2 catalyst at the Co/(Co + Mo) = 0.4 ratio is not shown since it presented similar signals. In these samples, sulfidation of Co and Mo was incomplete since the presence of oxysulfide and oxidic species was detected. However, the presence of oxysulfide molybdenum species indicated the transition of Mo^{6+} to Mo^{5+} during the sulfidation process. Despite this, sulfided phases, such as MoS_2, Co_9S_8, and CoMoS, were predominant on 10 wt.% and 15 wt.% Mo supported catalysts. The relative contributions of Mo and Co species from the data obtained from sulfide CoMo/AT2 catalysts at Co/(Co + Mo) = 0.2 and 0.4 ratios are presented in Tables 4 and 5.

Table 4. Binding energies of the Mo $3d_{5/2}$ contributions obtained for sulfided 15 wt.% Mo CoMo/AT2 catalysts at Co/(Co + Mo) = 0.2 and 0.4.

Catalysts	Mo^{4+}		Mo^{5+}		Mo^{6+}	
	BE (eV)	%	BE (eV)	%	BE (eV)	%
CoMo/AT2 (0.2)	228.76	69	230.40	19	232.80	12
CoMo/AT2 (0.4)	228.79	59	230.28	30	232.50	11

Table 5. Binding energies of the Co $2p_{3/2}$ contributions obtained for sulfided 15 wt.% Mo CoMo/AT2 catalysts at Co/(Co + Mo) = 0.2 and 0.4.

Catalyst	Co_9S_8		CoMoS		Co^{2+}	
	BE (eV)	%	BE (eV)	%	BE (eV)	%
CoMo/AT2 (0.2)	778.16	13	778.76	51	780.7	36
CoMo/AT2 (0.4)	778.30	19	778.80	44	780.8	37

Table 4 shows that MoS_2 species were the main phase on both catalysts (69%–59%), while oxysulfide and oxide species represented about 30% to 19% and 12% to 10%, respectively. It is possible to observe that a fraction of the MoS_2 phase decreased by 14% at high Co loadings, indicating that the sulfidation degree was higher for the Co/(Co + Mo) = 0.2 ratio than the Co/(Co + Mo) = 0.4. However, since the oxide Mo^{6+} species were the same, a partial sulfidation of Mo species occurred at high Co loadings due to the presence of 60% more oxysulfide species than at low Co concentrations. This could indicate that the excess of Co caused a decrease in the capacity of sulfidation of the Mo species due to the formation of the $CoMoO_4$ phase [72]. The Co species' contributions shown in Table 5 indicated that the mixed phase CoMoS represented 51% to 44%.

Besides, the Co_9S_8 and oxide phase contributed to 19% to 13% and about 37% of the Co supported on both catalysts. At Co/(Co + Mo) = 0.4, the amount of Co_9S_8 phase increased by 20%, whereas the CoMoS phase decreased by 15% and the oxide form was essentially the same. A high concentration of Co induced the formation of Co_9S_8 over the CoMoS phase due to the presence of $CoMoO_4$ species, which were difficult to sulfide. Therefore, the segregated Co^{Oh} was easily reduced and formed its sulfided phase. At low Co concentrations, Co was capable of occupying the octahedral sites of the MoS_2 phase. Thus, as this latter phase was more abundant, it led to an increase of the CoMoS phase concentration. Therefore, the Co/(Co + Mo) = 0.2 ratio was adequate for generation of the CoMoS phase, which is widely accepted to be the most active phase. Since the Co loading was different, a normalization of the fraction of each phase was carried out. Results of the concentration and mass fraction of the sulfided species are displayed in Table 6.

Table 6. Concentrations and atomic ratios of the Co and Mo species of the 15 wt.% Mo CoMo/AT2 catalyst at Co/(Co + Mo) = 0.2 and 0.4.

Catalysts	Concentration ($g_{metal} \cdot g_{cat}^{-1}$)				Fraction		
	MoS_2	Co_9S_8	CoMoS	Co^{2+}	$f_{CoMos/MoS2}$	$f_{CoMoS/Co9S8}$	$f_{CoMoS/Co2+}$
CoMo/AT2 (0.2)	0.104	0.004	0.014	0.010	0.14	3.92	1.42
CoMo/AT2 (0.4)	0.089	0.011	0.026	0.022	0.29	2.32	1.19

It is possible to observe that the concentration of sulfided and oxide Co species increased with the Co loading, whereas the MoS_2 concentration dropped. Nevertheless, the CoMoS/Co_9S_8 and CoMoS/Co^{2+} fractions decreased by 40% and 16%, respectively at Co/(Co + Mo) = 0.4. This means that as more Co was interacting with Mo at high Co concentrations, less MoS_2 could be formed. In other words, the formation of the CoMoS phase was limited due to the formation of $CoMoO_4$ and the sulfidation of the active phase was not complete. At low Co loadings, less Co_9S_8 and Co^{2+} were present on the catalyst than at high Co loadings. As Figure 10 indicates, more Co^{Oh} was present at the Co/(Co + Mo) = 0.2 ratio than at 0.4. Hence, these Co^{Oh} species interacted with Mo species to form the CoMoS phase. However, the Co concentration was not enough to promote all Mo species, and more MoS_2 was generated, i.e., a higher promotion of the active sites was achieved (see Table 2). In contrast, the excess Co provoked segregated species that could either form Co_9S_8 or Co oxide. Since, at high Co loadings, segregated Co oxide species in octahedral coordination were present on the surface, the sulfided Co phase tended to form. In comparison, the Co oxide species with tetrahedral coordination were difficult to sulfide and were more abundant at Co/(Co + Mo) = 0.4 than at Co/(Co + Mo) = 0.2.

The formation of the CoMoS phase provided electrophilic active sites that were selective to the hydrogenolysis route [5,73]. However, when the Co loading was increased, the HYD route did as well, due to the presence of more Co^{Oh} species. At Co/(Co + Mo) = 0.1, the production of benzene was higher than at the other ratios, indicating the presence of the CoMoS phase with hydrogenolysis sites. Nevertheless, the Co concentration was not enough to promote all MoS_2 slabs. Hence, a fraction of Co was segregated in the Co_9S_8 phase and a minor fraction interacted with the support. In this sense, the HYD route was mainly caused by metallic sites from unpromoted MoS_2 and Co_9S_8 phases than by CoMoS. The Co/(Co + Mo) = 0.2 ratio presented a higher catalytic activity and amount of CoMoS. This means that at this ratio, the amount of Co was adequate for direct interaction with Mo during the sulfidation process. The Co^{Oh} was available to promote the MoS_2 phase and generate the CoMoS phase over the other sulfided phases. At Co/(Co + Mo) > 0.2, activity decreased; however, the HYD route was enhanced. In these cases, there was enough Co to promote the MoS_2 phase. Nonetheless, the excess Co led to the production of a more segregated Co_9S_8 phase than at lower ratios. Hence, the metallic sites responsible for hydrogenation reactions increased their number. As more Co was added, more Co^{Th} and $CoMoO_4$ phase could have been formed, which were difficult to sulfide and consequently, sulfidation was not optimal.

3. Materials and Methods

3.1. *Support and Catalysts Synthesis*

3.1.1. Support Synthesis

Mixed oxide Al_2O_3–TiO_2, (Al/Ti = 2 labelled as AT2) support was synthesized by the sol–gel method as described in a previous work [34]. As organic precursors, tri-sec-butoxide (Al $(OCH(CH_3)$ $C_2H_5)_3$; Aldrich 99.9%, St. Louis, MI, USA) and titanium isopropoxide (Ti $(OCH_3H_7)_4$; Aldrich 98%, St. Louis, MI, USA) were employed. As a solvent, 2-propanol $(CH_3)_2CHOH$; Baker 99.5%, Ecatepec, Estado de Mexico, Mexico), and, as hydrolysis catalyst, nitric acid (HNO_3) were used. The nominal molar ratio used in all supports was 2-proponol:H_2O:alkoxide:HNO_3 = 325:100:5:1 [30]. 2-Propanol was cooled to 0 °C and under vigorous stirring, the theoretical amounts of Al and Ti were added. Then, HNO_3 aqueous solution was added dropwise. The obtained gel was aged for 24 h at 273 K. Subsequently, it was dried at 333 K. Finally, the dried gel was calcined for 3 h at 773 K with a rate of 3 K min^{-1}. AT2 support textural properties were: S_{BET} = 359 m^2 g^{-1}, V_p = 1.1 cm^3 g^{-1}, Dp = 7.7 nm, and PZC = 7.6, as previously reported by [32,34].

3.1.2. Catalyst Synthesis

The Al_2O_3–TiO_2 support was impregnated by the successive wetness impregnation method using an aqueous solution of ammonium heptamolybdate ($(NH_4)_6Mo_7O_{24}\cdot 4H_2O$; Aldrich 99.9%, St. Louis, MI, USA) and cobalt nitrate (($Co(NO_3)_2\cdot 6H_2O$; Aldrich 99%, St. Louis, MI, USA). The non-promoted Mo series were loaded at 5, 10, 15, and 20 wt.% Mo, while only Co catalyst was loaded at 2 wt.%. The promoted CoMo catalyst series were impregnated following four different molar Co/(Co + Mo) ratios: 0.1, 0.2, 0.3, and 0.4. For monometallic catalysts, the Mo (or Co) solution was impregnated on AT2 support and was macerated at room conditions for 12 h. After that, it was dried at 393 K and calcinated at 673 K for 5 h. For the promoted catalyst, the Mo calcined materials were impregnated with the cobalt solution and the heat treatment was repeated. Before XPS analysis and the HDO of phenol tests, calcined CoMo samples were sulfided ex-situ in a glass tube reactor with a 10 vol.-% H_2S/H_2 mixture at 673 K for 2 h. After this, sulfided catalysts were immediately immersed in dodecane to avoid oxidation from air.

3.2. *Materials Characterization*

3.2.1. Diffuse Reflectance Ultraviolet-Visible Spectroscopy

The diffuse reflectance UV-Vis (DRS UV-Vis) spectra of the synthesized support and promoted and unpromoted oxide catalyst series were recorded with a Lambda 35 spectrometer equipped with an integration sphere (Labsphere RSA-PE-20, North Sutton, NH, USA). The data acquisition was in the 200–1000 nm range with an interval of 0.5 nm and a scan speed of 240 nm^{-1}. The spectra were recorded in the reflectance mode for infinitely thick samples (R_∞) using the reflectance of MgO as a reference.

3.2.2. Laser Raman Spectroscopy

Laser Raman spectroscopy (LRS) of unpromoted Mo catalyst at 10 and 15 wt.% were analyzed with a Perkin Elmer GX Raman FT-IR (Waltham, MA, USA), equipped with an Nd: YAG (1064 nm) laser and InGaAs detector. The data acquisition was carried out with a laser power of 40 to 300 mW at the 3600 to 100 cm^{-1} Raman shift range with a resolution of 2 to 4 cm^{-1}.

3.2.3. Temperature Programmed Reduction

Temperature programmed reduction (TPR) experiments of the promoted and unpromoted oxide catalysts series were carried out with in an Altamira Instruments AMI-80 (Pittsburgh, PA, USA) apparatus provided with a thermal conductivity detector (TCD) interfaced to a data station. For each TPR test, 50 mg of catalyst precursor were set into a U-shaped quartz cell and pretreated in situ at 523 K for 1 h under 35 mL min^{-1} He flow. After this, the catalyst precursor was cooled to room temperature. TPR analysis was performed under a stream of 10 vol% of H_2/Ar, with a heating rate of 15 °C min^{-1} up to 1100 K. A moisture trap was used to avoid measurement interference.

3.2.4. X-Ray Photoelectron Spectroscopy (XPS)

The sulfided catalyst were analyzed in a K-alpha Thermo Fischer Scientific spectrometer equipped (Waltham, MA, USA) with a hemispherical electron analyzer and an Al Kα (hν = 1486.6 eV) X-ray source. The residual pressure was kept below 7×10^{-7} Pa during data acquisition. The binding energies (BEs) were referenced to the C 1s peak (284.9 eV) to account for the charging effects. The areas of the peaks were computed after fitting the experimental spectra to Gaussian/Lorentzian curves and removing the background (Shirley function). After that, surface atomic ratios were calculated from the peak area ratios normalized by the corresponding atomic sensitivity factors. The spectra were analyzed from the 222 to 244 eV region where the Mo 3d levels were located. The surface relative abundance percentages of the Mo and Co species were calculated according to the methodology proposed by Chen et al. [38]. Moreover, we estimated the relative amount of the CoMoS mixed phase, considering the Co species in this phase:

$$\left[Mo^{4+}\right](\%) = \frac{A_{Mo^{4+}}}{A_{Mo^{4+}} + A_{Mo^{5+}} + A_{Mo^{6+}}} \times 100, \tag{1}$$

$$[CoMoS](\%) = \frac{A_{CoMoS}}{A_{CoMoS} + A_{Co_9S_8} + A_{Co^{2+}}} \times 100, \tag{2}$$

$$C(MoS_2) = C(Mo)\left[Mo^{4+}\right]/100, \tag{3}$$

$$C(CoMoS) = C(Co)[CoMoS]/100, \tag{4}$$

$$f_{\frac{CoMoS}{MoS_2}} = \frac{C(CoMoS)}{C(MoS_2)}, \tag{5}$$

where A_{Mo4+}, A_{Mo5+}, A_{Mo6+}, A_{CoMoS}, A_{Co9S8}, and A_{Co2+} are the area of each species fitted from the Mo3d and Co2p XPS spectra; C(Mo) and C(Co) are the theoretical mass concentrations of Mo and Co per gram of oxidic catalyst ($g_{metal}/g_{catalyts}$); C(MoS_2) and C(CoMoS) are the mass concentrations of MoS_2 and CoMoS species per gram of catalyst; and f_{CoMoS}/MoS_2 is the mass ratio of the CoMoS and MoS_2 species. The C(Co_9S_8) and C(Co^{2+}) concentrations and mass ratios were quantified in the same way.

3.2.5. Catalytic Performance

To evaluate the catalytic performance of the Mo, Co, and CoMo sulfided catalyst, hydrodeoxygenation of phenol was carried out. HDO of phenol facilitates an understanding of the functionality and reaction mechanisms in catalytic tests since it is a relatively simple molecule and it constitutes the main component in the remnant of the HDO of guaiacol [40]. The reaction took place in a Parr Series 4540 high-pressure batch reactor (Parr Instrument Co., Moline, IL, USA) equipped with a Parr 4842 controller, mechanic impeller, wall-baffles, liquid and gas inlet/outlet gas valves, internal thermocouple, and pressure gauge. The reaction mixture consisted of phenol (500 ppm of oxygen) and CS_2 (100 ppm of sulfur) dissolved in 100 mL of n-dodecane and 0.1 g of freshly sulfide catalysts with a particle size between 150 and 180 μm. Catalyst sulfidation was carried out as was described in the XPS methodology. The reactor was pressurized up to 1.4 MPa with N_2 to flush air and to prevent oxidation of the sulfide catalysts. After this, the reactor was heated up to 593 K and kept an isothermal operation mode during the reaction. Then, N_2 was vented slowly, and hydrogen was introduced up to 5.5 MPa. The reactor operated in an isobaric mode during the reaction time with manual addition of H2 and vigorous agitation of 1000 rpm. The reaction time started from the incorporation of H_2. Small samples were collected (0, 10, 20, 30, 45, 60, 90, 120, 180, 240, and 300 min), ensuring that the sum of the volume samples was less than 5% of the initial volume. A gas chromatograph (Agilent 7820A, Santa Clara, CA, USA) equipped with a CP Sil-5 CB capillary column (100% dimethylpolysiloxane, 60 m × 0.32 mm) and a flame ionized detector (FID) were used for the quantification of the products' concentrations. The initial reaction rates, reagent mol transformed per time of reaction, and mass of a sulfided catalyst (mol $g^{-1}s^{-1}$) were compared.

4. Conclusions

In the unpromoted catalysts at high Mo loadings, more easily reduced Mo^{Oh} species were formed, and more active sites were present on these catalysts than at low Mo loadings. Nevertheless, a hardly reduced MoO_3 presence was detected at a high Mo content. On the other hand, the functionalities to produce benzene and cyclohexene did not change with the Mo content since all catalysts showed the same selectivity. This indicated that the active sites could cleave the C_{AR}–OH bond but could not properly hydrogenate the π-bonds of the cyclic compounds, i.e., S-edges sites played a more major role than Mo-edges.

Catalytic activity and selectivity were related to the Co^{Oh} content. A Co/(Co + Mo) = 0.2 ratio presented a maximum activity for these catalysts. At this Co concentration, Co^{Oh} was the main Co oxide coordination species. These Co^{Oh} species properly promoted the Mo oxide species and subsequently formed more CoMoS phase than at higher ratios. At a low Co content, it was insufficient to totally promote the Mo oxide species. Therefore, CoMoS phase formation was limited and the MoS_2 phase was predominant since benzene was the main product. Plus, Co may have generated Co^{Th} species that were difficult to reduce. At high ratios, the excess Co concentration led to the $CoMoO_4$ phase, and the formation of the CoMoS phase was inadequate. Hence, activity dropped compared with the Co/(Co + Mo) = 0.2 ratio. Moreover, the surplus of Co was segregated and formed Co^{Oh}, which transformed into the Co_9S_8 phase, leading to an enhancement of the HYD route. Finally, the segregated Co oxide species may have formed Co^{Th} as well; thus, catalytic activity decreased.

Author Contributions: C.E.S.-V. and O.U.V.-M. performed and analyzed the characterization experiments and data; J.A.T.-P. conceived, designed, and performed experiments, analyzed the data, and wrote the manuscript; J.A.d.l.R.H. contributed to writing—review of the manuscript, funding acquisition, and was the project administrator and laboratory chief.

Funding: This research was funded by Consejo Nacional de Ciencia y Tenología (CONACYT—Mexico), grant number 237857.

Acknowledgments: The authors are grateful to CONACYT for the financial support 237857 and for the scholarship of J.A. Tavizón-Pozos with number 221991. The authors acknowledge financial support from Instituto Politénico Nacional (Proyecto SIP 20196722) and CONACYT for the projects CB-2017-2018 #A1-S-32418 and the Cátedras-CONACYT number 216.

Conflicts of Interest: The authors declare no conflict of interest.

References

1. Pourzolfaghar, H.; Abnisa, F.; Wan Daud, W.M.A.; Aroua, M.K. Atmospheric Hydrodeoxygenation of Bio-Oil Oxygenated Model Compounds: A Review. *J. Anal. Appl. Pyrolysis* **2018**, *133*, 117–127. [CrossRef]
2. Li, X.; Chen, G.; Liu, C.; Ma, W.; Yan, B.; Zhang, J. Hydrodeoxygenation of Lignin-Derived Bio-Oil Using Molecular Sieves Supported Metal Catalysts: A Critical Review. *Renew. Sustain. Energy Rev.* **2017**, *71*, 296–308. [CrossRef]
3. Dabros, T.M.H.; Stummann, M.Z.; Høj, M.; Jensen, P.A.; Grunwaldt, J.-D.; Gabrielsen, J.; Mortensen, P.M.; Jensen, A.D. Transportation Fuels from Biomass Fast Pyrolysis, Catalytic Hydrodeoxygenation, and Catalytic Fast Hydropyrolysis. *Prog. Energy Combust. Sci.* **2018**, *68*, 268–309. [CrossRef]
4. Linck, M.; Felix, L.; Marker, T.; Roberts, M. Integrated Biomass Hydropyrolysis and Hydrotreating: A Brief Review. *Wiley Interdiscip. Rev. Energy Environ.* **2014**, *3*, 575–581. [CrossRef]
5. Furimsky, E. Hydroprocessing Challenges in Biofuels Production. *Catal. Today* **2013**, *217*, 13–56. [CrossRef]
6. Huber, G.W.; Iborra, S.; Corma, A. Synthesis of Transportation Fuels from Biomass Chemistry Catalysts and Enginneering. *Chem. Rev.* **2006**, *106*, 4044–4098. [CrossRef] [PubMed]
7. Li, X.; Zhang, J.; Liu, B.; Liu, J.; Wang, C.; Chen, G. Hydrodeoxygenation of Lignin-Derived Phenols to Produce Hydrocarbons over Ni/Al-SBA-15 Prepared with Different Impregnants. *Fuel* **2019**, *243*, 314–321. [CrossRef]
8. Jahromi, H.; Agblevor, F. Hydrogenation of Aqueous-Phase Catalytic Pyrolysis Oil to Liquid Hydrocarbons Using Multifunctional Nickel Catalyst. *Ind. Eng. Chem. Res.* **2018**, *57*, 13257–13268. [CrossRef]
9. Bui, V.N.; Laurenti, D.; Delichère, P.; Geantet, C. Hydrodeoxygenation of Guaiacol: Part II: Support Effect for CoMoS Catalysts on HDO Activity and Selectivity. *Appl. Catal. B Environ.* **2011**, *101*, 246–255. [CrossRef]
10. Şenol, O.İ.; Ryymin, E.-M.; Viljava, T.-R.; Krause, A.O.I. Effect of Hydrogen Sulphide on the Hydrodeoxygenation of Aromatic and Aliphatic Oxygenates on Sulphided Catalysts. *J. Mol. Catal. A Chem.* **2007**, *277*, 107–112. [CrossRef]
11. Bu, Q.; Lei, H.; Zacher, A.H.; Wang, L.; Ren, S.; Liang, J.; Wei, Y.; Liu, Y.; Tang, J.; Zhang, Q.; et al. A Review of Catalytic Hydrodeoxygenation of Lignin-Derived Phenols from Biomass Pyrolysis. *Bioresour. Technol.* **2012**, *124*, 470–477. [CrossRef] [PubMed]
12. Echeandia, S.; Arias, P.L.; Barrio, V.L.; Pawelec, B.; Fierro, J.L.G. Synergy Effect in the HDO of Phenol over Ni–W Catalysts Supported on Active Carbon: Effect of Tungsten Precursors. *Appl. Catal. B Environ.* **2010**, *101*, 1–12. [CrossRef]
13. Bui, V.N.; Laurenti, D.; Afanasiev, P.; Geantet, C. Hydrodeoxygenation of Guaiacol with CoMo Catalysts. Part I: Promoting Effect of Cobalt on HDO Selectivity and Activity. *Appl. Catal. B Environ.* **2011**, *101*, 239–245. [CrossRef]
14. Honkela, M.L.; Viljava, T.-R.; Gutierrez, A.; Krause, A.O.I. Chapter 11 Hydrotreating for Bio-Oil Upgrading. In *Thermochemical Conversion of Biomass to Liquid Fuels and Chemicals*; Crocker, M., Ed.; The Royal Society of Chemistry: London, UK, 2010; pp. 288–306.
15. Zhang, J.; Dong, K.; Luo, W. PdCl2-Catalyzed Hydrodeoxygenation of 5-Hydroxymethylfurfural into 2,5-Dimethylfuran at Room-Temperature using Polymethylhydrosiloxane as the Hydrogen Donor. *Chem. Eng. Sci.* **2019**, *201*, 467–474. [CrossRef]
16. Viljava, T.-R.; Komulainen, S.; Selvam, T.; Krause, A.O.I. Stability of CoMo/Al_2O_3 Catalysts: Effect of HDO Cycles on HDS. In *Studies in Surface Science and Catalysis: Hydrotreatment and Hydrocracking of Oil Fractions Proceedings of the 2nd International Symposium/7th European Workshop*; Delmon, B., Froment, G.F., Grange, P., Eds.; Elsevier: Antwerpen, Belgium, 1999; Volume 127, pp. 145–152. [CrossRef]
17. Massoth, F.E.; Politzer, P.; Concha, M.C.; Murray, J.S.; Jakowski, J.; Simons, J. Catalytic Hydrodeoxygenation of Methyl-Substituted Phenols: Correlations of Kinetic Parameters with Molecular Properties. *J. Phys. Chem. B* **2006**, *110*, 14283–14291. [CrossRef] [PubMed]
18. Badawi, M.; Paul, J.F.; Cristol, S.; Payen, E.; Romero, Y.; Richard, F.; Brunet, S.; Lambert, D.; Portier, X.; Popov, A.; et al. Effect of Water on the Stability of Mo and CoMo Hydrodeoxygenation Catalysts: A Combined Experimental and DFT Study. *J. Catal.* **2011**, *282*, 155–164. [CrossRef]
19. Wildschut, J.; Mahfud, F.H.; Venderbosch, R.H.; Heeres, H.J. Hydrotreatment of Fast Pyrolysis Oil Using Heterogeneous Noble-Metal Catalysts. *Ind. Eng. Chem. Res.* **2009**, *48*, 10324–10334. [CrossRef]

20. Lauritsen, J.V.; Besenbacher, F. Atom-Resolved Scanning Tunneling Microscopy Investigations of Molecular Adsorption on MoS2 and CoMoS Hydrodesulfurization Catalysts. *J. Catal.* **2015**, *328*, 49–58. [CrossRef]
21. Topsøe, H.; Clausen, B.S.; Massoth, F.E. *Hydrotreating Catalysis BT—Catalysis: Science and Technology*; Anderson, J.R., Boudart, M., Eds.; Springer Berlin Heidelberg: Berlin/Heidelberg, Germany, 1996; 269p.
22. Lauritsen, J.V.; Nyberg, M.; Nørskov, J.K.; Clausen, B.S.; Topsøe, H.; Lægsgaard, E.; Besenbacher, F. Hydrodesulfurization Reaction Pathways on MoS2 Nanoclusters Revealed by Scanning Tunneling Microscopy. *J. Catal.* **2004**, *224*, 94–106. [CrossRef]
23. Zhu, Y.; Ramasse, Q.M.; Brorson, M.; Moses, P.G.; Hansen, L.P.; Topsøe, H.; Kisielowski, C.F.; Helveg, S. Location of Co and Ni Promoter Atoms in Multi-Layer MoS_2 Nanocrystals for Hydrotreating Catalysis. *Catal. Today* **2016**, *261*, 75–81. [CrossRef]
24. Liu, B.; Liu, L.; Chai, Y.; Zhao, J.; Liu, C. Essential Role of Promoter Co on the MoS2 Catalyst in Selective Hydrodesulfurization of FCC Gasoline. *J. Fuel Chem. Technol.* **2018**, *46*, 441–450. [CrossRef]
25. Kim, D.S.; Segawa, K.; Soeya, T.; Wachs, I.E. Surface Structures of Supported Molybdenum Oxide Catalysts under Ambient Conditions. *J. Catal.* **1992**, *136*, 539–553. [CrossRef]
26. Ng, K.Y.S.; Gulari, E. Molybdena on Titania: I. Preparation and Characterization by Raman and Fourier Transform Infrared Spectroscopy. *J. Catal.* **1985**, *92*, 340–354. [CrossRef]
27. Cáceres, C.V.; Fierro, J.L.G.; Lázaro, J.; López Agudo, A.; Soria, J. Effect of Support on the Surface Characteristics of Supported Molybdena Catalysts. *J. Catal.* **1990**, *122*, 113–125. [CrossRef]
28. Breysse, M.; Afanasiev, P.; Geantet, C.; Vrinat, M. Overview of Support Effects in Hydrotreating Catalysts. *Catal. Today* **2003**, *86*, 5–16. [CrossRef]
29. Gutiérrez-Alejandre, A.; Ramírez, J.; Val, I.J.; Peñuelas-Galaz, M.; Sánchez-Neri, P.; Torres-Mancera, P. Activity of NiW Catalysts Supported on TiO_2-Al_2O_3 Mixed Oxides: Effect of Ti Incorporation Method on the HDS of 4,6-DMDBT. *Catal. Today* **2005**, *107–108*, 879–884. [CrossRef]
30. Escobar, J.; Antonio De Los Reyes, J.; Viveros, T. Nickel on TiO_2-Modified Al_2O_3 Sol–Gel Oxides: Effect of Synthesis Parameters on the Supported Phase Properties. *Appl. Catal. A Gen.* **2003**, *253*, 151–163. [CrossRef]
31. Núñez, S.; Escobar, J.; Vázquez, A.; de los Reyes, J.A.; Hernández-Barrera, M. 4,6-Dimethyl-Dibenzothiophene Conversion over Al2O3–TiO2-Supported Noble Metal Catalysts. *Mater. Chem. Phys.* **2011**, *126*, 237–247. [CrossRef]
32. Tavizón-Pozos, J.A.; Suárez-Toriello, V.A.; de los Reyes, J.A.; Guevara-Lara, A.; Pawelec, B.; Fierro, J.L.G.; Vrinat, M.; Geantet, C. Deep Hydrodesulfurization of Dibenzothiophenes Over NiW Sulfide Catalysts Supported on Sol–Gel Titania–Alumina. *Top. Catal.* **2016**, *59*, 241–251. [CrossRef]
33. Suárez-Toriello, V.A.; Santolalla-Vargas, C.E.; de los Reyes, J.A.; Vázquez-Zavala, A.; Vrinat, M.; Geantet, C. Influence of the Solution PH in Impregnation with Citric Acid and Activity of Ni/W/Al_2O_3 Catalysts. *J. Mol. Catal. A Chem.* **2015**, *404–405*, 36–46. [CrossRef]
34. Tavizón-Pozos, J.A.; Suárez-Toriello, V.A.; del Ángel, P.; de los Reyes, J.A. Hydrodeoxygenation of Phenol Over Sulfided CoMo Catalysts Supported on a Mixed Al_2O_3-TiO_2 Oxide. *Int. J. Chem. React. Eng.* **2016**, *14*, 1211–1223. [CrossRef]
35. Escobar, J.; Núñez, S.; Montesinos-Castellanos, A.; de los Reyes, J.A.; Rodríguez, Y.; González, O.A. Dibenzothiophene Hydrodesulfurization over PdPt/Al_2O_3–TiO_2. Influence of Ti-Addition on Hydrogenating Properties. *Mater. Chem. Phys.* **2016**, *171*, 185–194. [CrossRef]
36. Olguin, E.; Vrinat, M.; Cedeño, L.; Ramirez, J.; Borque, M.; López-Agudo, A. The Use of TiO_2—Al_2O_3 Binary Oxides as Supports for Mo-Based Catalysts in Hydrodesulfurization of Thiophene and Dibenzothiophene. *Appl. Catal. A Gen.* **1997**, *165*, 1–13. [CrossRef]
37. Ramírez, J.; Macías, G.; Cedeño, L.; Gutiérrez-Alejandre, A.; Cuevas, R.; Castillo, P. The Role of Titania in Supported Mo, CoMo, NiMo, and NiW Hydrodesulfurization Catalysts: Analysis of Past and New Evidences. *Catal. Today* **2004**, *98*, 19–30. [CrossRef]
38. Chen, W.; Long, X.; Li, M.; Nie, H.; Li, D. Influence of Active Phase Structure of CoMo/Al_2O_3 Catalyst on the Selectivity of Hydrodesulfurization and Hydrodearomatization. *Catal. Today* **2017**, *292*, 97–109. [CrossRef]
39. Wang, L.; Hall, W.K. On the Genesis of Molybdena-Alumina Catalyst. *J. Catal.* **1980**, *66*, 251–255. [CrossRef]
40. Kim, D.S.; Kurusu, Y.; Wachs, I.E.; Hardcastle, F.D.; Segawa, K. Physicochemical Properties of MoO3 TiO_2 Prepared by an Equilibrium Adsorption Method. *J. Catal.* **1989**, *120*, 325–336. [CrossRef]
41. Afanasiev, P.; Geantet, C.; Breysse, M. Preparation of High-Surface-Area Mo/ZrO2 Catalysts by a Molten Salt Method: Application to Hydrodesulfurization. *J. Catal.* **1995**, *153*, 17–24. [CrossRef]

42. Williams, C.C.; Ekerdt, J.G.; Jehng, J.M.; Hardcastle, F.D.; Wachs, I.E. A Raman and Ultraviolet Diffuse Reflectance Spectroscopic Investigation of Alumina-Supported Molybdenum Oxide. *J. Phys. Chem.* **1991**, *95*, 8791–8797. [CrossRef]
43. Xiong, G.; Li, C.; Feng, Z.; Ying, P.; Xin, Q.; Liu, J. Surface Coordination Structure of Molybdate with Extremely Low Loading on γ-Alumina Characterized by UV Resonance Raman Spectroscopy. *J. Catal.* **1999**, *186*, 234–237. [CrossRef]
44. Stencel, J.M. *Raman Spectroscopy for Catalysis*, 1st ed.; Davis, B., Ed.; Van Nostrand Reinhold: New York, NY, USA, 1990.
45. Lozano-Hernández, G.; Lozada-Ascencio, E.M.; Guevara-Lara, A. Support Composition Effect on Superficial Structures of Nickel and Molybdenum Oxides Supported on TiO_2-Al_2O_3 Mixed Oxides. *Rev. Mex. Ing. Química* **2006**, *5*, 311–320.
46. Aveston, J.; Anacker, E.W.; Johnson, J.S. Hydrolysis of Molybdenum(VI). Ultracentrifugation, Acidity Measurements, and Raman Spectra of Polymolybdates. *Inorg. Chem.* **1964**, *3*, 735–746. [CrossRef]
47. Platanitis, P.; Panagiotou, G.D.; Bourikas, K.; Kordulis, C.; Fierro, J.L.G.; Lycourghiotis, A. Preparation of Un-Promoted Molybdenum HDS Catalysts Supported on Titania by Equilibrium Deposition Filtration: Optimization of the Preparative Parameters and Investigation of the Promoting Action of Titania. *J. Mol. Catal. A Chem.* **2016**, *412*, 1–12. [CrossRef]
48. Vakros, J.; Lycourghiotis, A.; Voyiatzis, G.A.; Siokou, A.; Kordulis, C. CoMo/Al_2O_3-SiO_2 Catalysts Prepared by Co-Equilibrium Deposition Filtration: Characterization and Catalytic Behavior for the Hydrodesulphurization of Thiophene. *Appl. Catal. B Environ.* **2010**, *96*, 496–507. [CrossRef]
49. Malaibari, Z.O.; Croiset, E.; Amin, A.; Epling, W. Effect of Interactions between Ni and Mo on Catalytic Properties of a Bimetallic Ni-Mo/Al_2O_3 Propane Reforming Catalyst. *Appl. Catal. A Gen.* **2015**, *490*, 80–92. [CrossRef]
50. Arnoldy, P.; Franken, M.C.; Scheffer, B.; Moulijn, J.A. Temperature-Programmed Reduction of CoO $MoO_3Al_2O_3$ Catalysts. *J. Catal.* **1985**, *96*, 381–395. [CrossRef]
51. Bergwerff, J.A.; Visser, T.; Weckhuysen, B.M. On the Interaction between Co- and Mo-Complexes in Impregnation Solutions Used for the Preparation of Al2O3-Supported HDS Catalysts: A Combined Raman/UV–Vis–NIR Spectroscopy Study. *Catal. Today* **2008**, *130*, 117–125. [CrossRef]
52. Okamoto, Y.; Imanaka, T. Interaction Chemistry between Molybdena and Alumina: Infrared Studies of Surface Hydroxyl Groups and Adsorbed Carbon Dioxide on Aluminas Modified with Molybdate, Sulfate, or Fluorine Anions. *J. Phys. Chem.* **1988**, *92*, 7102–7112. [CrossRef]
53. Topsoe, N.Y.; Topsoe, H. FTIR Studies of Mo/Al_2O_3-Based Catalysts: I. Morphology and Structure of Calcined and Sulfided Catalysts. *J. Catal.* **1993**, *139*, 631–640. [CrossRef]
54. Besenbacher, F.; Brorson, M.; Clausen, B.S.; Helveg, S.; Hinnemann, B.; Kibsgaard, J.; Lauritsen, J.V.; Moses, P.G.; Nørskov, J.K.; Topsøe, H. Recent STM, DFT and HAADF-STEM Studies of Sulfide-Based Hydrotreating Catalysts: Insight into Mechanistic, Structural and Particle Size Effects. *Catal. Today* **2008**, *130*, 86–96. [CrossRef]
55. Lauritsen, J.V.; Kibsgaard, J.; Olesen, G.H.; Moses, P.G.; Hinnemann, B.; Helveg, S.; Nørskov, J.K.; Clausen, B.S.; Topsøe, H.; Lægsgaard, E.; et al. Location and Coordination of Promoter Atoms in Co- and Ni-Promoted MoS2-Based Hydrotreating Catalysts. *J. Catal.* **2007**, *249*, 220–233. [CrossRef]
56. Vrinat, M.; Letourneur, D.; Bacaud, R.; Harlé, V.; Jouguet, B.; Leclercq, C. CoMo/Al_2O_3 and CoMo/TiO_2-Al_2O_3 Catalysts in Hydrodesulfurization: Relationship between the Promoting Effect of Cobalt and the Nature of the Support. In *Studies in Surface Science and Catalysis: Hydrotreatment and Hydrocracking of Oil Fractions Proceedings ofthe 2nd International Symposium/7th European Workshop*; Delmon, B., Froment, G.F., Grange, P., Eds.; Elsevier: Antwerpen, Belgium, 1999; Volume 127, pp. 153–160.
57. Vakros, J.; Papadopoulou, C.; Lycourghiotis, A.; Kordulis, C. Hydrodesulfurization Catalyst Bodies with Various Co and Mo Profiles. *Appl. Catal. A Gen.* **2011**, *399*, 211–220. [CrossRef]
58. Montoya, J.A.; del Angel, P.; Viveros, T. The Effect of Temperature on the Structural and Textural Evolution of Sol-Gel AlO-TiO Mixed Oxides. *J. Mater. Chem.* **2001**, *11*, 944–950. [CrossRef]
59. Herrera, J.E.; Resasco, D.E. Loss of Single-Walled Carbon Nanotubes Selectivity by Disruption of the Co–Mo Interaction in the Catalyst. *J. Catal.* **2004**, *221*, 354–364. [CrossRef]

60. Vakros, J.; Bourikas, K.; Perlepes, S.; Kordulis, C.; Lycourghiotis, A. Adsorption of Cobalt Ions on the "Electrolytic Solution/γ-Alumina" Interface Studied by Diffuse Reflectance Spectroscopy (DRS). *Langmuir* **2004**, *20*, 10542–10550. [CrossRef] [PubMed]
61. Papadopoulou, C.; Vakros, J.; Matralis, H.K.; Voyiatzis, G.A.; Kordulis, C. Preparation, Characterization, and Catalytic Activity of CoMo/γ-Al_2O_3 Catalysts Prepared by Equilibrium Deposition Filtration and Conventional Impregnation Techniques. *J. Colloid Interface Sci.* **2004**, *274*, 159–166. [CrossRef]
62. Brito, J.L.; Barbosa, A.L. Effect of Phase Composition of the Oxidic Precursor on the HDS Activity of the Sulfided Molybdates of Fe(II), Co(II), and Ni(II). *J. Catal.* **1997**, *171*, 467–475. [CrossRef]
63. Papadopoulou, C.; Vakros, J.; Matralis, H.K.; Kordulis, C.; Lycourghiotis, A. On the Relationship between the Preparation Method and the Physicochemical and Catalytic Properties of the CoMo/γ-Al_2O_3 Hydrodesulfurization Catalysts. *J. Colloid Interface Sci.* **2003**, *261*, 146–153. [CrossRef]
64. Zhang, Z.; Zhang, Q.; Jia, L.; Wang, W.; Zhang, T.; Han, Y.; Tsubaki, N.; Tan, Y. Effects of Tetrahedral Molybdenum Oxide Species and MoOx Domains on the Selective Oxidation of Dimethyl Ether under Mild Conditions. *Catal. Sci. Technol.* **2016**, *6*, 2975–2983. [CrossRef]
65. Jermwongratanachai, T.; Jacobs, G.; Shafer, W.D.; Pendyala, V.R.R.; Ma, W.; Gnanamani, M.K.; Hopps, S.; Thomas, G.A.; Kitiyanan, B.; Khalid, S.; et al. Fischer–Tropsch Synthesis: TPR and XANES Analysis of the Impact of Simulated Regeneration Cycles on the Reducibility of Co/Alumina Catalysts with Different Promoters (Pt, Ru, Re, Ag, Au, Rh, Ir). *Catal. Today* **2014**, *228*, 15–21. [CrossRef]
66. Jacobs, G.; Ji, Y.; Davis, B.H.; Cronauer, D.; Kropf, A.J.; Marshall, C.L. Fischer–Tropsch Synthesis: Temperature Programmed EXAFS/XANES Investigation of the Influence of Support Type, Cobalt Loading, and Noble Metal Promoter Addition to the Reduction Behavior of Cobalt Oxide Particles. *Appl. Catal. A Gen.* **2007**, *333*, 177–191. [CrossRef]
67. Yamada, M.; Kasahara, S.; Kawahara, K. Interaction between Co Species and Mo Species in C0O-$M0O_3$/AI_2O_3. In *Studies in Surface Science and Catalysis: New Aspects of Spillover Effect in Catalysis*; lnui, T., Fujimoto, K., Uchijima, T., Masai, M., Eds.; Elsevier: Kyoto, Japan, 1993; Volume 77, pp. 349–352. [CrossRef]
68. Guo, C.; Wu, Y.; Wang, X.; Yang, B. Effect of the Support Calcination Temperature on Selective Hydrodesulfurization of TiO_2 Nanotubes Supported CoMo Catalysts. *J. Energy Chem.* **2013**, *22*, 517–523. [CrossRef]
69. Gandubert, A.D.; Krebs, E.; Legens, C.; Costa, D.; Guillaume, D.; Raybaud, P. Optimal Promoter Edge Decoration of CoMoS Catalysts: A Combined Theoretical and Experimental Study. *Catal. Today* **2008**, *130*, 149–159. [CrossRef]
70. Ninh, T.K.T.; Laurenti, D.; Leclerc, E.; Vrinat, M. Support Effect for CoMoS and CoNiMoS Hydrodesulfurization Catalysts Prepared by Controlled Method. *Appl. Catal. A Gen.* **2014**, *487*, 210–218. [CrossRef]
71. Laurenti, D.; Phung-Ngoc, B.; Roukoss, C.; Devers, E.; Marchand, K.; Massin, L.; Lemaitre, L.; Legens, C.; Quoineaud, A.-A.; Vrinat, M. Intrinsic Potential of Alumina-Supported CoMo Catalysts in HDS: Comparison between Γc, ΓT, and δ-Alumina. *J. Catal.* **2013**, *297*, 165–175. [CrossRef]
72. La Parola, V.; Dragoi, B.; Ungureanu, A.; Dumitriu, E.; Venezia, A.M. New HDS Catalysts Based on Thiol Functionalized Mesoporous Silica Supports. *Appl. Catal. A Gen.* **2010**, *386*, 43–50. [CrossRef]
73. Brorson, M.; Carlsson, A.; Topsøe, H. The Morphology of MoS2, WS2, Co–Mo–S, Ni–Mo–S and Ni–W–S Nanoclusters in Hydrodesulfurization Catalysts Revealed by HAADF-STEM. *Catal. Today* **2007**, *123*, 31–36. [CrossRef]

catalysts

MDPI

Article

Synthesis and Regeneration of Nickel-Based Catalysts for Hydrodeoxygenation of Beech Wood Fast Pyrolysis Bio-Oil

Caroline Carriel Schmitt [1,2,*], María Belén Gagliardi Reolon [1,3], Michael Zimmermann [1], Klaus Raffelt [1], Jan-Dierk Grunwaldt [1,4] and Nicolaus Dahmen [1]

1 Institute of Catalysis Research and Technology (IKFT), Karlsruhe Institute of Technology (KIT), 76344 Eggenstein-Leopoldshafen, Germany; magagliardi@itba.edu.ar (M.B.G.R.); michael.zimmermann@kit.edu (M.Z.); klaus.raffelt@kit.edu (K.R.); grunwaldt@kit.edu (J.-D.G.); nicolaus.dahmen@kit.edu (N.D.)

2 Institute of Microstructure Technology (IMT), Karlsruhe Institute of Technology (KIT), 76344 Eggenstein-Leopoldshafen, Germany

3 Instituto Tecnologico de Buenos Aires (ITBA), Buenos Aires C1106ACD, Argentina

4 Institute for Chemical Technology and Polymer Chemistry (ITCP), Karlsruhe Institute of Technology (KIT), 76131 Karlsruhe, Germany

* Correspondence: caroline.schmitt@partner.kit.edu

Received: 17 August 2018; Accepted: 8 October 2018; Published: 12 October 2018

Abstract: Four nickel-based catalysts are synthesized by wet impregnation and evaluated for the hydrotreatment/hydrodeoxygenation of beech wood fast-pyrolysis bio-oil. Parameters such as elemental analysis, pH value, and water content, as well as the heating value of the upgraded bio-oils are considered for the evaluation of the catalysts' activity and catalyst reuse in cycles of hydrodeoxygenation after regeneration. The reduction temperature, selectivity and hydrogen consumption are distinct among them, although all catalysts tested produce upgraded bio-oils with reduced oxygen concentration, lower water content and higher energy density. Ni/SiO_2, in particular, can remove more than 50% of the oxygen content and reduce the water content by more than 80%, with low coke and gas formation. The evaluation over four consecutive hydrotreatment reactions and catalyst regeneration shows a slightly reduced hydrodeoxygenation activity of Ni/SiO_2, mainly due to deactivation caused by sintering and adsorption of poisoning substances, such as sulfur. Following the fourth catalyst reuse, the upgraded bio-oil shows 43% less oxygen in comparison to the feedstock and properties comparable to the upgraded bio-oil obtained with the fresh catalyst. Hence, nickel-based catalysts are promising for improving hardwood fast-pyrolysis bio-oil properties, especially monometallic nickel catalysts supported on silica.

Keywords: hydrodeoxygenation; fast-pyrolysis bio-oil; nickel catalyst; upgrading

1. Introduction

The increase in global energy demand, depletion of fossil fuel reserves and climate change issues have drawn attention to renewable alternatives, particularly to biomass [1–3], considering its CO_2 neutrality for fuel applications and widespread availability [1,4]. Products such as heat, power, biomaterials, chemical compounds, and transportation fuels can be obtained from biomass [5]. Considering this purpose, thermochemical, chemical-catalytical or biological processes are used. Regarding the first category, combustion, gasification and pyrolysis are most common. Although considered the simplest way to convert biomass in either power or heat, combustion shows high emissions and ash generation [5]. Gasification is considered a very efficient method to obtain fuels, but it requires a high investment due to large-scale installations, storage and transportation [6].

Therewith, pyrolysis has been considered a promising process as it balances simple operation techniques with reasonable costs.

Pyrolysis is a thermochemical process in which the biomass is heated and converted in an inert atmosphere into a liquid fraction called bio-oil, a carbon-rich solid (biochar) and a mixture of non-condensable gases [7]. The bio-oil obtained has poorer physical and chemical characteristics if compared to liquid fossil fuels. The heating value is usually lower, only 40–50% compared to conventional fossil fuels (42–45 MJ/Kg), mainly due to the high oxygen and water content. Additionally, it shows high viscosity, low chemical stability and solid particles [8–10] due to incomplete solid separation or polymerization reactions during storage, for example. Carboxylic acids present in the bio-oil composition lead to high acidity (pH value around 2–3.7), resulting in a bio-oil with potentially corrosive properties. Furthermore, it is highly unstable during storage due to ongoing chemical reactions, resulting in larger molecules by polymerization, etherification and esterification [9], for example. Additionally, it is immiscible with fossil fuels and tends to undergo phase separation when stored for a long time. Considering these poor fuel properties, the direct application of bio-oil is limited to furnaces and boilers, being unsuitable for application in gas turbines, diesel engines and other applications without further treatment [11]. Concerning bio-oil production today, wood with low ash content is used, leading to relatively "well-natured" bio-oils. Using ash-rich feedstocks, bio-oil yield and quality is decreased, while the tendency for phase separation increases.

To improve these properties and obtain a product resembling diesel fuel, bio-oil requires an additional upgrading treatment. Upgraded bio-oil can then be used as feedstock for producing chemicals, such as phenols for resin production, additives for fertilizers and pharmaceutical industries, as well as flavoring agents in the food industry [12]. Regarding terms of energetic use, upgraded bio-oil might be used as feedstock in oil refineries and fuels in engines [13].

A variety of upgrading techniques already have been proposed, such as catalytic cracking, hydrodeoxygenation (HDO) and esterification in supercritical fluids [14]. Among them, HDO appears to be a propitious route, due to its flexibility with respect to the biomass feed, the good economy of the input materials, and its compatibility with refinery infrastructures [15]. HDO is a high-pressure catalytic treatment in which oxygen is removed by hydrogen resulting in water, which is environmentally benign [16]. Usually, sulfides, noble metals and transition metal catalysts are used [17]. Noble metals such as Pt, Pd and Ru have been evaluated widely for HDO and are often the first choice in hydrogenation reactions. Additionally, they have a low tendency to be poisoned by the sulfur present in the bio-oil [18]. Their relatively high costs, however, prevent them from being widely used. Recently, nickel-based catalysts have become more attractive, considering their lower price, availability, activity and reduced hydrogen consumption [10,19]. Jin et al. [20] evaluated a series of nickel-based catalysts on different supports (SiO_2, Al_2O_3, AC and SBA–15 mesoporous silica) for the HDO of anisole, used as a model compound. Boscagli et al. [21] investigated the HDO of the bio-oil light phase over a variety of nickel-based catalysts (NiCu/Al_2O_3, Ni/SiO_2, Ni/ZrO_2, Ni/TiO_2 and NiW/AC). Dongil et al. [22] studied the HDO of guaiacol over nickel-based catalysts, using different carbon-based supports.

The combination of nickel in bimetallic catalysts has also attracted attention for HDO, especially in combination with copper. Ardiyanti et al. [23] evaluated the application of NiCu at different loadings supported in δ–Al_2O_3 for upgrading of model compounds and fast-pyrolysis bio-oil. Dongil et al. [24] investigated the effect of Cu loading on nickel catalysts supported in carbon nanotubes over the HDO of guaiacol. Mortensen et al. [25] screened different catalysts, including NiCu/SiO_2 for phenol HDO and, more recently, Boscagli et al. [26] tested NiCu/Al_2O_3 for the HDO of phenol and bio-oils reusing the catalyst after a regeneration step.

The investigation of nickel and nickel–copper catalysts on SiO_2 and ZrO_2 supports with real feedstock (fast-pyrolysis bio-oil) is of interest, especially with respect to the lower acidity in comparison to Al_2O_3, a commonly studied support for HDO catalysts [27]. Many studies are focused on alumina-supported catalysts (Al_2O_3) [23,26] and upgrading applying model compounds [20,22] but supports with higher stability are required. The alumina-support is well known for its acidity, tendency for increased coke formation, low water tolerance, and conversion to boehmite, resulting in the oxidation and deactivation of the active metal [15,28,29]. According to He et al. [28], the selection of the support for HDO of bio-oils, must consider the resistance to the water content, the acidity of the supports to reduce coke formation, the porosity and its ability to keep the active metal dispersed for the activation of hydrogen. Hence, the investigation of different supports, such as SiO_2 and ZrO_2, appears interesting, especially when including catalyst regeneration in consecutive cycles of HDO-regeneration, evaluating the thermal stability of the catalyst [30]. Presently, only a few works have investigated the regeneration and evaluation of the reuse of the catalyst [26]. Most works, in fact, only consider one regeneration step and do not contemplate Ni catalysts [26,31,32]. It is an essential step to reduce costs, minimizing the waste generation at the same time helping to increase the reusability and recyclability of the catalysts, extending its lifetime [33]. Additionally, previous studies considered the HDO of model compounds whereas others considered the application of fast pyrolysis bio-oil. Usually, different temperature, pressure and reactor designs are used, which makes the comparison of the performance of different nickel-based catalysts difficult.

The current work synthesizes, characterizes and evaluate four nickel-based catalysts for a multi-phase fast-pyrolysis bio-oil upgrading. Supports with higher stability (SiO_2 and ZrO_2) are selected. The catalyst with the best performance is then reused in subsequent HDO-regeneration steps, resulting in four consecutive reactions. Finally, the performance and the catalytic activity along the HDO-regeneration steps are assessed and discussed.

2. Results

2.1. Characterization of the Synthesized Catalysts

The results obtained from the temperature programmed reduction (H_2–TPR), in Figure 1, were useful to identify the catalysts' reduction temperatures before the hydrodeoxygenation (HDO) reactions. The H_2–TPR profile for Ni/SiO_2 showed a clear peak at 350 °C, while the peak for Ni/ZrO_2 is found between 350–400 °C, in agreement with literature [21,34,35]. Since the reduction of bulk Ni oxide occurs around 400–450 °C [36,37], the reduction temperature of Ni/SiO_2 and Ni/ZrO_2 was set to 500 °C to ensure a full reduction before hydrodeoxygenation (HDO) reactions. The addition of Cu to Ni [36,37], as well as the higher loading of Ni [38,39] seems to promote the reduction of Ni oxide, as for both bimetallic NiCu catalysts the temperature of reduction is lower compared to monometallic Ni catalysts. Concerning NiCu/SiO_2, it occurs at 300 °C, while for NiCu/ZrO_2 it was closer to 200 °C. Different peaks are present in the H_2–TPR profiles, attributed to bulk nickel oxide reduction, reduction of Cu(II) to Cu(0) (below 250 °C), as well as reduction of bimetallic NiCu species which, according to Ardiyanti et al. [40], should occur approximately in the range of 290–390 °C. Regarding the current catalysts, most of the reduction was observed at lower temperatures. Consequently, the reduction of NiCu catalyst was set at 350 °C.

Powder X-ray diffraction (XRD), provided information about the crystalline structure of the catalyst (Figure 2). Both SiO_2-supported catalysts show similar XRD patterns. The reflections located at 37.25° and 43.29° indicate the presence of Ni oxide (NiO) in the calcined catalysts. Additionally, metallic Ni was identified due to the reflections at 44.49°, 51.85°, 76.38°, 92.93° and 98.44°. Following the reduction, reflections attributed to NiO disappeared, remaining just metallic Ni reflections. A similar behavior can be seen for Ni/ZrO_2 and NiCu/ZrO_2 (Figure 2). Both NiO and metallic Ni are present in the calcined catalyst. Subsequently, the reduction reflections of NiO are no longer observed for Ni/ZrO_2 and showed a reduced intensity for NiCu/ZrO_2. Reflections attributed to copper were not observed in the bimetallic catalysts, which can be a result of high dispersion of the metal, as well as low concentration [21,23].

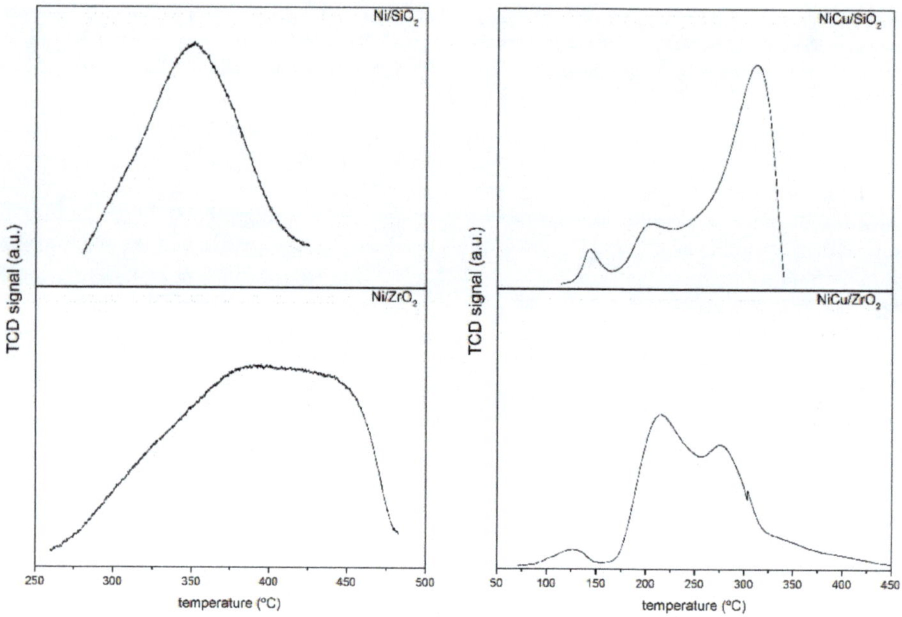

Figure 1. Temperature programmed reduction profile for the nickel-based catalysts. TCD: Thermal conductivity detector; a.u.: arbitrary units.

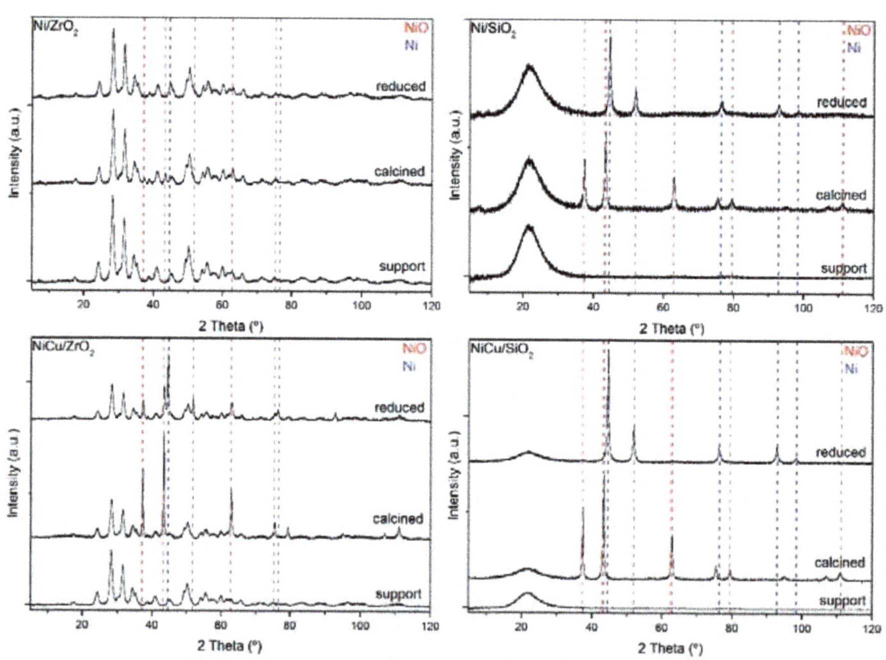

Figure 2. X-ray powder diffraction of the freshly synthesized catalysts, Ni/ZrO_2, Ni/SiO_2, $NiCu/ZrO_2$ and $NiCu/SiO_2$. a.u.: arbitrary units.

The crystallite sizes were estimated using the Scherrer equation. The crystallite size was 17.7 nm for Ni/SiO_2, whereas for $NiCu/SiO_2$, the value was estimated to be 21.4 nm. $NiCu/ZrO_2$ showed a crystallite size of 43.3 nm and the crystallite size of Ni/ZrO_2 was estimated at 9.7 nm.

The metal concentration, as well as the specific surface area, is compiled in Table 1. The Ni/SiO_2 catalyst displays the highest specific surface area (215 m^2/g), while $NiCu/ZrO_2$ shows the lowest (50 m^2/g). Usually, SiO_2-supported catalysts show higher surface areas in comparison to ZrO_2-supported catalysts [41]. The BET surface area of the catalysts, as well as the micropore area and volume, is reduced with the addition of Cu and higher nickel loading. This behavior was also observed by Dongil et al. [24] and Zhang et al. [42]. Furthermore, no micropores were observed in the zirconia-supported catalysts.

Table 1. BET surface area, pore area, volume, diameter, and metal content in the freshly synthesized catalysts.

Properties	Ni/SiO_2	Ni/ZrO_2	$NiCu/SiO_2$	$NiCu/ZrO_2$
BET surface area (m^2/g)	215	65	156	50
Micropore area (m^2/g)	24	0	12	0
Micropore volume (cm^3/g)	0.009	0	0.004	0
Pore diameter (nm)	17.1	12.7	17.1	10.2
Ni (wt.%)	7.9	8.0	27.9	27.3
Cu (wt.%)	-	-	3.2	3.1

2.2. *Hydrotreatment Reactions*

2.2.1. Upgraded Bio-Oil Yields and Properties

The HDO reactions with Ni/SiO_2 showed the highest yield of upgraded bio-oil (49.36 wt.%), and the lowest yields of the aqueous phase (35.57 wt.%) and solids (Table 2). These tendencies were not clear for the remaining catalysts. The lowest yield of upgraded bio-oil (45.32 wt.%) was obtained with Ni/ZrO_2, while $NiCu/ZrO_2$ showed the highest production of solids (0.32 wt.%) and aqueous phase (43.52 wt.%). The highest production of gas was obtained with Ni/ZrO_2 (4.54 wt.%) whereas the lowest was obtained with $NiCu/SiO_2$ (3.56 wt.%).

Table 2. Mass balance, elemental analysis and physicochemical properties of the upgraded bio-oils obtained with fresh Ni-based catalysts.

Properties	Ni/SiO_2	Ni/ZrO_2	$NiCu/SiO_2$	$NiCu/ZrO_2$
Mass balance				
Upgraded bio-oil (wt.%)	49.36 ± 0.07	45.32 ± 0.03	49.03 ± 0.03	46.39 ± 0.04
Aqueous phase (wt.%)	35.57 ± 0.07	42.72 ± 0.01	41.45 ± 0.01	43.52 ± 0.03
Gas (wt.%)	4.29 ± 0.06	4.54 ± 0.05	3.56 ± 0.07	3.87 ± 0.03
Solids (wt.%)	0.23 ± 0.11	0.31 ± 0.06	0.31 ± 0.22	0.32 ± 0.29
Losses (wt.%)	10.54 ± 0.17	7.10 ± 0.05	5.65 ± 0.47	5.90 ± 0.49
Upgraded bio-oil (dry basis; wet basis)				
Carbon (wt.%)	73.15 ± 0.06; 69.75 ± 0.21	72.22 ± 0.01; 68.10 ± 0.21	72.37 ± 0.58; 66.80 ± 0.85	72.15 ± 1.01; 67.25 ± 0.21
Hydrogen (wt.%)	8.42 ± 0.07; 8.55 ± 0.07	8.25 ± 0.01; 8.40 ± 0.01	8.51 ± 0.31; 8.70 ± 0.28	8.55 ± 0.31; 8.70 ± 0.28
Oxygen (wt.%)	17.86 ± 0.05; 21.30 ± 0.01	19.46 ± 0.15; 23.00 ± 0.14	19.00 ± 0.91; 24.10 ± 1.13	20.35 ± 0.01; 23.60 ± 1.41
Nitrogen (wt.%)	0.32 ± 0.01; 0.30 ± 0.01	0.32 ± 0.01; 0.30 ± 0.01	0.32 ± 0.01; 0.30 ± 0.01	0.32 ± 0.01; 0.30 ± 0.01
Physicochemical properties				
H_2O (wt.%)	4.85 ± 0.07	5.10 ± 0.01	7.30 ± 0.42	5.95 ± 0.35
HHV (MJ/kg)	31.18 ± 0.08	30.85 ± 0.02	29.86 ± 0.26	30.27 ± 0.31
pH value	3.55 ± 0.35	3.05 ± 0.35	3.35 ± 0.07	3.20 ± 0.14
Density (g/cm^3)	1.11 ± 0.01	1.12 ± 0.01	1.09 ± 0.01	1.11 ± 0.01

The elemental composition of the upgraded bio-oils by the different nickel-based catalysts is presented in Table 2. The concentration of carbon increases in all upgraded bio-oil in comparison to the concentration in the feed (57.31 wt.% dry basis) [43]. The highest carbon content (73.15 wt.% dry basis) was obtained applying Ni/SiO_2. The remaining oils upgraded with the other catalysts displaying a concentration of around 72.25 wt.% in dry and 67.38 wt.% in wet basis, respectively. The hydrogen concentration was slightly higher in the upgraded bio-oils in comparison to the initial feed, while the oxygen content was reduced in comparison to the original beech wood bio-oil (35.84 wt.% in dry basis). The upgraded bio-oil over Ni/SiO_2 shows the lowest oxygen concentration (17.86 wt.% dry basis), followed by NiCu/SiO_2 (19.00 wt.% dry basis), Ni/ZrO_2 (19.46 wt.% dry basis) and NiCu/ZrO_2 (20.35 wt.% dry basis).

The water content in the upgraded bio-oils was reduced significantly from 26.77 wt.% in the original feed to values between 4.85 wt.% and 7.30 wt.% in the upgraded bio-oils. More than 70% of the water content in the original oil was removed by the hydrotreatment. The upgraded bio-oil over Ni/SiO_2 showed the lowest water content (4.85 wt.%), while NiCu/SiO_2 had the highest value (7.30 wt.%). The opposite was observed for the upgraded aqueous phase. While the upgraded aqueous phase obtained with Ni/SiO_2 showed 74.35 wt.% of water, the aqueous phase obtained with NiCu/SiO_2 showed 67.15 wt.% of H_2O (Table S1). This is because the water removed from the upgraded bio-oils mainly was concentrated in the aqueous phase.

The higher heating value (HHV) also changed by the reactions. In the feedstock, this value was 24.33 MJ/kg, whereas, for the upgraded bio-oils, the HHV increased to values ranging from 29.86 MJ/kg to 31.18 MJ/kg in the case of SiO_2 supported catalysts. A slight increase in the pH value was observed, except for the upgraded bio-oil with Ni/ZrO_2. The density of the upgraded bio-oils decreased after the hydrotreatment. Comparing the density of the upgraded bio-oils with the density of the heavy phase, for example, the value was reduced from 1.19 g/cm^3 to values between 1.09 g/cm^3 and 1.12 g/cm^3. The upgraded bio-oil over NiCu/SiO_2 displayed the lowest density (1.09 g/cm^3), while the highest corresponded to Ni/ZrO_2 (1.12 g/cm^3).

Further information regarding the upgraded bio-oils composition was obtained by proton nuclear magnetic resonance (^{1}H-NMR), depicted in Figure 3. Table 3 shows the assignment of chemical groups to the integration range of the spectra [44]. The aqueous phases are compared to the light phase (LP) of the feedstock, whereas the upgraded bio-oil is compared to the heavy phase (HP) of the feedstock.

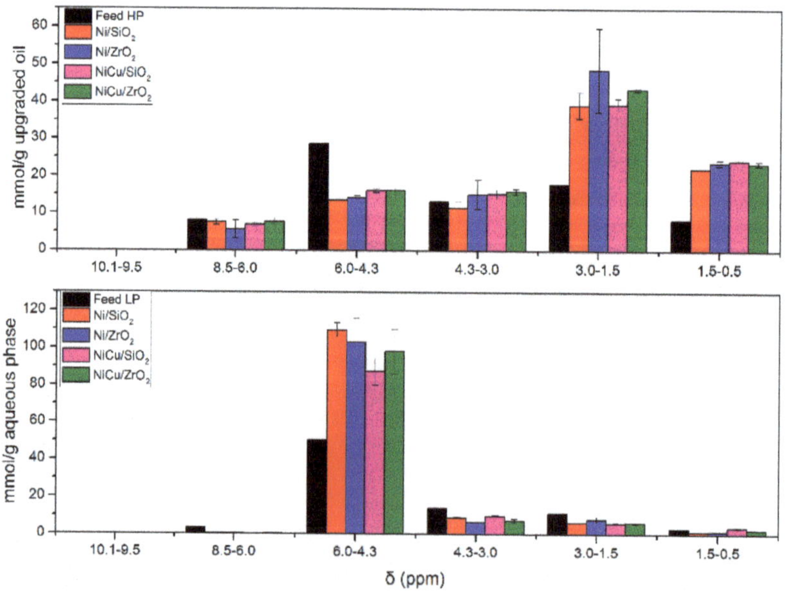

Figure 3. ^{1}H-NMR spectra integrals of the bio-oil components, HP (**top**) and LP (**bottom**), in contrast to the products (upgraded bio-oil and aqueous phase) obtained by different catalysts.

Table 3. Integration ranges of ^{1}H-NMR spectra and their corresponding proton assignment [44].

Integration Range of ^{1}H-NMR Spectra	Proton Assignment
10.1–9.5	Aldehydes
8.5–6.0	(Hetero-)aromatics
6.0–4.3	Carbohydrates, water, O-H exchanging groups
4.3–3.0	Alcohols, ethers, alkenes
3.0–1.5	α proton to carboxylic acid or keto-groups, α proton to unsaturated groups
1.5–0.5	Alkanes

The signals for aldehydes (10.1–9.5 ppm) are very small in the feed, below 0.1 mmol/g sample in both LP and HP. They were not observed in the upgraded products as they are very reactive even at mild conditions and, therefore, are hydrogenated to alcohols [45]. The main signals for the upgraded bio-oil were found in the region of α-protons to carboxylic acid or keto-groups and α-protons to unsaturated groups (3.0–1.5 ppm).

The aromatic (8.5–6.0 ppm) were concentrated mostly in the upgraded bio-oil and almost absent in the aqueous phase (7.0 versus 0.2 mmol/g sample on average). No significant differences among catalysts were observed. A slight increasing tendency could be observed for the concentration of alcohols, ethers, and alkenes (4.3–3.0 ppm), although it was not significantly different among all tested catalysts. The concentration of protons in this region was more abundant in the upgraded bio-oil (14.5 mmol/g sample on average) in comparison to the aqueous phase (8.2 mmol/g sample on average). The abundance of alkanes (1.5–0.5 ppm) almost triples in comparison to the feed (heavy phase) and was significantly higher in the upgraded bio-oil than in the aqueous phase (23.4 versus 2.2 mmol/g sample).

The accumulation of water in the aqueous phase is also confirmed by the ^{1}H-NMR measurements, considering that the main signal obtained for the aqueous phase was found in the carbohydrates, water, and O-H exchanging groups (6.0–4.3 ppm). The high concentration of protons in this region is attributed to the removal of water from the bio-oil [46,47], in agreement with Karl–Fisher results. On the other hand, the proton in this region decreases in all upgraded bio-oils, especially for Ni/SiO_2, which produced the upgraded bio-oil with the lowest water concentration (Table 3).

To identify the main compounds in the upgraded bio-oils, as well as to investigate differences in selectivity among the catalysts tested, the upgraded bio-oils were analyzed qualitatively by gas chromatography-mass spectrometer (GC-MS), depicted in Figure 4 and Table 4. The chromatograms of the upgraded bio-oils are discussed in comparison to the chromatograms of the heavy phase (feedstock), available in the Supplementary Material (Figure S3). The chromatograms of the upgraded aqueous phases, as well as the light phase of the feedstock, are also available as Supplementary Material (Figures S3 and S4).

Typical compounds were observed in the upgraded bio-oils, such as carboxylic acids, ketones, phenolic compounds and others [48]. The main reaction pathways identified and later discussed are available in Figure S6.

A variety of ketones, especially but not limited to cyclic forms, were identified in all the upgraded bio-oils, such as 2-pentanone, cyclopentanone, 3-methyl-cyclopentanone, 2-ethyl-cyclohexanone, cycloheptanone and others, in agreement with Boscagli et al. [46] and Ardiyanti et al. [23]. They are attributed as products of sugar conversion and its derivatives.

Aromatic compounds initially present in the feedstock also were observed in the upgrade-oil, such as guaiacol (2-methoxy-phenol), phenol and 4-ethyl-2-methoxy-phenol. Acetic acid also was present in the feedstock as well as in the upgraded products.

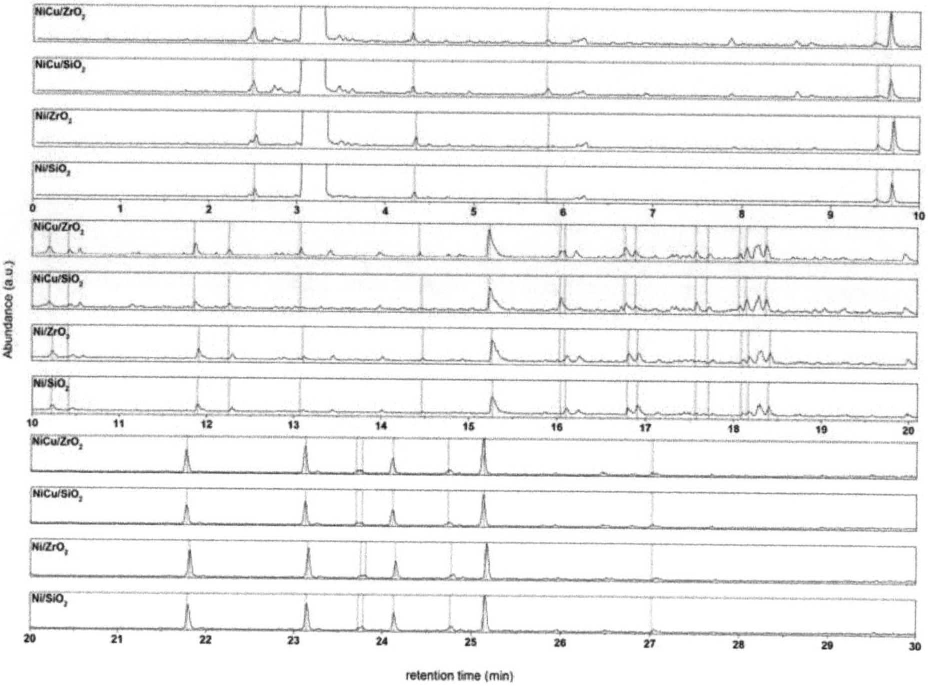

Figure 4. Chromatograms of the upgraded bio-oils over different nickel-based catalysts.

Table 4. Retention time of the main compounds in the upgraded bio-oil identified by GC-MS.

Retention Time (min)	Compound
2.51	Acetic acid, methyl ester
4.32	2-pentanone
5.82	1-propanol
9.46	Cyclopentanone
9.64	2-methyl-cyclopentanone
10.20	3-methyl-cyclopentanone
10.42	(R)-3-methyl-cyclohexanone
11.84	2-ethyl-cyclopentanone
12.23	Cycloheptanone
13.08	2-methyl-2-propanol
14.47	6-hepten-1-ol
15.23	Acetic acid
16.04	Tetrahydro-2-furanmethanol
16.10	2,3,4-trimethyl-cyclopent-2-ene-1-one
16.77	Propanoic acid
16.87	2,3-dimethyl-2-cyclopenten-1-one
17.57	Propylene glycol
18.07	Dihydro-5-methyl-2(3*H*)-furanone
18.13	1,2-ethanediol
18.40	γ-butyrolactone
19.05	3-pentanol
21.74	2-methoxy-phenol
23.09	2-methoxy-4-methyl-phenol
23.67	4-methyl-phenol
23.70	Phenol
24.07	4-ethyl-2-methoxy-phenol
24.72	2-ethyl-phenol
25.1	2-methoxy-4-propyl-phenol
27.02	2-propyl-phenol

After the upgrading reaction, compounds initially present in the bio-oil mixtures, such as furfural, were completely converted for all the catalysts tested. The same was observed for eugenol and compounds with GC retention times higher than 30 min, such as vanillin. Although very similar compositions of all upgraded bio-oils, some differences in selectivity were observed for the catalysts. 1-propanol (5.82 min) was observed only in the bio-oils upgraded with bimetallic catalysts, as well as the peak at 16.04 min, attributed to tetrahydro-2-furanmethanol (tetrahydrofurfuryl alcohol), also observed in the bio-oils upgraded with NiCu catalysts. It shows that furfural is completely hydrogenated to tetrahydrofurfuryl alcohol over these catalysts [49]. Earlier studies observed the selectivity of copper-containing catalysts for furfural hydrogenation, particularly at high temperatures [50]. Furfural also can be converted to cyclopentanone [48,49] and 2-pentanone [8,51], both identified in the products, as well as other cyclopentanones [52].Compounds such as methane, identified in the gas phase and later discussed, also can be derived from furfural conversion; the conversion of furfural to furfuryl alcohol and later to furan leads to methane formation [53]. Furthermore, the peak at 17.57 min, attributed to propylene glycol, was observed only in the upgraded bio-oils as well as in the aqueous phases (Figure S5) upgraded with NiCu catalysts. Propylene glycol can be obtained from hydrogenation of hydroxyacetone [54], which was converted completely after the HDO reactions (see peak at 12.46 min Figure S3). While copper-containing catalysts seem to favor the production of propylene glycol in comparison to other catalysts [55], nickel catalysts seems to follow a different pathway. Resulting from the C-O bond cleavage of propylene glycol [54], 1-propanol was observed in the oils upgraded with bimetallic catalysts. Furthermore, compounds initially absent in the feedstock, such as 2-methoxy-4-propyl-phenol, resulting from the hydrogenation of the double bond of eugenol, were identified in the upgraded bio-oil [47]. A peak at a retention time of 26.9 min was observed in the upgraded aqueous phases obtained with bimetallic catalysts, although the identification of the compound was not possible.

Differences among the feed (LP) and upgraded products (aqueous phases), as well as among the products obtained with different catalysts, were observed. For example, 2-methyl-propanol (13.08 min) was observed in all upgraded aqueous phases, except for $NiCu/SiO_2$. The peak attributed to tetrahydro-2-furanmethanol increased significantly, mainly in the aqueous phases obtained by bimetallic catalysts, in the same way as observed in the upgraded bio-oils. Propanoic acid (16.77 min) was observed for all aqueous phases.

2.2.2. Hydrogen Consumption and Gaseous Products

The hydrogen consumption was considered for the evaluation of the different catalysts. $NiCu/SiO_2$ presented the highest consumption (239.3 NL/kg bio-oil), followed by $NiCu/ZrO_2$ (201.6 NL/kg bio-oil), Ni/SiO_2 (186.2 NL/kg bio-oil) and Ni/ZrO_2 (181.9 NL/kg bio-oil). Yin et al. [35] reported, higher hydrogenation activity can be seen when adding copper to Ni catalysts, due to changes in the catalytic activity and selectivity, favoring some hydrogenation reactions [56]. The hydrogen consumption can be correlated with the H/C molar ratio. A higher H/C molar ratio and lower O/C molar ratio indicates an upgraded bio-oil with improved properties [15]. The catalysts evaluated resulted in H/C ratios between 1.37 (Ni/ZrO_2) to 1.53 ($NiCu/SiO_2$), showing a tendency between the hydrogen uptake and the H/C molar ratio. The same tendency was not observed for the O/C ratio. The lowest ratio was observed using Ni/SiO_2 (0.18) whereas the highest was for $NiCu/ZrO_2$ (0.28). This indicates that a higher consumption of H_2 does not reflect in the HDO of the feedstock.

Although a very similar total gas production (Figure 5) was obtained for all tested catalysts, Ni/ZrO_2 showed the highest total gas production (1.06 mol/kg bio-oil), followed by Ni/SiO_2 (0.99 mol/kg bio-oil), $NiCu/ZrO_2$ (0.90 mol/kg bio-oil) and $NiCu/SiO_2$ (0.86 mol/kg bio-oil). The total gas production was determined mainly by the production of CO_2, the most abundant gas product for all four catalysts, in agreement with other studies [35,40,41,46]. Decarboxylation of carboxylic acids can result in CO_2 formation [46]. Carbon monoxide, methane, and other gases such as propane, propene, ethane and ethene were formed in smaller amounts.

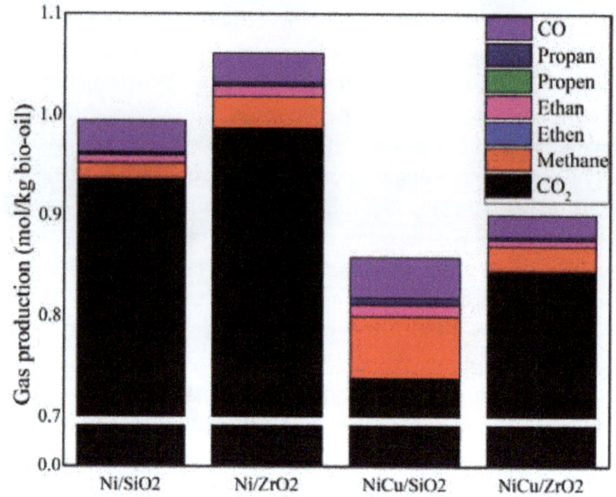

Figure 5. Production and composition of the gas phase for the nickel-based catalysts tested.

Monometallic Ni catalysts produced the highest amount of CO_2, followed by NiCu/ZrO_2 and, later, by NiCu/SiO_2. According to Gallakota et al. [14], in an ideal scenario all C atoms should be converted to hydrocarbons without CO_2 formation. Nevertheless, since the total gas production of the four catalysts represented around 4 wt.% (see mass balance Table 2), the produced amounts of CO_2 are, in general, quite reduced in these four cases. The formation of CO_2 might indicate lower hydrogen consumption [57]. This effect was observed when the CO_2 production was compared to the hydrogen consumption—both variables were inversely proportional (Figure S2), as reported by Boscagli et al. [46]. This tendency only was noticed in the CO_2 production, but not for the remaining produced gases.

The formation of methane was observed mainly for NiCu/SiO_2. The smallest concentration of methane was obtained for Ni/SiO_2. Methane resulted from the hydrogenation of carbohydrates, acetic acid decomposition, cleavage of C-C bonds of alcohols or even from methoxy groups demethylation [57]. It is important to highlight that the higher methane production was observed with NiCu/SiO_2, the catalyst that showed the highest H_2 consumption. This is an indication of hydrocracking of the molecules and excessive hydrogen consumption [35,58].

CO formation was more abundant in the SiO_2-supported catalysts. NiCu/SiO_2 displayed the highest amount, followed by Ni/SiO_2, Ni/ZrO_2 and NiCu/ZrO_2. The formation of CO can be a result of the C-O cleavage of different groups, such as carboxylic acids, aldehydes, and alcohols. The loss of C via CO (decarbonylation), is less advantageous than the loss of CO_2 (decarboxylation), considering more O is removed per mole of lost carbon [57].

2.2.3. Catalysts Characterization

The spent catalysts were evaluated mainly in terms of metal leaching, XRD, sintering, carbon deposition as well as surface area and composition. Inductively Coupled Plasma Emission Spectroscopy (ICP-OES) was further used to evaluate the metal content present in the upgraded aqueous phases and determine the metal leaching. Ni/SiO_2 presented 0.8% of Ni leached while using Ni/ZrO_2 0.43% of Ni was washed into the aqueous phase. Both NiCu catalysts showed 0.16% of Ni leached while NiCu/SiO_2 and NiCu/ZrO_2 displayed 0.08% and 0.04% of Cu leached, respectively. Similar leaching levels also were found by Boscagli et al. [26] for a reaction at 240 °C compared to a NiCu/Al_2O_3 catalyst. The extent of leaching seems to correlate with the support. The lowest proportions of metals leached were observed in ZrO_2-supported catalysts. Reported previously, ZrO_2 seems to be stable in the harsh reaction conditions [28,40,42,59]. Due to the severity of the HDO reactions, the structure, morphology and texture of the catalysts might be affected [40], therefore, the spent catalysts were analyzed by XRD after the reactions. No significant differences were observed between the X-ray diffraction patterns of the fresh and spent catalysts, except for NiCu/ZrO_2. The small diffractions attributed to NiO disappeared in the spent catalysts, indicating further nickel reduction. The XRD patterns are available in Figure S1. The crystallite sizes were calculated for the spent catalysts. The crystallite size of NiCu/SiO_2 increased from 21.4 nm to 43.0 nm after the reaction. Such an increase in the crystallite size was previously observed [60] and is an indication of sintering, which might result in loss of the catalyst activity [61]. Conversely, the crystallite size of Ni/ZrO_2, Ni/SiO_2 and NiCu/ZrO_2 remained in the same range observed for the fresh catalysts. The composition of selected particles was analyzed by EDX (Energy Dispersive X-ray spectroscopy) and the results obtained for the fresh and spent catalysts are presented in Table 5.

Table 5. Composition of selected particles of different Ni-based catalysts (fresh and spent) obtained by EDX.

Catalyst		Ni (wt.%)	Cu (wt.%)	Si (wt.%)	Zr (wt.%)	C (wt.%)	S (wt.%)	Ca (wt.%)	Fe (wt.%)
Ni/SiO_2	Fresh	11.2	-	38.6	-	3.6	-	-	-
	Spent	7.3	-	44.9	-	12.1	0.2	-	-
Ni/ZrO_2	Fresh	8.6	-	-	60.9	6.4	-	-	-
	Spent	6.3	0.7	-	45.5	26.3	0.1	0.1	1.3
NiCu/SiO_2	Fresh	30.0	2.9	29.2	-	3.2	-	-	-
	Spent	26.9	2.4	18.2	-	26.7	0.2	-	-
NiCu/ZrO_2	Fresh	55.8	6.3	-	20.0	4.0	-	-	-
	Spent	44.0	4.5	-	16.4	24.3	0.1	0.1	0.7

Carbon deposition was observed for all catalysts after HDO, with lower concentration observed in Ni/SiO_2. Poisoning substances, such as sulfur and calcium, were observed near the detection limit in the spent forms. Sulfur appeared in all spent catalysts, whereas calcium and iron were observed only in zirconia-supported catalysts. The Ni proportion over the catalysts' surfaces decreased, attributed mainly to the carbon deposits of higher molecular weight polymerization products [35]. The specific surface areas of the spent catalysts were reduced in comparison to the fresh ones, especially for the silica-supported catalysts. The specific surface area of Ni/SiO_2 was reduced to 46 m^2/g, a sharp reduction in comparison to the original surface area (215 m^2/g). NiCu/SiO_2 showed a reduction from 156 m^2/g to 36 m^2/g. It can be attributed to the fact that the pores probably were blocked by carbonaceous deposition. The micropore area of Ni/SiO_2 was reduced to 8 m^2/g, whereas the micropore area of NiCu/SiO_2 was reduced to 3 m^2/g. The reduction in the surface area of the zirconia-supported catalyst was less significant. Ni/ZrO_2 showed a reduction to 57 m^2/g (original: 65 m^2/g), whereas NiCu/ZrO_2 was reduced from 50 to 36 m^2/g. The reduction of the surface area is attributed to deposition of carbon in all cases, as already documented by many authors [41,62–64].

Considering that SEM-EDX (Scanning Electron Microscopy/Energy Dispersive X-ray spectroscopy) provides the elemental mapping of only selected regions [65], further composition analysis was performed measuring the active metals and poisoning substances in the bulk catalyst (Table 6).

Agreeing with EDX results, an increase in the calcium concentration over both zirconia-supported catalysts was observed. Sulfur, considered a very persistent poisoning substance of nickel [26,66], increased in all catalysts tested after the reaction, with slightly lower concentrations at Ni/SiO_2. Interestingly, no differences were observed in the XRD spectra, such as nickel sulfide formation [66]. The concentration of active metals (Ni and Cu) was reduced in the spent forms in comparison to the fresh catalyst. This behavior correlates to the carbon deposition. Accompanying a higher concentration of carbon, which initially is absent in the fresh catalysts, the average concentration of nickel and copper in the spent catalyst decreases. Furthermore, leaching also can play a role in the reduction of active metal in the catalyst, as discussed previously. The carbon concentration obtained by elemental analysis shows the same tendencies as observed with the EDX measurements.

Table 6. Metal content and poisoning substances on the catalyst before and after the reaction.

Catalyst		Ca * (wt.%)	S * (wt.%)	Mg * (wt.%)	Ni * (wt.%)	Cu * (wt.%)	C ** (wt.%)
Ni/SiO_2	Fresh	0.03	0.01	0.01	7.94	-	-
	Spent	0.03	0.12	0.01	7.05	0.01	4.2
Ni/ZrO_2	Fresh	0.01	0.01	-	8.02	-	-
	Spent	0.06	0.13	0.01	6.94	0.07	5.6
$NiCu/SiO_2$	Fresh	0.02	0.01	0.01	27.90	3.25	-
	Spent	0.02	0.13	0.01	25.90	2.76	8.6
$NiCu/ZrO_2$	Fresh	0.01	0.01	-	27.30	3.10	-
	Spent	0.05	0.14	0.01	25.25	2.72	7.1

* Results obtained by ICP-OES. ** Results obtained by elemental analysis.

2.3. Cycles of HDO and Regeneration: Catalyst and Product Behavior

Ni/SiO_2 was selected for further consecutive HDO reactions as soon as the original reactions with the fresh catalyst were carried out. Key parameters were considered for selection of this catalyst for consecutive HDO-regeneration cycles: HDO activity, water and carbon concentration in the upgraded bio-oil, hydrogen consumption as well as solid formation. The upgraded bio-oil obtained with Ni/SiO_2 showed the highest carbon content and the lowest oxygen content in comparison to other catalysts, an indication of the improvement of the oil quality (higher energy density and better chemical stability). Additionally, upgrading with Ni/SiO_2 resulted in an upgraded bio-oil with the lowest water concentration, low hydrogen consumption, lowest amount of solids formed, highest HHV and pH (Table 2). Finally, the smallest production of methane also was obtained for this catalyst (Figure 5). Therefore, Ni/SiO_2 was selected to carry out the following regeneration steps.

Along the cycles of HDO-regeneration, the catalyst was characterized in the intermediate steps of the process: spent, calcined, and reduced form. Only a small amount of sample was available for characterization, considering that the catalyst needed to be reused in the following experiments. Due to that reason, two techniques which require low sample amounts showing meaningful results were selected to monitor the catalyst along the cycles—SEM-EDX and XRD. The upgraded products were characterized using the same techniques as described previously.

Looking at the SEM-EDX results, it was possible to follow the main changes occurring on the catalyst surface (Figure 6). The fresh catalyst showed a good dispersion of nickel particles over the support surface (Figure S7 fresh a). Along the cycles, sintering of the nickel particles was observed. Specific regions were analyzed by EDX to estimate the composition. While the fresh catalyst mainly was composed of nickel, silica and a low concentration of carbon, the spent catalysts clearly showed a small concentration of sulfur (near the detection limit) located around the nickel particles. Interestingly, in the regions with very low nickel concentration or even absent of nickel, no sulfur was identified. Al and Fe were also observed in the EDX spectrum, although in very small concentrations (below 0.1 wt.%). Aluminum was observed in small concentrations (support composition) but the authors cannot discard the possibility of contamination during the removal of the catalyst from the autoclave (aluminum paste is used to seal the reactor).

Considering that EDX measurements can be made either as element mapping or point analysis, both methods were used. Further compilation of images, as well as the mapping of the catalyst surface along the cycles, is available in the Supplementary Material (Figures S7 and S8 and Table S3).

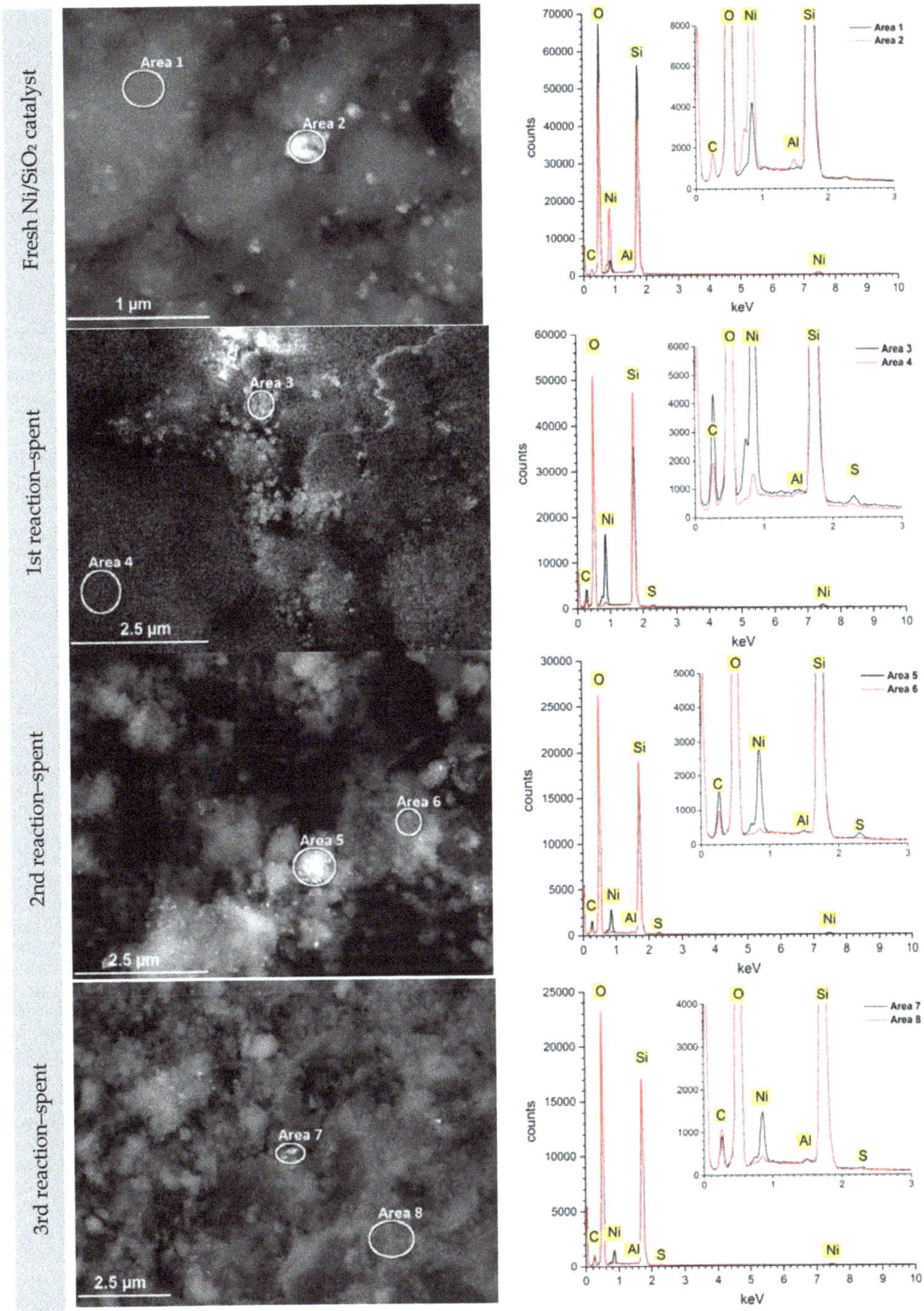

Figure 6. *Cont.*

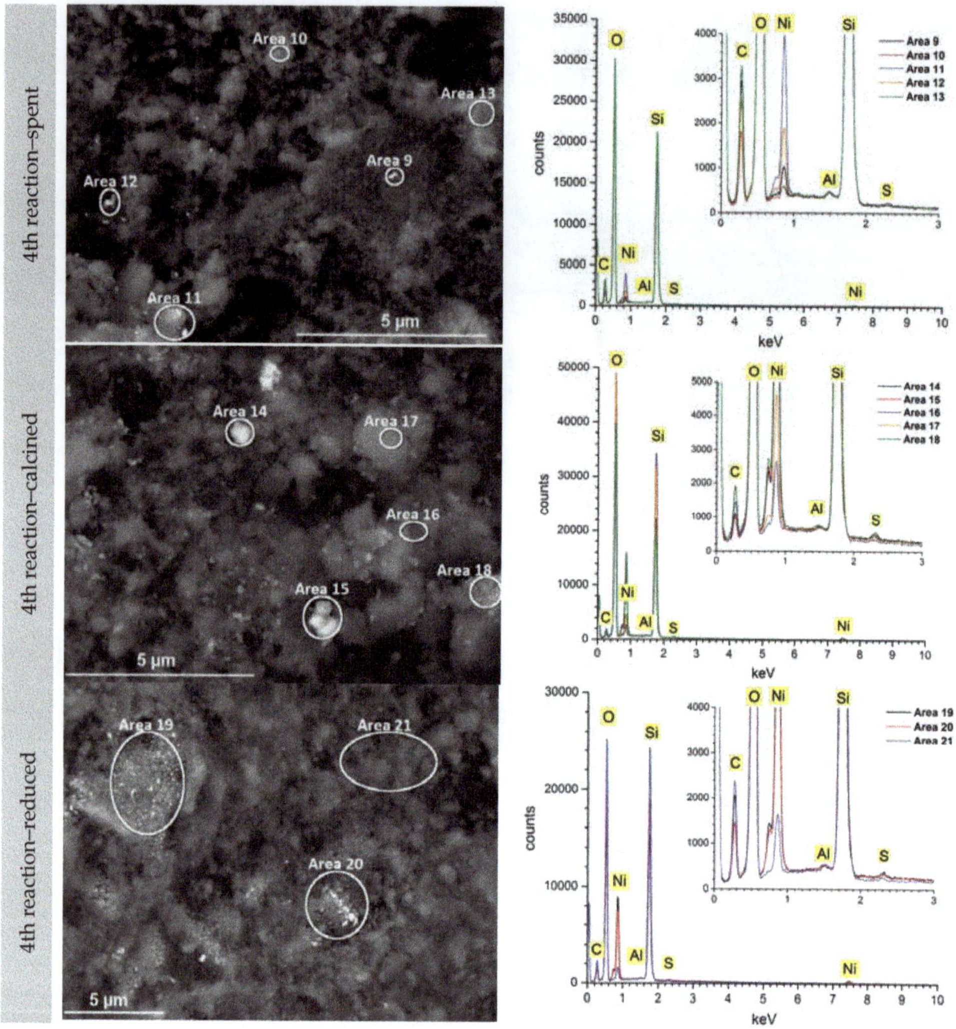

Figure 6. SEM-EDX images and spectra of Ni/SiO_2 along the consecutive cycles.

Following the fourth reaction, the catalyst was analyzed by SEM-EDX in three different situations—spent, after the calcination and, finally, after the reduction step. Although punctual analyses were performed (Table 7), some trends could be observed. Regarding carbon, an expected lower concentration was observed after the calcination step (spectrum 14–16), in comparison to the spent catalyst (spectrum 9–13). Furthermore, the sulfur remained over the reduced catalyst, mainly in the regions with higher nickel concentration, indicating a persistent adsorption. Further structural investigation along the cycles was obtained by XRD and, in this case, the catalyst was analyzed along each new regeneration cycle, in the spent, calcined, and reduced forms (Figure 7). Metallic Ni reflection was observed in the fresh, spent, and reduced catalysts along the consecutive reactions. Due to the calcination step, reflections attributed to NiO were observed which disappeared after the reduction. Considering the low concentration of sulfur over the spent catalyst surface, reflections attributed to Ni_3S_2 as found by other authors [26,67,68] were not observed, even after the fourth reaction.

Table 7. SEM-EDX scan composition of selected regions of Ni/SiO_2 over consecutive cycles.

Spectrum	C (wt.%)	Si (wt.%)	S (wt.%)	Ni (wt.%)
1—Fresh Ni/SiO$_2$	4.1	41.5	-	3.1
2—Fresh Ni/SiO$_2$	2.6	24.9	-	44.9
3—1st reaction (spent)	8.4	24.5	0.3	37.1
4—1st reaction (spent)	6.0	42.8	-	1.2
5—2nd reaction (spent)	8.2	29.3	0.4	19.8
6—2nd reaction (spent)	5.8	38.6	<0.1	0.6
7—3rd reaction (spent)	5.5	35.4	0.1	13.6
8—3rd reaction (spent)	7.61	38.1	-	-
9—4th reaction (spent)	14.1	32.5	-	4.2
10—4th reaction (spent)	10.1	37.0	-	1.0
11—4th reaction (spent)	14.2	27.5	0.1	17.4
12—4th reaction (spent)	12.3	32.8	-	7.7
13—4th reaction (spent)	14.7	32.8	-	0.7
14—4th reaction (calcined)	2.4	22.3	0.1	41.6
15—4th reaction (calcined)	2.3	21.0	0.2	45.2
16—4th reaction (calcined)	4.6	38.7	<0.1	1.4
17—4th reaction (calcined)	5.7	34.7	0.1	7.1
18—4th reaction (calcined)	4.5	20.6	0.3	37.5
19—4th reaction (reduced)	9.1	23.6	0.2	40.4
20—4th reaction (reduced)	6.8	28.9	0.2	33.8
21—4th reaction (reduced)	12.6	37.7	0.1	5.0

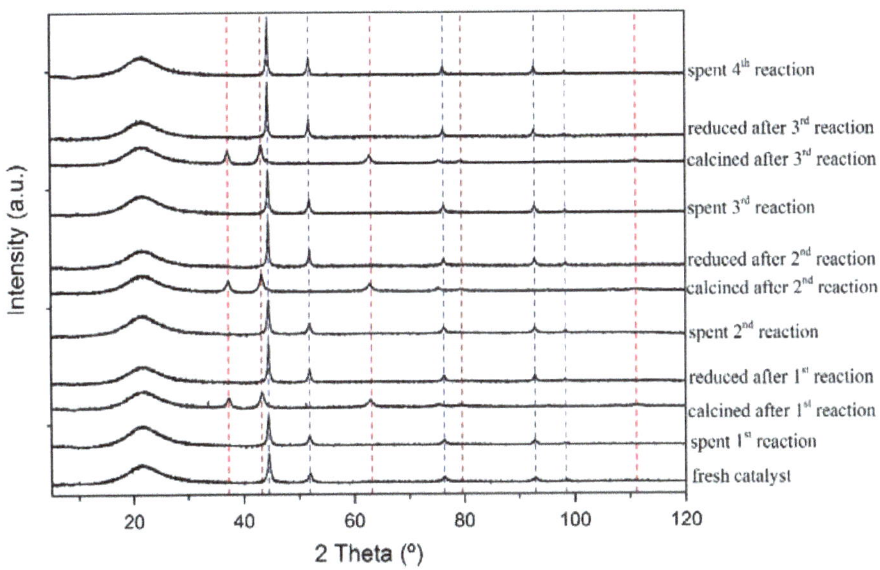

Figure 7. XRD patterns of Ni/SiO_2 catalyst along consecutive reactions and regeneration steps. Dashed blue lines refer to Ni and dashed red lines refer to NiO.

An increase in the crystallite sizes could be observed, as also identified by other authors [26]. Ni/SiO_2 showed initially a crystallite size of 17.7 nm. Following subsequent regenerations, the crystallite size reached 37.3 nm (fourth reaction), due to particle sintering [45], in agreement with the SEM-EDX observations.

Due to the small amount of sample available, the BET surface area was measured only after the fourth reaction (spent catalyst). The original catalyst had a surface area of 215 m^2/g of catalyst, reduced to 46 m^2/g after the first reaction and later to 39 m^2/g after the fourth reaction (spent catalyst), slightly lower compared to the first use. Furthermore, the amount of nickel leached to the light phase was calculated. It could be noticed that along the cycles the amount of nickel leached was being reduced as follows: during the 1st reaction 0.8% of nickel was leached, followed by 0.6% in the 2nd, 0.12% in the 3rd reaction and, finally, in the 4th only 0.10% of nickel was leached.

The physicochemical properties of the upgraded bio-oils over consecutive reactions also were determined (Table 8). The carbon content remained above 72 wt.%, although a slight decrease was observed. The water concentration slightly increased after the second reaction, remaining in the same range over the following reactions (around 5.4 wt.%). The HHV and pH remained in the same range as observed in the reaction performed with the original catalyst. The oxygen concentration increased after the third reaction, reaching 20.35 wt.% after the fourth reuse of the catalyst. Similar tendencies were found by Boscagli et al. [26] evaluating the performance of regenerated catalysts, although, in their study, only one regeneration was considered and a different catalyst ($NiCu/Al_2O_3$) was used. As the number of consecutive reactions increased, the hydrogen consumption decreased, mainly between the first and second reuse. A reduction of 17.18% in the consumption of hydrogen between the original reaction and the first regeneration, 6.28% between the first and the second, and just 0.66% between the second and third regenerations was observed. The lower consumption of hydrogen indicates a decline in the hydrotreatment activity, which can be a result of the sintering, leaching and due to the poisoning substances [61] observed by XRD, ICP-OES and EDX measurements.

Table 8. Physicochemical properties, elemental analysis (in dry basis) of the upgraded bio-oil, hydrogen consumption and total gas production over consecutive HDO reactions and catalyst regeneration.

Property	1st Reaction	2nd Reaction	3rd Reaction	4th Reaction
Carbon (wt.%)	73.15	73.54	72.37	72.15
Hydrogen (wt.%)	8.42	8.25	8.51	8.55
Oxygen (wt.%)	17.86	17.74	19.00	20.35
Nitrogen (wt.%)	0.32	0.32	0.32	0.32
H_2O (wt.%)	4.85	5.4	5.5	5.3
HHV (MJ/kg)	31.18	30.92	31.11	31.26
pH value	3.6	2.8	3.5	3.4
H_2 consumption (NL/kg)	186.21	154.22	144.53	143.58
Gas production (mol/kg)	0.99	1.23	1.31	1.35

Furthermore, as the number of consecutive reactions increased, higher amounts of CO_2 were produced (Figure 8). Discussed previously, the consumption of hydrogen seems to be inversely proportional to the amount of CO_2 produced; Additionally, as the number of consecutive reactions increased, the higher the amounts of CO, propene, ethane and ethene were produced, although less significant in comparison to CO_2. Methane remained constant along the cycles.

The changes over consecutive reactions also were monitored by ^{1}H-NMR (Figure 9). Some trends observed for the original reaction also were observed along the cycles. The signal for aldehydes (10.1–9.5 ppm), already absent in the first reaction due to its high reactivity at low temperatures [45], was not observed in any of the phases for all regeneration stages. The highest signal for the upgraded bio-oil was found in the α protons to unsaturated, carboxylic acids and keto-groups (3.0–1.5 ppm) while, for the aqueous phase, it was in the water, O-H exchanging and carbohydrate groups (6.0–4.3 ppm). Aromatics (8.5–6.0 ppm) mostly were concentrated in the upgraded bio-oil and almost absent in the aqueous phase (8.3 mmol/g sample versus 0.2 mmol/g sample on average). Alkenes, alcohols, and ethers (4.3–3.0 ppm) were concentrated in the upgraded bio-oil (10.6 mmol/g sample versus 6.3 mmol/g in the aqueous phase on average). The same was observed for alkanes (1.5–0.5 ppm), 22.3 mmol/g bio-oil versus 1.1 mmol/g aqueous phase on average. It could be seen that the concentration of aromatics remained similar along the reactions. The values also were similar as observed for the feedstock (HP). A similar trend was observed for the carbohydrates, water, and O-H exchanging groups. The increase in this region can be attributed to the higher water concentration in the upgraded bio-oils, as well as a lower conversion of compounds belonging to this region. As the number of reactions increased, the concentration of alcohols, ether and alkenes in the upgraded bio-oils decreased. The α proton to unsaturated groups in the upgraded bio-oils was much higher in comparison to the feedstock. An increase of protons was observed in the first reaction, dropping in the following reactions. Alkanes were also in higher concentration compared to the feedstock and the values were quite similar among all the reactions. Conversely, within the aqueous phase, the opposite tendency was observed for the carbohydrates, water and O-H exchanging groups. A decline along the regenerations was seen, probably due to the increase of the water content in the upgraded bio-oil (Figure 9). The signal for alcohols, ethers and alkenes also dropped, from 8.7 mmol/g after the first reaction to 5.2 mmol/g after the fourth regeneration. The signals for α proton to carboxylic acid or keto-groups, α proton to unsaturated groups, and for alkanes displayed a very small reduction after each reuse of the catalyst.

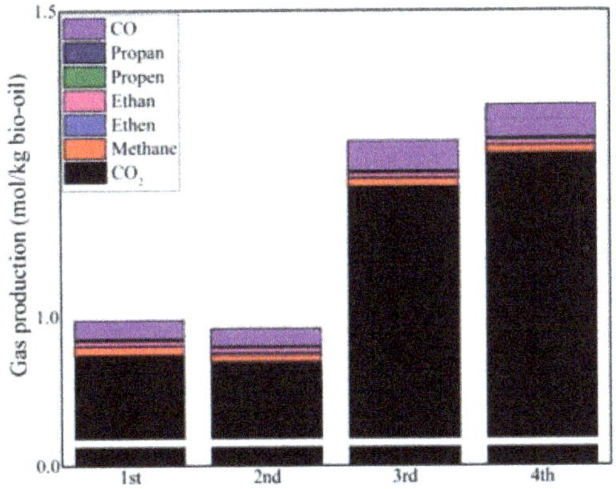

Figure 8. Production and composition of the gas phase obtained for the cycles of hydrotreatment-regeneration.

Further qualitative investigation was performed with GC-MS (Figure 10). The composition of the upgraded bio-oils along the cycles was very similar, although some changes could be observed. The intensity of the peaks attributed to cycloheptanone (12.23 min) and 2-methyl-2-propanol (13.08 min) was reduced along the cycles. The same was observed for two more substances: dihydro-5-methyl-2-(3*H*)-furanone was being reduced until disappeared in the fourth reaction while a visible reduction in the γ-butyrolactone peak (18.4 min) was visible.

The dehydration of C6 sugars, such as glucose, results in compounds such as hydroxymethylfurfural [69]. The following conversion of hydroxymethylfurfural resulted in 5-(Hydroxymethyl)dihydro-2(3*H*)-furanone which can be further converted to γ-valerolactone (Dihydro-5-methyl-2(3*H*)-furanone) through hydrogenation and direct deoxygenation [70]. Conversely, γ-butyrolactone is obtained through hydrogenation of 2(5*H*)-furanone [70]. These sugar derivative compounds then can be further converted to ketones and alcohols [28,69]. Based on these results, a correlation between ^{1}H-NMR and GC-MS could be proposed to explain these findings: Considering the higher signal obtained for the protons belonging to the carbohydrates, water and O-H exchanging groups, it can be assumed that the upgraded bio-oils along the cycles showed higher concentration of not just of water, but possibly also sugars, in comparison to the first reaction. Furthermore, based on the lower intensity of the GC-MS peaks of sugar derivative compounds, such as γ-valerolactone, γ-butyrolactone as well as 2-methyl-2-propanol, it is assumed that the conversion of sugars along the upgrading cycles is reduced. The peak intensity of 2-ethyl-cyclopentanone in the light phases (Figure S5) is also reduced along the cycles. It also is evidence of lower conversion of sugars along the cycles, considering that this compound can be obtained from a compound with similar structure to furfural [71].

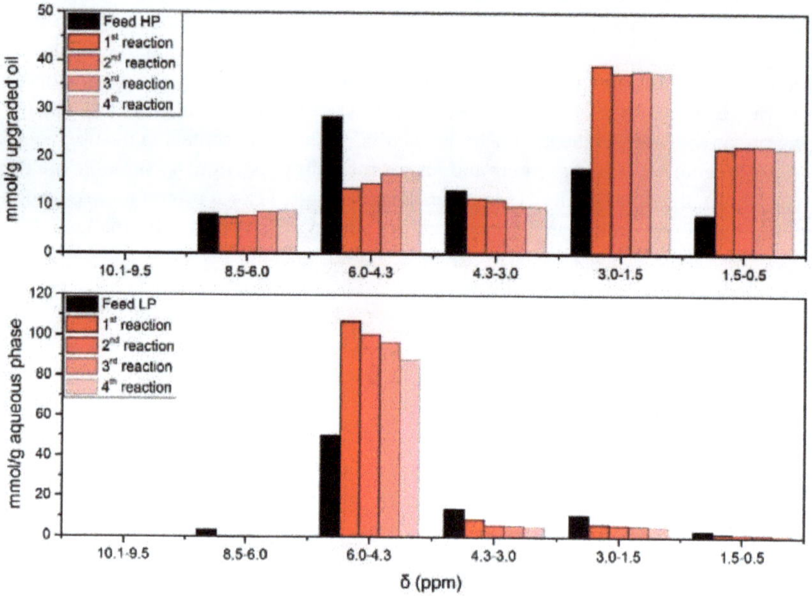

Figure 9. ^{1}H-NMR spectra integrals of upgraded bio-oil and HP (**top**) and aqueous phase and LP (**bottom**) along the HDO-regeneration cycles.

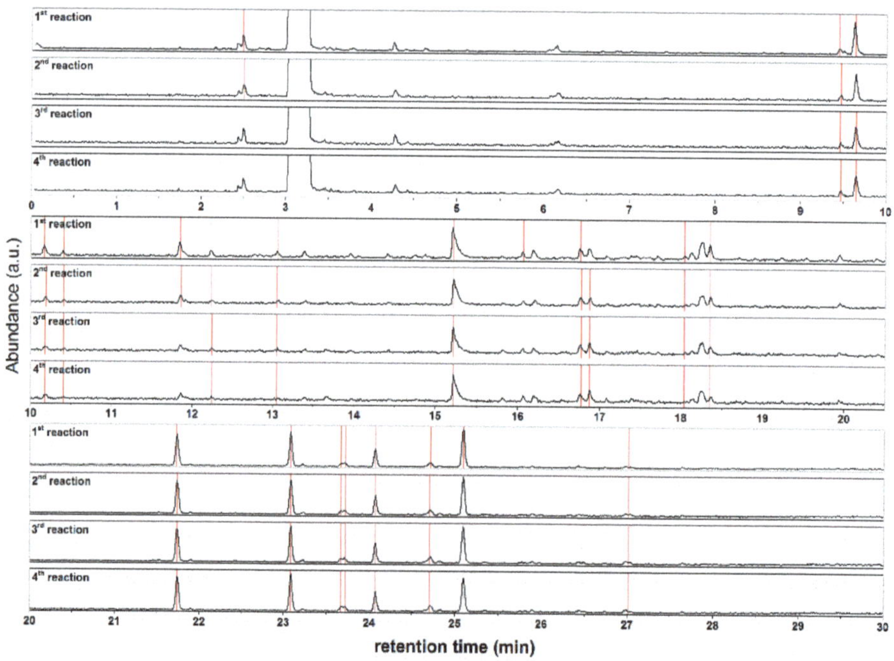

Figure 10. GC-MS upgraded bio-oil along the regeneration cycles.

3. Discussion

The synthesis and evaluation of nickel-based catalysts showed differences among the catalysts evaluated. The temperature programmed reduction (H_2-TPR) showed that the catalysts, with the addition of copper as a promoter and higher loading of nickel, had lowered the reduction temperature of nickel oxides, which is in agreement with previous studies [30,38].

The evaluation of the catalysts for hydrodeoxygenation (HDO) reactions showed that upgraded bio-oil obtained with Ni/SiO_2 showed the best properties in terms of low oxygen concentration, low water concentration and high HHV. The higher HDO activity for this catalyst can be correlated with its higher surface area in comparison to the other catalysts tested in this study, which could be beneficial to increase the dispersion of the active components, resulting in a more active catalyst [20,64]. Coincidently, the ZrO_2-supported catalysts showed a lower HDO activity, which might be attributed to the lower surface area of this support [42]. Similar to the current study's findings, Dongil et al. [24] observed lower HDO activity for NiCu catalysts in comparison to monometallic Ni catalysts. The authors attributed the lower guaiacol HDO to the larger particle size, the presence of NiO particles as well as to copper particles located at the nickel active sites. Furthermore, a higher crystallite size of bimetallic catalysts, as observed for both NiCu catalysts, also might play a role in the lower HDO activity; the higher crystallite size decreases the number of step/corner sites, which, according to Mortensen et al. [15], are more active for breaking C-O bonds.

Coke formation is another important parameter for the selection of the catalyst. Observed for both Ni/SiO_2 and $NiCu/SiO_2$, the volume of the micropores was reduced significantly after the reactions. Coke is known for blocking the pores, covering the catalyst active sites, resulting in partial or even complete loss of activity [61], therefore, it is considered one of the main causes of deactivation in HDO reactions [8,15,28]. Considering that the catalyst acidity is connected directly to coke formation [8], the lower coke formation observed with Ni/SiO_2 could be related to the low acidity of silica [28]. Due to the amphoteric nature of ZrO_2, reduced coke formation would be expected [28]; However, SiO_2 was more resistant to coke formation. The addition of Cu also seems to contribute to the slightly higher amount of coke deposited on the bimetallic catalysts. Stated by Zhang et al. [42], the addition of copper increases the acidity of the catalyst, which can result in higher coke deposition [27], as observed. Hence, the lowest coke deposition observed for Ni/SiO_2 can be related to the lower acidity compared to $NiCu/SiO_2$ and higher resistance of SiO_2 in comparison to ZrO_2.

Reactive compounds, such as aldehydes, were converted completely after the upgrading. The reduction of these very reactive compounds results in a more stable oil [15,29]. The lower conversion of aromatic compounds over all the catalysts evaluated also is interesting; if further upgrading is intended, aiming at fuel production, the presence of aromatic compounds could result in a high octane number gasoline [29].

Different selectivity among mono and bimetallic catalysts was attributed mostly to the addition of copper [69], with a minor contribution of nickel loading [22]. Furfural, also a very reactive compound [72], mainly seemed to be hydrogenated to tetrahydrofurfuryl alcohol over NiCu catalysts. Additionally, the presence of propylene glycol leads to the conclusion that hydroxyacetone was mainly hydrogenated to this compound over bimetallic catalysts. Concurring with other authors, the addition of copper seems to increase the hydrogenation [24], increasing the hydrogen consumption. Furthermore, the higher hydrogen consumption observed for NiCu catalysts is in agreement with literature [21,23] giving evidence that the addition of a second metal can increase the hydrogenation activity of the catalyst [69]. Stated by Mortensen et al. [15], the H/C and O/C is used to evaluate the quality of the upgraded product. A higher H/C ratio is intended, whereas a lower O/C ratio is desired. In this case, it was observed that the higher H/C ratio for $NiCu/SiO_2$ agrees with the higher consumption of H_2 observed for this catalyst, but it was not reflected in the O/C ratio. The lowest O/C was obtained for Ni/SiO_2 (0.18), being much lower in comparison to the O/C ratio of the feed (0.47). It is important to note higher hydrogenation results in higher consumption of hydrogen [57,73], but it is not necessarily reflected in the reduction of the oxygen content. Moreover, the higher consumption of hydrogen can also result in higher methane formation [35], considering that during the hydrocracking the consumption of hydrogen is higher compared to hydrotreating [74]. This behavior was observed for $NiCu/SiO_2$. The catalyst showed not just the higher hydrogen consumption, but also a higher methane formation. The desired catalyst should be able to remove the larger amount of oxygen with minimal hydrogen consumption [24] as hydrogen consumption, bio-oil yield and catalyst deactivation are among the most important parameters to be considered in the HDO process [75]. Ni/SiO_2, therefore, was considered the more appropriate catalyst in terms of H_2 consumption and HDO activity.

After the reaction, compounds such as calcium (only over ZrO_2) and sulfur were observed on the spent catalysts by two different analytical techniques, Inductively Coupled Plasma Emission Spectroscopy (ICP-OES) and Scanning Electron Microscopy/Energy Dispersive X-ray spectroscopy (SEM-EDX). Calcium, observed in higher concentrations in comparison to sulfur over the zirconia-supported catalysts, on the one hand acts as a poisoning agent, reducing the mobility and re-dispersion of the active metal centers over the support and, on the other hand, can reduce sintering, due to the reduction in the atom mobility, resistance to dissociation and migration [61].

Despite the detection of sulfur on the catalyst's surface, structural changes were not observed after the reactions by Powder XRD, as observed in previous investigations. Mortensen et al. [66] observed a reflection at approximately $2\theta = 45.2°$, attributed to NiS during the evaluation of the influence of sulfur over the conversion of guaiacol with Ni/ZrO_2. Boscagli et al. [26] observed the formation of Ni_3S_2, which was persistent to regeneration and changed the catalyst structure. The concentration of sulfur in the feed, in this case, varies significantly in both works. While Mortensen et al. used a model mixture containing 0.05 wt.% of sulfur, Boscagli et al. used feedstocks with a higher concentration of sulfur (light bio-oil phase with 0.05 wt.% of sulfur and straw bio-oil obtained at 450 °C with 0.3 wt.% of sulfur). The high concentration of sulfur resulted in a spent catalyst with a much higher concentration of sulfur on the catalyst (0.6–2.0 wt.%, SEM-EDX), in comparison to the current study's findings (feed sulfur concentration = 0.012 wt.% and spent catalysts $\leq$0.1–0.4 wt.%, SEM-EDX). Even without structural changes, sulfur is one of the most persistent poisons for nickel catalysts. It is irreversibly chemisorbed and responsible for blocking the reaction-adsorption active site, modifying electronically the neighbor atom of metals and, thereby, reducing the ability to adsorb and dissociate H_2, at the same time influencing the diffusion or reactants, blocking their contact with the active site [61].

Along with the consecutive reactions with Ni/SiO_2, comparable carbon and hydrogen concentrations and a slightly higher oxygen concentration was observed in the upgraded bio-oils. The oxygen concentration reached20.34 wt.% after the fourth reaction. It was higher compared to the first reaction (17.86 wt.%) but still much lower compared to the feed (35.84 wt.%, dry basis). It gives evidence of a low rate of deactivation and the possibility to reuse the catalyst. The increase in the oxygen concentration can be correlated with the reduction in the H_2 uptake, in agreement with lower HDO activity of Ni/SiO_2 in comparison to the fresh catalyst [76]. The current authors assumed that the reduced activity can be correlated mainly to sintering, poisoning and coke deposition (although easily removed during the calcination step), as the leaching was negligible. The crystallite size increased over the cycles in the same proportion that the HDO activity was reduced along the consecutive reactions. The fresh catalyst showed a crystallite size of 17.7 nm, reaching 37.3 nm after the fourth reuse. Thus, the number of active surfaces available were reduced with the increase of the crystallite size [61]. Furthermore, the lower H_2 uptake can also result in a higher amount of CO_2 formation, as observed along the cycles, considering the higher the hydrogen uptake, the lower the CO_2 production [46]. Curiously, the sulfur concentration remained approximately constant along the cycles. Furthermore, no differences were observed in the XRD diffractions. The current authors concluded that sulfur was strongly adsorbed on the catalyst, affecting surface-sensitive reactions, resulting only in partial loss of activity of the catalytic surface [61].

The results obtained by ^{1}H-NMR showed that the concentration of protons in the region related to water and carbohydrates increased along the cycles in the upgraded bio-oils. Moreover, the gas chromatography-mass spectrometer (GC-MS), showed that the peak of some sugar derivatives became smaller over the consecutive cycles. It was assumed that the catalytic conversion of sugars was possibly affected by the lower activity of Ni/SiO_2 along the cycles. This assumption corroborates the results of GC-MS, from which the intensity of sugar derivative compounds (γ-valerolactone γ-butyrolactone 2-ethyl-cyclopentanone 2-methyl-2-propanol) have been reduced as the number of cycles increased. Considering the assumption that the conversion of sugars through hydrogenation is reduced with the increase in the number of consecutive reactions, higher amounts of coke formation could be expected, considering that the thermal polymerization of the sugar fraction can lead to increased char deposition [35,46,77] over the cycles. The results of SEM-EDX showed a slightly higher carbon deposition over the spent catalyst after the fourth reaction.

Since poisoning substances might affect some specific reactions [61], the investigation of model compound conversions as well as the effect of poisonings, over single compounds, could contribute to the understanding of the selectivity changes observed along the cycles. Considering the difficulty for regeneration of sulfur-poisoned catalysts, due to the harsh conditions required (700 °C in steam) [61], its influence over the conversion of model compounds should be investigated in detail. Since deactivation mechanisms are difficult to be monitored in batch experiments [66], continuously operated reactors are more appropriate for this investigation.

4. Experimental

4.1. Catalyst Syntheses

The four nickel-based catalysts, Ni/SiO_2, Ni/ZrO_2, $NiCu/SiO_2$ and $NiCu/ZrO_2$, were synthesized by wet impregnation as follows. Both supports (silica and zirconia from Alfa Aesar, Haverhill, MA, USA) were milled to 0.125–0.250 mm and added to the metal solution in a ratio of 1:10. The metal precursors were $Ni(NO_3)_2.6H_2O$ (Sigma–Aldrich, St. Louis, MI, USA) and $Cu(NO_3)_2{\cdot}2.5H_2O$ (Alfa Aesar, Haverhill, MA, USA). The water was evaporated at 35 °C, 45 mbar, 100 rpm in a rotary evaporator (Hei-VAP Advantage ML/G3) and the formed catalyst was dried for 12 h at 105 °C. The catalysts were then calcined at 450 °C for 4 h after reaching the set point, with a heating ramp of 10 °C/min (Thermolyne F6010) and later reduced in a 25% H_2/N_2 flow of 3 L/min, heating ramp of 5 K/min, during 4 h. The reduction temperatures were defined by temperature programmed reduction (H_2-TPR) experiments (see Sections 2.1 and 4.5). Regarding the monometallic Ni catalyst, the metal concentration in the catalyst was defined in 8.6 wt.% [78] while, for the bimetallic catalysts, the metal concentration was defined in 28 wt.% of Ni and 3.5 wt.% Cu [23].

4.2. Beech Wood Fast-Pyrolysis Bio-Oil

The experiments were carried out with a beech wood fast-pyrolysis bio-oil (BTG Group, Enschede, The Netherlands). Following intentional aging (24 h, 80 °C) two phases were observed. The final oil was then composed by 41 wt.% of heavy phase (HP) and 59 wt.% of light phase (LP), respectively. Both phases were separated, and previously characterized separately, as presented elsewhere [43]. The main physicochemical properties and elemental analyses are presented (Table 9).

Table 9. Physicochemical properties and elemental analysis of the HP and LP of aged beech wood fast pyrolysis bio-oil [43].

Properties	Beechwood Fast-Pyrolysis Bio-Oil	
	HP	LP
Carbon (wt.%) (wet basis; dry basis)	54.3; 63.7	34.3; 52.4
Hydrogen (wt.%) (wet basis; dry basis)	7.5; 6.9	8.3; 6.8
Nitrogen (wt.%) (wet basis; dry basis)	<0.2	<0.2
Oxygen (wt.%) (wet basis; dry basis)	38.2; 29.2	57.3; 39.9
Sulfur (wt.%) (wet basis; dry basis)	0.009;0.011	0.015;0.023
H_2O (wt.%)	14.5	35.3
HHV (MJ/kg) (wet basis; dry basis)	23.1; 27.3	14.1; 22.4
pH value	3	2.7
Density (g/cm^3)	1.19	1.17

4.3. HDO Reactions

The HDO (hydrodeoxygenation) experiments were conducted in a 200 mL volume batch reactor. More details about the reactor are given in the Supplementary Material. The conditions selected for this study were fixed to 325 °C and 80 bar of H_2 based on the current authors' previous investigation [47]. Fifty grams of bio-oil (composed of the mixture of the LP and HP, in the proportion advised in Section 4.2) and 5 wt.% of catalyst in relation to the amount of the bio-oil were added to the autoclave. The reactor was closed and purged with N_2 for 5 min and pressurized with H_2 (Air liquid Alphagaz 2, purity 6.0) at ambient temperature. The stirrer was switched to 1000 rpm and the heating program was started at a rate of 5 °C/min. The global reaction time, including the heating ramp, was 120 min. When this time was reached, the reaction was quenched first using a flow of compressed air and, later, in a cold water bath with ice, until reaching ambient temperature. Two experiments were performed for each set of conditions and are presented as an average.

Following the reaction, the final pressure was recorded for hydrogen consumption determination and the gas phase was collected for gas chromatography analysis. The mixture of spent catalyst, coke/char, upgraded bio-oil and aqueous phase was collected, centrifuged for 40 min at 7000 rpm in a Heraeus Biofuge Stratos centrifuge (Thermo Fischer Scientific, Waltham, MA, USA) and then separated for further characterization. The H_2 consumption was calculated using the ideal gas law, as the difference of the moles of hydrogen loaded to the reactor and the remaining moles after the hydrotreatment, from the pressure before and after the reaction and the gas composition determined by gas chromatography [40].

4.4. Products Characterization

The gas composition was determined by collecting and analyzing the gaseous fraction by gas chromatography. A 100 μL sample was injected at 250 °C (split 28:1) in an Agilent 7890A (two detectors: thermal conductivity and flame ionization detector, TCD and FID respectively) equipped with two columns: Restek 79,096 Hayesep Q and Restek Molsieve 5A. The oven temperature was programmed as follows: the initial temperature was set to 50 °C maintained for 10 min; increased to 90 °C at a heating rate of 3 °C per minute and then increased to 150 °C at a heating rate of 20 °C per minute, maintained at this temperature for 16 min and finally heated to 230 °C at 50 °C per minute and kept for 10 min.

The liquid products (aqueous phase and upgraded bio-oil) were characterized using the same methodology, except the higher heating value (HHV), which was not determined for the aqueous phases. The pH values were measured with a pH-meter from Metrohm. The HHV was determined using a calorimeter IKA C5000 control and the water content using a volumetric Karl–Fischer titrator from Metrohm (Titrando 841, titration reagents Composite 5 and dry Methanol). Carbon, hydrogen, and nitrogen content were measured using a micro-elemental analyzer Elementar Vario el Cube. The content of oxygen was estimated by the difference.

Quantitative ^{1}H-NMR (proton nuclear magnetic resonance) was employed to characterize the functional groups in the product molecules, based on the number of protons in the corresponding ^{1}H-shift range [44]. ^{1}H-NMR spectra were recorded at 25 °C on a Bruker Biospin spectrometer, equipped with a 5.47 T magnet (^{1}H frequency 250 Hz). Sample preparation consisted of 0.1 g of either upgraded bio-oil or aqueous phase and their dilution in 0.7 g of deuterated methanol (CD_3OD) containing TMSP-d4 [3-(trimethylsilyl)-2,2,3,3-tetradeuteropropionic acid sodium salt] as an internal standard (0.1 g TMSP in 50 mL CD_3OD). Subsequently, the samples were centrifuged (removal of particles not solubilized) and placed in NMR tubes.

The liquid samples also were analyzed qualitatively using a gas chromatography mass spectrometer (GC-MS) HPG1800A, with a Restek stabilwax column. Prior to measurement, the samples were diluted in methanol and filtrated using a 0.25 μm filter. A 1 μL sample was injected at 250 °C, with a split of 1:20. The oven temperature was programmed as follows: the initial temperature was set to 40 °C, maintained for 5 min; increased to 300 °C with a heating rate of 20 °C per minute and maintained for 20 min at this condition.

4.5. Catalysts Characterization

The concentration of metal in the catalysts was analyzed by ICP-OES (Inductively Coupled Plasma Emission Spectroscopy Agilent 725 Spectrometer). The sample was dissolved using a mixture of HNO_3 (2 mL), HCl (6 mL) and H_2O_2 (0.5 mL) and digested in a microwave oven for 45 min at 240 °C.

To determine the reduction temperature profile of the active metal, an Autochem HP 2950 (Micrometrics, Ottawa, ON, Canada) was used for the temperature programmed reduction by hydrogen (H_2-TPR). The measurements were performed at a heating rate of 1 K/min until 500 °C and 5% H_2 in Ar at 30 mL/min. The samples were dried in a 30 mL/min flow of Ar at 300 °C for 3 h before the measurement.

The total specific surface area of the catalyst was determined by nitrogen physisorption with a Belsorp Mini II at 77 K and calculated by applying the BET theory in the fitting rate between 0.05–0.30 p/p0 (12 points). Powder XRD was measured using an X'Pert PRO MPD instrument (PANalytical GmbH, Nuremberg, BY, Germany) equipped with a Cu anode (Cu Kα 1.54060 Å). The XRD patterns were recorded in a 2θ range between 5–120° (1 h, step size 0.017°). The average crystallite size was estimated using the Scherrer equation (shape factor K = 0.9) after correcting the instrumental line broadening.

Leaching of the catalyst was monitored by analyzing the aqueous phase by ICP-OES. Sample preparation involved the filtration with a 0.2 µm polytetrafluoroetylen (PTFE) filter membrane of the produced aqueous phase after each reaction.

To identify the elements present on the particles of fresh and spent catalysts, SEM/EDX (Scanning Electron Microscopy/Energy Dispersive X-ray spectroscopy) was applied. The equipment for this technique was a GeminiSEM 500, Zeiss, software SmartSEM Version 6.01, with a thermal Schottky field-emitter cathode. An energy dispersive X-ray spectrometer X-Max^N from Oxford with a silicon drift detector (80 mm^2 and resolution of 127 eV) was employed for the quantitative analysis of micro areas and the distribution of the elements, in addition to the software Aztec 3.3. The C, H, N and S of the spent catalysts were measured by a micro-elemental analyzer Elementar Vario el Cube. The solid over the spent catalysts (coke) was calculated by the carbon concentration over the spent catalyst (determined by elemental analysis), considering oxygen concentration negligible [46]. More detail is given in the Supplementary Materials (Equation (S1)).

4.6. Catalyst Regeneration

Based on the experimental results, such as oxygen content, water concentration, solid and gas production as well as upgraded bio-oil yield, one of the catalysts was selected for further application in cycles of HDO and regeneration. The cycles each consisted of an HDO reaction, calcination of the spent catalyst (as described in Section 4.1), followed by a reduction and a subsequent HDO reaction. Altogether, the catalyst was used four times. To evaluate the behavior of the catalyst over the consecutive uses, the spent catalyst was analyzed through SEM-EDX and XRD between the regenerations: EDX was performed for the spent and reduced catalysts, whereas XRD was performed for the spent, calcined, and reduced catalysts. Furthermore, the upgraded products also were characterized along the cycles, as described in Section 4.4.

5. Conclusions

Four nickel-based catalysts on different supports were synthesized and evaluated for the HDO of a multi-phase beech wood fast-pyrolysis bio-oil. The bimetallic catalysts showed lower reduction temperature, attributed to the addition of copper, and higher metal loading. Furthermore, NiCu catalysts presented higher consumption of hydrogen and different selectivity toward the conversion of compounds such as furfural, compared to monometallic catalysts. Upgraded bio-oils with reduced concentration of oxygen, lower water concentration and higher carbon content were obtained after the HDO reactions. Ni/SiO_2, in particular, showed the highest HDO activity, reducing more than 50% of the oxygen content and more than 80% of the water content, thus selected for application in cycles of regeneration-reaction. During the consecutive reactions, the activity of the catalyst decreased, attributed mainly to sintering and poisoning by sulfur, as coke was removed easily during the regeneration steps. Lower hydrogen consumption and higher carbon dioxide production were observed in comparison to the reactions when applying the fresh catalyst. Correlating the results obtained by ^1H-NMR and GC-MS, it was possible to observe that compounds known as sugar derivatives were being reduced along the consecutive reactions while the concentration of protons in the region attributed to water, carbohydrates and O-H exchange increased. The partial loss of activity seemed to lower the conversion of sugars.

Accompanying that, the catalysts evaluated seemed to be suitable for HDO of hardwood fast-pyrolysis bio-oil, especially Ni/SiO_2. Further investigation will be addressed, considering a detailed investigation of model compound conversions and the effect of sulfur and calcium over HDO of single compounds. The influence of sintering over the activity and selectivity will also be investigated. Due to the limitation of the batch reactor for deactivation mechanism studies, the application of a trickle bed reactor in future studies is intended.

Supplementary Materials: The following are available online at http://www.mdpi.com/2073-4344/8/10/449/s1.

Author Contributions: C.C.S. design the experiments, performed part of the experiments, analyzed the data and wrote the paper; M.B.G.R. performed most of the experiments and analyzed the data; M.Z. contributed with the microscopy experiments; J.-D.G., K.R. and N.D. contributed with the discussion, reviewed and supervised the work.

Funding: This research received no external funding.

Acknowledgments: The authors are grateful to the Bioeconomy Graduate Program—BBWForwerts, Brazilian National Council for Science and Technology (CNPQ) and BeMundus for the financial support. The authors are also grateful to Herman Köhler, Pia Griesheimer, Petra Janke, Jessica Heinrich, and Simon Wodarz for the support with analytical techniques, Oliver Schade for the support with the catalyst reduction oven and to the assistant bachelor student (Hiwi student) Mouhannad Moulla, for the activities developed during his internship which supported the development of this work.

Conflicts of Interest: The authors declare no conflict of interest.

References

1. Kan, T.; Strezov, V.; Evans, T.J. Lignocellulosic biomass pyrolysis: A review of product properties and effects of pyrolysis parameters. *Renew. Sustain. Energy Rev.* **2016**, *57*, 1126–1140. [CrossRef]
2. García-Olivares, A.; Solé, J.; Osychenko, O. Transportation in a 100% renewable energy system. *Energy Convers. Manag.* **2018**, *158*, 266–285. [CrossRef]
3. Balsalobre-Lorente, D.; Shahbaz, M.; Roubaud, D.; Farhani, S. How economic growth, renewable electricity and natural resources contribute to CO_2 emissions? *Energy Policy* **2018**, *113*, 356–367. [CrossRef]
4. Guedes, R.E.; Luna, A.S.; Torres, A.R. Operating parameters for bio-oil production in biomass pyrolysis: A review. *J. Anal. Appl. Pyrolysis* **2018**, *129*, 134–149. [CrossRef]
5. Bridgwater, A.V. Renewable fuels and chemicals by thermal processing of biomass. *Chem. Eng. J.* **2003**, *91*, 87–102. [CrossRef]
6. El Bassam, N. *Handbook of Bioenergy Crops. A Complete Reference to Species, Development and Applications*; Routledge: Abingdon-on-Thames, UK, 2010; Volume 18, ISBN 978-1-84407-854-7.
7. Sharma, A.; Pareek, V.; Zhang, D. Biomass pyrolysis—A review of modelling, process parameters and catalytic studies. *Renew. Sustain. Energy Rev.* **2015**, *50*, 1081–1096. [CrossRef]
8. Si, Z.; Zhang, X.; Wang, C.; Ma, L.; Dong, R. An Overview on Catalytic Hydrodeoxygenation of Pyrolysis Oil and Its Model Compounds. *Catalysts* **2017**, *7*, 169. [CrossRef]
9. Czernik, S.; Johnson, D.K.; Black, S. Stability of wood fast pyrolysis oil. *Biomass Bioenergy* **1994**, *7*, 187–192. [CrossRef]

10. Olbrich, W.; Boscagli, C.; Raffelt, K.; Zang, H.; Dahmen, N.; Sauer, J. Catalytic hydrodeoxygenation of pyrolysis oil over nickel-based catalysts under H_2/CO_2 atmosphere. *Sustain. Chem. Process.* **2016**, *4*, 9. [CrossRef]
11. Li, X.; Gunawan, R.; Wang, Y.; Chaiwat, W.; Hu, X.; Gholizadeh, M.; Mourant, D.; Bromly, J.; Li, C.-Z. Upgrading of bio-oil into advanced biofuels and chemicals. Part III. Changes in aromatic structure and coke forming propensity during the catalytic hydrotreatment of a fast pyrolysis bio-oil with Pd/C catalyst. *Fuel* **2014**, *116*, 642–649. [CrossRef]
12. Balat, M. An overview of the properties and applications of biomass pyrolysis oils. *Energy Sources Part A Recover. Util. Environ. Eff.* **2011**, *33*, 674–689. [CrossRef]
13. Bridgwater, A.V. Review of fast pyrolysis of biomass and product upgrading. *Biomass Bioenergy* **2012**, *38*, 68–94. [CrossRef]
14. Gollakota, A.R.K.; Reddy, M.; Subramanyam, M.D.; Kishore, N. A review on the upgradation techniques of pyrolysis oil. *Renew. Sustain. Energy Rev.* **2016**, *58*, 1543–1568. [CrossRef]
15. Mortensen, P.M.; Grunwaldt, J.-D.; Jensen, P.A.; Knudsen, K.G.; Jensen, A.D. A review of catalytic upgrading of bio-oil to engine fuels. *Appl. Catal. A Gen.* **2011**, *407*, 1–19. [CrossRef]
16. Furimsky, E. Catalytic hydrodeoxygenation. *Appl. Catal. A Gen.* **2000**, *199*, 147–190. [CrossRef]
17. Song, H.; Gong, J.; Song, H.; Li, F. Applied Catalysis A: General A novel surface modification approach for synthesizing supported nickel phosphide catalysts with high activity for hydrodeoxygenation of benzofuran. *Appl. Catal. A Gen.* **2015**, *505*, 267–275. [CrossRef]
18. Elliott, D.C. Biofuel from fast pyrolysis and catalytic hydrodeoxygenation. *Curr. Opin. Chem. Eng.* **2015**, *9*, 59–65. [CrossRef]
19. Ewald, S.; Standl, S.; Hinrichsen, O. Characterization of nickel catalysts with transient methods. *Appl. Catal. A Gen.* **2018**, *549*, 93–101. [CrossRef]
20. Jin, S.; Xiao, Z.; Li, C.; Chen, X.; Wang, L.; Xing, J.; Li, W.; Liang, C. Catalytic hydrodeoxygenation of anisole as lignin model compound over supported nickel catalysts. *Catal. Today* **2014**, *234*, 125–132. [CrossRef]
21. Boscagli, C.; Raffelt, K.; Zevaco, T.A.; Olbrich, W.; Otto, T.N.; Sauer, J.; Grunwaldt, J.-D. Mild hydrotreatment of the light fraction of fast-pyrolysis oil produced from straw over nickel-based catalysts. *Biomass Bioenergy* **2015**, *83*, 525–538. [CrossRef]
22. Dongil, A.B.; Ghampson, I.T.; García, R.; Fierro, J.L.G.; Escalona, N. Hydrodeoxygenation of guaiacol over Ni/carbon catalysts: Effect of the support and Ni loading. *RSC Adv.* **2016**, *6*, 2611–2623. [CrossRef]
23. Ardiyani, A.R.; Khromova, S.A.; Venderbosch, R.H.; Yakovlev, V.A.; Heeres, H.J. Catalytic hydrotreatment of fast-pyrolysis oil using non-sulfided bimetallic Ni-Cu catalysts on a δ-Al_2O_3 support. *Appl. Catal. B Environ.* **2012**, *117–118*, 105–117. [CrossRef]
24. Dongil, A.B.; Bachiller-Baeza, B.; Rodriguez-Ramos, I.; Fierro, J.L.G.; Escalona, N. Effect of Cu loading on Ni/carbon nanotubes catalyst for hydrodeoxygenation of guaiacol. *RSC Adv.* **2016**, *3*, 26658–26667. [CrossRef]
25. Mortensen, P.M.; Grunwaldt, J.-D.; Jensen, P.A.; Jensen, A.D. Screening of catalysts for hydrodeoxygenation of phenol as a model compound for bio-oil. *ACS Catal.* **2013**, *3*, 1774–1785. [CrossRef]
26. Boscagli, C.; Yang, C.; Welle, A.; Wang, W.; Behrens, S.; Raffelt, K.; Grunwaldt, J.-D. Effect of pyrolysis oil components on the activity and selectivity of nickel-based catalysts during hydrotreatment. *Appl. Catal. A Gen.* **2017**, *544*, 161–172. [CrossRef]
27. Saidi, M.; Samimi, F.; Karimipourfard, D.; Nimmanwudipong, T.; Gates, B.C.; Rahimpour, M.R. Upgrading of lignin-derived bio-oils by catalytic hydrodeoxygenation. *Energy Environ. Sci.* **2014**, *7*, 103–129. [CrossRef]
28. He, Z.; Wang, X. Hydrodeoxygenation of model compounds and catalytic systems for pyrolysis bio-oils upgrading. *Catal. Sustain. Energy* **2012**, *1*, 28–52. [CrossRef]
29. Elliott, D.C. Historical developments in hydroprocessing bio-oils. *Energy Fuels* **2007**, *21*, 1792–1815. [CrossRef]
30. Bykova, M.V.; Ermakov, D.Y.; Kaichev, V.V.; Bulavchenko, O.A.; Saraev, A.A.; Lebedev, M.Y.; Yakovlev, V.A. Applied Catalysis B: Environmental Ni-based sol–gel catalysts as promising systems for crude bio-oil upgrading: Guaiacol hydrodeoxygenation study. *Appl. Catal. B Environ.* **2012**, *113–114*, 296–307. [CrossRef]
31. Roldugina, E.A.; Naranov, E.R.; Maximov, A.L.; Karakhanov, E.A. Hydrodeoxygenation of guaiacol as a model compound of bio-oil in methanol over mesoporous noble metal catalysts. *Appl. Catal. A Gen.* **2018**. [CrossRef]

32. Esmaeili, J.; Rahimpour, F. Regeneration of spent nickel catalyst from hydrogenation process of edible oils: Heat treatment with hydrogen injection. *Int. J. Hydrogen Energy* **2017**, *42*, 24197–24204. [CrossRef]
33. Ng, A.; Lup, K.; Abnisa, F.; Mohd, W.; Wan, A. A review on reactivity and stability of heterogeneous metal catalysts for deoxygenation of bio-oil model compounds. *J. Ind. Eng. Chem.* **2017**, *56*, 1–34. [CrossRef]
34. Koike, N.; Hosokai, S.; Takagaki, A.; Nishimura, S.; Kikuchi, R.; Ebitani, K.; Suzuki, Y.; Oyama, S.T. Upgrading of pyrolysis bio-oil using nickel phosphide catalysts. *J. Catal.* **2016**, *333*, 115–126. [CrossRef]
35. Yin, W.; Hendrikus, R.; Alekseeva, M.V.; Bernardes, M.; Heeres, H.; Khromova, A.; Yakovlev, V.A.; Cannilla, C.; Bonura, G.; Frusteri, F.; et al. Hydrotreatment of the carbohydrate-rich fraction of pyrolysis liquids using bimetallic Ni based catalyst: Catalyst activity and product property relations. *Fuel Process. Technol.* **2018**, *169*, 258–268. [CrossRef]
36. Rodríguez, J.C.; Marchi, A.J.; Borgna, A.; Romeo, E.; Monzón, A. Gas Phase Selective Hydrogenation of Acetylene. Importance of the Formation of Ni-Co and Ni-Cu Bimetallic Clusters on the Selectivity and Coke Deposition. *Stud. Surf. Sci. Catal.* **2001**, *139*, 37–44. [CrossRef]
37. De Rogatis, L.; Montini, T.; Lorenzut, B.; Fornasiero, P. NixCuy/Al_2O_3 based catalysts for hydrogen production. *Energy Environ. Sci.* **2008**, *1*, 501–509. [CrossRef]
38. Li, X.; Cheng, H.; Liang, G.; He, L.; Lin, W.; Yu, Y.; Zhao, F. Effect of Phosphine Doping and the Surface Metal State of Ni on the Catalytic Performance of Ni/Al_2O_3 Catalyst. *Catalysts* **2015**, *5*, 759–773. [CrossRef]
39. Jahromi, H.; Agblevor, F.A. Hydrodeoxygenation of pinyon-juniper catalytic pyrolysis oil using red mud-supported nickel catalysts. *Appl. Catal. B Environ.* **2018**, *236*, 1–12. [CrossRef]
40. Ardiyanti, A.R.; Khromova, S.A.; Venderbosch, R.H.; Yakovlev, V.A.; Melián-Cabrera, I.V.; Heeres, H.J. Catalytic hydrotreatment of fast pyrolysis oil using bimetallic Ni-Cu catalysts on various supports. *Appl. Catal. A Gen.* **2012**, *449*, 121–130. [CrossRef]
41. Wang, Z.; Zeng, Y.; Lin, W.; Song, W. In-situ hydrodeoxygenation of phenol by supported Ni catalyst—Explanation for catalyst performance. *Int. J. Hydrogen Energy* **2017**, *42*, 21040–21047. [CrossRef]
42. Zhang, X.; Wang, T.; Ma, L.; Zhang, Q.; Yu, Y.; Liu, Q. Characterization and catalytic properties of Ni and NiCu catalysts supported on ZrO_2–SiO_2 for guaiacol hydrodeoxygenation. *Catal. Commun.* **2013**, *33*, 15–19. [CrossRef]
43. Carriel Schmitt, C.; Boscagli, C.; Rapp, M.; Raffelt, K.; Dahmen, N. Characterization of light and heavy phase of pyrolysis-oil from distinct biomass for further upgrading reactions. In Proceedings of the 25th European Biomass Conference and Exhibition, Stockholm, Sweden, 12–15 June 2017. [CrossRef]
44. Mullen, C.A.; Strahan, G.D.; Boateng, A.A. Characterization of various fast-pyrolysis bio-oils by NMR spectroscopy. *Energy Fuels* **2009**, *23*, 2707–2718. [CrossRef]
45. Yin, W.; Kloekhorst, A.; Venderbosch, R.H.; Bykova, M.V.; Khromova, S.A.; Yakovlev, V.A.; Heeres, H.J. Catalytic hydrotreatment of fast pyrolysis liquids in batch and continuous set-ups using a bimetallic Ni–Cu catalyst with a high metal content. *Catal. Sci. Technol.* **2016**, *6*, 5899–5915. [CrossRef]
46. Boscagli, C.; Raffelt, K.; Grunwaldt, J.-D. Reactivity of platform molecules in pyrolysis oil and in water during hydrotreatment over nickel and ruthenium catalysts. *Biomass Bioenergy* **2017**, *106*, 63–73. [CrossRef]
47. Schmitt, C.C.; Raffelt, K.; Zimina, A.; Krause, B.; Otto, T.; Rapp, M.; Grunwaldt, J.-D.; Dahmen, N. Hydrotreatment of Fast Pyrolysis Bio-oil Fractions Over Nickel-Based Catalyst. *Top. Catal.* **2018**, *61*, 1769–1782. [CrossRef]
48. Oh, S.; Choi, H.S.; Choi, I.-G.; Choi, J.W. Evaluation of hydrodeoxygenation reactivity of pyrolysis bio-oil with various Ni-based catalysts for improvement of fuel properties. *RSC Adv.* **2017**, *7*, 15116–15126. [CrossRef]
49. Nakagawa, Y.; Tamura, M.; Tomishige, K. Catalytic Conversions of Furfural to Pentanediols. *Catal. Surv. Asia* **2015**, *19*, 249–256. [CrossRef]
50. Resasco, D.E.; Sitthisa, S.; Faria, J.; Prasomsri, T.; Ruiz, M.P. *Furfurals as Chemical Platform for Biofuels Production*; Kubička, D., Kubičková, I., Eds.; Research S.: Kerala, India, 2011; Volume 661, ISBN 9788130804620.
51. Mariscal, R.; Maireles-Torres, P.; Ojeda, M.; Sádaba, I.; López Granados, M. Furfural: A renewable and versatile platform molecule for the synthesis of chemicals and fuels. *Energy Environ. Sci.* **2016**, *9*, 1144–1189. [CrossRef]
52. Feng, J.; Yang, Z.; Hse, C.Y.; Su, Q.; Wang, K.; Jiang, J.; Xu, J. In situ catalytic hydrogenation of model compounds and biomass-derived phenolic compounds for bio-oil upgrading. *Renew. Energy* **2017**, *105*, 140–148. [CrossRef]

53. Agblevor, F.A.; Jahromi, H. Aqueous phase synthesis of hydrocarbons from furfural reactions with low molecular weight biomass oxygenates. *Energy Fuels* **2018**, *32*, 8552–8562. [CrossRef]
54. Vispute, T. Pyrolysis Oils: Characterization, Stability Analysis, and Catalytic Upgrading to Fuels and Chemicals. Ph.D. Thesis, University of Massachussets Amherst, Amherst, MA, USA, 2011.
55. Deutsch, K.L. Copper Catalysts in the C-O Hydrogenolysis of Biorenewable Compounds. Ph.D. Thesis, Iowa State University, Ames, IA, USA, 2012.
56. Khromova, S.A.; Smirnov, A.A.; Bulavchenko, O.A.; Saraev, A.A.; Kaichev, V.V.; Reshetnikov, S.I.; Yakovlev, V.A. Anisole hydrodeoxygenation over Ni-Cu bimetallic catalysts: The effect of Ni/Cu ratio on selectivity. *Appl. Catal. A Gen.* **2014**, *470*, 261–270. [CrossRef]
57. French, R.J.; Stunkel, J.; Black, S.; Myers, M.; Yung, M.M.; Iisa, K. Evaluate impact of catalyst type on oil yield and hydrogen consumption from mild hydrotreating. *Energy Fuels* **2014**, *28*, 3086–3095. [CrossRef]
58. Huynh, T.M. Development of Novel Bimetallic Nickel-Cobalt Catalysts for Hydrodeoxygenation of Bio-Oil Producing a Co-Feed for a Standard Refinery Unit. Ph.D. Thesis, University of Rostock, Rostock, Germany, 2015.
59. Ardiyanti, A.R.; Gutierrez, A.; Honkela, M.L.; Krause, A.O.I.I.; Heeres, H.J. Hydrotreatment of wood-based pyrolysis oil using zirconia-supported mono- and bimetallic (Pt, Pd, Rh) catalysts. *Appl. Catal. A Gen.* **2011**, *407*, 56–66. [CrossRef]
60. Van Haasterecht, T.; Ludding, C.C.I.; De Jong, K.P.; Bitter, J.H. Toward stable nickel catalysts for aqueous phase reforming of biomass-derived feedstock under reducing and alkaline conditions. *J. Catal.* **2014**, *319*, 27–35. [CrossRef]
61. Argyle, M.D.; Bartholomew, C.H. Heterogeneous Catalyst Deactivation and Regeneration: A Review. *Catalysts* **2015**, *5*, 145–269. [CrossRef]
62. Oh, S.; Hwang, H.; Seok, H.; Weon, J. The effects of noble metal catalysts on the bio-oil quality during the hydrodeoxygenative upgrading process. *Fuel* **2015**, *153*, 535–543. [CrossRef]
63. Li, X.; Luo, X.; Jin, Y.; Li, J.; Zhang, H.; Zhang, A.; Xie, J. Heterogeneous sulfur-free hydrodeoxygenation catalysts for selectively upgrading the renewable bio-oils to second generation biofuels. *Renew. Sustain. Energy Rev.* **2018**, *82*, 3762–3797. [CrossRef]
64. Chen, S.; Miao, C.; Luo, Y.; Zhou, G.; Xiong, K.; Jiao, Z. Study of catalytic hydrodeoxygenation performance of Ni catalysts: Effects of prepared method. *Renew. Energy* **2018**, *115*, 1109–1117. [CrossRef]
65. González-Cobos, J.; de Lucas-Consuegra, A. A Review of Surface Analysis Techniques for the Investigation of the Phenomenon of Electrochemical Promotion of Catalysis with Alkaline Ionic Conductors. *Catalysts* **2016**, *6*, 15. [CrossRef]
66. Mortensen, P.M.; Gardini, D.; de Carvalho, H.W.P.; Damsgaard, C.D.; Grunwaldt, J.-D.; Jensen, P.A.; Wagner, J.B.; Jensen, A.D. Catalysis Science & Technology of sulfur, potassium, and chlorine in the feed. *Catal. Sci. Technol.* **2014**, *4*, 3672–3686. [CrossRef]
67. Hou, L.; Bu, Q.; Li, S.; Wang, D.; Xie, T. Ni_3S_2-Decorated TiO_2 nanotube arrays as effective photoanodes for photoelectrochemical water splitting. *RSC Adv.* **2016**, *6*, 99081–99087. [CrossRef]
68. Xiong, X.; Zhao, B.; Ding, D.; Chen, D.; Yang, C.; Lei, Y.; Liu, M. One-step synthesis of architectural Ni_3S_2 nanosheet-on-nanorods array for use as high-performance electrodes for supercapacitors. *NPG Asia Mater.* **2016**, *8*, e300. [CrossRef]
69. Alonso, D.M.; Wettstein, S.G.; Dumesic, J.A. Bimetallic catalysts for upgrading of biomass to fuels and chemicals. *Chem. Soc. Rev.* **2012**, *41*, 8075–8098. [CrossRef] [PubMed]
70. Sanna, A.; Vispute, T.P.; Huber, G.W. Hydrodeoxygenation of the aqueous fraction of bio-oil with Ru/C and Pt/C catalysts. *Appl. Catal. B Environ.* **2015**, *165*, 446–456. [CrossRef]
71. Gunawan, R.; Li, X.; Lievens, C.; Gholizadeh, M.; Chaiwat, W.; Hu, X.; Mourant, D.; Bromly, J.; Li, C.Z. Upgrading of bio-oil into advanced biofuels and chemicals. Part I. Transformation of GC-detectable light species during the hydrotreatment of bio-oil using Pd/C catalyst. *Fuel* **2013**, *111*, 709–717. [CrossRef]
72. Sitthisa, S.; An, W.; Resasco, D.E. Selective conversion of furfural to methylfuran over silica-supported NiFe bimetallic catalysts. *J. Catal.* **2011**, *284*, 90–101. [CrossRef]
73. Jahromi, H.; Agblevor, F.A. Hydrotreating of guaiacol: A comparative study of Red mud-supported nickel and commercial Ni/SiO_2-Al_2O_3 catalysts. *Appl. Catal. A Gen.* **2018**, *558*, 109–121. [CrossRef]
74. Narayanasamy, L.; Murugesan, T. Degradation of Alizarin Yellow R using UV/H_2O_2 Advanced Oxidation Process. *Environ. Sci. Technol.* **2014**, *33*, 482–489. [CrossRef]

75. Srifa, A.; Faungnawakij, K.; Itthibenchapong, V.; Viriya-Empikul, N.; Charinpanitkul, T.; Assabumrungrat, S. Production of bio-hydrogenated diesel by catalytic hydrotreating of palm oil over $NiMoS_2/\gamma\text{-}Al_2O_3$ catalyst. *Bioresour. Technol.* **2014**, *158*, 81–90. [CrossRef] [PubMed]
76. Elliott, D.C.; Hart, T.R.; Neuenschwander, G.G.; Rotness, L.J.; Olarte, M.V.; Zacher, A.H.; Solantausta, Y. Catalytic hydroprocessing of fast pyrolysis bio-oil from pine sawdust. *Energy Fuels* **2012**, *26*, 3891–3896. [CrossRef]
77. Venderbosch, R.H.; Ardiyanti, A.R.; Wildschut, J.; Oasmaa, A.; Heeres, H.J. Stabilization of biomass-derived pyrolysis oils. *J. Chem. Technol. Biotechnol.* **2010**, *85*, 674–686. [CrossRef]
78. Zhao, H.Y.; Li, D.; Bui, P.; Oyama, S.T. Hydrodeoxygenation of guaiacol as model compound for pyrolysis oil on transition metal phosphide hydroprocessing catalysts. *Appl. Catal. A Gen.* **2011**, *391*, 305–310. [CrossRef]

catalysts

MDPI

Article

Thermal and Catalytic Pyrolysis of Dodecanoic Acid on SAPO-5 and Al-MCM-41 Catalysts

Carolina Freitas [1], Marizania Pereira [2], Damari Souza [2], Noyala Fonseca [3], Emerson Sales [1,3], Roger Frety [1,3], Camila Felix [4], Aroldo Azevedo Jr. [5] and Soraia Brandao [1,2,*]

1 UFBA—Instituto de Química, Universidade Federal da Bahia, Rua Barão de Jeremoabo, 147, Ondina, Salvador CEP 40170-115, BA, Brazil; carolina.c_freitas@hotmail.com (C.F.); andradesales.emerson@gmail.com (E.S.); ro_fre@hotmail.fr (R.F.)
2 UFBA—Escola Politécnica, Universidade Federal da Bahia, Rua Professor Aristides Novis, 2, Federação, Salvador CEP 40210-630, BA, Brazil; msena.pet@gmail.com (M.P.); damari.eng@hotmail.com (D.S.)
3 LABEC—Laboratório de Bioenergia e Catálise, Escola Politécnica, Universidade Federal da Bahia, Rua Professor Aristides Novis, 2, Federação, Salvador CEP 40210-630, BA, Brazil; noyala_nscf@hotmail.com
4 IFBA—Instituto Federal da Bahia, Rua Vital Brazil, Pitanguinha, s. n, Simões Filho, Salvador CEP 43700-000, BA, Brazil; camila_rib@yahoo.com.br
5 UFRB—Universidade Federal do Recôncavo da Bahia, Av. Centenário, 697, Feira de Santana CEP 44042-280, BA, Brazil; aroldo.ufrb@gmail.com
* Correspondence: soraia.ufba@gmail.com; Tel.: +55-71-3283-6882 (ext. 6889)

Received: 25 February 2019; Accepted: 27 April 2019; Published: 3 May 2019

Abstract: In this study, dodecanoic acid was decomposed during fast pyrolysis experiments either thermally or in the presence of SAPO-5 and Al-MCM-41catalysts. The catalysts were synthesized by a hydrothermal route and subsequently characterized by XRD, TPD-NH_3, and TGA, and dodecanoic acid was characterized by TGA and DSC. Analysis of the post-pyrolysis products was performed online by gas chromatography coupled with mass spectrometry (GC-MS). The results from pyrolysis at 650 °C indicated that the nature of the catalysts strongly influences the composition of the products. Linear alkenes were standard products for all pyrolysis experiments, but with Al-MCM-41, various alkene isomers with a linear and cyclic structure formed, as well as saturated and aromatic hydrocarbons. As a whole, Al-MCM-41 led to a much higher dodecanoic acid conversion and higher deoxygenation than SAPO-5. As these catalysts present small differences in strong acid site density, the difference in the global conversion of dodecanoic acid could be attributed to textural characteristics such as pore volume and surface area. In this case, the textural properties of the SAPO-5 are much lower when compared to Al-MCM-41 and, due to a lower accessibility of the reactant molecule to the acidic sites of SAPO-5, partially blocked for fatty acid molecules by the considerable amount of amorphous material, as detected by XRD.

Keywords: fast pyrolysis; SAPO-5; Al-MCM-41; dodecanoic acid

1. Introduction

The global energy matrix is still based on non-renewable sources and fossil fuels. The use of these sources generates pollution and emissions of greenhouse gases, leaving environmental problems for future generations. Since the second half of the 20th Century, with oil crises, rising fuel demand, and growing environmental awareness, research into alternative energy sources has grown. Renewable sources are now much favored due to their full availability, biodegradability, and low cost [1].

Vegetable oils have a large amount of triacylglycerides (about 90%) and a lower amount of free fatty acids, mono-, and di-glycerides (8–10%). Among the acids present in the various oily compounds, saturated fatty acids, such as palmitic (C16:0) and stearic (C18:0), and unsaturated, such as linoleic

(C18:2) and oleic (C18:1), are the most common, but other acids, such as lauric (C12:0) and myristic (C14:0), exist in specific oils and fats [2].

Brazil has several sources for the supply of oily compounds, such as soybean (whose predominant fatty acid is linoleic acid), babassu (lauric acid), and animal fat (stearic acid). Research into oils from other biomasses has been carried out, for example, castor oil, palm oil pulp, palm kernel almond, babassu coconut kernel, sunflower seeds, and other raw materials [3]. The country is the fourth largest producer of coconut in the world with almost three million tons of fruit per year, and the State of Bahia is the largest producer. Coconut and coconut oil are relevant natural sources of saturated fats, based mainly on dodecanoic acid (C12:0), also known as lauric acid. Coconut oil is obtained from the pulp of mature fresh coconut (*Cocos nucifera* L.) and contains more than 80% of saturated fatty acids, such as caproic, caprylic, capric, lauric, myristic, palmitic, and stearic acids. Coconut oil has a concentration of lauric acid above 40% [4].

Pyrolysis or thermal cracking of triglycerides is a well-known method for producing renewable fuels (Bressler, Maher). Pyrolysis operates in the absence of external O_2, providing high yields in liquids when performed at elevated temperatures (400–600 °C), with vapor residence times lower than 5 s and heating rates between 10 and 200 °C/s [2]. Pyrolysis of triglycerides can be summarized by Reaction 1:

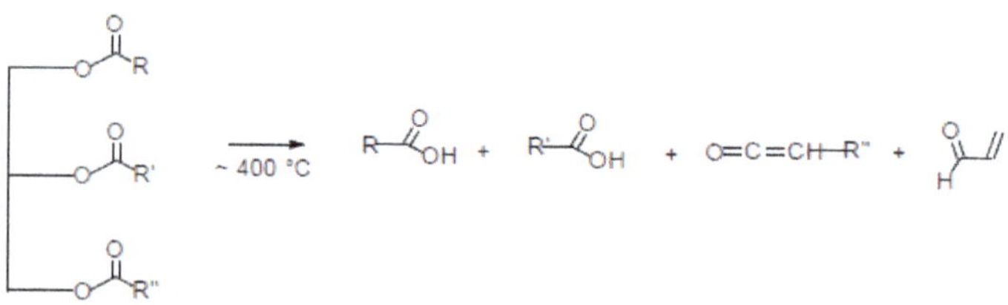

Reaction 1

Thermal cracking produces saturated, unsaturated linear hydrocarbons, or both, via the deoxygenation of carboxylic acids, which can occur through two distinct reaction paths: decarboxylation and decarbonylation. In decarboxylation, the carboxyl group of the fatty acid is removed, forming a saturated hydrocarbon and CO_2 (Reaction 2). The decarbonylation consists of the removal of the carbonyl of the fatty acid forming CO, H_2O, and olefins (Reaction 3) [1].

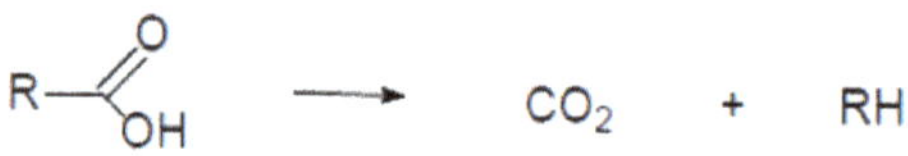

Reaction 2

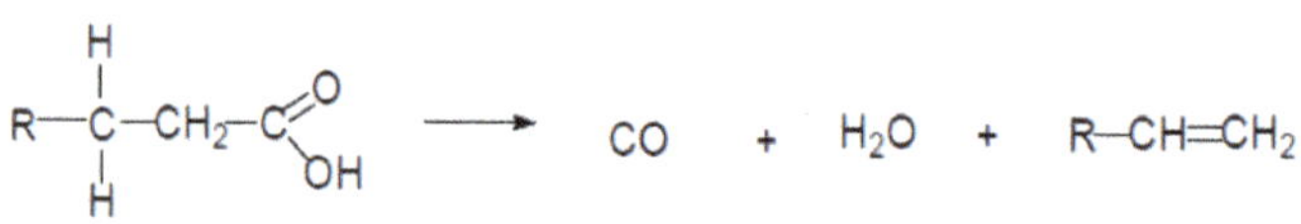

Reaction 3

In addition to the formation of the primary pyrolysis products, many parallel and sequential transformations can occur, leading to a series of oxygenated and deoxygenated molecules with a smaller or higher number of carbon atoms, compared to the initial carbon chains, such as olefins, paraffins, ketones, aldehydes, and aromatics, among others [5].

One way of making pyrolysis more efficient is to promote the reaction in the presence of a suitable catalyst. In recent years, studies dealing with the catalytic conversion of oily biomass feeds, through thermochemical reactions, have reported impressive results, and many reviews have summarized the positive and negative aspects of such transformations [6–8]. Further, current research has been done into the use of multi-element and multi-functional catalysts and complex mixtures of feeds in coprocessing strategies [9–12]. However, much essential work is still needed to understand the complexity of the reaction mechanism sequences, and therefore, studies with simpler systems, or model molecules, must be performed.

The interest of the petrochemical and chemical industries in the use of molecular sieves is due to the application of these types of materials in catalytic cracking, isomerization, and alkylation. Molecular sieves have a combination of important properties, such as surface area, large ion exchange capacity, and strong acidity, with high thermal and hydrothermal stability. Due to the regularity of their crystalline channels, molecular sieves allow higher selectivity of the reagents, products, and transition states, on a molecular scale [13]. Catalysts based on molecular sieves with different crystalline structures can induce changes in the product distribution, in the conversion of bio-oils and triglycerides. An example is the several zeolites, such as ZSM-5, Y, and beta, and silicoaluminophosphates studied in bio-oil cracking [3].

The SAPO-5 catalyst has one-dimensional 12-membered straight channel rings. It is a crystalline and microporous material, presenting excellent thermal and hydrothermal stability, and its acidity reaches intermediate values between the zeolite and the AlPO (aluminophosphate molecular sieves). Due to its topology, SAPO-5 provides potential catalytic applications, such as catalytic cracking, isomerization, and alkylation reactions [14].

The molecular sieve MCM-41 belongs to the class of materials known as M41S (Mobil 41 Synthesis). They are Si-based materials with highly-ordered mesopores and a high surface area. MCM-41, consisting of pure silica, presents low acidity. The incorporation of aluminum into the MCM-41 structure increases the number of acid sites, improving catalytic activity [15].

In this work, molecular sieves SAPO-5 and Al-MCM-41 were synthesized and evaluated as catalysts in the fast pyrolysis of dodecanoic acid, a simple model molecule of fatty compounds. The differences between thermal and catalytic pyrolysis, as well as the distribution of the products formed were analyzed.

2. Results and Discussion

Although the textural properties of the present samples were not determined, other samples of SAPO-5 and Al-MCM-41 using the same synthesis procedures have been prepared by some of us. In these cases, surface areas of SAPO-5 and Al-MCM-41 ranged from 247–261 and 485–510 m^2/g, respectively, whereas pore volumes were in the order of 0.19 and 0.85 cm^3/g, respectively. Therefore, Al-MCM-41 presented a higher surface area and a much larger pore volume than SAPO-5 [16].

2.1. X-ray Diffraction

Figure 1a shows the diffractogram of the SAPO-5 molecular sieve. The material presented a characteristic diffractogram of the AFI structure, similar to that found in the literature [17]. The presence of the intense halo at about $2\theta = 15–35°$ was associated with some extra frame silica.

The diffractogram of the Al-MCM-41 catalyst shown in Figure 1b presented a strong reflection at $2\theta = 2.1°$, characteristic of the MCM-41 structure. The two peaks of lower intensity and greater width observed at $2\theta = 3–4°$ suggest a loss of homogeneity and hexagonal symmetry of the crystalline structure due to the incorporation of aluminum in the structure of MCM-41 [18–20].

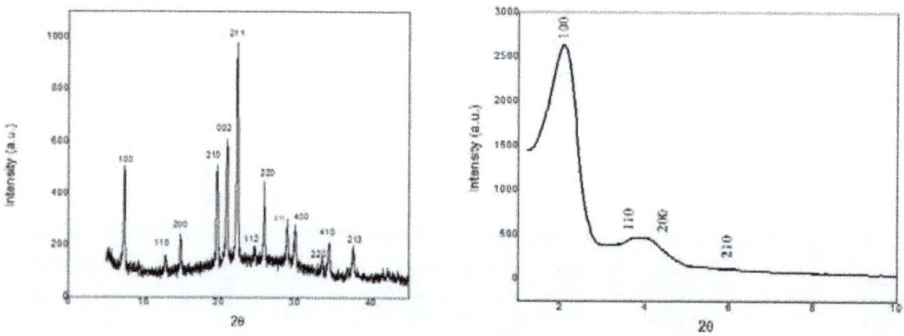

(a) (b)

Figure 1. X-ray diffraction of SAPO-5 (**a**); X-ray diffraction of Al-MCM-41 (**b**)

2.2. Thermodesorption of Ammonia

Thermodesorption (TPD) of ammonia is often used to estimate the acidity of the molecular sieves and establish strong correlations both for the extent of the reactions and product selectivity [21]. TPD-NH_3 profiles of the SAPO-5 and Al-MCM-41 catalysts in Figure 2a,b, respectively, are generally divided into three temperature ranges: (i) below 300 °C is associated with weak acid sites; (ii) between 300 and 500 °C with acidic sites of moderate strength; and (iii) above 500 °C corresponding to strong acid sites [22].

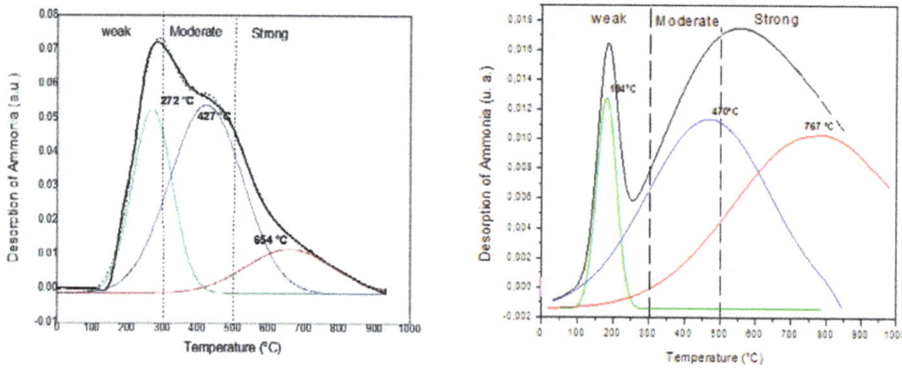

(a) (b)

Figure 2. TPD ammonia analysis of SAPO-5 (**a**); TPD ammonia analysis of Al-MCM-41(**b**).

Table 1 presents the concentrations, in mmol/g, of the acid sites derived from the deconvolution of the experimental NH_3 thermodesorption curves, using a Gaussian distribution function.

Table 1. Concentration ($\times 10^{-2}$ mmol/g) of SAPO-5 and Al-MCM-41 acid sites.

Catalysts	Weak	Moderate	Strong
SAPO-5	1.04	1.94	0.54
Al-MCM-41	0.11	0.55	0.41

For the SAPO-5 catalyst, there was a predominance of weak and moderate acid sites with maximum desorption temperature respectively at 272 °C and 427 °C, in addition to the presence of a peak ascribed to strong acid sites, at a temperature of 654 °C, with a lower concentration. Al-MCM-41 had three desorption peaks at maximum temperatures of 184 °C, 470 °C, and 767 °C, respectively.

The highest concentration, in mmol/g, found for the Al-MCM-41 catalyst was for moderate acid sites. The desorption peak at a low temperature (<300 °C) probably refers to weak Brønsted acid sites, which are bound to aluminum with tetrahedral coordination (AlO^{4-}). Thus, a high temperature (>500 °C) desorption may be ascribed to ammonia adsorbed in Lewis acid sites on the surface of the molecular sieve, created by the dehydroxylation of the aluminum atoms substituting silicon atoms in the structure [21,22].

2.3. Thermogravimetric Analysis

The profiles of the thermogravimetric analysis curves of the Al-MCM-41 and SAPO-5 samples are shown respectively in Figure 3a,b.

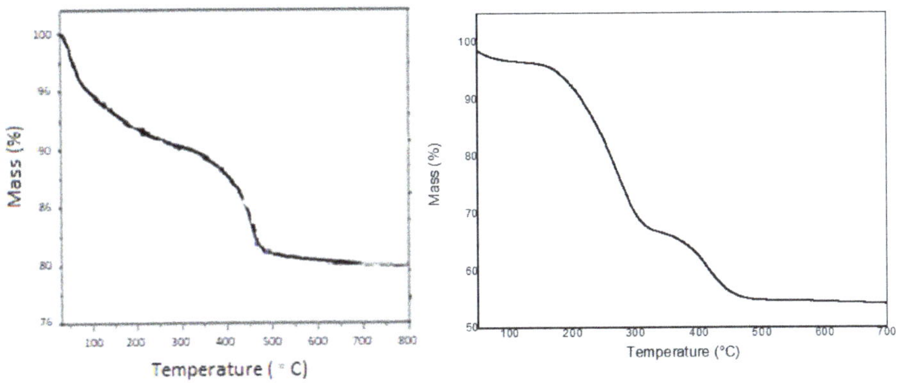

(**a**) (**b**)

Figure 3. TGA profile of SAPO-5 catalyst (**a**); TGA profile of Al-MCM-41 catalyst (**b**).

In Figure 3a, the thermogravimetric analysis of the SAPO-5 catalyst is presented, in which three regions of mass loss can be observed. The first 8% mass loss event, from room temperature up to 200 °C, was associated with the desorption of adsorbed water on the outer surface and occluded in the pores. The second step had a loss of 10% in the range of 200 °C–470 °C, with a maximum rate from 440 °C–470 °C, which was ascribed to the desorption of triethylamine (template) occluded in the channels. Finally, the third region, between 470 and 750 °C, with a loss of 2%, was related to the desorption of the occluded channel surfactant and to the decomposition of the protonated amine (TEA^+) [23].

In Figure 3b, the first mass loss event (2%) was related to the loss of water physically adsorbed in the pores of the Al-MCM-41 catalyst. The second mass loss event (33%) was attributed to the degradation of the organic molecules of the template CTMABr (hexadecyltrimethylammonium bromide). The third event (10%) was associated with the removal of the residual template and also water resulting from the secondary condensation of the silanol groups, according to the literature [24,25].

The TGA/DTG curves of pure lauric acid, Figure 4, exhibited a region of significant mass loss (95%) between 150 °C and 250 °C, followed by small mass losses up to 600 °C. The excessive loss of mass was attributed to the volatilization of lauric acid and the losses at higher temperatures to progressive degradation of non-volatile residues.

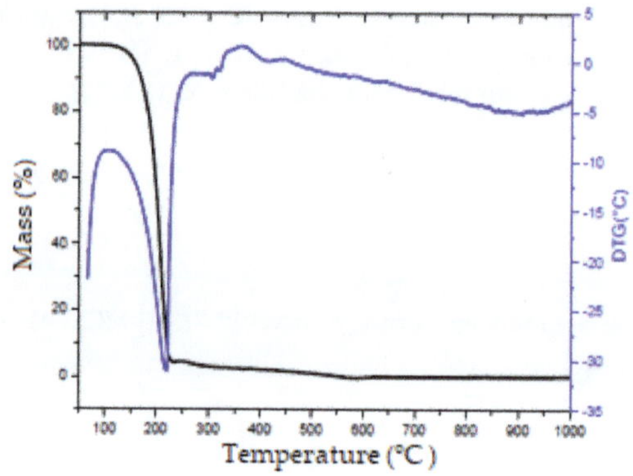

Figure 4. TGA/DTG profile of lauric acid.

2.4. Differential Scanning Calorimetry

The DSC profile of pure lauric acid, Figure 5, shows a succession of endothermic and exothermic events. The endothermic phenomenon appearing at around 50 °C, not associated with mass loss, according to Figure 4, was related to the fusion of lauric acid. At about 200 °C, a second weak endothermal phenomenon was observed, probably related to the initial volatilization of lauric acid, quickly followed by an exothermic event linked to the processes of cracking and autoxidation, possibly associated with the degradation of the fatty acid. The results obtained by the DSC and TGA/DTG analyses suggest that the lauric acid was fully converted to the vapor phase below 250 °C and transformed into products between 250 and 600 °C.

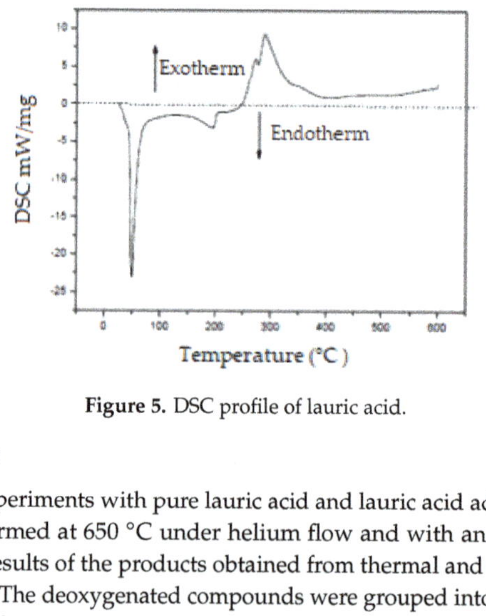

Figure 5. DSC profile of lauric acid.

2.5. Fast Pyrolysis Reaction

The fast pyrolysis experiments with pure lauric acid and lauric acid adsorbed on both molecular sieve catalysts was performed at 650 °C under helium flow and with an estimated heating rate of 1000 °C s^{-1}. The global results of the products obtained from thermal and thermo-catalytic pyrolysis are shown in Figures 6–8. The deoxygenated compounds were grouped into different product families to facilitate the analysis of the products. The remaining reactant is also indicated in Figures 6 and 7, as the observed conversions were 9% and 12%, respectively. When using the Al-MCM-41 catalyst (Figure 8), no lauric acid was detected after pyrolysis, indicating a total conversion.

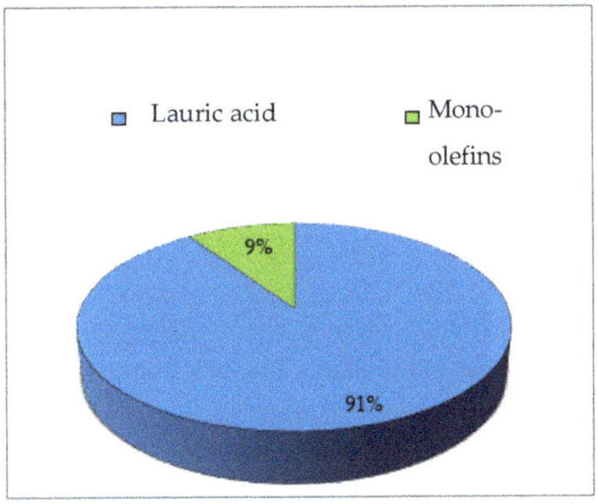

Figure 6. Compounds identified after fast pyrolysis of pure lauric acid at 650 °C.

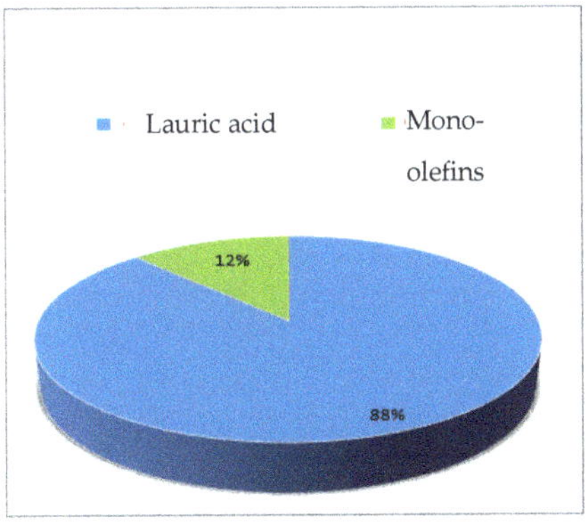

Figure 7. Compounds identified after fast pyrolysis of lauric acid at 650 °C in the presence of SAPO-5.

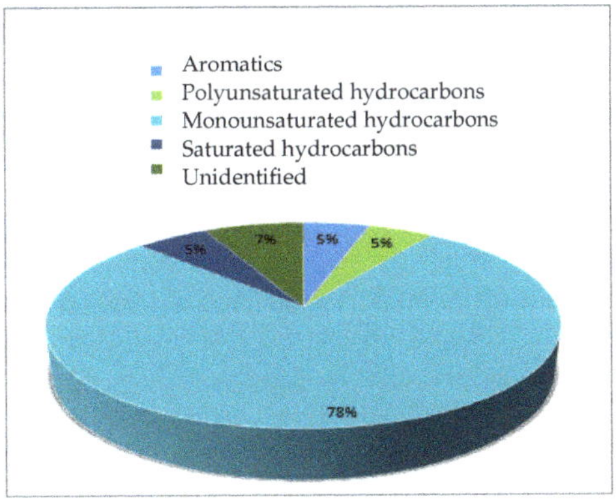

Figure 8. The yield of fast pyrolysis products of lauric acid at 650 °C in the presence of Al-MCM-41.

In thermal pyrolysis, limited deoxygenation of lauric acid occurs via the decarbonylation route, mainly giving α-olefins (C_3 to C_{11}) [8], as observed in Figure 6. Pyrolysis of fatty acids can proceed through the breaking of C-C and C-O bonds. Two competing routes can explain the presence of products with fewer than ten carbon atoms: (i) deoxygenation of the starting acid, followed by breaking of the C-C bond to produce hydrocarbon radicals; and (ii) cracking of the C-C bond of the chain, followed by deoxygenation of a shorter chain carboxylic acid. With saturated compounds, such as lauric acid, the first route is favored [16,24].

In the catalytic pyrolysis of the acid adsorbed on SAPO-5 (Figure 7), a slightly larger number of α-olefins (from C_3–C_{11}) were formed, which consisted of potential feedstock to produce high octane gasoline. A similar study with the same catalyst (SAPO-5), using palmitic acid, i.e., a saturated fatty acid with a carbon chain longer than lauric acid, can be found in [16]. In both cases, the pyrolysis products presented great similarity since they showed a high distribution of linear mono-unsaturated hydrocarbons.

For pyrolysis of lauric acid in the presence of Al-MCM-41 catalysts (Figure 8), C_3–C_{11} olefins (68%) including isomers with internal C=C bonds, both in cis and trans configuration, cyclic mono-unsaturated molecules, and methylated molecules, were identified. Cyclic and alkylated cyclic saturated products (14%) and aliphatic alkanes (4%), together with mono-aromatics (benzene, toluene, and p-xylene) (4%), were also found. The formation of both aromatic and saturated products can be attributed to the acidity of the Al-MCM-41 catalyst, which should promote hydrogen transfer reactions. The hydrogen liberated during aromatic formation could be in part used to saturate olefin-based products and possibly helped the alkylation reaction.

Figure 9 presents the normalized distribution of the α-olefins obtained (an area greater than 0.5% of the total area) after thermal pyrolysis of lauric acid at 650 °C and catalytic pyrolysis of lauric acid with SAPO-5 at the same temperature. The SAPO-5 catalyst did not alter the distribution of the products significantly, compared to the thermal pyrolysis. Only the conversion was slightly higher (12% and 9%, as seen in Figures 6 and 7), which is not visible in the figure because of normalization.

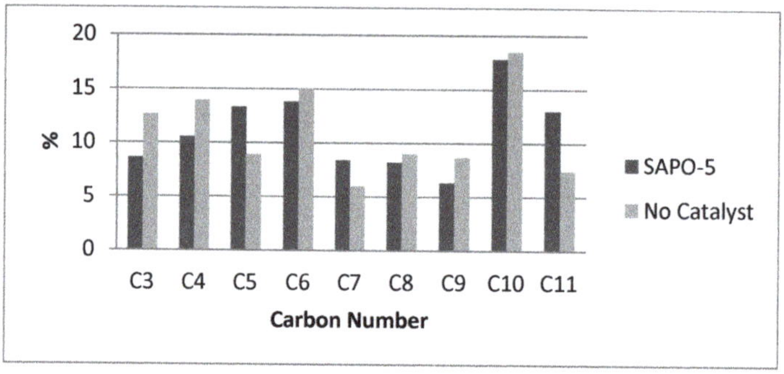

Figure 9. Normalized distribution of the α-olefins obtained after pyrolysis of lauric acid at 650 °C: (■) SAPO-5 catalyst and (■) no catalyst.

Table 2 presents the C6 (6 isomers) and C11 (13 isomers) olefins identified after dodecanoic acid pyrolysis at 650 °C in the presence of Al-MCM-41, limiting the identification of the products to those with a relative area greater than 0.5% of the pyrogram total area. Among the main products, linear, cyclic, and alkylated olefins were identified, as well as cis and trans isomers.

When considering the complete family of monounsaturated molecules obtained in the presence of Al-MCM-41, represented in Figure 10, the amount of isomers does not increase continuously with chain length; C_8 olefins have both a lower number of identified isomers (two) and the most moderate content. It suggests a potential application for gasoline and kerosene fuels, after partial hydrogenation.

Table 2. C_6 and C_{11} mono-unsaturated main products observed during dodecanoic acid pyrolysis at 650 °C using Al-MCM-41 catalyst.

Carbon Number	Retention Time (min)	Mono-Unsaturated
C6	1.686	2-Pentene, 4-methyl-
	1.785	1-Hexene
	1.883	3-Hexene, (E)-
	1.918	3-Hexene, (Z)-
	1.995	2-Pentene, 3-methyl-, (Z)-
	2.268	Cyclopentene, 1-methyl-
C11	8.383	4-Decene, 3-methyl-, (E)-
	8.426	2-Decene, 4-methyl-, (Z)-
	8.538	5-Undecene, (E)-
	8.608	4-Undecene, (Z)-
	8.671	2-Decene, 7-methyl-, (Z)-
	8.699	5-Undecene, (Z)-
	8.769	1-Decene, 5-methyl-
	8.825	4-Undecene, (e)-
	8.846	2-Decene, 6-methyl-, (Z)-
	8.916	3-Undecene, (Z)-
	8.973	3-Undecene, (E)-
	9.064	2-Undecene, (Z)-
	9.169	2-Undecene, (E)-

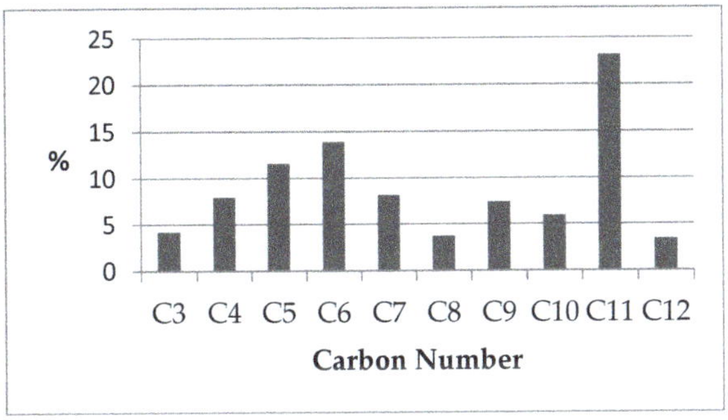

Figure 10. Mono-unsaturated isomers' distribution from pyrolysis of lauric acid at 650 °C, in the presence of Al-MCM-41.

2.6. Comparison between the SAPO-5 and Al-MCM-41 Catalytic Behavior

The extent of lauric acid conversion between the three pyrolysis reactions was different: 9% was observed during thermal pyrolysis, 12% when lauric acid was adsorbed on the SAPO-5 catalyst, and about 100% when the fatty acid was adsorbed on Al-MCM-41 catalyst. The results are consistent and close to those found in the literature regarding the pyrolysis of saturated fatty acids [4]. However, the results obtained for pyrolysis of unsaturated fatty acids on Al-MCM-41-based catalyst, similar to those in the present work, showed that deoxygenation was not as effective as for saturated acids [19].

An essential point of the discussion is linked to the low conversion of lauric acid observed when using SAPO-5 catalyst compared to Al-MCM-41. The TPD of NH_3 profiles are different for these two catalysts, but small differences in the number of strong acid sites were observed; this alone cannot explain the significant differences in conversion between both molecular sieves. Specific surface area and pore volume of SAPO-5 were much lower than the values for Al-MCM-41. Moreover, a large amount of amorphous material in SAPO-5 was revealed by XRD analysis, and this can cause lower accessibility of the dodecanoic acid molecule to acid sites in SAPO-5 catalyst. These two last characteristics are probably responsible for the observed differences in performances between both catalysts. The access of large lauric acid molecules to the acidic sites located inside the porous structure of SAPO-5 could be limited for two main reasons: (i) a lower mean diameter of SAPO-5 pores compared to that of Al-MCM-41; (ii) the pore mouth in SAPO-5 can be partially blocked by the amorphous species. In other words, only a small proportion of SAPO-5 can interact with dodecanoic acid, limiting the conversion of the fatty acid during pyrolysis experiments.

3. Materials and Methods

3.1. Catalyst Preparation

The synthesis of the SAPO-5 catalyst was carried out following the methodology found in [4], and some modifications were made to adjust the Si/Al ratio. Pseudoboehmite and orthophosphoric acid (85%, Sigma-Aldrich) were used as sources of Si and P, respectively. Then, the triethylamine (Sigma-Aldrich, St. Louis, MO, USA) template was added to tetraethyl orthosilicate (98%, Sigma-Aldrich, St. Louis, MO, USA) and the surfactant hexadecyltrimethylammonium bromide (Sigma-Aldrich, St. Louis, MO, USA), in solution with hexanol (Sigma-Aldrich, St. Louis, MO, USA). Initially, 19.56 mL of orthophosphoric acid (previously diluted in 20 mL of deionized water) were placed in a beaker containing 22.4 g of pseudoboehmite (dispersed in 63 mL of H_2O); this system was kept under stirring for two hours. After that, 38.64 mL of triethylamine were added, and stirring was continued for a further two hours. After this time, a solution prepared by the addition of 4.82 mL of TEOS (Tetraethyl Orthosilicate) + 78.82 mL of hexanol + 3.8 g of CTMABr (hexadecyltrimethylammonium bromide) was added and again stirred for another two hours. The mixture was partitioned into autoclaves and put into an oven preheated to 170 °C, where it remained for 24 h. The material obtained after this period was centrifuged, washed with alcohol and distilled water, respectively, and again routed to a drying oven at 100 °C for six hours.

A sample of Al-MCM-41 with a silica/alumina (SAR) ratio of 60 was prepared. Commercial silica (Aerosil 200) and sodium aluminate were used as sources of Si and Al, respectively. As a guideline of the mesoporous structure, hexadecyltrimethylammonium bromide (CTMABr) was used in the form of 25% *w*/*w* aqueous solution. Sodium hydroxide solution (50% *w*/*w*) was used to maintain the basic pH. The molar composition of the synthesis gel was: 1 SiO_2: 0.025 Al_2O_3: 0.08 Na_2O: 0.3 CTMA: 26 H_2O.

The samples were prepared from suspensions and solutions named as A, B, C. In Suspension A, 20.2 g of silica (SiO_2) were mixed in 38.3 g of distilled water. For Solution B, 2.8 g of sodium hydroxide (NaOH) were added to 2.8 g of distilled water, and in Solution C, 36.8 g of hexadecyltrimethylammonium bromide (CTMABr) were dissolved in 110.5 g of distilled water.

In Solution B, the amount of 0.9 g of sodium aluminate and then Solution C were added under vigorous and constant stirring until complete dissolution of the sodium aluminate. Solution A was then added under vigorous stirring at room temperature. The mixture was then placed in autoclaves without stirring and heated at 150 °C for 48 h.

Then were added 0.9 g of sodium aluminate to Solution B and then Solution C added under vigorous and constant stirring until complete dissolution of the sodium aluminate. Suspension A was submitted to vigorous stirring at room temperature. The mixture was then placed in autoclaves without stirring and heated at 150 °C for 48 h.

After this procedure, the product was washed with distilled water separated using a centrifuge and oven dried at 100 °C for 4 h. After this step, the material was calcined to remove the remaining template in the porous structure.

Al-MCM-41 catalyst was calcined under an inert atmosphere (N_2, 30 mL/min) to 370 °C for 1 h 30 min to remove the remaining template in the porous structure. After this step, the N_2 flux was replaced by a synthetic air flow (30 mL/min) maintaining the temperature of 550 °C for over ten hours for remaining burning of the template.

The SAPO-5 catalyst was calcined in an inert atmosphere (N_2, 100 mL/min) to 450 °C (10 °C/min), remaining for one hour at this temperature, to remove the remaining template in the structure. After this stage, the N_2 was replaced by synthetic air (100 mL/min), maintained at a temperature of 550 °C for 5 h to burn the coke formed with the decomposition of the template.

3.2. *Catalyst Characterization*

3.2.1. X-ray Diffraction

The X-ray diffraction measurements were conducted on a Shimadzu apparatus, model XRD-6000, using Cu Kα radiation (λ = 1.5418 Å) (Shimadzu, Kobe, Japan). The diffractograms were collected in a range from 1–50°, with a scanning speed of 0.25° min^{-1}. They were obtained at a constant power source of 40 kV and 30 mA, at room temperature. The powder samples were analyzed without any previous treatment.

3.2.2. Temperature Programmed Desorption of Ammonia (TPD-NH_3)

The measurement of the density and strength of the acid sites of the samples was carried out using the temperature programmed desorption (TPD-NH_3). Initially, the ammonia was adsorbed at room temperature for 1 h. The samples were then heated at a rate of 10 °C/min to 150 °C for 1 h under He flow to remove the poorly-adsorbed molecules. The ammonia desorption step was performed under a constant helium flow of 25 mL/min to 700 °C, using a heating rate of 10 °C/min. The equipment used was the Chemisorb 2720, Pulse Chemisorption System, Micromeritics, equipped with a thermal conductivity detector (TCD), a quartz reactor, and gas supply system.

3.2.3. Thermogravimetric Analysis

The samples (2 mg of mass) were heated with a temperature ramp of 10°C/min from ambient temperature to 1000 °C in a Shimadzu apparatus, Model TGA-50 (Shimadzu, Kobe, Japan), under a constant flow of 50 mL/min of nitrogen.

3.3. *Catalytic Test*

The pyrolysis experiments were performed on a multi-shot pyrolyzer model EGA/PY-3030D (Frontier Laboratories LTD, Fukushima, JPN) connected online with a GC-MS-5799 (Agilent, Santa Clara, CA, USA). In the thermal fast pyrolysis, 1.1 milligrams of pure lauric acid were used, whereas in the catalytic pyrolysis tests, 1.1 mg of the lauric acid/catalyst mixture were used.

The mixture of the lauric acid with the catalyst was carried out with the addition of the fatty acid to the catalyst in the proportion 1:10 (m/m), by mechanical mixing. Subsequently, the mixture was heated to 60 °C, a temperature higher than the melting point of lauric acid (43 °C), and continuously homogenized. This treatment allowed the lauric acid to migrate inside the pores of the catalysts. After this step, the mixture still in powder form was stored away from moisture, then added to the titanium sample holder, and covered with quartz wool.

The samples were dropped by gravity into the pyrolysis furnace preheated to 650 °C under a helium atmosphere and remained under these conditions for 20 s before an aliquot of the gaseous atmosphere was introduced into the analytical system. For the chromatographic analysis of the products, an HP-5MS column was used with a heating ramp comprising an initial temperature of 40 °C for 3 minutes, followed by a ramp of 20 °C/min to 320 °C, with the temperature maintained for 20 min. The rate of helium flow through the column was 2 mL/min. The temperature of the pyrolyzer interface was 320 °C. The products obtained by GC/MS were identified based on the retention times and by comparison with the standard mass spectral fragments of the National Institute of Standards and Testing database.

4. Concluding Remarks

The presence of the catalysts modified the quantity and distribution of the products. The fast pyrolysis in the presence of catalysts based on mesoporous and microporous molecular sieves allowed the production of olefins in the gasoline range. Using both thermal and catalytic pyrolysis with the SAPO-5 catalyst, the conversion of dodecanoic acid was limited, 9% and 12%, respectively, and the main products were terminal olefins (C_3–C_{11}). In the presence of the Al-MCM-41 catalyst, the total conversion of the lauric acid was observed, and the main products were internal and alkylated olefins, fundamental for both chemical and petrochemical industries.

Author Contributions: S.B. has conceived the project, funding acquisition, designed the experiments, analyzed the data and written the paper. C.F. (Carolina Freitas), M.P., D.S. and N.F. have designed and performed the experiments, analyzed the data and written the paper. E.S., R.F., C.F. (Camila Felix) and A.A.J. have designed the experiments, analyzed the data and written the paper.

Funding: FAPESB (Bahia Research Foundation), project DTE 0043/2011, and CNPq, project 552367/2011-7 funded this research.

Acknowledgments: The authors are grateful to CNPq, FAPESB, and CAPES for scholarships and financial support.

Conflicts of Interest: The authors declare no conflict of interest.

References

1. De M. Araújo, A.M.; de O. Lima, R.; Gondim, A.D.; Diniz, J.; Di Souza, L.; de Araujo, A.S. Thermal and catalytic pyrolysis of sunflower oil using AlMCM-41. *Renew. Energy* **2017**, *101*, 900–906.
2. Badoga, S.; Ganesan, A.; Dalai, A.K.; Chand, S. Effect of synthesis technique on the activity of CoNiMo tri-metallic catalyst for hydrotreating of heavy gas oil. *Catal. Today* **2017**, *291*, 160–171. [CrossRef]
3. Rodrigues, T.O.; Rousset, P.; do Vale, A.T.; Broust, F. Bioóleo: Uma Alternativa para Valorização Energética da Biomassa. *Rev. Bras. Energ.* **2011**, *17*, 39–56.
4. Maher, K.D.; Bressler, D.C. Pyrolysis of Triglyceride Materials for the Production of Renewable Fuels and Chemicals. *Bioresour. Technol.* **2007**, *98*, 2351–2368. [CrossRef]
5. Wu, P.; Jiang, X.; Jin, Q.; Liu, F.; Zhang, J.; Zhu, Y.; Xia, L.; Shao, T.; Wang, K.; Li, T. Production of jet fuel range biofuels by catalytic transformation of triglycerides based oils. *Fuel* **2017**, *188*, 205–211. [CrossRef]
6. Sonthalia, A.; Kumar, N. Hydroprocessed vegetable oil as a fuel for transportation sector: A review. *J. Energy Inst.* **2019**, *82*, 1–17. [CrossRef]
7. Bezergianni, S.; Dimitriadis, A.; Kikhtyanin, O.; Kubicka, D. Refinery co-processing of renewable feeds. *Prog. Energy Comb. Sci.* **2018**, *68*, 29–64. [CrossRef]
8. Kubicka, D.; Kikhtyanin, O. Opportunities for zeolites in biomass up-grading: Lessons from the refining and petrochemical industry. *Catal. Today* **2015**, *243*, 10–22. [CrossRef]
9. Merdun, H.; Sezgin, I.V. Products distribution of catalytic co-pyrolysis of greenhouse vegetable waste and coal. *Energy* **2018**, *162*, 953–963. [CrossRef]
10. Jahromi, H.; Agblevor, F.A. Hydrodeoxygenation of Aqueous-Phase Catalytic Pyrolysis Oil to Liquid Hydrocarbons using Multifunctional Nickel catalysts. *Ind. Eng. Chem. Res.* **2018**, *57*, 13257–13268. [CrossRef]
11. Castille, A.; Bessette, C.; Thomas, F.; Eternad, M. Sustainable hydrocarbons production via simultaneous condensation-hydrodeoxygenation of propionic acid with furfural over red mud-supported noble metal catalysts. *Catal. Commun.* **2019**, *121*, 5–10. [CrossRef]
12. Wang, Y.; Dai, L.; Fan, L.; Cao, L.; Zhou, Y.; Zhao, Y.; Liu, Y.; Ruan, R. Catalytic co-pyrolysis of waste vegetable oil and high density polyethylene for hydrocarbon fuel production. *Waste Manag.* **2017**, *61*, 276–282. [CrossRef]
13. Abdelfattah, M.S.H.; Abu-Elyazeed, O.S.M.; Abdelazeem, M.A. On biodiesels from castor raw oil using catalytic pyrolysis. *Energy* **2018**, *143*, 950–960. [CrossRef]
14. Nash, C.P.; Ramanathan, A.; Ruddy, D.A.; Behl, M.; Gjersing, E.; Griffin, M.; Zhu, H.; Subramaniam, B.; Schaidle, J.A.; Hensley, J.E. Mixed alcohol dehydration over Brønsted and Lewis acidic catalysts. *Appl. Catal. A Gen.* **2016**, *510*, 110–124. [CrossRef]
15. Karnjanakom, G.; Suriya-umporn, S.; Bayu, T.; Kongparakul, A.; Samart, S.; Fushimi, C.; Abudula, C.; Guan, A. High selectivity and stability of Mg-doped Al-MCM-41 for in-situ catalytic upgrading fast pyrolysis bio-oil. *Energy Convers. Manag.* **2017**, *142*, 272–285. [CrossRef]

16. De Azevedo, A.F., Jr. Desenvolvimento de Catalisadores a Base de SAPO-5 Para Avaliação na Reação de Fast Pirólise de Compostos Graxos. Ph.D. Thesis, Universidade Federal da Bahia (UFBA), Salvador, Brazil, 24 Auguest 2014.
17. Bandyopadhyay, R.; Bandyopadhyay, M.; Kubota, Y.; Sugi, Y. Synthesis of $AlPO_4$ Molecular Sieves with AFI and AEL Structures by Dry-Gel Conversion Method and Catalytic Application of Their SAPO Counterparts on Isopropylation of Biphenyl. *J. Porous Mat.* **2002**, *9*, 83–95. [CrossRef]
18. Vieira, S.S. Produção de Biodiesel via Esterificação de Ácidos Graxos Livres Utilizando Catalisadores Heterogêneos Ácidos. Ph.D. Thesis, Universidade Federal de Lavras (UFLA), Lavras, Brazil, 24 February 2011.
19. Yu, F.; Gao, L.; Wang, W.; Zhang, G.; Ji, J. Bio-fuel production from the catalytic pyrolysis of soybean oil over Me-Al-MCM-41 (Me = La, Ni or Fe) mesoporous materials. *J. Anal. Appl. Pyrolysis* **2013**, *104*, 325–329. [CrossRef]
20. Barbosa, F.A.; dos Santos, A.C.B.; da Silva, M.I.P.; Stumbo, A.M. Resistance to poisoning by nitrogen compounds of NiMo/Al-MCM-41 hydrocracking catalysts. *Catal. Today* **2004**, *98*, 109–113. [CrossRef]
21. Liu, C.W.; Hu, W.W.; Yang, B.H.; Tong, Y.; Zhu, D.M.; Zhang, L.F.; Zhao, R. N Study on the effect of metal types in (Me)-Al-MCM-41 on the mesoporous structure and catalytic behavior during the vapor-catalyzed co-pyrolysis of pubescens and LDPE. *Appl. Catal. B Environ.* **2013**, *129*, 202–213. [CrossRef]
22. Intana, T.; Föttinger, K.; Rupprechter, G.; Kongkachuichay, P. Physicochemical properties of Cu loaded onto core-shell Al-MCM-41: Effect of loading methods. *Colloids Surf. A Physicochem. Eng. Asp.* **2015**, *467*, 157–165. [CrossRef]
23. Benítez-Guerrero, M.; López-Beceiro, J.; Sánchez-Jiménez, P.E.; Pascual-Cosp, J. Comparison of thermal behavior of natural and hot-washed sisal fibers based on their main components: Cellulose, xylan and lignin. TG-FTIR analysis of volatile products. *Thermochim. Acta* **2014**, *581*, 70–86. [CrossRef]
24. Zafar, R.; Watson, J.S. Adsorption of tetradecanoic acid on kaolinite minerals: Using fast pyrolysis to characterize the catalytic efficiency of clay mineral adsorbed fatty acids. *Chem. Geol.* **2017**, *471*, 111–118. [CrossRef]
25. Souza, M.J.B.; Marinkovic, B.A.; Jardim, P.M.; Araujo, A.S.; Pedrosa, A.M.G.; Souza, R.R. HDS of thiophene over CoMo/AlMCM-41 with different Si/Al ratios. *Appl. Catal. A Gen.* **2007**, *316*, 212–218. [CrossRef]

catalysts

MDPI

Article

Dibenzothiophene Hydrodesulfurization over P-CoMo on Sol-Gel Alumina Modified by La Addition. Effect of Rare-Earth Content

José Escobar [1,*], María C. Barrera [2], Jaime S. Valente [1], Dora A. Solís-Casados [3], Víctor Santes [4], José E. Terrazas [2] and Benoit A.R. Fouconnier [2]

1 Instituto Mexicano del Petróleo, Eje Central Lázaro Cárdenas 152, San Bartolo Atepehuacan, G.A. Madero, Cd. de México 07730, México; jsanchez@imp.mx

2 Facultad de Ciencias Químicas-Centro de Investigación en Recursos Energéticos y Sustentables, Universidad Veracruzana, Campus Coatzacoalcos, Av. Universidad km. 7.5, Col. Santa Isabel, Coatzacoalcos, Veracruz 96538, México; mcbdgavilan@gmail.com (M.C.B.); eterrazas@uv.mx (J.E.T.); broger@uv.mx (B.A.R.F.)

3 Universidad Autónoma del Estado de México. Centro Conjunto de Investigación en Química Sustentable, km. 14.5 Carretera Toluca-Atlacomulco, Toluca, Estado de México 50200, México; solis_casados@yahoo.com.mx

4 Departamento de Biociencias e Ingeniería, Centro Interdisciplinario de Investigaciones y Estudios en Medio Ambiente y Desarrollo (CIIEMAD), Instituto Politécnico Nacional, Ciudad de México C.P. 07340, México; vsantes@ipn.mx

* Correspondence: jeaguila@imp.mx

Received: 11 March 2019; Accepted: 9 April 2019; Published: 13 April 2019

Abstract: Alumina-lanthana (La at 1, 3, or 5 wt%) supports were prepared by sol-gel from Al alkoxide sol where $La(NO_3)_3$ was added. Annealed (550 °C) xerogels were characterized by N_2 physisorption, thermal analysis (TG-DTA), X-ray diffraction (XRD), scanning electron microscopy- energy dispersive spectroscopy (SEM-EDS), CO_2-adsorption studied in IR region, Raman and ultraviolet-vis (UV-vis) spectroscopies. The texture of amorphous binary matrices of high La dispersion was adequate to applications in catalysts for middle distillates hydrodesulfurization (HDS). Generally, the amount and strength of surface basic sites increased with La content in solids. Mo (at 2.8 at. nm^{-2}) and Co (at Co/(Co+Mo) = 0.3) were deposited over carriers by one-pot simultaneous impregnation in the presence of PO_4^{3-} (P_2O_5/(NiO+MoO_3) = 0.2 mass ratio). Calcined (400 °C) Co-Mo-P impregnated precursors had decreased basicity as to that of corresponding carriers, suggesting strong La-deposited species interaction. As La content in carriers increased Mo=O Raman stretching vibrations shifted to lower wave-numbers (949 to 935 cm^{-1}) suggesting octahedral molybdates coordination change to tetrahedral. Although La at the lowest concentration (1 wt%) enhanced dibenzothiophene, HDS (~38% higher as to the Al_2O_3-supported formulation) desulfurization was significantly diminished at augmented content. Presence of hardly sulfidable tetrahedral Mo originated during impregnation at basic conditions in pores of La-modified carriers seemed to dictate observed behavior. Rare earth content in formulations enhanced selectivity to biphenyl.

Keywords: hydrodesulfurization; CoMo/Al_2O_3; basic additive; lanthanum

1. Introduction

Improved catalytic formulations of enhanced properties in hydrodesulfurization (HDS) reactions that allow for compliance with ever stricter environmental regulations on S content in internal combustion engine fuels remains a challenging research field. In this context, one of the most relevant lines of investigation involves preparation of improved catalyst supports, where conventional active phases used in formulations applied at the industrial hydrotreating scale (sulfided Mo or W promoted by either Co or Ni) could have enhanced properties, including better dispersion and sulfidability or more efficient promoter integration [1–3]. In this regard, the effects of using carriers with augmented surface basicity, as to that of conventionally used alumina, still remains a matter of debate. Some reports have shown that over carriers of increased basicity, by alkaline [4] or alkaline-earth metal [5] species addition, catalysts of enhanced selectivity to desulfurization and limited hydrogenating properties could be obtained. That effect has been attributed to several factors, for instance, decreased surface acidity by basic agent addition [4] or formation of oxidic species (magnesium molybdate and NiO-MgO solid solution, [5]) that resulted after sulfiding in MoS_2 (from $MgMoO_4$, for example) phases of increased slab length and diminished saturation ability. However, decreased activity for organo-S species desulfurization (thiophene) was also observed [4], although in a lower degree, as to that registered for alkenes hydrogenation. Formation of K-decorated and K-intercalated MoS_2 phases supported on SBA-15 mesoporous materials of enhanced activity and selectivity to methanethiol from a H_2S-CH_4 mixture have been recently reported [6]. According to the authors, electron donation from K atoms could enhance the electron density of sulfided Mo species that facilitated formation of increased amounts of Mo-coordinatively unsaturated sites (CUS).

Environmentally friendly lanthanum is a cheap rare earth that has been added to catalyst carriers in order to increase their thermal resistance [7] and surface basicity [8]. For instance, decreased acidity of alumina-La oxides contributed to augmented stability in ethanol conversion, due to diminished coke formation [9]. Also, due to that diminished acidity, corresponding rare-earth-modified Al_2O_3 had enhanced selectivity to ethylene (instead of to carbonaceous species) at high ethanol conversion [10]. Augmented basicity provoked by La addition also resulted in improved activity in a rare-earth-modified Ni/alumina formulation tested in CO_2 methanation, that fact being originated by stronger carbon dioxide adsorption as surface carbonates that acted as so-called reactant reservoirs [11]. Higher stability of La-modified HZSM-5 zeolite catalysts in methyl mercaptan decomposition, as to that of the non-doped material, has been attributed to decreased Brønsted acidity [12].

Regarding lanthanum-containing formulations applied as HDS catalysts carriers, to our best knowledge, the information available is rather scarce [13,14]. Diminished activity in thiophene HDS of sulfided CoMo/alumina catalysts modified by La has been attributed to enhanced proportions of isolated MoO_4^{2-} refractory and less-sulfidable species adsorbed on lanthana domains [14]. From their studies on molybdenum supported on La-modified alumina, Massoth et al. [13] found that strong interaction of rare earth oxidic domains with Al_2O_3 resulted in LaO monolayer dispersion. At high La loadings, however, a second lanthanum oxide layer could be formed right over the first, the latter being partially sulfidable under 10% H_2S/H_2 treatment (400 °C, 2 h). Higher activity, as to that of non-doped alumina-supported sulfided Mo catalysts, in both thiophene conversion and hexene saturation was found for formulations of high-La content ($\geq$ 10 wt%). Different trends in those reactions, as a function of rare-earth concentration in tested formulations, were related to electronic support effects, which affected those reactions in distinctive ways. Also, a diminished proportion of disintegration reactions was observed, decreased cracking being originated by the neutralization of surface acidic sites over alumina, due to the basic agent deposition. Considering those contradictory results, the influence of basic La as a carrier additive on HDS catalyst properties deserves further investigation.

Respecting HDS catalysts' active phases (as opposed to corresponding supports), Chevrel phases ($M_xMo_6S_8$, where M could be La) have been synthesized and tested in HDS reaction schemes [15]. Different to conventional MoS_2-based hydrodesulfurization catalysts, where Mo is in 4^+ oxidation state, in Chevrel phases Mo is in either 2^+ or 2.666^+, depending on the type of second metal M. In general, HDS properties of Chevrel phases are not as good as those of MoS_2-based catalysts, which explains why studies on those species (and particularly on those where La is the second metal) are very scarce.

In this work, lanthana modified-alumina at various rare-earth contents (1, 3, and 5 wt%) is used as CoMo-based HDS catalyst support. Textural, structural, and surface properties of carriers and corresponding impregnated materials, in an oxidic phase, were characterized through various techniques. Sulfided materials were tested in dibenzothiophene (most representative organo-S species present in middle distillates) hydrodesulfurization, in batch reactor at conditions close to those used in commercial-scale hydrotreaters aimed at diesel fuel production from oil-derived feedstocks. The activity trends found were correlated to physicochemical characteristics of studied catalysts. Some advances from this investigation have been recently reported [16].

2. Results and Discussion

2.1. N_2 Physisorption

N_2 physisorption isotherms (Figure S1) of alumina and La-modified supports (see Section 3.1. *Materials Synthesis* for samples nomenclature) were intermediate between types II and IV, according to IUPAC (International union of pure and applied chemistry) classification [17], suggesting mesoporous materials. All solids had type H1 hysteresis characterizing solids with porous networks of uniform size and shape [17], where capillary condensation started at ~$P/P_s = 0.5$.

From Table 1, La addition was reflected in slightly increased (~10%) surface area (S_{gBET}) as to that of the pristine alumina carrier. More noticeable improvements were observed in both pore volume (V_p) and average pore diameter, that effect being more evident in materials at a higher rare earth content (in ALa3, 40% and 53% higher, respectively, as to sol-gel alumina). The texture of prepared binary matrices was adequate for carriers of catalysts applied in oil-derived middle distillate hydrotreatment aimed at diesel fuel production [18].

Table 1. Textural properties (as determined by low-temperature N_2 physisorption) of ALaz supports and corresponding CoMo-impregnated oxidic materials.

Sample	S_{gBET} ($m^2\ g^{-1}$)	V_p ($cm^3\ g^{-1}$)	D_p [a] (nm)	S_{gBJHa} ($m^2\ g^{-1}$)	S_{gBJHd} ($m^2\ g^{-1}$)
A	308	0.45	5.9	347	435
ALa1	334	0.55	6.6	403	485
ALa3	335	0.69	8.3	423	504
ALa5	326	0.60	7.3	419	505
CM/A	199	0.43	8.6	240	277
CM/ALa1	135	0.26	7.8	191	240
CM/ALa3	211	0.43	8.2	252	291
CM/ALa5	141	0.41	11.7	268	323

[a] $4 \times V_p \times S_{gBET}^{-1}$ [19].

Considering Co-Mo-P oxidic impregnated materials (see corresponding isotherms in Figure S2), CM/A had ~35% lower surface area as to that of the Al_2O_3 carrier, commensurate with non-porous oxidic phase loading (~32%), which suggests well-dispersed deposited phases. On the other hand, La-modified supports had significant textural losses after Co-Mo-P impregnation (~60% diminished S_g in CM/ALa1, as to that of the corresponding carrier), pointing to a partial collapse of the binary support matrix. Strandberg anions (phosphopentamolybdates), present in highly acidic impregnating solutions (see Section 3.1. *Materials synthesis*), could probably be decomposed under basic conditions in the interior of pores of rare earth modified materials (as indeed reported in Section 2.6. *Raman spectroscopy*), releasing PO_4^{3-} anions. Due to their strong affinity, lanthanum atoms in the solid matrices could strongly interact with those phosphate anions [20,21] being partially leached from the carriers, thus provoking the mentioned partial porous network destruction. Additionally, it has been reported [14] that, when impregnated through originally acidic solutions (as in our case), Mo could partially extract lanthanum atoms deposited over an alumina matrix, forming defined La molybdate domains. It seemed that particle growth could cause an enhanced average pore size (Table 1, as determined by Gurvich´s law [19]) of impregnated Co-Mo-P materials over La-doped carriers. Others [22] have also found that, during HDS catalysts preparation, impregnating conditions could be crucial in preserving textural properties of corresponding supports, avoiding then their collapse. Expectably, that fact could be reflected in improved catalytic activity.

Pore size distributions (PSD) were determined through Barrett-Joyner-Halenda (BJH) methodology. In order to determine which PSD profile, obtained from an either adsorption or desorption data branch, better fits the actual ones, corresponding cumulative surface area values were compared to those obtained by the BET (Brunauer-Emett-Teller method, Table 1). Values from adsorption branch data (S_{gBJHa}) were closest to the BET ones as to those obtained from desorption data (S_{gBJHd}). Thus, PSD profiles from the former data were deemed the most suitable in describing the actual porous networks. PSD maxima of La-containing sol-gel carriers shifted to higher diameters as to that of non-modified alumina (~6.6 nm), ALa3 showing the maximum average value (~8.5 nm) (Figure S3a). Considering Co-Mo-P impregnated samples (Figure S3b), the maxima of CM/A and CM/ALa5 profiles shifted to larger diameters, as to those of corresponding supports (from 6.6 to 8.6 nm and from 6.4 to 8.1 nm, respectively). The amount of pores with diameters ≤ 5 nm, in the PSD of all carriers, strongly diminished after Co-Mo-P deposition.

2.2. *Thermal Analyses (Thermogravimetric and Differential Thermal Analyses, TG and DTA)*

Thermogravimetric (TG) and differential thermal analyses (DTA) profiles of various carriers at different La contents were similar, with just small differences (not shown) in either weight losses (TG) or signal intensity (DTA). For example, corresponding plots for ALa5 are included in Figure 1. From room temperature to 550 °C, weight losses at a more or less constant rate were observed. The first part of the TGA curve (until approximately 140 °C) could be related to the evaporation of physisorbed water and alcohol from the sol-gel technique used during materials synthesis [23] (see Section 3.1. *Materials synthesis*). Further losses until 550 °C could be originated by the elimination of both nitrates (from La salt used) and organic remains from the Al alkoxide utilized. The endothermal signal centered at ~140 °C in the ALa5 DTA profile corresponded to aforementioned evaporation of water, alcohol, and structural hydroxyls. Meanwhile, the exothermal peak, at ~220 °C, could be provoked by NO_3^- anion decomposition [24]. The intensity of this signal increased with La content in mixed formulations. The strong exothermal peak, centered at 277 °C, could be provoked by organic residua (from Al alkoxide) combustion [23]. Due to crystallization phenomena (to any Al_2O_3, LaO or La_2O_3 defined phases), no further signals were observed at more severe treatment conditions. In full agreement with our findings, it has been reported [25] that lanthanum addition during sol-gel alumina synthesis contributes to obtaining highly stable binary matrices delaying crystallization to the gamma phase, with defined La phases detected just after high-temperature (1000 °C) annealing.

2.3. *X-ray Difracction (XRD)*

Diffractograms of various supports at various compositions (not shown) showed no crystalline phases. The amorphousness of carriers suggested homogeneous materials, where La was both well-integrated and well-dispersed in alumina matrices. In Figure 2, diffractograms of the oxide at the highest La content and the corresponding Co-Mo-P impregnated material are shown. Both patterns did not show any diffraction peak, pointing to either well-dispersed deposited molybdates on the support surface or the presence of disordered amorphous Mo domains, undetectable by the used technique [26].

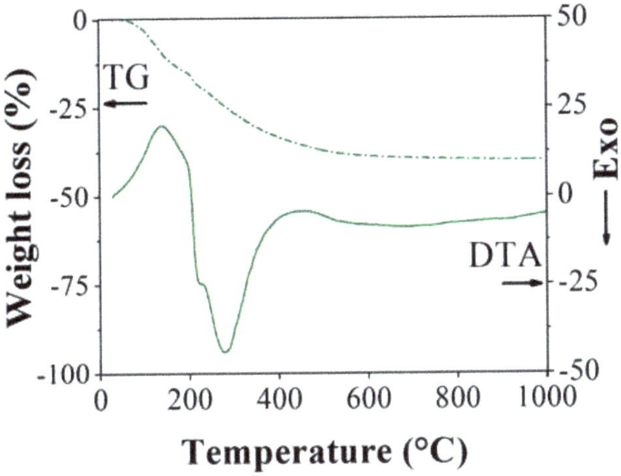

Figure 1. Thermal analysis (TG and DTA) profiles of the ALa5 dried carrier.

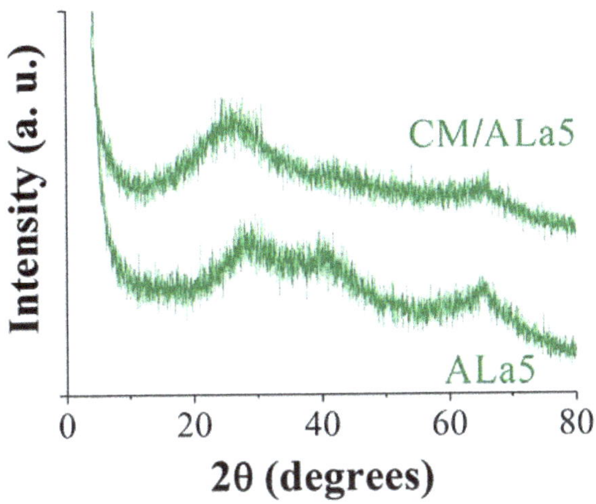

Figure 2. X-ray diffraction patterns of ALa5 support and corresponding P-doped CoMo impregnated oxidic material.

2.4. Chemical Analysis (Scanning Electron Microscopy-Energy Dispersive Spectroscopy (SEM-EDS)

The chemical composition of the supports and the various impregnated materials prepared were determined from EDS (energy dispersive spectroscopy, equipment attached to a SEM, scanning electron microscope, with back-scattered electrons detector). Corresponding micrographs of La-modified carriers and Co-Mo-P impregnated materials are included in Figures S4 and S5, respectively.

No important differences in sample morphology were observed at this stage, all solids being constituted by faceted particles. Regarding Co-Mo-P impregnated solids, morphology of corresponding carriers is preserved (note higher mixrograph magnification in former samples). In Figure 3, representative EDS profiles of samples ALa5 and CM/ALa5 are shown. From Table 2, actual La loading (wt%) in sol-gel carriers nicely corresponded to nominal values, suggesting homogeneous binary matrices.

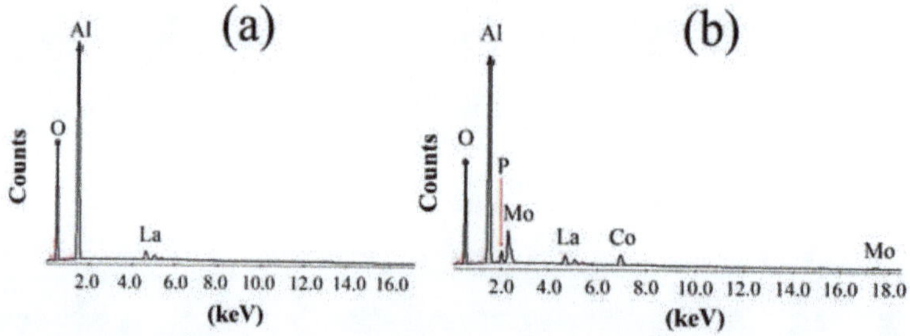

Figure 3. EDS profiles of ALa5 support and corresponding oxidic P-doped CoMo impregnated material. (**a**) ALa5; (**b**) CM/ALa5.

Table 2. Chemical analysis (by SEM-EDS) of La-modified sol-gel alumina supports used.

ALa1	Wt%	At%
O_K	49.82	63.14
Al_K	48.78	36.66
La_L	1.4	0.2
Total	100	100
ALa3		
O_K	48.55	62.75
Al_K	47.91	36.72
La_L	3.54	0.53
Total	100	100
ALa5		
O_K	48.12	63.15
Al_K	46.27	36.01
La_L	5.6	0.85
Total	100	100

Regarding Co-Mo-P impregnated solids, phosphorus content in CM/A and CM/ALa1 was very similar (Table S1), being slightly lower in the rest of samples. Additionally, Co and Mo contents in samples were alike in all impregnated samples.

2.5. Surface Bascity (CO_2-Adsorption Studied In IR Region, CO_2-FTIR)

The surface basic properties of the studied materials were estimated by integrating the area of peaks registered in the mid-infrared region, between 1800–1200 cm^{-1}, observed after CO_2 room-temperature adsorption [27], followed by thermo-desorption at various temperatures (Table 3). CO_2 has been applied as a probe in determining the density and strength of surface basic sites of La-containing materials [28,29]. Different surface species could be originated by carbon dioxide adsorption, namely, unidentate and bidentate carbonates and bicarbonates [30] (Scheme S1). Carbonates, related to CO_2 adsorbed on surface sites with intermediate and high basic strengths, respectively, requires of surface basic oxygen atoms [31]. Unidentate carbonates, characterized by symmetric and asymmetric O-C-O stretching at 1360–1400 cm^{-1} and 1510–1560 cm^{-1}, respectively, could be originated on isolated low-coordination surface O^{2-} anions in corners or edges. On the other hand, bidentate carbonates that could be formed on Lewis acid-base pairs (M-O^{2-} pair site, where M is La or Al cation, Scheme S1) are characterized by absorptions due to symmetric and asymmetric O-C-O stretching at 1320–1340 cm^{-1} and 1610–1630 cm^{-1}, respectively [31]. Finally, bicarbonates involve CO_2 adsorption on low-strength basic surface hydroxyls [31], showing a C-OH bending mode at 1220 cm^{-1} and symmetric and asymmetric O-C-O stretching vibrations at 1480 cm^{-1} and 1650 cm^{-1}, respectively (Scheme S1). We arbitrarily classified the strength of various types of sites according to the temperature at which they could retain CO_2 (T_d as weak (200 °C < T_d), medium (200 °C > T_d < 400 °C) and strong (T_d > 400 °C).

Table 3. Relative surface basicity (CO_2-FTIR, integrated area of peaks in the 1200–1800 cm^{-1} region) of supports used and corresponding Co-Mo-P oxidic impregnated materials. T_d: Desorption temperature. ND: Non-determined.

T_d (°C)	A	ALa1	ALa3	ALa5	CM/A	CM/ALa1	CM/ALa3	CM/ALa5
25	71	79	96	108	36	30	42	41
100	21	32	54	38	27	16	34	31
200	15	26	44	29	22	7	24	24
300	14	26	42	29	19	6	22	21
400	12	26	41	27	15	5	20	21
500	11	17	33	25	0	ND	13	12

The number of surface basic sites of alumina augmented, by La addition, being that especially evident in samples of higher rare-earth concentration (see, for instance, values from room temperature spectra of ALa3 and ALa5, Table 3). Cui et al. [13] found a linear relationship between the amount of adsorbed CO_2 and La content (until ~20 wt%) in sulfided Mo/alumina samples prepared by rare-earth nitrate impregnation, although non-doped Mo/Al_2O_3 had some surface basicity. In addition, the strength of sites as a function of CO_2, retained after a progressively increased desorption temperature, was clearly enhanced in aforementioned solids. Most of the sites on sol-gel alumina, ALa1 and ALa5, were weak, mainly forming bicarbonates. In the opposite, ALa3 contained the strongest surface basic sites (those related to unidentate carbonates formation) among the studied materials, preserving around the third part of the total number of sites retaining CO_2 at room temperature and after treatment at 500 °C. Additionally, the relative amount of strong sites was three-fold to those over pristine alumina. In full agreement, it has been reported [32] that, on La-modified Al_2O_3 at contents close to monolayer (~5.12 at. La^{3+} nm^{-2} [33]) population of strong sites retaining CO_2, was significantly enhanced.

For Co-Mo-P impregnated samples, the amount of adsorbed CO_2 was clearly diminished as to those of corresponding carriers, with the sole exception of the alumina-supported solid. In this case, significantly decreased basicity (~50%) was registered just in the case of weak sites, which could be related to the well-known strong interaction between molybdate anions and the most basic hydroxyl groups on Al_2O_3 surface [34] (those related to bicarbonates formation during CO_2 adsorption experiments). Additionally, CM/ALa1 lost around 62% of sites adsorbing CO_2 at room temperature. The observed trend was even more marked regarding strong sites (~80% loss). Those facts could probably be related to strong deposited phases-La interaction. In the opposite, CM/ALa5 retained between 72%–83% of medium strength sites (mainly related to bidentate carbonates) of those on corresponding support. Due to strong Co [35], Mo [14] and P [20,21] interaction with La, it is probable that after impregnation the rare-earth could be lixiviated, to some extent, from the sol-gel matrix, with that fact explaining the partial textural collapse (Table 1). According to Cui et al. [13], during impregnation, Mo could preferably interact with La domains then with the alumina carrier. In addition, Ledford et al. [35] found that, in materials of La/Al $\geq$ 0.0075 atomic ratio (as in the case of our carriers with 3 and 5 wt% La), a Co-La phase could be formed that, being reflected in, lessened cobalt reducibility. The effects of lanthanum extraction could be more significant in the sample of the lowest rare earth content (CM/ALa1), rationalizing the important surface basicity lost. Conversely, the Co-Mo-P impregnated samples of enhanced La concentration had around 33% more sites retaining CO_2 at 400 °C as to the alumina-supported solid, pointing out the retention of mid-strength basic sites.

2.6. Raman Spectroscopy

Al_2O_3 does not show any Raman band due to the low polarizability of light atoms and the ionic character of Al-O bonds [36]. Characteristic peaks of lanthana at 104, 191 [37], 310, 350, and 415 cm^{-1} [38] were not evidenced, pointing to the absence of La_2O_3 definite domains in our oxidic carriers. Raman signals at 225, 359, 911 (shoulder), and 949 cm^{-1} were observed in the CM/A spectrum

(Figure 4a). The former two signals could be assigned to Mo-O-Mo and O-Mo-O vibrations [39]. Meanwhile, those at higher wavenumbers were related to Mo=O bands. They were ascribed to Anderson species $Al(OH)_6Mo_6O_{18}^{3-}$ [40], whose presence suggested partial support dissolution during contact with the acidic Co-Mo-P impregnating solution (pH ~2.5). Considering the Mo/P ratio in those solutions (~2.0) and their pH, $H_2P_2Mo_5O_{23}^{4-}$ anions (two PO_4 tetrahedrons surrounded by five interconnected octahedral MoO_6 species) could, very probably, be preferentially present [40,41]. Although, due to their stability, those Strandberg heteropolyanions ($H_xP_2Mo_5O_{23}^{(6-x)-}$) [42] could exist in various protonation states over a wide pH range [41], they could be decomposed to phosphates and molybdates owing to their strong interaction with both basic hydroxyls and Al^{3+} coordinatively unsaturated (CUS) Lewis acid sites on alumina surfaces [43].

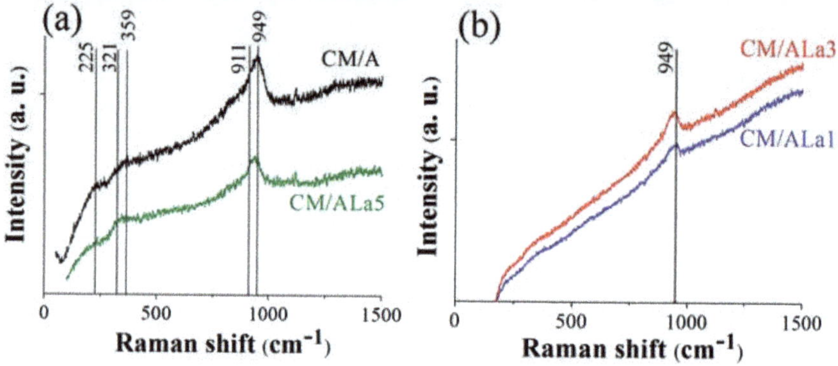

Figure 4. Raman spectra of oxidic Co-Mo-P impregnated materials. (**a**) CM/A and CM/ALa5; (**b**) CM/ALa1 and CM/ALa3.

In spite of some fluorescence interference that decreased sensitivity of the technique used, CM/ALa1 and CM/ALa3, Raman spectra (Figure 4b) evidenced that the signal related to M=O stretching vibrations progressively shifted to a lower wavelength as the rare earth concentration increased, being registered at 935 cm^{-1} for CM/ALa5 (as compared to 949 cm^{-1} observed in CM/A spectrum, Figure 4a). That fact, accompanied by the apparition of a signal at 321 cm^{-1} (more evident in CM/ALa5), strongly suggested the co-existence of both octahedral and tetrahedral Mo species in La-containing samples. In full agreement, Payen et al. [14] reported that the Mo/ La ratio determined the type of oxomolybdenum entities adsorbed on lanthanum-modified alumina. Monomeric tetrahedral species (MoO_4^{2-}) could be preferably formed by an interaction with La-OH groups, whereas polymeric octahedral molybdates could be deposited on Al_2O_3 domains. For materials at a high La concentration (Mo/La~1), just tetrahedral molybdates were observed [14]. In the same line, tetrahedral Mo species were mainly found [44] on basic magnesium-modified sol-gel alumina supports. It seemed that the high point of zero charge values of carriers with basic properties determined that behavior, as in contact with water those solids could produce high pH impregnating conditions where MoO_4^{2-} species are favored. Additionally, coordination state modifications of Mo^{6+} species, from octahedral to tetrahedral (as determined by corresponding Raman shifts), by basic additive doping of alumina-supported unpromoted Mo (Mg, [5]) and CoMo formulations have been reported by others (K [45], Mg-Li [46]).

Definite domains (although not large enough to be detectable by X-ray diffraction) of lanthanum molybdate have also been registered [14] by impregnating La-modified alumina with acidic solutions prepared from ammonium heptamolybdate. In our case, contributions of characteristic bands at 915 and 940 cm^{-1} could not be discarded, although, due to the low intensity of the registered peaks, they could remain masked. If existent, $La_2(MoO_4)_3$-like domains could also be responsible for tetrahedral molybdenum presence [47]. According to the peaks' intensity in spectra, in Figure 4a,b, it seemed that samples CM/ALa1 and CM/ALa5 were the ones with the highest Mo dispersion (although with a high proportion of tetrahedral species). Small signals, at 1119 cm^{-1}, could be related to C-C bond stretching vibrations [48], probably originated by organic remains from either aluminum alkoxide used during carrier synthesis or due to acetate utilized during Co impregnation.

2.7. UV-Vis Spectroscopy

Being an insulator alumina does not absorb in ultraviolet-visible (UV-vis) region [49]. Absorptions in the 267–306 nm range (more notable for ALa1, Figure 5) could be related to charge transfer from O^{2-} to framework La^{3+} ions [50]. Regarding Co-Mo-P impregnated materials, the $O^{-2} \rightarrow Mo^{6+}$ charge transfer transition bands of tetrahedral ($Mo_{(Th)}{}^{6+}$) and octahedral ($Mo_{(Oh)}{}^{6+}$) molybdenum species were centered at ~280 and ~314 nm, respectively [51] (Figure 6). That fact suggests that, although basic pH conditions that prevailed during Co-Mo-P impregnation of rare-earth-containing carriers strongly contributed to formation of isolated monomeric $Mo_{(Th)}{}^{6+}$ species (Figure 4a,b), $Mo_{(Oh)}{}^{6+}$ species co-existed with them, as already suggested by our Raman spectroscopy studies (Section 2.6). As proposed by Payen et al. [14], $Mo_{(Th)}{}^{6+}$ species, related to molybdenum in strong interaction with the support, could be related to La domains, whereas $Mo_{(Oh)}{}^{6+}$ ones could be deposited on alumina surfaces. As the former species were absent in the impregnating solutions (containing octahedral Mo^{6+} heteropolyanions) their presence strongly pointed to the decomposition of the Strandberg species through their interaction with basic OH groups and CUS (Lewis sites) on the Al_2O_3 carrier surface [43]. As aforementioned, basic conditions prevalent in the pores of La-modified carriers could also play a major role in tetrahedral Mo^{6+} species formation. The bathochromic shift of the $Mo_{(Oh)}{}^{6+}$ low-energy absorption edge strongly suggested an augmented oxomolybdates polymerization degree [52] in La-containing formulations as to that in CM/A following the order: CM/ALa1 > CM/ALa3 > CM/ALa5 > CM/A. The diminished surface area of lanthanum-modified CM materials (Table 1) could contribute to lowered octahedral Mo species dispersion. Indeed, a shoulder at ~370 nm, whose intensity increased according to the order CM/ALa1 > CM/ALa3 > CM/ALa5, pointed to the formation of MoO_3 domains [53].

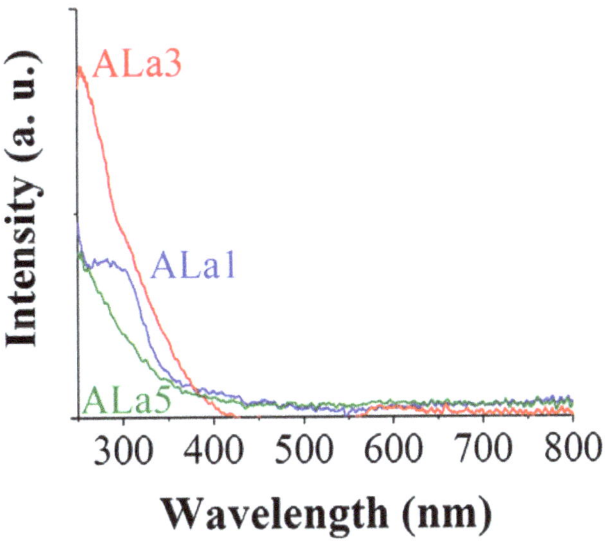

Figure 5. UV-vis spectra of La-modified alumina supports.

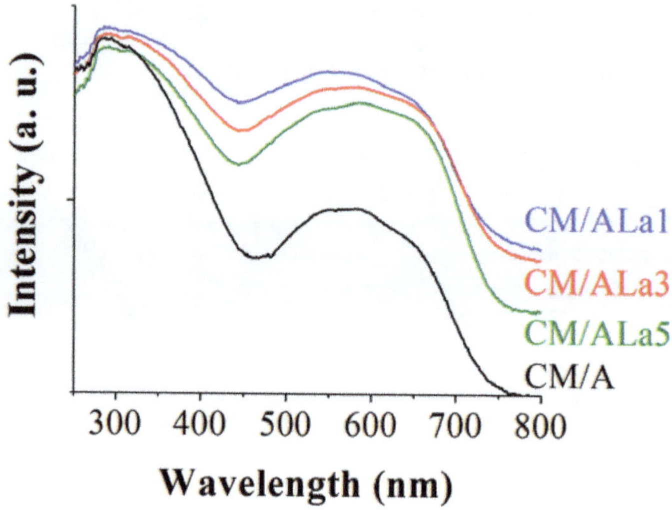

Figure 6. UV-vis spectra of Co-Mo-P impregnated oxidic formulations on La-modified Al_2O_3. Alumina-supported material included as reference.

The triplet band related to $^4A_2(F) \rightarrow {}^4T_1(P)$ transitions of tetrahedral cobalt $Co_{(Td)}{}^{2+}$ at 540, 580, and 625 nm [51] was augmented in rare earth-modified solids, but with no definite trend regarding La concentration. Those cobalt species could be originated in lanthanum cobaltate-like entities. Interestingly, the enhanced proportion of tetrahedral Co^{2+} in alumina-supported CoMo oxidic formulations has been observed in the past by progressively increasing the basic additives concentration (Mg and Li) [46], with augmented $CoAl_2O_4$ formation being invoked in that case. For La-containing samples, shoulders at ~495 and ~563 nm could be related to octahedral Co^{2+} species cobalt ($Co_{(Oh)}{}^{2+}$) [54]. These signals progressively increased in the order CM/ALa1 > CM/ALa3 > CM/ALa5, being essentially absent in the alumina-supported formulation. Additionally, a small shoulder at 518 nm (probably from $^4T_{1g}(F) \rightarrow {}^4T_1g(P)$ electronic transition) suggested $Co_{(Oh)}{}^{2+}$ entities [55] in the CM/ALa1 spectrum.

2.8. HDS Reaction Test

La content in supports of sulfided Co-Mo-P (see 3.3. HDS Reaction Test) catalysts clearly affected their dibenzothiophene HDS activity (Figure 7). The formulation with the lowest rare-earth content had the highest pseudo first order kinetic constant value (~38% enhanced as to that of the non-doped Al_2O_3-supported material). However, significant diminution in HDS properties was observed as lanthanum concentration augmented in tested materials. Payen et al. [14] reported progressive diminution in thiophene HDS activity as La content in alumina- supported CoMo formulations increased. Very probably, enhanced proportions of refractory tetrahedral $MoO_4{}^{2-}$ (see Section 2.6. *Raman spectroscopy*), as rare earth content in formulations augmented, could be primarily responsible for that behavior. Those species are characterized by being hardly sulfidable under the used conditions [14], then precluding MoS_2 phase formation.

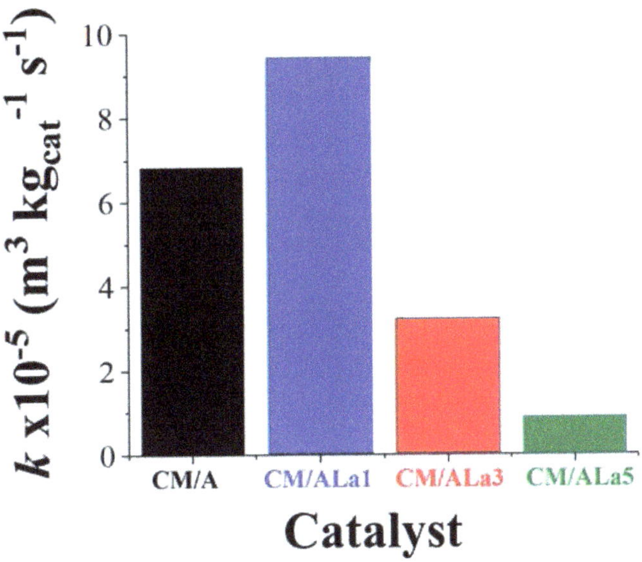

Figure 7. Pseudo first order kinetic constant (DBT HDS) of various tested catalysts. Batch reactor, *n*-hexadecane as solvent, P = 5.67 MPa, T = 320 °C, 107 rad s^{-1} (1030 rpm) mixing speed.

The increased amount of MoO_4^{2-} species by the addition of basic agents in alumina-supported HDS catalysts has been reported in the past [14,56]. For instance, Malet et al. [56] found augmented proportions of hardly-reducible tetrahedral molybdenum entities in Mo/TiO_2 solids doped with Na (through NaOH impregnation) at various concentrations. The decreased reducibility of tetrahedral Mo^{6+} species on Al_2O_3-supported oxidic Mo and CoMo formulations modified by basic agents (Mg and Li) was confirmed by temperature-programmed reduction studies [5,46]. Regarding sulfided alumina-supported CoMo formulations modified by K addition (2.7 wt%), Maugé et al. [4] observed (by FTIR) decreased intensity in bands related to adsorbed CO as to that over the counterpart with a non-doped Al_2O_3 carrier. That could mean a lower proportion of sulfided phases, although the authors mainly attributed that to adsorption sites poisoning provoked by augmented sulfided phase electron density that could be reflected in diminished CO adsorption strength, under tested conditions (−173 °C). Similar to our case, those authors identified enhanced surface basicity of K-doped solids by formation of hydrogen carbonates and bidentate carbonates (weak and medium strength sites, respectively [30]), as studied by CO_2 adsorption followed by FTIR). Decreased conversion in the 2-methylthiophene in *n*-heptane was observed in the case of K-doped CoMo/alumina catalysts when tested in a model FCC (fluid catalytic cracking) naphtha mixture also containing 2,3-dimethylbut-2-ene and orthoxylene. Regarding unpromoted sulfided alumina-supported Mo catalysts, Cui et al. [13] observed unaltered activity in thiophene HDS for materials at low La content (2 wt%), which was followed by decreased desulfurating properties in the solid modified with 5 wt% rare earth loading. Regarding the effect of basic additives on dibenzothiophene (DBT) HDS activity of alumina-supported materials, strongly decreased activity was observed [44] in the case of NiMo formulations with a MgO-modified (5 mol%) Al_2O_3 carrier. As the HDS activity of sulfided Mo has often been related to surface acidity [57], it seems reasonable to assume that diminished Brønsted or Lewis ones (or both) in formulations doped with basic additives could provoke limited organo-S species conversion.

The high activity in DBT HDS of our sulfided alumina-supported Co-Mo-P formulation modified with lanthanum at the lowest content could result from a combination of factors. Our UV-vis results (Figure 6) suggested that, in CM/ALa1, extant octahedral molybdenum species were at lower interaction with the support. Although, according to Raman characterization (Figure 4), amount of refractory tetrahedral Mo entities augmented in that formulation, as to those over the non-doped material with an Al_2O_3 carrier, the proportion of those monomeric species seemed to be not large enough to significantly affect proportion of sulfidable phases. In addition, the increased proportion of octahedral Co^{2+} (as also observed by UV-vis, Figure 6) in the corresponding oxidic solid could be beneficial as that species is considered a precursor of cobalt that could efficiently form the highly active "CoMoS" phase during catalyst activation by sulfiding (see Section 3.3. *HDS Reaction Test*) [58]. A linear relationship between $Co_{(Oh)}{}^{2+}$ concentration in oxidic materials and HDS activity, in both gas-phase thiophene conversion and liquid phase real feedstock (vacuum gas oil and deasphalted oil) desulfurization, was found in that case. Interestingly, improved DBT HDS activity (as to that of corresponding non-modified catalysts) was observed in CoMo/alumina modified by basic agent addition, but just at a low doping agent content (1.2 wt% Mg or that solid with additional 4 wt% potassium loading) [59]. However, and similarly to our case, HDS activity significantly diminished by augmenting basic species concentration in corresponding sulfided catalysts. Additionally, Cao et al. [60] reported decreased DBT HDS activity on magnesium modified (20 wt% MgO) CoMo/Al_2O_3 materials.

Very interestingly, it has been reported [61] that activity in the SO_2 conversion to sulfur, using coal gas as a reductant, of La-modified sulfided (at similar conditions to those used in the present work, Section 3.3. *HDS Reaction Test*) Co-Cu/γ-Al_2O_3 catalysts was significantly improved over formulations at low rare earth content, as to that of the reference non-doped formulation. Activity was augmented by adding 0.5 wt% rare earth, the maximum being found at 1 wt% La. However, diminished SO_2 conversion was observed by augmenting dopant concentration (4 wt%). The authors attributed that to excessive La_2O_3 loading that could accumulate around crystals of active components, generating an undesired phase change by embedding corresponding particles. Those facts clearly show that La-modified sulfided catalysts at low rare-earth contents deserve further studies, not just in HDS reaction schemes. It is also worth mentioning that recently [62] we reported on the increased activity in naphthalene conversion of Pt (1 wt%) supported on alumina modified with a basic additive (Mg in this case) at low content, as compared to the non-doped formulation with the Al_2O_3 carrier. However, and similarly to what found during the present investigation, activity diminished in materials with a higher magnesium content (8 wt%).

Regarding selectivity to various products, DBT conversion could be carried out through direct desulfurization (DDS, to biphenyl, BP) and hydrogenation (HYD, firstly to hydrodibenzothiophenes, HDBT's) [63] reaction pathways (Scheme S2). At our reaction conditions (see Section 3.3. *HDS Reaction Test*), bicyclohexyl (BCH) saturation, from both aromatic rings, was observed just in trace amounts. Taking into account that BP hydrogenation to cyclohexylbenzene (CHB) could be strongly inhibited by DBT competitive adsorption under our HDS conditions, all CHB produced must have predominantly come from the HYD route through the sulfur removal of partially saturated HDBTs [64]. Selectivity to BP was clearly enhanced with La content in sulfided formulations at isoconversion (x ~17%) (Figure 8). Previous studies in HDS of DBT [65,66] and benzothiophene (BT) [67] strongly evidenced that an augmented DDS/HYD ratio could be related to more efficient MoS_2 promotion by proper Co integration, then enhanced "CoMoS" phase formation. In this direction, Kaluža et al. [67] proposed that the amount of dihydrobenzothiophene (DHBT, formed from BT aromatic ring partial saturation) that could be then subsequently eliminated by C-S hydrogenolysis could be a measure of relative hydrogenation/hydrogenolysis selectivity. During their studies on CoMo sulfided formulations with various carriers, it was found that a very high promotion degree (17.9, over MgO-supported catalyst) was accompanied by a significantly decreased DHBT formation.

In the case of CoMo materials supported on alumina modified by basic agents, significantly augmented DDS/HYD selectivity in DBT HDS has been reported [46] in the past. For instance, by adding magnesia (0.05 mol ratio) to an Al_2O_3 carrier CoMo catalysts producing strongly increased (100% enhancement at 30% conversion) DDS/HYD ratio were obtained. Even more, selectivity to biphenyl was further progressively augmented by additional lithium doping of the aforementioned formulation. Others [44] have also found decreased DBT transformation through the HYD route over sulfided NiMo catalysts supported on alumina modified by 5 mol% magnesia.

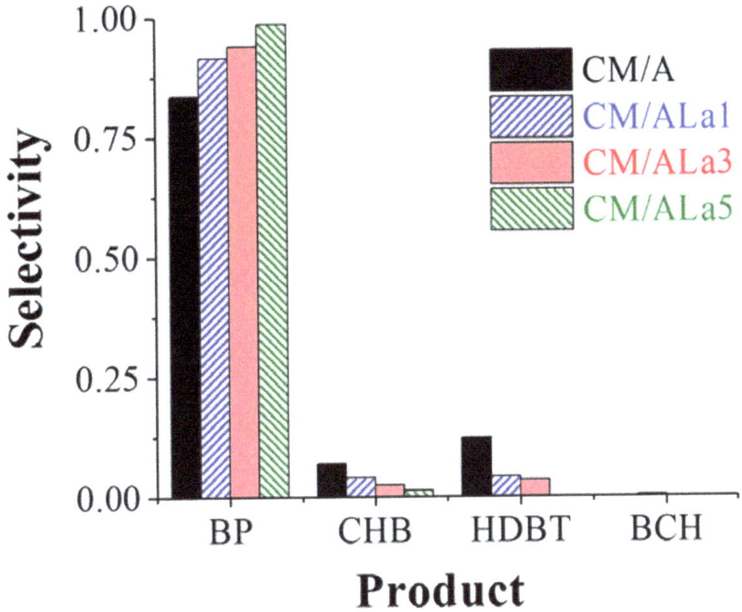

Figure 8. Selectivity to various products (DBT HDS, ~17% conversion), of various tested catalysts. HDBT: hydrodibenzothiophenes; BP: biphenyl; CHB: cyclohexylbenzene; BCH: bicyclohexyl. Batch reactor, *n*-hexadecane as solvent, P = 5.67 MPa, T = 320 °C, 107 rad s^{-1} (1030 rpm) mixing speed.

Following this line of reasoning, the enhanced DDS/HYD ratio, in our case, could be originated by well-promoted (by Co) MoS_2 slab edges, whose number could be restricted due to limited Mo^{6+} sulfidation provoked by extant tetrahedral species. That could justify low DBT HDS activity of formulations of higher La content where cobalt could yet be properly integrated to molybdenum sulfide edges. In those catalysts, however, a certain proportion of the sulfided promoter could be as partially segregated Co_xS_y domains (of marginal HDS activity [66]) due to the reduced number of MoS_2 layer edges, where they could be properly incorporated to produce the highly active "CoMoS phase".

Well-dispersed molybdenum sulfide phases (thus composed of short slabs) could be very efficient in aromatic ring hydrogenation [68]. HYD sites have been related [46] to corner sites in Co-MoS_2 slabs, whose proportion decreases as slab length grows. Those longer molybdenum sulfide layers could be originated by oxidic molybdates of lower dispersion, similarly to that proposed by Halachev et al. [69] in the case of alumina-supported P-modified NiW materials where the ratio W=O/W-O-W, considered as a measure of dispersion of oxidic tungsten species, was related to naphthalene hydrogenation activity of corresponding sulfided formulations. Solids with a higher proportion of terminal M=O bonds resulted (after sulfiding) in catalysts with improved dispersion, and, thus, with enhanced saturation properties. In the opposite, materials with diminished dispersion resulted in decreased hydrogenating capabilities by generating longer MoS_2 with an increased edge sites/corner sites ratio. In this regard, it should be mentioned that, in our case, La-addition resulted in materials with lower oxidic Mo^{6+} phases dispersion (Figure 6). Interestingly, sulfided CoMo/alumina catalysts with limited dispersion (longer slabs and augmented stacking) have also shown [70] enhanced HDS of thiophenic species, accompanied by decreased olefins hydrogenation in the selective hydrotreating of FCC naphtha focused on lowering octane number losses.

Bataille et al. [65] proposed that the main role of the promoter in DBT HDS is to enhance the reaction rate through the DDS pathway, or the C-S bond breakage activity in general. The involved mechanism could consist of attacking a hydrogen atom, in a β position relative to the S atom in a DBT molecule, by a basic sulfur anion. Primarily, both reaction routes (DDS and HYD) require a dihydrodibenzohiophene intermediate [65]. Cleaving the C-S bond to produce two phenyl rings species (BP) could be carried out by hydrogenating one of the double bonds (in any of the aromatic rings) in the sulfur atom vicinity, resulting in the dihydrogenated DBT. Then, the C-S bond in the partially saturated ring could be opened by an elimination process. The second C-S bond cleavage, leading to BP, could possibly occur through a similar mechanism [65].

As the promoter (either Ni or Co) increases electron density of basic S anions, they could favor C-S bonds cleavage. The possibility of electron-rich alumina-lanthana supports contributing to increased sulfur anion basicity (then enhancing the DDS reaction pathway), as recently reported in the case of K-doped MoS_2 phases applied in methanethiol synthesis from a H_2S-CH_4 mixture [6], remains a question that deserves to be answered. Mechanistic studies on DBT HDS over La-modified alumina supports are clearly needed, as information on reaction routes and kinetic parameters could provide valuable parameters that could be useful in investigations focused on applying that kind of catalyst under more realistic conditions in the presence of oil-derived middle distillates [71].

Regarding CoMo alumina formulations modified by La at various loadings, Payen et al. [14] found a similar promotion degree in the thiophene HDS as to that observed over a non-modified solid with an Al_2O_3 carrier. Different trends, as to those reported by others, for corresponding catalysts doped with basic agents (in BT HDS, for instance [67]) could be originated in distinctive rate-limiting steps in tested reactions. Unfortunately, no data on selectivity through DDS and HYD reaction pathways were provided in that case [14].

From the results of the present investigation we consider that sulfided CoMo formulations, supported on alumina doped with La at low concentration (up to 1 wt%), clearly deserve deeper studies. These ongoing investigations will be the subject of further reports.

Finally, the development of HDS catalysts with enhanced activity and increased DDS/HYD selectivity ratio (as our CM/ALa1 formulation) results is promising considering their limited consumption of expensive hydrogen, mainly in the case of H_2-constrained refineries.

3. Experimental

3.1. Materials Synthesis

Al_2O_3-La_2O_3 mixed oxides at three compositions (1, 3, and 5 wt%, ALa1, ALa3, and ALa5, respectively) were prepared by a sol-gel method. First, aluminum alkoxide was dissolved at 80 °C in corresponding alcohol (ROH/alkoxide = 60), under stirring and refluxing being kept during 2 h. Then, water and HNO_3 (3 N) hydrolysis catalysts [23] were added (H_2O/alkoxide= 1, HNO_3/alkoxide = 0.03, respectively), the mixture being kept under stirring and refluxing during approximately 1 more h until sol formation. Pertinent amounts of $La(NO_3)_3{\cdot}6H_2O$ in an alcoholic solution were added to the prepared alumina sol. Additional amounts of distillated water were supplied until total aluminum alkoxide hydrolysis. The mixtures were kept under stirring and reflux during 1 more h at room temperature until gelification. The obtained gels were aged, dried (~100 °C), and further calcined (550 °C, 4 h).

Over annealed mixed oxides Mo and Co were deposited (at 2.8 atoms Mo nm^{-2} and (Co/(Co+Mo)) = 0.3, respectively). Pore-filling simultaneous impregnation was carried out over previously dried supports (120 °C, 2 h) with an acidic solution (pH ~2.5) prepared from digestion (at ~80 °C, in water and under vigorous stirring, 4 h) of MoO_3 99.5 wt% (PQM) in the presence of H_3PO_4 85.3 wt% (Tecsiquim, Mexico City, Mexico). H_2O in excess (typically 250 ml of starting solution to impregnate 5 g of carrier) was used to accelerate molybdenum salt dissolution and digestion. A yellow transparent solution was obtained after hydrolysis. $(CH_3COO)_2Ni{\cdot}4H_2O$ (Sigma-Aldrich, Darmstadt, Germany) was then added, with the stirring and processing temperature being maintained for 2 h. A P_2O_5/(NiO+MoO_3) = 0.2 mass ratio was fulfilled [72]. A transparent dark wine solution was thus finally obtained. The solution volume was then reduced by evaporation until reaching the suitable one for pore-filling impregnation of the given mass of support. The described one-pot impregnation method was chosen as it constitutes a readily scalable methodology during catalyst preparation at the commercial scale. After impregnation materials were left aging overnight at room temperature for the diffusion of the impregnated species into the carrier porous network. Then, the materials were dried (2 h at 120 °C) and further calcined (400 °C for 5 h). Impregnated solids were identified by using the CM/ALaz key, where "z" represents La loading in corresponding supports.

3.2. Materials Characterization

Textural properties of prepared materials were determined by N_2 physisorption (at −198 °C, nitrogen saturation temperature at Mexico City barometric pressure) in a Micromeritics (Norcross, GA, USA) ASAP 2000 apparatus after ultra-high vacuum (133.32×10^{-5} Pa) degassing at 300 °C for 2 h to eliminate adsorbed molecules. The surface area (BET, Brunauer-Emett-Teller method), pore volume, and average pore size of the materials (BJH, Barrett-Joyner-Halenda method) were determined from the corresponding isotherms.

Thermal analyses (thermogravimetric and differential thermal analysis (TG and DTA, respectively) of dried (non-calcined) samples were carried out with a Netzsch Thermische Analyze (Burlington, MA, USA) STA 409 EP apparatus, under a static air atmosphere. The crystallographic order of the studied samples was determined by powder X-ray diffraction (XRD, Siemens (Munich, Germany) D-500 Kristalloflex, copper anode, $Cu_{K\alpha}$ radiation, λ = 0.15406 nm, 35 kV, 25 mA), in the 5–80° 2θ range.

Scanning electron microscopy pictures and chemical analysis by energy dispersive spectroscopy (EDS) of the prepared samples were performed in an environmental scanning electron microscope XL30 with an attached energy dispersive X-ray spectroscope (EDAX, Berwyn, PA, USA).

The surface basicity of alumina, alumina-lanthana supports, and the corresponding impregnated materials were characterized by CO_2 adsorption studied in the mid-infrared region. As an acid probe, carbon dioxide interacts with basic catalysts' surfaces, forming bidentate carbonates, unidentate carbonates, and bicarbonates (Scheme S1, [30]), depending on surface basic site strength. Previously, to corresponding studies, all materials (carefully weighed) were submitted to thermal treatments (500 °C), under an inert atmosphere, to eliminate adsorbed species. Samples were cooled down to room temperature with CO_2 being then fed to the system (at room temperature). Once samples were saturated, the flow of probe gas was suspended, with excess CO_2 being flushed by an inert gas stream. The corresponding spectra were acquired by using a Nicolet (Waltham, MA, USA) Magna 560 FTIR spectrometer at 4 cm^{-1} resolution (50 scans) in diffuse reflectance mode. The analyzed samples were treated at various progressively higher temperatures (25–500 °C) and detected peaks in the 1800–1200 cm^{-1} wavenumbers infrared range (corresponding to formed surface bicarbonates and carbonates) were then registered and integrated.

The Raman spectra of various impregnated samples were obtained using a Jobin Yvon Horiba (Northampton, UK) T64000 spectrometer, equipped with a CCD camera detector. As an excitation source, the 514.5 nm line of a Spectra Physics 2018 Argon/Krypton ion laser system was focused through an Olympus (Tokyo, Japan) BX41 microscope equipped with a 100× magnification objective. Laser power on the sample never exceeded 5 mW to avoid thermal effects on the samples studied.

3.3. HDS Reaction Test

Prepared catalysts were tested in sulfided form (see below) by using a batch reactor, as the main goal was to determine the effect of La content on the HDS properties of otherwise conventional P-modified $CoMo/Al_2O_3$ catalysts. That approach allowed a simple, rapid comparison among the catalytic activity of various studied materials [44,46], although commercial-scale HDS applications in refineries are carried out in fixed-bed plug-flow reactors. Sulfided catalysts were obtained by submitting impregnated precursors to treatment at 400 °C (heating rate 6 °C min^{-1}), under H_2/H_2S (Praxair, Mexico

City, Mexico) flow at 50/6 (ml min^{-1})/(ml min^{-1}) during 2 h. The HDS activity of sulfided catalysts was studied in a tri-phasic slurry batch reactor (Parr 4575, Moline, IL, USA). The reaction mixture was prepared by dissolving ~0.3 g of dibenzothiophene (representing S-bearing species present in oil-derived middle distillates) in 100 cm^3 of *n*-hexadecane (98 mass % and 99+ mass %, respectively, both from Aldrich, Darmstadt, Germany). *N*-hexadecane (cetane) well-represented the hydrocarbons extant in aforementioned real feedstocks from which diesel fuel is produced. Approximately ~0.2 g of sieved sulfided catalyst (80–100 Tyler mesh, ~0.165 mm average particle diameter) were also added, avoiding contact with atmospheric air, precluding sulfates formation. The HDS test reactions were carried out at T = 320 °C, P = 5.67 MPa, and 1000 rpm (~105 rad s^{-1}, mixing speed). Catalysts' particle size and operating conditions were carefully chosen to avoid internal/external diffusional limitations [73]. Liquid samples were taken from the reactor, periodically, then analyzed by gas chromatography (Agilent (Santa Clara, CA, USA) 6890N with flame ionization detector and Econo-Cap-5 capillary column length: 30 m, 0.53 mm diameter, and film thickness of 1.2 μm, from Alltech (Nicholasville, KY, USA). The HDS kinetic constants were calculated assuming pseudo-first order model kinetics referred to DBT concentration and zero order with respect to excess H_2, as follows:

$$k = \frac{-\ln(1-x)}{t_r} \tag{1}$$

where x is DBT conversion and t_r refers to reaction time. The k values for various catalysts were normalized by considering the reaction volume and the mass of the catalyst used (k expressed in $m^3\ kg_{cat}{}^{-1}\ s^{-1}$). It should be mentioned that H_2S, a byproduct from S-C bond scission reactions, contributed in maintaning tested catalysts in sulfided form and then avoiding their reduction to corresponding metallic phases, characterized by being prone to sintering [74].

4. Conclusions

Alumina-lantana (1, 3, or 5 wt% La) mixed oxides of suitable texture, to be applied as supports of catalysts for hydrotreatment of oil-derived middle-distillates, were prepared by a sol-gel method. In general, the amount and strength of surface basic sites increased with rare-earth content in binary carriers. La at low content (1 wt%) was beneficial in sulfided P-doped CoMo formulations where enhanced dibenzothiophene hydrodesulfurization (HDS) activity (~38%), as to that over the Al_2O_3-supported counterpart, was observed. However, increased La concentration in supports was detrimental on HDS properties. Hardly sulfidable tetrahedral Mo species, originated during impregnation at basic conditions, in pores of La-modified carriers seemed to dictate that behavior. Augmented rare earth concentrations in mixed supports of sulfided CoMo catalysts favored dibenzothiophene conversion through the direct desulfurization (DDS) reaction pathway to biphenyl, over the hydrogenation (HYD) one. That fact, presumably originated in enhanced MoS_2 promotion by efficient Co integration, seems to be promising in the development of alumina-supported HDT formulations at low La content of limited expensive hydrogen consumption.

Supplementary Materials: The following are available online at http://www.mdpi.com/2073-4344/9/4/359/s1, Figure S1: N_2 adsorption isotherms (at −198 °C) of alumina support (A) and La-modified carriers at various rare earth contents (ALaz). Closed symbols: adsorption branch; open symbols: desorption branch. Figure S2: N_2 physisorption isotherms (at −198 °C) of oxidic P-doped CoMo materials impregnated over sol-gel alumina (A) and corresponding La-modified (ALaz) carriers. Closed symbols: adsorption branch; open symbols: desorption branch. Figure S3: Pore size distributions of various prepared supports (a) and oxidic Co-Mo-P impregnated materials (b), as calculated by Barrett-Joyner-Halenda methodology with data from adsorption branch of corresponding N_2 adsorption isotherms. Figure S4: SEM micrographs of La-modified supports. At 1000× magnification, back-scattered electrons detector. (a) ALa1; (b) ALa3; (c) ALa5. Figure S5: SEM micrographs of Co-Mo-P impregnated oxidic samples on alumina and La-modified supports. At 4000× magnification, back-scattered electrons detector. (a) CM/A; (b) CM/ALa1 (c) CM/ALa3; (d) CM/ALa5. Table S1: SEM-EDS chemical analysis of Co-Mo-P impregnated oxidic materials prepared. Scheme S1: Absorption bands in the infrared region of CO_2 species adsorbed on basic sites [from ref. S1]. Scheme S2: Dibenzothiophene HDS reaction network over sulfided $CoMo/Al_2O_3$ [from ref. S2]. HDBT's: hydrodibenzothiophenes; BP: biphenyl; CHB: cyclohexylbenzene; BCH: bicyclohexyl.

Author Contributions: Funding acquisition, J.E.; Investigation, J.E., M.C.B., J.S.V., D.A.S.-C., V.S., J.E.T., and B.A.R.F.; Project administration, J.E.; Validation, J.E. and M.C.B.; Writing, review and editing, J.E. and M.C.B.

Funding: 117086 SENER-CONACYT-Hidrocarburos grant and Y.00105 Project from IMP.

Acknowledgments: J. Escobar acknowledges financial support from IMP (Y.00105) and SENER-CONACYT-Hidrocarburos (115086) fund.

Conflicts of Interest: The authors declare no conflict of interest.

References

1. Xu, J.; Guo, Y.; Huang, T.; Fan, Y. Hexamethonium bromide-assisted synthesis of CoMo/graphene catalysts for selective hydrodesulfurization. *Appl. Catal. B Environ.* **2019**, *244*, 385–395. [CrossRef]
2. Yang, L.; Peng, C.; Fang, X.; Cheng, Z.; Zhou, Z. Hierarchically macro-mesoporous Ni-Mo/Al_2O_3 catalysts for hydrodesulfurization of dibenzothiophene. *Catal. Commun.* **2019**, *121*, 68–72. [CrossRef]
3. Zhang, C.; Brorson, M.; Li, P.; Liu, X.; Liu, T.; Jiang, Z.; Li, C. CoMo/Al_2O_3 catalysts prepared by tailoring the surface properties of alumina for highly selective hydrodesulfurization of FCC gasoline. *Appl. Catal. A Gen.* **2019**, *570*, 84–95. [CrossRef]
4. Mey, D.; Brunet, S.; Canaff, C.; Maugé, F.; Bouchy, C.; Diehl, F. HDS of a model FCC gasoline over a sulfided CoMo/Al_2O_3 catalyst: Effect of the addition of potassium. *J. Catal.* **2004**, *227*, 436–447. [CrossRef]
5. Klimova, T.; Solís Casados, D.; Ramírez, J. New selective Mo and NiMo HDS catalysts supported on Al_2O_3-MgO(x) mixed oxides. *Catal. Today* **1998**, *43*, 135–146. [CrossRef]
6. Lu, J.; Luo, Y.; He, D.; Xu, Z.; He, S.; Xie, D.; Mei, Y. An exploration into potassium (K) containing MoS_2 active phases and its transformation process over MoS_2 based materials for producing methanethiol. *Catal. Today* **2019**. [CrossRef]
7. Alphonse, P.; Faure, B. Thermal stabilization of alumina modified by lanthanum. *Microporous Mesoporous Mater.* **2014**, *196*, 191–198. [CrossRef]
8. Lu, J.; Hao, H.; Zhang, L.; Xu, Z.; Zhong, L.; Zhao, Y.; He, D.; Liu, J.; Chen, D.; Pu, H.; et al. The investigation of the role of basic lanthanum (La) species on the improvement of catalytic activity and stability of HZSM-5 material for eliminating methanethiol-(CH_3SH). *Appl. Catal. B Environ.* **2018**, *237*, 185–197. [CrossRef]
9. Garbarino, G.; Vijayakumar, R.P.P.; Riani, P.; Finocchio, E.; Busca, G. Ethanol and diethyl ether catalytic conversion over commercial alumina and lanthanum-doped alumina: Reaction paths, catalyst structure and coking. *Appl. Catal. B Environ.* **2018**, *236*, 490–500. [CrossRef]
10. Garbarino, G.; Wang, C.; Valsamakis, I.; Chitsazan, S.; Riani, P.; Finocchio, E.; Flytzani-Stephanopoulos, M.; Busca, G. Acido-basicity of lanthana/alumina catalysts and their activity in ethanol conversion. *Appl. Catal. B Environ.* **2017**, *200*, 458–468. [CrossRef]
11. Garbarino, G.; Wang, C.; Cavattoni, T.; Finocchio, E.; Riani, P.; Flytzani-Stephanopoulos, M.; Busca, G. A study of Ni/La-Al_2O_3 catalysts: A competitive system for CO_2 methanation. *Appl. Catal. B Environ.* **2018**. [CrossRef]
12. He, D.; Zhao, Y.; Yang, S.; Mei, Y.; Yu, J.; Liu, J.; Chen, D.; He, S.; Luo, Y. Enhancement of catalytic performance and resistance to carbonaceous deposit of lanthanum (La) doped HZSM-5 catalysts for decomposition of methyl mercaptan. *Chem. Eng. J.* **2018**, *336*, 579–586. [CrossRef]
13. Cui, J.-W.; Massoth, F.E.; Topsøe, N.Y. Studies of Molybdena-Alumina Catalysts XVIII. Lanthanum-Modified Supports. *J. Catal.* **1992**, *136*, 361–377.
14. Blanchard, P.; Payen, E.; Grimblot, J.; Le Bihan, L.; Poulet, O.; Loutaty, R. Preparation of Co-Mo-based hydrodesulphurization catalysts: Characterizations of deposited species on lanthanum modified γ-alumina. *J. Mol. Catal. A Chem.* **1998**, *135*, 143–153. [CrossRef]
15. Afanasiev, P.; Bezverkhyy, I. Ternary transition metals sulfides in hydrotreating catalysis. *Appl. Catal. A Gen.* **2007**, *322*, 129–141. [CrossRef]
16. Escobar, J.; Barrera, M.C.; Solís-Casados, D.A.; Santes, V.; Sánchez-Valente, J.; Terrazas, J.E. Hidrodesulfuración de dibenzotiofeno sobre CoMo/Al_2O_3-La sol-gel. Effecto del contenido de tierra rara. In Proceedings of the Livro de atas XXVI Congresso Ibero-americano de Catálise (CICAT 2018), Coimbra, Portugal, 9–14 September 2018; Gomes, H., Silva, A., Machado, B., Ribeiro, F., Fonseca, I., Faria, J., Pereira, M., Rocha, R., Eds.; Abstract number: P-IR21. Sociedade Portuguesa de Química: Coimbra, Portugal, 2018; pp. 1070–1075.

17. Leofanti, G.; Padovan, M.; Tozzola, G.; Venturelli, B. Surface area and pore texture of catalysts. *Catal. Today* **1998**, *41*, 207–219. [CrossRef]
18. Dicks, A.L.; Ensell, R.L.; Phillips, T.R.; Szczepura, A.K.; Thorley, M.; Williams, A.; Wragg, R.D. A study of relationships between pore size distribution, hydrogen chemisorption, and activity of hydrodesulphurisation catalysts. *J. Catal.* **1981**, *72*, 266–273. [CrossRef]
19. Gurvich, L. Physico-chemical attractive force. *J. Russ. Phys. Chem. Soc.* **1915**, *47*, 805–827.
20. Ou, E.; Zhou, J.; Mao, S.; Wang, J.; Xia, F.; Min, L. Highly efficient removal of phosphate by lanthanum-doped mesoporous SiO_2. *Colloid Surf. A* **2007**, *308*, 47–53. [CrossRef]
21. Zhang, L.; Wan, L.; Chang, N.; Liu, J.; Duan, C.; Zhou, Q.; Li, X.; Wang, X. Removal of phosphate from water by activated carbon fiber loaded with lanthanum oxide. *J. Hazard. Mater.* **2011**, *190*, 848–855. [CrossRef]
22. Kaluža, L.; Gulková, D.; Vít, Z.; Zdražil, M. High-activity MgO-supported CoMo hydrodesulfurization catalysts prepared by non-aqueous impregnation. *Appl. Catal. B Environ.* **2015**, *162*, 430–436. [CrossRef]
23. Escobar, J.; De Los Reyes, J.A.; Viveros, T. Influence of the Synthesis Additive on the Textural and Structural Characteristics of Sol-Gel Al_2O_3-TiO_2. *Ind. Eng. Chem. Res.* **2000**, *39*, 666–672. [CrossRef]
24. Chaturvedi, S.; Dave, P.N. Review on Thermal Decomposition of Ammonium Nitrate. *J. Energ. Mater.* **2013**, *31*, 1–26. [CrossRef]
25. Barrera, A.; Fuentes, S.; Viniegra, M.; Avalos-Borja, M.; Bogdanchikova, N.; Campa-Molina, J. Structural properties of Al_2O_3-La_2O_3 binary oxides prepared by sol-gel. *Mater. Res. Bull.* **2007**, *42*, 640–648. [CrossRef]
26. Escobar, J.; Barrera, M.C.; De Los Reyes, J.A.; Cortés, M.A.; Santes, V.; Gómez, E.; Pacheco, J.G. Effect of Mo and Co loading in HDS catalysts supported on solvo-thermally treated ZrO_2-TiO_2 mixed oxide. *Catal. Today* **2008**, *133–135*, 282–291. [CrossRef]
27. Hattori, H. Heterogeneous Basic Catalysis. *Chem. Rev.* **1995**, *95*, 537–558. [CrossRef]
28. Bálsamo, N.; Mendieta, S.; Heredia, A.; Crivello, M. Nanoclays as dispersing precursors of La and Ce oxide catalysts to produce high-valued derivatives of biodiesel by-product. *Mol. Catal.* **2019**. [CrossRef]
29. Bernal, S.; Díaz, J.A.; García, J.R.; Rodríguez-Izquierdo, J.M. Study of some aspects of the reactivity of La_2O_3 with CO_2 and H_2O. *J. Mater. Sci.* **1985**, *20*, 537–541. [CrossRef]
30. Morterra, C.; Ghiotti, G.; Boccuzzi, F.; Coluccia, S. An infrared spectroscopic investigation of the surface properties of magnesium aluminate spinel. *J. Catal.* **1978**, *51*, 299–313. [CrossRef]
31. Di Cosimo, J.I.; Diez, V.K.; Xu, M.; Iglesia, E.; Apesteguia, C.R. Structure and Surface and Catalytic Properties of Mg-Al Basic Oxides. *J. Catal.* **1998**, *78*, 499–510. [CrossRef]
32. Boukha, Z.; Fitian, L.; López-Haro, M.; Mora, M.; Ruiz, J.R.; Jiménez-Sanchidrián, C.; Blanco, G.; Calvino, J.J.; Cifredo, G.A.; Trasobares, S.; et al. Influence of the calcination temperature on the nano-structural properties, surface basicity, and catalytic behavior of alumina-supported lanthana samples. *J. Catal.* **2010**, *272*, 121–130. [CrossRef]
33. Bettman, M.; Chase, R.E.; Otto, K.; Weber, W.H. Dispersion Studies on the System La_2O_3/γ-$A1_2O_3$. *J. Catal.* **1989**, *117*, 447–454. [CrossRef]
34. Kraus, H.; Prins, R. Proton NMR Investigations of Surface Hydroxyl Groups on Oxidic Mo-P/γ-Al_2O_3 Catalysts. *J. Catal.* **1996**, *164*, 260–267. [CrossRef]
35. Ledford, J.S.; Kim, Y.-M.; Houalla, M.; Proctor, A.; Hercules, D.M. Surface Analysis of Lanthanum-modified Cobalt Catalysts. *Analyst* **1992**, *117*, 323–327. [CrossRef]
36. Wachs, I.E. Raman and IR studies of surface metal oxide species on oxide supports: Supported metal oxide catalysts. *Catal. Today* **1996**, *27*, 437–455. [CrossRef]
37. Scheithauer, M.; Knözinger, H.; Vannice, M.A. Raman Spectra of La_2O_3 Dispersed on γ-Al_2O_3. *J. Catal.* **1998**, *178*, 701–7005. [CrossRef]
38. Dedov, A.G.; Loktev, A.S.; Moiseev, I.I.; Aboukais, A.; Lamonier, J.F.; Filimonov, I.N. Oxidative coupling of methane catalyzed by rare earth oxides: Unexpected synergistic effect of the oxide mixtures. *Appl. Catal. A Gen.* **2003**, *245*, 209–220. [CrossRef]
39. Chen, J.; Mi, J.; Li, K.; Wang, X.; Dominguez Garcia, E.; Cao, Y.; Jiang, L.; Oliviero, L.; Mauge, F. The role of citric acid in preparing highly active $CoMo/Al_2O_3$ catalyst: From aqueous impregnation solution to active site formation. *Ind. Eng. Chem. Res.* **2017**, *56*, 14172–14181. [CrossRef]
40. Nicosia, D.; Prins, R. ^{31}P MAS NMR and Raman study of a Co(Zn)MoP/γ-Al_2O_3 HDS catalyst precursor containing triethylene glycol. *J. Catal.* **2005**, *234*, 414–420. [CrossRef]

41. Bergwerff, J.A.; Visser, T.; Leliveld, B.R.G.; Rossenaar, B.D.; de Jong, K.P.; Weckhuysen, B.M. Envisaging the physicochemical processes during the preparation of supported catalysts: Raman microscopy on the impregnation of Mo onto Al_2O_3 extrudates. *J. Am. Chem. Soc.* **2004**, *126*, 14548–14556. [CrossRef]
42. Catita, L.; Quoineaud, A.-A.; Espinat, D.; Pichon, C.; Delpoux, O. Application of magnetic resonance imaging and raman imaging to study the impact of phosphorus in impregnation of hydrotreatment catalysts. *Appl. Catal. A Gen.* **2017**, *547*, 164–175. [CrossRef]
43. Kraus, H.; Prins, R. Composition of impregnation solutions and wet impregnated Mo-P/γ-Al_2O_3 catalysts as investigated by ^{31}P and ^{95}Mo NMR. *J. Catal.* **1995**, *164*, 251–259. [CrossRef]
44. Vázquez-Garrido, I.; López-Benítez, A.; Berhault, G.; Guevara-Lara, A. Effect of support on the acidity of NiMo/Al_2O_3-MgO and NiMo/TiO_2-Al_2O_3 catalysts and on the resulting competitive hydrodesulfurization/hydrodenitrogenation reactions. *Fuel* **2019**, *236*, 55–64. [CrossRef]
45. Ishutenko, D.; Nikulshin, P.; Pimerzin, A. Relation between composition and morphology of K(Co)MoS active phase species and their performances in hydrotreating of model FCC gasoline. *Catal. Today* **2016**, *271*, 16–27. [CrossRef]
46. Solís-Casados, D.A.; Escobar-Alarcón, L.; Klimova, T.; Escobar-Aguilar, J.; Rodríguez-Castellón, E.; Cecilia, J.A.; Morales-Ramírez, C. Catalytic performance of CoMo/Al_2O_3-MgO-Li(*x*) formulations in DBT hydrodesulfurization. *Catal. Today* **2016**, *271*, 35–44. [CrossRef]
47. Sheik Saleem, S.; Aruldhas, G. Raman and Infrared Spectra of Lanthanum Molybdate. *J. Solid State Chem.* **1982**, *42*, 158–162. [CrossRef]
48. Frank, C.J.; Redd, D.C.B.; Gansler, T.S.; McCreery, R.L. Characterization of Human Breast Biopsy with Near-IR Raman Spectroscopy Specimens. *Anal. Chem.* **1994**, *66*, 319–326. [CrossRef]
49. Ramírez, J.; Macías, G.; Cedeño, L.; Gutiérrez-Alejandre, A.; Cuevas, R.; Castillo, P. The role of titania in supported Mo, CoMo, NiMo, and NiW hydrodesulfurization catalysts: Analysis of past and new evidences. *Catal. Today* **2004**, *98*, 19–30. [CrossRef]
50. Liu, Y.; Liu, X.; Zhao, L.; Lyu, Y.; Xu, L.; Rood, M.J.; Wei, L.; Liu, Z.; Yan, Z. Effect of lanthanum species on the physicochemical properties of La/SAPO-11 molecular sieve. *J. Catal.* **2017**, *347*, 170–184. [CrossRef]
51. Torres-Mancera, P.; Ramírez, J.; Cuevas, R.; Gutiérrez-Alejandre, A.; Murrieta, F.; Luna, R. Hydrodesulfurization of 4,6-DMDBT on NiMo and CoMo catalysts supported on B_2O_3-Al_2O_3. *Catal. Today* **2005**, *107–108*, 551–558. [CrossRef]
52. Weber, R.S. Effect of local structure on the UV-visible absorption edges of molybdenum oxide clusters and supported molybdenum oxides. *J. Catal.* **1995**, *151*, 470–474. [CrossRef]
53. Sreedhara, M.B.; Ramakrishna Matte, H.S.S.; Govindaraj, A.; Rao, C.N.R. Synthesis, Characterization, and Properties of Few-Layer MoO_3. *Chem. Asian J.* **2013**, *8*, 2430–2435. [CrossRef] [PubMed]
54. Makshina, E.V.; Zhilinskaya, E.A.; Siffert, S.; Mazo, G.N.; Aboukaïs, A.; Grünert, W.; Romanovsky, B.V. Nanostructured lanthanum cobaltate: Oxidation and coordination states of Co atoms. *J. Exp. Nanosci.* **2009**, *5*, 427–437. [CrossRef]
55. Xie, L.L.; Gao, Q.M.; Li, Q.H. Nanoporous metal phosphate CoVSB-1 catalyst for oxidation of styrene with H_2O_2. In *From Zeolites to Porous MOF Materials—The 40th Anniversary of International Zeolite Conference. Proceedings of the 15th International Zeolite Conference, Beijing, China, 12–17 August 2007*; Xu, R., Gao, Z., Chen, J., Yan, W., Eds.; Elsevier B.V.: Amsterdam, The Netherlands, 2007; pp. 1338–1343.
56. Martin, C.; Martin, I.; Rives, V.; Malet, P. Changes in the Struture of TiO_2-supported Molybdena Induced by Na-doping. *J. Catal.* **1994**, *147*, 465–475. [CrossRef]
57. Topsøe, N.Y.; Topsøe, H. FTIR Studies of Mo/Al_2O_3-Based Catalysts: II. Evidence for the Presence of SH Groups and Their Role in Acidity and Activity. *J. Catal.* **1993**, *139*, 641–651. [CrossRef]
58. Morales, A.; Ramírez de Agudelo, M.M. Promoter Role of Octahedral Co (and Ni) in Modified Co(Ni)Mo-Al_2O_3 Catalysts for Hydrodesulfurization Reactions. *Appl. Catal.* **1986**, *23*, 23–34. [CrossRef]
59. Solís-Casados, D.; Escobar, J.; García Orozco, I.; Klimova, T. Effect of Potassium Content on the Performance of CoMo/Al_2O_3-MgOK_2O(*x*) Catalysts in Hydrodesulfurization of Dibenzothiophene. *Ind. Eng. Chem. Res.* **2011**, *50*, 2755–2761. [CrossRef]
60. Wu, L.; Jiao, D.; Wang, J.; Chen, L.; Cao, F. The role of MgO in the formation of surface active phases of CoMo/Al_2O_3-MgO catalysts for hydrodesulfurization of dibenzothiophene. *Catal. Commun.* **2009**, *11*, 302–305. [CrossRef]

61. Ge, T.; Zuo, C.; Chen, H.; Muhammad, Y.; Wei, L.; Li, C. Catalytic Activity and Molecular Behavior of Lanthanum Modified $CoS_x/\gamma\text{-}Al_2O_3$ Catalysts for the Reduction of SO_2 to Sulfur in Smelter Off-Gas Using $CO\text{-}H_2$ Mixture as Reductant. *Ind. Eng. Chem. Res.* **2019**. [CrossRef]
62. Escobar, J.; Barrera, M.C.; Santes, V.; Terrazas, J.E. Naphthalene hydrogenation over Mg-doped Pt/Al_2O_3. *Catal. Today* **2017**, *296*, 197–204. [CrossRef]
63. Houalla, M.; Nag, N.K.; Sapre, A.V.; Broderick, D.H.; Gates, B.C. Hydrodesulfurization of dibenzothiophene catalyzed by sulfided $CoO\text{-}MoO_3/\gamma\text{-}Al_2O_3$: The reaction network. *AIChEJ* **1978**, *24*, 1015–1021. [CrossRef]
64. Hensen, E.J.M.; Kooyman, P.J.; van der Meer, Y.; van der Kraan, A.M.; de Beer, V.H.J.; van Veen, J.A.R.; van Santen, R.A. The relation between morphology and hydrotreating activity for supported MoS_2 particles. *J. Catal.* **2001**, *199*, 224–235. [CrossRef]
65. Bataille, F.; Lemberton, J.L.; Michaud, P.; Pérot, G.; Vrinat, M.; Lemaire, M.; Schulz, E.; Breysse, M.; Kasztelan, S. Alkyldibenzothiophenes hydrodesulfurization-promoter effect, reactivity, and reaction mechanism. *J. Catal.* **2000**, *191*, 409–422. [CrossRef]
66. Wu, Z.; Whiffen, V.M.L.; Zhu, W.; Wang, D.; Smith, K.J. Effect of Annealing Temperature on $Co\text{-}MoS_2$ Nanosheets for Hydrodesulfurization of Dibenzothiophene. *Catal. Lett.* **2014**, *144*, 261–267. [CrossRef]
67. Kaluža, L.; Gulková, D.; Vít, Z.; Zdražil, M. Effect of support type on the magnitude of synergism and promotion in CoMo sulphide hydrodesulphurisation catalyst. *Appl. Catal. A Gen.* **2007**, *324*, 30–35. [CrossRef]
68. Wu, H.; Duan, A.; Zhao, Z.; Qi, D.; Li, J.; Liu, B.; Jiang, G.; Liu, J.; Wei, Y.; Zhang, X. Preparation of NiMo/KIT-6 hydrodesulfurization catalysts with tunable sulfidation and dispersion degrees of active phase by addition of citric acid as chelating agent. *Fuel* **2014**, *130*, 203–210. [CrossRef]
69. Halachev, T.; Atanasova, P.; Lopez Agudo, A.; Arias, M.G.; Ramirez, J. Activity of $P\text{-}Ni\text{-}W/Al_2O_3$ catalysts with varying phosphorus content in the hydrogenation of naphthalene. *Appl. Catal. A Gen.* **1996**, *136*, 161–175. [CrossRef]
70. Wang, T.; Fan, Y.; Wang, X.; Chou, L.; Lin, H. Selectivity enhancement of CoMoS catalysts supported on tri-modal porous Al_2O_3 for the hydrodesulfurization of fluid catalytic cracking gasoline. *Fuel* **2015**, *157*, 171–176. [CrossRef]
71. Jaf, Z.N.; Altarawneha, M.; Miran, H.A.; Jiang, Z.T.; Dlugogorski, B.Z. Hydrodesulfurization of Thiophene over $\gamma\text{-}Mo_2N$ catalyst. *Mol. Catal.* **2018**, *459*, 21–30. [CrossRef]
72. Fujikawa, T.; Chiyoda, O.; Tsukagoshi, M.; Idei, K.; Takehara, S. Development of a high activity HDS catalyst for diesel fuel: From basic research to commercial experience. *Catal. Today* **1998**, *45*, 307–312. [CrossRef]
73. Escobar, J.; Gutiérrez, A.; Barrera, M.C.; Colín, J.A. NiMo/alumina hydrodesulphurization catalyst modified by saccharose: Effect of addition stage of organic modifier. *Can. J. Chem. Eng.* **2016**, *94*, 66–74. [CrossRef]
74. Shalagina, A.E.; Aleshina, G.I.; Startsev, A.N. Hydrogen occluded in sulfide catalysts: Evidence from the temperature programmed heating Technique. *React. Kinet. Catal. Lett.* **2003**, *78*, 91–98. [CrossRef]

catalysts

MDPI

Article

Assessment of Ag Nanoparticles Interaction over Low-Cost Mesoporous Silica in Deep Desulfurization of Diesel

Rafael V. Sales [1], Heloise O. M. A. Moura [1], Anne B. F. Câmara [1], Enrique Rodríguez-Castellón [2], José A. B. Silva [1], Sibele B. C. Pergher [1], Leila M. A. Campos [3], Maritza M. Urbina [4], Tatiana C. Bicudo [1] and Luciene S. de Carvalho [1,*]

1 Energetic Technologies Research Group, Institute of Chemistry and School of Science and Technology, Federal University of Rio Grande do Norte, Natal 59078-900, Brazil

2 Dpto. de Química Inorgánica, Cristalografía y Mineralogía, Facultad de Ciencias, Universidad de Málaga, 29071 Málaga, Spain

3 Chemical Engineering Post-Graduate Program, Salvador University (UNIFACS), Salvador 40140-110, BA, Brazil

4 Chemical Engineering, Center for Technology, Federal University of Alagoas, Maceió 57072-900, AL, Brazil

* Correspondence: luciene_car@hotmail.com; Tel.: +55-849-8828-5261

Received: 5 July 2019; Accepted: 18 July 2019; Published: 30 July 2019

Abstract: Chemical interactions between metal particles (Ag or Ni) dispersed in a low-cost MCM-41^M produced from beach sand amorphous silica and sulfur compounds were evaluated in the deep adsorptive desulfurization process of real diesel fuel. N_2 adsorption-desorption isotherms, X-ray diffraction (XRD), X-ray photoelectron spectroscopy (XPS), high-resolution transmission electron microscopy (HRTEM) and scanning transmission electron microscopy coupled to energy-dispersive X-ray spectroscopy (STEM-EDX) were used for characterizing the adsorbents. HRTEM and XPS confirmed the high dispersion of Ag nanoparticles on the MCM-41 surface, and its chemical interaction with support and sulfur compounds by diverse mechanisms such as π-complexation and oxidation. Thermodynamic tests indicated that the adsorption of sulfur compounds over Ag(I)/MCM-41^M is an endothermic process under the studied conditions. The magnitude of $\Delta H°$ (42.1 kJ/mol) indicates that chemisorptive mechanisms govern the sulfur removal. The best fit of kinetic and equilibrium data to pseudo-second order ($R^2 > 0.99$) and Langmuir models ($R^2 > 0.98$), respectively, along with the results for intraparticle diffusion and Boyd's film-diffusion kinetic models, suggest that the chemisorptive interaction between organosulfur compounds and Ag nanosites controls sulfur adsorption, as seen in the XPS results. Its adsorption capacity (q_m = 31.25 mgS/g) was 10 times higher than that obtained for pure MCM-41^M and double the q_m for the Ag(I)/MCM-41^C adsorbent from commercial silica. Saturated adsorbents presented a satisfactory regeneration rate after a total of five sulfur adsorption cycles.

Keywords: desulfurization; chemisorption; MPI silica; MCM-41; Ag nanoparticles; XPS assessment

1. Introduction

The demand for clean energy has attracted great attention in recent years and has generated an increasing focus on sulfur compounds removal from fuels [1]. The combustion of these compounds results in liberating sulfur oxides (SO_x), which poison catalytic converters and are corrosive to fuel cell electrodes [2], in addition to polluting the environment; therefore, governments worldwide have adopted standards to decrease sulfur concentration to 10 ppm for diesel fuel [3]. The current technique for sulfur removal in oil refineries is the highly expensive hydrodesulfurization (HDS), performed at hard conditions (>573 K, >4 MPa) using sophisticated Co-Mo/Al_2O_3 or Ni-Mo/Al_2O_3 catalysts. However, some recalcitrant organosulfur such as thiophene derivatives are not removed by this method [4].

Methods based on membrane separation [5], catalytic oxidation [3], biological desulfurization [6], and adsorption [1,7–11] have been developed for desulfurizing fuels. Among these techniques, adsorptive desulfurization stands out due to its simplicity, efficiency and low-cost according to the solid chosen as the adsorbent. Activated carbon [1], MOFs [9], KIT-6 [10], SBA-15 [8], and MCM-41 [7,11] are currently being investigated for sulfur adsorption from liquid fuels, among other materials.

These molecular sieves, especially MCM-41, are extensively applied as supports for adsorbents and heterogeneous catalysts due to their high surface area, ordered structure, and uniform pore size [12]. In general, heterogeneous micrometric size catalysts present low catalytic activity due to slow diffusion of reagents [13]. Therefore, active metal nanoparticles (NPs), which have high catalytic activity due to their high surface-volume ratio and chemical reactivity, have been incorporated into solid supports for increasing their efficiency, recyclability and stability, thus providing more sustainable processes [14–16]. When unsupported, the high surface energy of nanocatalysts increases system instability and leads to aggregation, thus resulting in the loss of its catalytic properties. The use of molecular sieve supports is fundamental to prevent the undesirable aggregation and loss of nanocatalysts [17,18].

Moreover, MCM-41 has gained attention for its flexibility regarding the silica feedstock for synthesis. Recently, it has been produced from greener sources of silica such as coal fly ash (CFA) [12], rice husk ash [19], iron ore residues [20], wheat stem ash (WSA) [21], and MPI silica from beach sand [22,23], used in this study. This material has been functionalized by adding active sites to its surface for promoting new chemical interactions in the support-site-adsorbate system and improving the adsorptive treatment [24]. Researchers have observed that adsorbents functionalized with transition metals are capable of capturing aromatic sulfur compounds refractory to HDS processes via π-complexation [25,26]. Furthermore, Ag species dispersed onto metal and semi-metal oxide adsorbents generate materials described as high activity catalysts for oxidation of several toxic compounds such as formaldehyde [27], benzene [28], toluene [29] and carbon monoxide [24], among others.

The Ag(0) sites present high reactivity both by adsorbing and activating atmospheric O_2 at ambient conditions (300 K, 1 atm) [30] and by increasing the mobility of lattice oxygens from the adsorbent structure. This increase in mobility can occur by bridging bonds between Ag(0) and the metal oxide molecules [28], but it is also reported by covalent bonding in the case of a more oxidized state of silver as Ag(I), and in a cation exchange of hydrolyzed silica silanol groups (Si-O-H) [31]. Thus, an instrumental technique such as X-ray photoemission spectroscopy (XPS) is required to identify the chemical species in the sample and to indicate what kind of interactions could be involved in impregnation and chemisorptions processes. This analysis is able to capture chemical changes in the adsorbent surface after metal deposition and adsorption by recording displacements in binding energies of the atoms and ions involved in the mechanisms [32].

This paper proposes a low-cost method for mitigating sulfur content in diesel fuel based on the adsorption of organosulfur compounds over Ag and Ni impregnated MCM-41 supports from beach sand silica, in a complementary process to the hydrodesulfurzation (HDS) applied in refineries with a greener and effective approach. Furthermore, high-efficiency characterization techniques such as X-ray photoelectron spectroscopy (XPS) and high-resolution transmission electron microscopy (HRTEM) were used to enable the more accurate analysis of the different species and interaction mechanisms in the metal-support-organosulfur system, and were associated with a deep evaluation of the kinetic and thermodynamic features.

2. Results and Discussion

2.1. Preliminary Sulfur Adsorption Results

The results of sulfur adsorption in columns for $AgNO_3$/MCM-41 produced from both MPI and commercial silicas interacted more strongly with organosulfur compounds and reached higher desulfurization results. The adsorbent generated from MPI silica presented lower performance in comparison to the commercial material; however, its efficiency is significant and it stands out for being extracted from a renewable and low-cost source. There was no relevant variation in adsorption by changing metal concentration from 2% to 8%. Taking into consideration the performance and cost-benefits, the concentration of 2% salts in MCM-41^M was selected for the next steps of characterization and adsorption studies. The sulfur adsorption results for all materials applied in this research are depicted in Figure S1 of the Supplementary Materials. The XRD patterns for the adsorbents with 8% metal (Ni or Ag) in salt and oxide forms are presented in Figure S2 of the Supplementary Materials.

2.2. Characterization of the Adsorbents

Low-angle XRD patterns of the samples are shown in Figure 1a. The presence of diffraction peaks indexed by (100), (110) and (200) confirms the well-ordered structure with hexagonal *p6mm* symmetry, indicating a relative perfect structure of MCM-41 samples [33]. Less ordered materials were obtained after impregnation; however, the deposition of transition metals in the adsorbent via wet impregnation did not compromise the ordered porous structure of MCM-41^M.

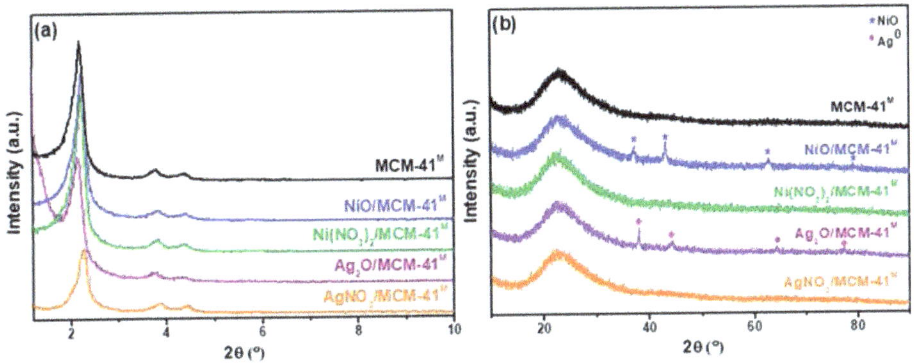

Figure 1. (**a**) Low-angle and (**b**) wide-angle X-ray diffractograms of the adsorbents.

Wide-angle XRD patterns of all samples (Figure 1b) exhibited broad diffraction peaks centered at 22.8°, originated from amorphous silica (MPI silica). $AgNO_3$/MCM-41^M and $Ni(NO_3)_2$/MCM-41^M adsorbents have not shown any reflection referent to deposited species, supporting the idea that they are highly dispersed and have a high interaction with the mesoporous support [34]. Refraction peaks corresponding to the cubic phase of NiO at 2θ of 37.16, 43.31, 62.88 and 79.31 (JCPDS 01-073-1523) and the face-centered-cubic (FCC) phase from metallic Ag at 2θ of 38.16, 44.30, 64.30 and 77.44 (JCPDS 04-0783) were detected in NiO/MCM-41^M and Ag_2O/MCM-41^M, respectively (Figure 1b).

HRTEM micrographs for pure MCM-41^M (Figure 2a,b) confirm the hexagonal high-organized pore matrix, typical of this material, as seen in XRD patterns [35]. The $AgNO_3$/MCM-41^M sample was analyzed by STEM-EDX mapping images and HRTEM. The STEM-EDX mapping profile for deposited species (Figure 2c–e) shows highly uniform distribution of silver nanoparticles (red spots), with some agglomerations, and silica particles were also observed in green. According to a HRTEM micrograph (Figure 2f), the MCM-41^M mesostructure was preserved after its modification with Ag^+ ions.

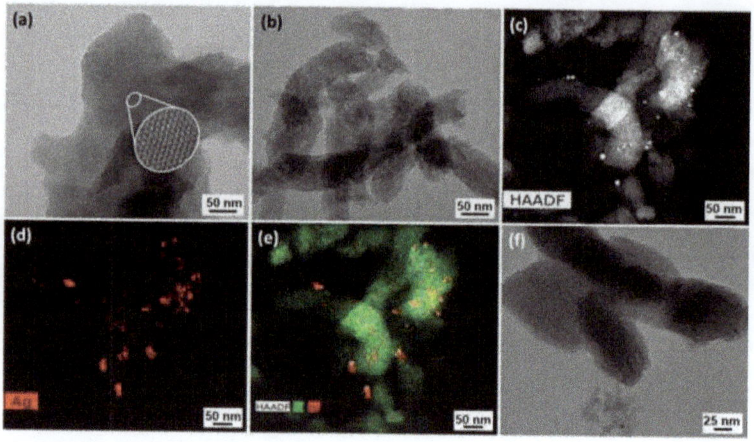

Figure 2. (**a**,**b**) HRTEM micrograph for pure MCM-41^M; (**c**) STEM, (**d**) EDX, (**e**) STEM-EDX and (**f**) HRTEM analysis for $AgNO_3$/MCM-41^M.

The N_2 adsorption-desorption isotherms and pore size distribution are depicted in Figure 3. The isotherms for all the samples are type IV according to the IUPAC classification, which is characteristic of MCM-41 [36]. As depicted in Figure 3a, the N_2 adsorption–desorption isotherms for all the samples were correspondent to type IV according to the IUPAC classification, which is associated to mesoporous materials with H3-type hysteresis loop due to capillary condensation [34]. The adsorption-desorption in intermediary relative pressures ($0.3 < P/P_0 < 0.4$) correspond to N_2 capillary condensation of uniform mesopores. The metal incorporation to the mesoporous sieve has not affected the isotherms profile. A slight inflection for the Ag_2O/MCM-41^M sample can be noticed, which suggests a less-organized mesoporous structure [11], as observed in the XRD data shown in Figure 1a. All samples presented an increase in adsorbed volume due to the macropore filling resulting from interparticle spaces and untransformed amorphous silica in the relative pressure range 0.90–0.99 [35].

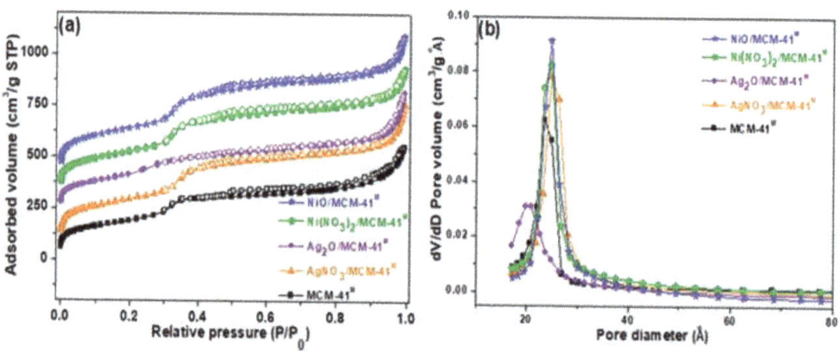

Figure 3. (**a**) N_2 adsorption-desorption isotherms at 77 K (**b**) and the corresponding BJH adsorption pore size distributions for the adsorbents.

The pore size distribution (Figure 3b) estimated by the BJH method indicates that the materials have uniform mesoporosity with a maximum distribution in the range of 23.54–24.82 Å. An exception was the Ag_2O/MCM-41^M sample, which exhibited a smaller maximum pore size centered at about 20.20 Å and heterogeneous distribution [4]. In this case, Ag_2O species are possibly deposited inside the pores, while the species were preferentially deposited outside the pore system for the other materials.

Textural parameters of the adsorbents are listed in Table 1 and presented a reduction in specific area (S_{BET}) and pore volume (V_P) compared to the initial silica material, indicating that the deposited particles caused the occlusion of some pores, reducing the amount of adsorbed N_2. In addition, the embedded species are not porous, which reduces porosity and consequently the specific area of the materials [37].

Table 1. Textural properties of the studied samples.

Adsorbent	S_{BET} [1] (m^2/g)	V_P [2] (cm^3/g)	Maximun Pore Diameter (Å)
MCM-41^M	722	0.79	23.54
$AgNO_3$/MCM-41^M	656	0.74	24.82
Ag_2O/MCM-41^M	704	0.71	20.20
$Ni(NO_3)_2$/MCM-41^M	641	0.57	24.70
NiO/MCM-41^M	703	0.71	24.75

[1] Specific surface area calculated by the BET method. [2] Total pore volume recorded at P/P_0 = 0.993.

XPS analysis was employed to obtain information about the surface composition, species identification and chemical state of the transition metal over MCM-41^M surface. The XPS spectrum of the pure MCM-41^M sample is shown in Figure 4. The binding energy (BE) of the Si 2p orbital in all the samples was about 103.3 eV, which is characteristic of mesoporous silicates such as MCM-41. Additionally, a single peak centered at 532.7 eV would be assigned to the O 1 s photoemission of oxygen atoms from the siliceous support [38].

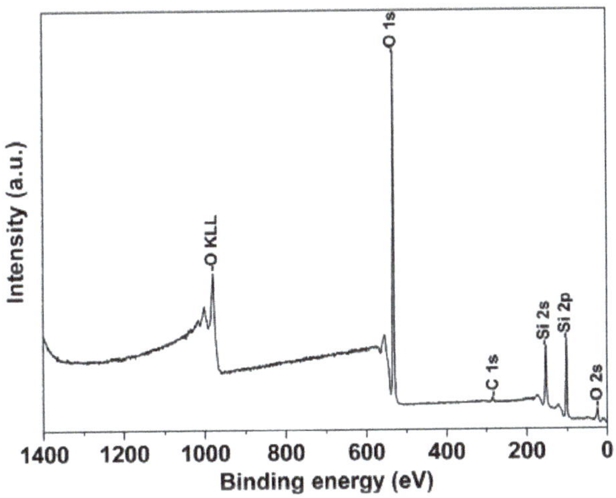

Figure 4. XPS spectra for pure MCM-41^M.

For $AgNO_3$/MCM-41^M (Figure 5a), broad peaks in between the characteristic BE for Ag(I) species (367.5 and 373.8 eV) and Ag (0) nanoparticles (369.7 and 375.9 eV) are observed for the Ag $3d_{5/2}$ and $3d_{3/2}$ orbitals at 368.1 and 374.1 eV, respectively [39]. This phenomenon suggests the interaction of impregnated $AgNO_3$ Ag(I) ions with other atoms such as adsorbed atmospheric oxygen (Ag-O_{ads}), lattice oxygen (Ag-O_{lat}) and silanol groups (Ag-O-Si) of MCM-41^M, now presenting a BE state in between the ionic and metallic forms ($Ag^{\delta+}$, $0 < \delta < 1$) [31,40].

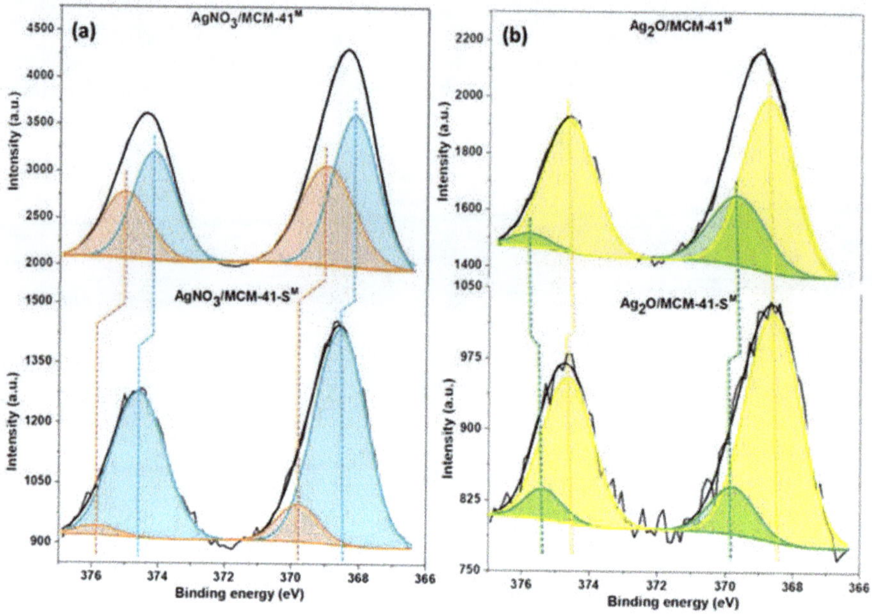

Figure 5. XPS spectra for (**a**) $AgNO_3$/MCM-41^M and (**b**) Ag_2O/MCM-41^M before and after desulfurization.

Smaller peaks at 369.0 and 375.0 eV are assigned to the typical BE region for Ag(0) nanoparticles highly dispersed in the support [30,41]. The presence of the metallic electronic state of silver may be related to the photolysis of the $AgNO_3$ molecule in the presence of light; furthermore, it is known in the literature that some oxygenated metal-adsorbents can undergo autoreduction processes in the presence of lattice oxygen and water [42]. Ag(0) can also interact with oxygen (Ag-O_{ads} and Ag-O_{lat}). All these oxygenated species (Ag-O) are highly active for oxidation of aromatic pollutants [34] and even for weak π-complexation interactions with aromatic organosulfurs, as well as the non-bonding Ag(0) nanodomains [36].

After sulfur adsorption was performed with model diesel, the displacement to higher BE values indicates more positive electronic density of Ag species [27]. The increase in BEs of oxygenated Ag species (368.4 and 374.5 eV) can be assigned to oxygen loss after oxidation reactions, while the broad peaks which emerged at 369.4 and 375.6 eV are ascribed to the formation of an Ag-S-R bridging configuration via π-complexation (S-R = DBT) performed by Ag(0) and also by active Ag-O species [26,43]. In this mechanism, Ag(0) ([Kr]$4d^{10}5s^1$) and Ag(I) ([Kr]$4d^{10}5s^0$) species are able to form bonds with their empty or semi-filled *s* orbitals and their *d* orbitals can retrograde electronic density to the π anti-ligands (π*) of aromatic organosulfur rings [25].

For the sample that was calcined in an oxidizing atmosphere, the XPS spectra (Figure 5b) show broad peaks at 368.5 and 374.6 eV, which can be ascribed to the production of stable Ag_2O species that can emerge in this BE region [44]; since the spectral resolution for all XPS analysis was 0.8 eV, significant peak separation was not possible. Highly dispersed Ag nanodomains were also observed at 369.4 and 375.9 eV. The spectra recorded after this material was used for desulfurization shows that the Ag_2O/Ag-O peak BE remained almost unchanged; however, the increase in Ag(0) BE (369.8 and 375.4 eV) demonstrates the same phenomenon observed for the non-calcined sample, where metallic and active silver species were complexed by DBT molecules.

The atomic concentrations in the adsorbents surface were measured by the XPS technique and are summarized in Table 2. Ni, C, O and Si contents in $Ni(NO_3)_2$/MCM-41^M and NiO/MCM-41^M adsorbents were similar; however, the Ag-impregnated sample after calcination presented some significant variations in the atomic concentrations, with a increase in the C atomic percentage. As the penetration depth of XPS radiation is about 10 nm [44], X-rays are only able to reach the atoms near the surface, and the higher amount of carbon in this region promotes a decrease in other atoms' concentrations, as seen for silver and oxygen. The $AgNO_3$/MCM-41^M adsorbent presented higher sulfur concentrations after adsorption tests among all samples.

Table 2. Atomic concentrations (%) in the adsorbents surface and Si/metal molar ratio.

Samples	Atomic Concentrations (%)					Molar Ratios		After Desulfurization (%)
	C 1s	O 1s	Si 2p	Ag 3d	Ni 2p	Si/Ag	Si/Ni	S 2p
MCM-41^M	2.24	69.59	28.18					
$AgNO_3$/MCM-41^M	1.94	68.05	29.34	0.67		43.79		1.87
Ag_2O/MCM-41^M	5.66	64.58	29.43	0.32		91.97		1.78
$Ni(NO_3)_2$/MCM-41^M	4.75	66.27	26.38		0.54		48.85	1.62
NiO/MCM-41^M	3.75	67.10	26.94		0.56		48.10	1.64

2.3. Kinetic Study

Sulfur adsorption rate performed with real diesel on $AgNO_3$/MCM-41^M and pure MCM-41^M was determined via kinetic analysis. Contact time effect data presented in Figure 6a show that the organosulfur contaminants were rapidly adsorbed by $AgNO_3$/MCM-41^M (about 5 min of contact) and reached equilibrium in approximately 120 min. No significant adsorption of these compounds was observed after 5 min for the pure MCM-41^M, which may be related to the fewer available active sites [45]. Figure 6 shows the fitting and Table 3 brings the summary of the corresponding adsorption parameters and determination coefficients (R^2).

Table 3. Kinetic parameters fitted to pseudo-first order, pseudo-second order, Elovich, intraparticle diffusion and Boyd models. Experimental conditions: room temperature and pressure, 500 ppm sulfur real diesel.

Adsorbent	q_e, exp (mg/g)	Pseudo-First Order			Pseudo-Second Order		
		k_1 (min^{-1})	q_e, cal (mg/g)	R^2	k_2 (g/(mg min))	q_e, cal (mg/g)	R^2
$AgNO_3$/MCM-41^M	7.26	0.005	2.05	0.814	0.010	7.31	0.998
MCM-41^M	1.64	0.002	0.79	0.516	0.013	1.60	0.890

Adsorbent	Elovich Model			Intraparticle Diffusion			Boyd	
	α (mg/g min)	β (g/mg)	R^2	k_{id} (mg/(g min))	C (mg/g)	R^2	B [1]	R^2
$AgNO_3$/MCM-41^M	45.257	1.426	0.977	0.126	4.654	0.894	0.007	0.977
MCM-41^M	0.786	5.432	0.710	0.034	0.709	0.691	0.771	0.710

[1] B is the Boyd's constant, which is equivalent to the angular coefficient of Boyd's plot ($t \times B_t$).

Experimental results of sulfur removal on $AgNO_3$/MCM-41^M adsorbent were better fitted to the pseudo-second order model, with $R^2 > 0.99$ and similar calculated and experimental q_e values (7.31 mg/g and 7.26 mg/g, respectively), followed by Elovich ($R^2 = 0.977$) and pseudo-first order ($R^2 = 0.814$) models, respectively. The best data fitting to the pseudo-second order and Elovich models is a clue that the chemisorption process is the rate determining step of adsorption [46]. Furthermore, the higher value of parameter α obtained by Elovich equation (Table 3) suggests that there is a strong affinity between the recalcitrant compounds containing S and the active sites of $AgNO_3$/MCM-41^M adsorbent, strengthening the hypothesis that this adsorption is highly influenced by Ag sites and its different active forms (Ag(0) and Ag-O), and mainly occurs via π-complexation and oxidation [47,48]. The slower adsorption rate after the initial minutes may be a reflection of concurrent kinetic mechanisms [46]. For pure MCM-41^M, the experimental results were better fitted to the pseudo-second order model, with $R^2 > 0.89$ and calculated q_e (1.60 mg/g) close to the experimental result (1.64 mg/g). Low R^2 values indicate that the pseudo-first-order and Elovich kinetic models did not adequately describe the adsorption process performed by this sample.

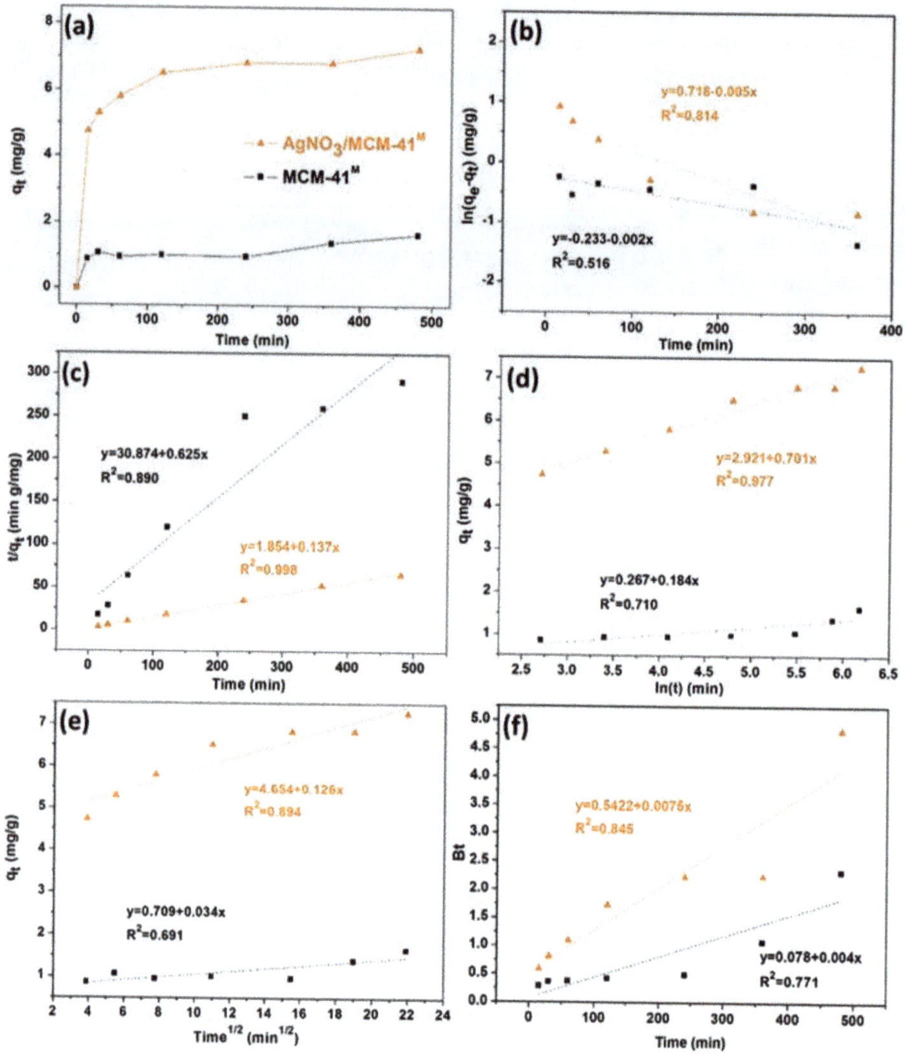

Figure 6. Plots of (**a**) contact time effect, (**b**) pseudo-first order, (**c**) pseudo-second order, (**d**) Elovich, (**e**) intraparticle diffusion and (**f**) Boyd models, where ▲ represents the kinetic data obtained for AgMCM-41^M and ■ represents the kinetic data obtained for MCM-41^M.

Considering molecular structures, critical diameters of sulfur compounds in model diesel are smaller than those in commercial diesel [49], whose main composition is listed in Table S1 of the Supplementary Materials. Thus, performing desulfurization tests with the real fuel is the best way to evaluate the performance of the adsorbents. For the effective adsorption of larger molecules in terms of physisorption, not only should the pore size of the material be at least larger than the critical diameter of the adsorbate, but it should also be large enough to reduce diffusional resistance during adsorption [50]. Therefore, the high sulfur adsorption performed by $AgNO_3$/MCM-41 can also be related to its larger pore diameter (24.82 Å) in relation to the other synthesized impregnated materials (as shown in Table 1), favoring adsorption due to both the more effective reduction of the diffusional resistance of aromatic organosulfurs (whose critical diameters are shown in Table S2 of the Supplementary Materials) and the stronger chemical interactions promoted by the more active Ag species over this adsorbent.

This fact can be verified by the results of intraparticle diffusion and film diffusion of Boyd models. Since the data fitting curves for intraparticle diffusion did not cross the axes origin (Figure 6e), it can be assumed that intrapore diffusion is not the limiting step of the adsorption process, and therefore other mechanisms must be acting simultaneously. This is confirmed by the Boyd model (Figure 6f), since the graphs have non-zero intercepts. The non-linearity of Boyd model graphs indicates that pore diffusion is not the rate control step and that there is a mass transfer resistance step in the outer film [51]. Thus, although both intraparticle diffusion and film diffusion are involved in the adsorptive process, these physical phenomena do not govern desulfurization.

2.4. Adsorption Equilibrium Isotherms

Langmuir, Freundlich, Temkin, and Dubinin-Radushkevich equations were used to fit the experimental sulfur adsorption results with real diesel. The isotherm parameters and the correlation coefficients (R^2) obtained from the linear fitting of experimental data are summarized in Table 4. The non-linear adjustments and variations in the separation factor (R_L) as a function of initial sulfur concentration (C_0) are displayed in Figures 7 and 8, respectively. Based on R^2 values, Langmuir and Freundlich mathematical models show good fitting to the adsorption equilibrium data for all the studied adsorbents. Experimental adsorption results for MCM-41^M were better adjusted to the Freundlich model ($R^2 = 0.9904$) in comparison to the Langmuir model ($R^2 = 0.9802$), indicating that the process preferably occurs through physisorption [52]. However, as shown in Figure 7, the adsorption on materials modified with transition metals better fit the Langmuir model ($R^2 > 0.98$), suggesting that the physisorption mechanism was preferably substituted by chemisorption [26].

The values of parameter n in the Freundlich isotherm are higher than 1 for all adsorbents, indicating favorable adsorption. The Temkin model ($R^2 < 0.93$) could be used to describe the experimental data. D-R model ($R^2 < 0.47$) was not able to fit the S-compounds adsorption processes performed in this work.

For the Langmuir isotherm, R_L reached values between 0 and 1 (Figure 8), indicating that the sulfur adsorption is favorable. The maximum adsorption capacity of S-compounds (q_m) for this model, in mg/g, was estimated in the following descending order: $AgNO_3$/MCM-41^M > Ag_2O/MCM-41^M > $Ni(NO_3)_2$/MCM-4^M > NiO/MCM-41^M > MCM-41^M. This result is a clue that $AgNO_3$/MCM-41^M was able to promote the best adsorption performance among the tested adsorbents, considering the same study conditions. MCM-41^M modification with Ag species from $AgNO_3$ can offer a high quantity of new active sites over the silica surface to chemically interact with sulfur compounds [53].

The adsorption equilibrium results of the materials synthesized from commercial silica are shown in Table 4 and Figure 7f–h. As expected, the experimental sulfur adsorption data of pure MCM-41^C were better fitted to the Freundlich model, and the results of the samples impregnated with non-calcined salts were better described by the Langmuir mathematical model. $AgNO_3$/MCM-41^C synthesized from commercial silica obtained the highest maximum adsorptive capacity among them (q_m = 15.41 mg/g), approximately half of the content adsorbed by $AgNO_3$/MCM-41^M synthesized from amorphous silica MPI, emphasizing the higher sulfur adsorption efficiency of the renewable material even when compared to other adsorbents applied for desulfurization in the literature (Table S3 of the Supplementary Materials).

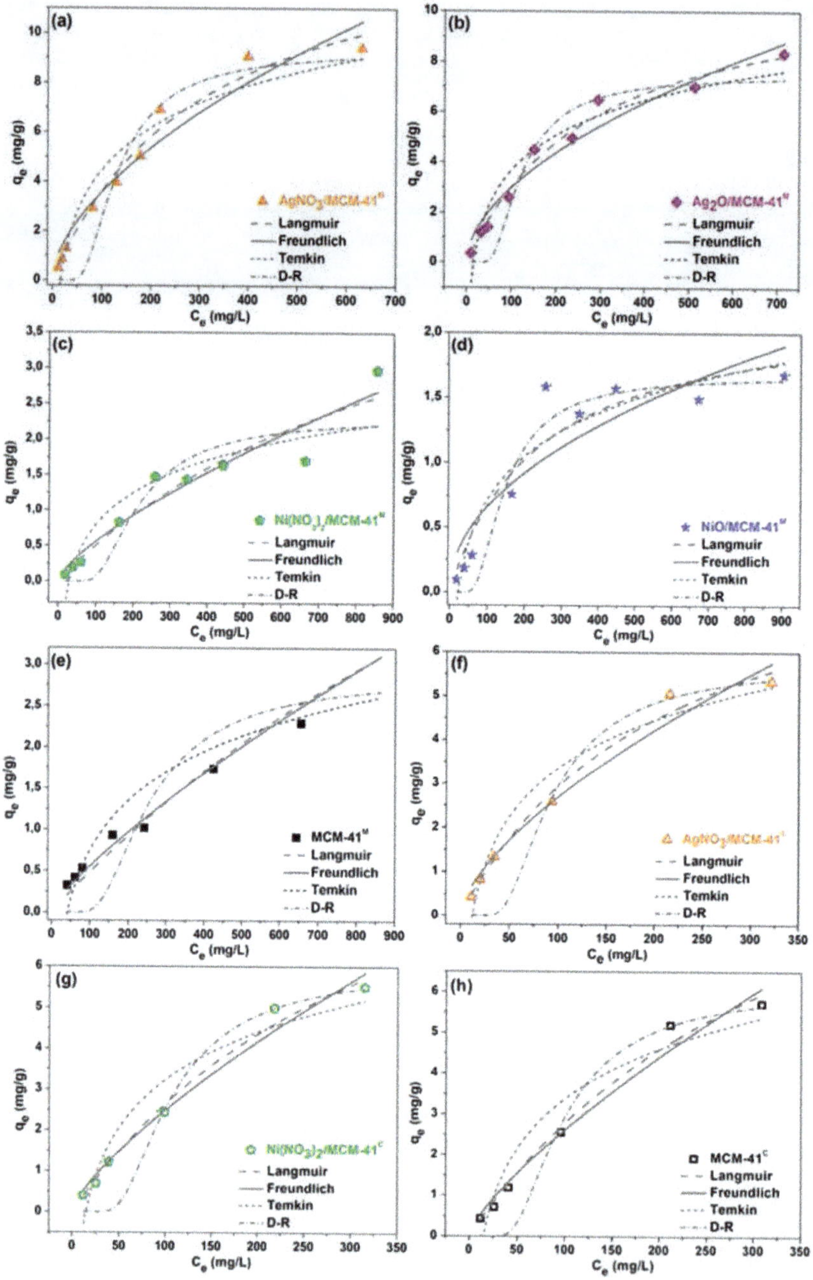

Figure 7. Non-linear adjustments of Langmuir, Freundlich, Temkin and Dubinin-Radushkevich (D-R) equilibrium isotherms of S-compounds adsorption over (**a**) $AgNO_3$/MCM-41^M, (**b**) Ag_2O/MCM-41^M, (**c**) $Ni(NO_3)_2$/MCM-41^M, (**d**) NiO/MCM-41^M, (**e**) MCM-41^M, (**f**) $AgNO_3$/MCM-41^C, (**g**) $Ni(NO_3)_2$/MCM-41^C and (**h**) MCM-41^C. Experimental conditions: room temperature and pressure, real diesel with sulfur concecntration ranging from 25 to 1000 ppm.

Table 4. Parameters and determination coefficients for Langmuir, Freundlich, Temkin and Dubinin-Radushkevich (D-R) isotherm models.

(a) Adsorbents Obtained from MPI Silica

Isotherm	Parameter	Adsorbent				
		$AgNO_3$/MCM-41^M	Ag_2O/MCM-41^M	$Ni(NO_3)_2$/MCM-41^M	NiO/MCM-41^M	MCM-41^M
Langmuir	K_L (L/mg)	0.0013	0.0025	0.0005	0.0009	0.0028
	q_m (mg/g)	31.25	13.95	8.97	5.38	3.06
	R^2	0.9888	0.9973	0.9954	0.9969	0.9802
Freundlich	K_F	0.0812	0.0823	0.0084	0.0106	0.0213
	n	1.27	1.35	1.16	1.25	1.37
	R^2	0.9719	0.9633	0.9461	0.9215	0.9904
Temkin	α	0.062	0.063	0.034	0.044	0.024
	B	2.457	2.006	0.658	0.482	0.863
	R^2	0.9249	0.9267	0.8343	0.8919	0.8575
D-R	q_s	302.808	115.492	4.214	3.369	3.777
	B	0.0002	0.0001	0.0001	0.00009	0.0004
	E	50.000	70.711	70.711	74.535	35.355
	R^2	0.4471	0.3171	0.3432	0.4556	0.4632

(b) Adsorbents Obtained from Commercial Silica

Isotherm	Parameter	Adsorbent		
		$AgNO_3$/MCM-41^C	$Ni(NO_3)_2$/MCM-41^C	MCM-41^C
Langmuir	K_L (L/mg)	0.0025	0.0035	0.0051
	q_m (mg/g)	15.41	9.51	7.25
	R^2	0.9941	0.9916	0.9814
Freundlich	K_F	0.0757	0.0511	0.0516
	n	1.31	1.20	1.19
	R^2	0.9827	0.9924	0.9925
Temkin	α	0.082	0.067	0.068
	B	1.599	1.694	1.761
	R^2	0.9553	0.9324	0.9238
D-R	q_s	34.046	28.994	30.883
	B	0.0001	0.0001	0.00009
	E	70.711	70.711	74.536
	R^2	0.4745	0.3969	0.3705

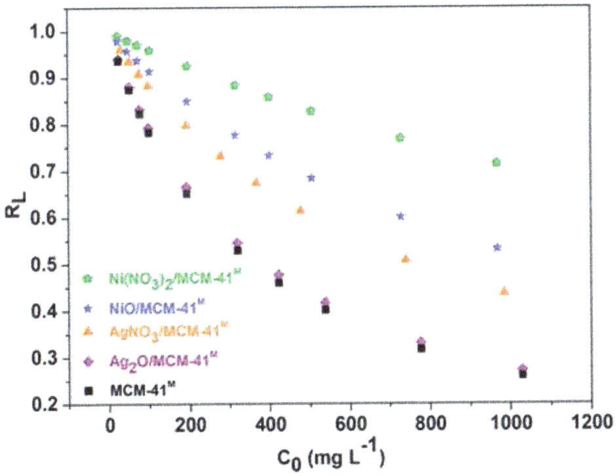

Figure 8. Plots of R_L versus initial concentrations.

2.5. Adsorption Thermodynamics

A thermodynamic assessment was performed for the sulfur adsorption over the $AgNO_3$/MCM-41^M material. The properties observed at 298, 308 and 318 K are given in Table 5. The values of ΔH° and ΔS° can be obtained from the slope and intercept of Van't Hoff plot of ln(K_L) against the reciprocal of T (1/T, K^{-1}) (Figure 9). The Langmuir model fitting for sulfur adsorption results with this material at 298, 308 and 318 K is depicted in Figure S4 of the Supplementary Materials.

Table 5. Thermodynamic parameters for sulfur adsorption on $AgNO_3$/MCM-41^M. Experimental conditions: room pressure, real diesel with sulfur concentration ranging from 25 to 500 ppm.

Adsorbent	T (K)	K_L (L/g) [1]	ΔG° (kJ/mol)	ΔH° (kJ/mol)	ΔS° (kJ/K mol)
$AgNO_3$/MCM-41^M	298	1.94×10^{-3}	15.528	42.161	0.089
	308	3.13×10^{-3}	14.634		
	318	5.69×10^{-3}	13.740		

[1] K_L is the Langmuir equilibrium constant calculated by the equation $\frac{C_e}{q_e} = \frac{C_e}{q_m} + \frac{1}{q_m K_L}$.

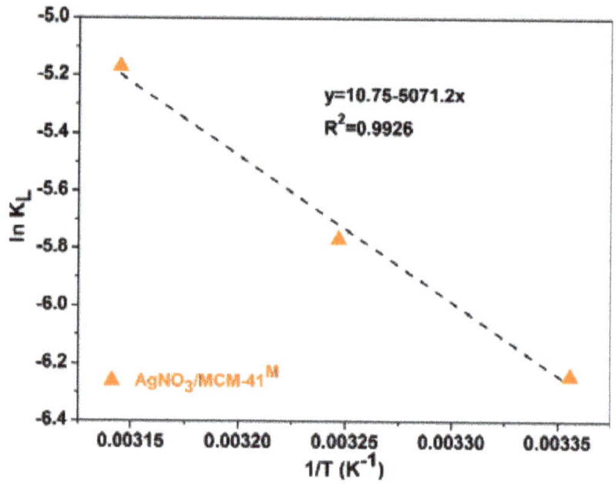

Figure 9. Van't Hoff plot.

The positive value for ΔH° indicates an endothermic process and its magnitude reaches the region defined for chemisorption in literature (≥40 kJ/mol) [54,55], thus suggesting predominantly chemical interactions between S-compounds and $AgNO_3$/MCM-41^M adsorbent. Although the process is not spontaneous, it can be noticed that there is a tendency of increasing spontaneity by increasing the adsorption temperature, since the total free energy (ΔG°) of the system decreases. Finally, the positive entropy (ΔS°) confirms the affinity of the adsorbent to S-compounds [46,56].

The data from the kinetic, equilibrium and thermodynamic tests reinforce the results observed in XPS spectra (Figure 5) for $AgNO_3$/MCM-41^M and Ag_2O/MCM-41^M materials before and after desulfurization, where stable Ag_2O species apparently did not interact with DBT molecules and only bridging bonds occurred between Ag (0) sites and the organosulfur rings (Ag-S-R), while active Ag-O species and Ag(0) nanodomains concurrently participate for adsorbing the contaminants via oxidation reactions and bridging interactions via π-complexation on the non-calcined adsorbent, visibly increasing the process efficiency. The same pattern occurs for Ni-impregnated MCM-41^M, where the non-calcined sample presents higher sulfur adsorption capacity than the calcined material; however, their results are not significant in comparison to those obtained with Ag species.

The sulfur adsorption results obtained for the MCM-41^M and $AgNO_3$/MCM-41^M materials regenerated by eluting a chloroform/hexane/acetone solution in the 40:30:30 ratio are depicted in Figure 10. Both materials presented a satisfactory reuse rate with just a slight decrease in sulfur adsorption after five regeneration cycles, thus indicating that the methodology was successful in removing the adsorbed organosulfurs without compromising their adsorption efficiency.

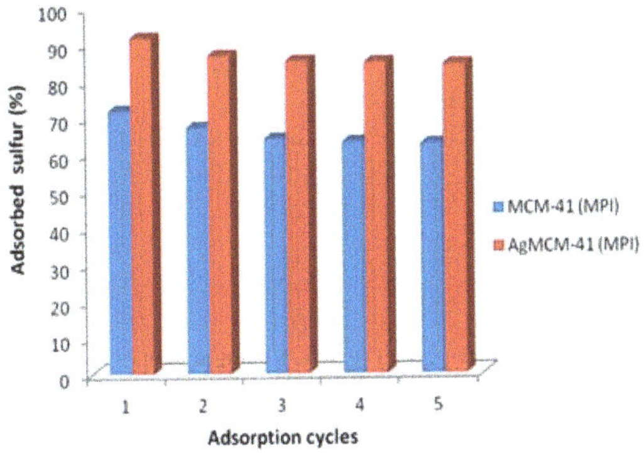

Figure 10. Sulfur adsorption results for the regenerated materials. Experimental conditions: room temperature and pressure.

3. Materials and Methods

3.1. Synthesis

Amorphous silica (MPI) was synthesized from beach sand following a new methodology described in our previous publication [23]. This silica presents isotherm type III, typical of some porous solids with meso and macroporosity, 33.54 m^2/g surface area (S_{BET}), 130.3 Å pore diameter and 0.18 cm^3/g pore volume. In addition, this material presents a chemical composition of 96.05% SiO_2 and main impurities such as K_2O, Fe_2O_3 and CaO with average contents of 1.84%, 1.12% and 0.81%, respectively.

MCM-41 was synthesized by the hydrothermal method described in [22,23] using MPI silica, with some modifications. The reaction system was composed of two initial solutions: (I) a mixture containing 14.75 g of MPI silica, 4.66 g of sodium hydroxide (PA, Vetec|Sigma-Aldrich, Duque de Caxias, RJ, Brazil) and 105.0 mL of distillated water, which was stirred for 2 h at 333 K; and (II) a mixture containing 21.68 g of cetyltrimethylammonium bromide (CTAB) (98.0%, Vetec|Sigma-Aldrich, Duque de Caxias, RJ, Brazil) and 105.0 mL of distillated water, stirred for 1 h at room temperature. Then, mixture (I) was added to (II) and stirred for 1 h at room temperature. The final gel had a molar composition of 1.0 CTAB:4.0 SiO_2:2.0 NaOH:200.0 H_2O, and was added into a teflon autoclave at 373 K for 96 h. The pH was adjusted daily with a 30% acetic acid (99.8%, Proquímios, Bangu, RJ, Brazil) solution until achieving a range of 9.5–10.0 pH. Then, the gel was washed with distilled water and 37% HCl (PA, Proquímios, Bangu, RJ, Brazil) solution in ethanol (99.5%, NEON, São Paulo, SP, Brazil) at 2%. The material was recovered by vacuum filtration and then dried at 393 K for 2 h. Finally, the solid was calcinated in air flow at 823 K with 278 K/min heating rate for 5 h. MCM-41 was also synthesized with commercial silica (silica gel 60, Macherey-Nagel, Düren, Germany), MCM-41^C, by following the same methodology for comparison with MCM-41^M in sulfur adsorption tests.

3.2. Adsorbent Modification

The mesoporous MCM-41 silica was modified with Ag^+ and Ni^{2+} cations from silver nitrate (99.8%, Vetec|Sigma-Aldrich, Duque de Caxias, RJ, Brazil) and hexa-hydrated nickel nitrate (99.999%, Sigma-Aldrich, St. Louis, MO, USA) salts via wet impregnation. Ethanolic solutions of $AgNO_3$ (0.03 mol/L for 2% (*w*/*w*) and 0.12 mol/L for 8% (*w*/*w*) concentrations) and $Ni(NO_3)_2{\cdot}6H_2O$ (0.05 mol/L for 2% (*w*/*w*) and 0.22 for 8% (*w*/*w*)) salts were prepared and added to 3.0 g of calcinated MCM-41. The functionalizations were performed in rotary evaporators at 120 rpm for 1 h at room temperature and 3.5 h under progressive heating (283 K/30 min) in a water bath until the solvent completely evaporated.

The modified adsorbents were dried in an oven at 393 K for 2h [57]. Next, 50% of the material was separated and denominated $AgNO_3$/MCM-41 and $Ni(NO_3)_2$/MCM-41, while the other 50% was calcinated at 773 K during 4 h in an oven at 278 K/min heating rate and denominated Ag_2O/MCM-41 and NiO/MCM-41.

3.3. Adsorbent Characterization

X-ray diffraction analyses (XRD) were performed using a Bruker D2 Phaser (Bruker AXS, Madison, WI, USA) with CuKα radiation (λ = 1.5406 Å), 30 kV filament, 10 mA current, Ni filter and a LYNXEYE detector. The XRD patterns were obtained in the range of 1°–10° (low-angle) and 10°–90° (wide-angle) 2θ.

The textural parameters were evaluated via nitrogen adsorption–desorption isotherms at 77 K, as determined by an automatic ASAP 2420 system from Micrometrics (Micrometrics, Norcross, GA, USA). Prior to the measurements, the samples were previously degassed at 473 K and 10^{-4} mbar. The pore size distribution was calculated by applying the Barrett–Joyner–Halenda (BJH) method to the desorption branch of the N_2 isotherm.

MCM-41^M and $AgNO_3$/MCM-41^M were evaluated by high-resolution transmission electron microscopy (HRTEM) using a Philips CCCM 200 Supertwin-DX4 microscope. Scanning transmission electron microscopy (STEM) analysis and mapping data were recorded in a Helios Nanolab 650 (FEI, Brno, Czech Republic) instrument with a high-angle annular dark field (HAADF) detector, at 200 kV and 200 nA. The microanalysis was carried out with energy dispersive X-ray (EDAX) spectroscopy Super-X system provided with four X-ray detectors and an X-FEG beam.

X-ray photoelectron spectra (XPS) was collected using a Physical Electronics PHI 5700 (Physical Electronic, Minneapolis, MN, USA) spectrometer with non-monochromatic Al Kα radiation (95.2 W, 15 kV, and 1486.6 eV) and a multi-channel detector. C1s peaks were used as an inner standard calibration peak at 284.8 eV. The Multipack software version 9.6.0.15 was used for data analysis. The recorded spectra were fitted using Gaussian–Lorentzian curves to more accurately determine the binding energies of the different element core levels.

3.4. Real and Model Fuels

Diesel fuel samples containing 1234.9 ppm (high-sulfur) and 5.1 ppm (low-sulfur) sulfur were kindly donated by the Clara Camarão Potiguar Refinery (Guamaré/RN, Brazil). This material was used for equilibrium, kinetic and thermodynamic evaluation of sulfur adsorption tests performed in a batch system. Moreover, model diesel solutions prepared with dibenzothiophene (98.0%, Sigma-Aldrich, Co., St. Louis, MO, USA) and n-decane (99.0%, Sigma-Aldrich, Co., St. Louis, MO, USA) in a concentration of around 2000 ppm sulfur were used in a set of adsorption tests in fixed bed columns, exclusively for evaluating the composition of the adsorbents after desulfurization via XPS analysis, since real diesel has many interferents in its composition and would generate noisy spectra.

3.5. Preliminary Adsorption Tests

All materials prepared in this work were subjected to initial sulfur adsorption tests with real diesel fuel samples (1234.9 ppm sulfur). The tests were performed in adsorption columns (0.61 cm × 30.0 cm) with downward flow. Each adsorbent was added to the glass columns up to 9.0 cm in height (about 0.5 g of material), and a volume of 5.0 mL of diesel was kept constant during the experiments. The first aliquots (0.3 mL) were collected and the sulfur content data were recorded using a total sulfur analyzer via ultraviolet fluorescence (UVF) spectrometry (Antek Multitek, PAC, L.P, Houston, TX, USA).

3.6. Batch Adsorption

The adsorptive desulphurization tests were performed using the finite bath method at ambient temperature and pressure. An amount of 0.3 g of adsorbent and 8.0 mL of diluted real diesel prepared by a solution of the high and low-sulfur samples were placed in erlenmeyer flasks under constant stirring (100 rpm) in a stirring table (SL 180/DT, Solab, Piracicaba, SP, Brazil). Different proportions of real diesel samples were used to produce the solutions with distinct sulfur concentrations which were applied in each test. The solids were separated by centrifugation and the residual sulfur contents in the liquid phase were collected for quantification following ASTM D5453 methodology. Adsorption effectivity was measured by the UVF sulfur analyzer and the sulfur adsorption capacity per gram of adsorbent at equilibrium (q_e) was calculated using the following expression:

$$q_e = (C_i - C_e)V/W \tag{1}$$

where V is the diesel solution volume (L), W is the adsorbent mass (g) and Ci and Ce are the initial sulfur content in the fuel (mg/L) and at equilibrium, respectively.

3.6.1. Kinetic Tests

Kinetic tests were performed for MCM-41^M and $AgNO_3$/MCM-41^M samples following the methodology described for batch adsorption using a real diesel solution with 500 ppm sulfur content. The supernatant fluid was collected at predetermined times between 5 and 180 min of stirring. The generated models are used to explain the adsorption mechanism characteristics. Pseudo-first-order (Equation (2)), pseudo-second-order (Equation (3)), Elovich (Equation (4)), intra-particle diffusion (Equation (5)) and the film-diffusion model of Boyd (Equation (6)) were used to analyze the experimental data using the following equations [46]:

$$\ln(q_e - q_t) = lnq_e - k_1 t \tag{2}$$

$$\frac{t}{q_t} = \frac{1}{k_2 q_e^2} + \frac{t}{q_e} \tag{3}$$

$$q_t = \frac{1}{\beta} ln\alpha\beta + \frac{1}{\beta} lnt \tag{4}$$

$$q_t = k_{id} t^{\frac{1}{2}} + C \tag{5}$$

$$F = 1 - \frac{6}{\pi^2} \sum_{n=1}^{\infty} \frac{1}{n^2} \exp\left(-n^2 B_t\right) \tag{6}$$

where q_t and q_e are the amounts of sulfur adsorbed (mg/g) at time t (min) and at equilibrium, respectively, k_1 is the pseudo-first-order constant (min^{-1}), k_2 is the pseudo-second-order adsorption constant (mg/g min), the parameter α is the initial adsorption rate of Elovich equation (mg/g min) and β is the desorption constant (g/mg) and k_{id} is the rate constant of the intra-particle diffusion models (mg/g min$^{1/2}$). F is fractional uptake (q_t/q_e) at a certain time and B_t is a mathematical function of F calculated by Equations (7) and (8) [58]:

$$\text{For } F > 0.85,\ B_t = f(F) = -0.4977 - \ln(1 - F) \tag{7}$$

$$\text{For } F < 0.85,\ B_t = f(F) = \left(\sqrt{\pi} - \sqrt{\pi - \sqrt{\frac{\pi^2 F}{3}}}\right)^2 \tag{8}$$

3.6.2. Adsorption Equilibrium Tests

The evaluation of sulfur adsorption equilibrium was performed in a batch system during 24 h with samples of real diesel containing sulfur concentrations in the 25–1000 ppm range. Langmuir, Freundlich, Temkin and Dubinin-Radushkevich mathematical models were used to fit the experimental data of sulfur adsorption in order to explain the adsorption on the adsorbent/adsorbate system, as well as predict their equilibrium parameters. The adsorption equilibrium of MCM-41^{C} produced from commercial silica and impregnated with non-calcined salts was also evaluated for comparison with the renewable material. The Langmuir model (Equations (9) and (10)) was used to quantify and contrast the performance of various adsorbents [59]:

$$q_e = \frac{q_m K_L C_e}{1 + K_L C_e} \tag{9}$$

$$\frac{C_e}{q_e} = \frac{1}{q_m K_L} + \frac{C_e}{q_m} \tag{10}$$

where C_e (g/L) and q_e (mg/g) are the equilibrium concentration of the adsorbate and adsorption capacity of the adsorbent, respectively; Langmuir q_m is a constant expressing the maximum absorption of the adsorbate (mg/g) and K_L is the Langmuir constant (L/g), also related to the adsorption energy and the affinity of the adsorbent. Dimensionless constant commonly known as a separation factor (R_L) defined by Weber and Chakravorti (1974) [60], can be given by the following equation:

$$R_L = 1/(1 + K_L C_0) \tag{11}$$

where C_0 in mg/L is the initial concentration of the analytes and K_L in L/mg is Langmuir constant. $R_L >$ 1 indicates an unfavorable process, $R_L = 1$ indicates linearity, $0 < R_L < 1$ indicates a favorable process and $R_L = 0$ signifies an irreversible adsorption.

The Freundlich model (Equations (12) and (13)) describes non-ideal and reversible adsorption, not being restricted to the formation of a monolayer [61]. This empirical model can be applied to the adsorption of multiple layers, with a non-uniform distribution of adsorption heat and affinities of the heterogeneous surface:

$$q_e = K_F C_e^{\frac{1}{n}} \tag{12}$$

$$ln q_e = ln K_F + \frac{1}{n} ln C_e \tag{13}$$

The Freundlich expression is an exponential equation and therefore assumes that the adsorbent concentration in the surface increases with the concentration of adsorbate. With the use of this expression, an infinite amount of adsorption can theoretically occur [62]. The equation is widely used in heterogeneous systems, where K_F (L/g) and n are characteristic constants of Freundlich, which indicate the adsorption capacity and adsorption intensity, respectively, where values of n in the range $1 < n < 10$ indicate favorable adsorption.

The Temkin isotherm model takes into consideration the effects of indirect adsorbate/adsorbate interactions which suggest that the adsorption heat of all adsorbed molecules in the layer decreases linearly with coverage [47]. The model can be expressed according to Equation (14):

$$q_e = \beta ln \alpha C_e \tag{14}$$

where β is related to the heat of adsorption (J/mol) and α is the Temkin isotherm constant (L/g). The Temkin linearized equation is given in Equation (15):

$$q_e = \beta ln \alpha + \beta ln C_e \tag{15}$$

The Dubinin-Radushkevich (D-R) isotherm expresses the adsorption mechanism on a heterogeneous and porous surface with variable parameters [63]. The D-R model is described by Equation (16), and the linearized equation is presented in Equation (17):

$$q_e = q_s e^{-\beta \varepsilon^2} \tag{16}$$

$$ln q_e = ln q_s - \beta \varepsilon^2 \tag{17}$$

where q_s is the theoretical saturation capacity (mg/g), β is the D-R constant (mol^2/kJ^2) and ε is the Polanyi potential, and can be expressed as follows:

$$\varepsilon = RTln(1 + 1/C_e) \tag{18}$$

where R is the universal gas constant (8.314 J/mol K), and T is the absolute temperature (K).

The β constant is related to the mean free energy of adsorption which is computed through Equation (19):

$$E = 1/\sqrt{2\beta} \tag{19}$$

where E is the mean adsorption energy (kJ/mol). The values of $E < 40$ kJ/mol implies physical adsorption while $E > 40$ kJ/mol suggests chemical adsorption [64].

Origin 2018 software (Origin Labs) was used to adjust the linearized equations of the models for calculating the equilibrium parameters.

3.6.3. Sulfur Adsorption Thermodynamics

The effect of temperature in S-compounds adsorption was studied by varying the temperature from 298, 308 and 318 K. The obtained thermodynamics parameters describe the variation or the transformation of a system. The standard free energy ($\Delta G°$, kJ/mol), standard enthalpy change ($\Delta H°$, kJ/mol) and standard entropy change ($\Delta S°$, kJ/mol K) can be calculated from the following Equations [54]:

$$\Delta G° = -RTln(K_L) = \Delta H° - T\Delta S° \tag{20}$$

$$ln(K_L) = -\frac{\Delta H°}{RT} + \frac{\Delta S°}{R} \tag{21}$$

where K_L is the Langmuir isotherm constant, R is the gas constant (8.314 J/K mol) and T is temperature in Kelvin.

3.6.4. Regeneration Tests

The regeneration methodology consisted of packing 1 g of residual adsorbent collected from the sulfur adsorption tests in a column with ~10 mm diameter and 50 cm length, forming a fixed bed with about 8 cm adsorbent. Next, 25 mL of a chloroform/hexane/acetone solution in a ratio of 40:30:30 (chosen after some preliminary tests using various proportions) at 323 K were added to the column. When the entire solution was completely eluted, the washed adsorbent was heated to 343 K in a rotary evaporator for 1 h to evaporate the solvent residue and replaced in a column for a new adsorption test. This evaluation was repeated for 5 regeneration cycles. A process flow diagram of the overall methodology is presented in Scheme 1.

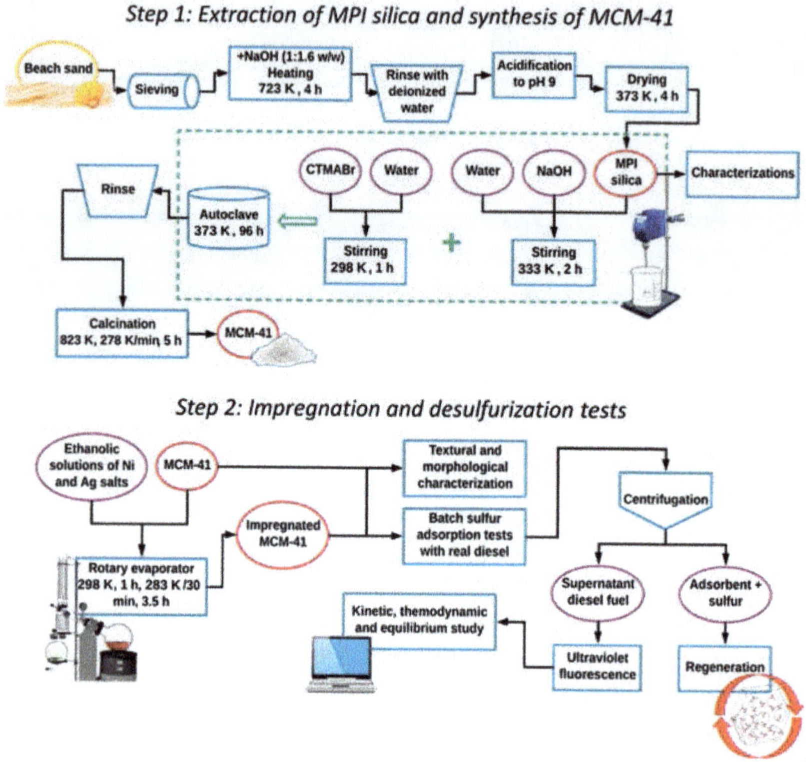

Scheme 1. Process flow diagram for the methodology used in this work.

4. Conclusions

The MCM-41^M adsorbent obtained from MPI silica presented the typical characteristics of this material, and its modification with metallic ions did not alter the material structure. Preliminary desulfurization results indicated better adsorption performance of $AgNO_3$/MCM-41^M and Ag_2O/MCM-41^M samples in comparison to the Ni-impregnated materials. High-resolution images of STEM-EDX and XPS spectra showed the presence of highly dispersed Ag on MCM-41^M support in the forms of Ag(0) nanoparticles and active Ag-O species, which were able to interact with organosulfur rings via oxidation and π-complexation in the non-calcined adsorbent ($AgNO_3$/MCM-41^M).

The results from the kinetic, equilibrium and thermodynamic tests reinforced the phenomena observed in XPS data and suggest that the chemisorptive interaction between organosulfur compounds and active Ag nanosites controls sulfur adsorption. The maximum adsorption capacity calculated by the Langmuir equation was about ten times higher in comparison to pure MCM-41^M and double the q_m for the $AgNO_3$/MCM-41^C sample from commercial silica. Thus, the $AgNO_3$/MCM-41^M material synthesized from MPI silica presents significant efficiency for adsorptive desulfurization and can be used as a complementary process to the expensive HDS method for its low-cost, sustainability and efficient removal of recalcitrant organosulfur compounds. Furthermore, the adsorbents presented a satisfactory regeneration rate by using a chloroform/hexane/acetone mixture (40:30:30 ratio) after a total of five sulfur adsorption cycles. This allows for efficient reuse of this material with certain longevity, ensuring even more reduction in operating costs.

Supplementary Materials: The following are available online at http://www.mdpi.com/2073-4344/9/8/651/s1, Figure S1: Results for preliminary sulfur adsorption tests, Figure S2: XRD results for the adsorbents with 8% metal,

Figure S3: Langmuir model fitting for sulfur adsorption results over $AgNO_3$/MCM-41^M material at different temperatures, Table S1: Sulfur compound distribution in a straight run diesel oil sample (Hua et al., 2003), Table S2: Critical diameter data for the main aromatic organosulfurs in real diesel, Table S3: Sulfur adsorption performance of some materials studied in literature for comparison.

Author Contributions: Conceptualization, R.V.S. and L.S.d.C.; methodology, R.V.S., A.B.F.C. and J.A.B.S.; formal analysis, E.R.-C., L.M.A.C., M.M.U. and S.B.C.P.; investigation, R.V.S., L.S.d.C., H.O.M.A.M., and A.B.F.C.; resources, L.M.A.C., J.A.B.S. and T.C.B.; data curation, E.R.-C. and S.B.C.P.; writing—original draft preparation, R.V.S., H.O.M.A.M., A.B.F.C., J.A.B.S. and L.S.d.C.; writing—review and editing, R.V.S., H.O.M.A.M., A.B.F.C., E.R.-C. and L.S.d.C.; supervision, R.V.S. and L.S.d.C.; project administration, L.S.d.C.; funding acquisition, S.B.C.P. and E.R.-C.

Funding: This research was funded by the Coordenação de Aperfeiçoamento de Pessoal de Nível Superior (CAPES—Brazil), grant number 001.

Acknowledgments: The authors thank the Energetic Technologies Laboratory (LTEN) and the Clara Camarão Potiguar Refinery (RPCC—Petrobras) for the diesel samples. The authors thank the Coordenação de Aperfeiçoamento de Pessoal de Nível Superior (CAPES—Brazil). ERC thanks to Ministerio de Ciencia, Innovación y Universidades, Project RTI2018-099668-B-C22 and FEDER funds.

Conflicts of Interest: The authors declare no conflict of interest.

References

1. Saleh, T.A.; Sulaiman, K.O.; AL-Hammadi, S.A.; Dafalla, H.; Danmaliki, G.I. Adsorptive desulfurization of thiophene, benzothiophene and dibenzothiophene over activated carbon manganese oxide nanocomposite: With column system evaluation. *J. Clean. Prod.* **2017**, *154*, 401–412. [CrossRef]
2. Pouladi, B.; Fanaei, M.A.; Baghmisheh, G. Optimization of oxidative desulfurization of gas condensate via response surface methodology approach. *J. Clean. Prod.* **2019**, *209*, 965–977. [CrossRef]
3. Andevary, H.H.; Akbari, A.; Omidkhah, M. High efficient and selective oxidative desulfurization of diesel fuel using dual-function [Omim]FeCl4 as catalyst/extractant. *Fuel Process. Technol.* **2011**, *185*, 8–17. [CrossRef]
4. Subhan, F.; Aslam, S.; Yan, Z.; Ikram, M.; Rehman, S. Enhanced desulfurization characteristics of Cu-KIT-6 for thiophene. *Microporous Mesoporous Mater.* **2014**, *199*, 108–116. [CrossRef]
5. Yang, D.; Yang, S.; Jiang, Z.; Yu, S.; Zhang, J.; Pan, F.; Cao, X.; Wang, B.; Yang, J. Polydimethyl siloxane–graphene nanosheets hybrid membranes with enhanced pervaporative desulfurization performance. *J. Membr. Sci.* **2015**, *487*, 152–161. [CrossRef]
6. Etemadi, N.; Sepahy, A.A.; Mohebali, G.; Yazdian, F.; Omidi, M. Enhancement of bio-desulfurization capability of a newly isolated thermophilic bacterium using starch/iron nanoparticles in a controlled system. *Int. J. Biol. Macromol.* **2018**, *120*, 1801–1809. [CrossRef]
7. Song, L.; Duan, Z.; Zhu, L.; Zhou, Y.; Xiang, Y.; Xia, D. Selective functionalization of external and internal surface of MCM-41 for adsorptive desulfurization. *J. Porous Mater.* **2016**, *23*, 1181–1187. [CrossRef]
8. Subhan, F.; Aslam, S.; Yan, Z.; Zhen, L.; Ikram, M.; Ullah, R.; Etim, U.J.; Ahmad, A. Ammonia assisted functionalization of cuprous oxide within confined spaces of SBA-15 for adsorptive desulfurization. *Chem. Eng. J.* **2018**, *339*, 557–565. [CrossRef]
9. Zhao, Z.; Zuhra, Z.; Qin, L.; Zhou, Y.; Zhang, L.; Tang, F.; Mu, C. Confinement of microporous MOF-74(Ni) within mesoporous γ-Al2O3 beads for excellent ultra-deep and selective adsorptive desulfurization performance. *Fuel Process. Technol.* **2018**, *176*, 276–282. [CrossRef]
10. Aslam, S.; Subhan, F.; Yan, Z.; Xing, W.; Zeng, J.; Liu, Y.; Ikram, M.; Rehman, S.; Ullah, R. Rapid functionalization of as-synthesized KIT-6 with nickel species occluded with template for adsorptive desulfurization. *Microporous Mesoporous Mater.* **2015**, *214*, 54–63. [CrossRef]
11. Teymouri, M.; Samadi-Maybodi, A.; Vahid, A.; Miranbeigi, A. Adsorptive desulfurization of low sulfur diesel using palladium containing mesoporous silica synthesized via a novel in-situ approach. *Fuel Process. Technol.* **2013**, *116*, 257–264. [CrossRef]
12. Sikarwar, P.; Kumar, U.K.A.; Gosu, V.; Subbaramaiah, V. Catalytic oxidative desulfurization of DBT using green catalyst (Mo/MCM-41) derived from coal fly ash. *J. Environ. Chem. Eng.* **2018**, *6*, 1736–1744. [CrossRef]
13. Shokouhimehr, M.; Kim, T.; Jun, S.W.; Shin, K.; Jang, Y.; Kim, B.H.; Kim, J.; Hyeon, T. Magnetically separable carbon nanocomposite catalysts for efficient nitroarene reduction and Suzuki reactions. *Appl. Catal. A Gen.* **2014**, *476*, 133–139. [CrossRef]

14. Ahadi, A.; Rostamnia, S.; Panahi, P.; Wilson, L.D.; Kong, Q.; An, Z.; Shokouhimehr, M. Palladium comprising dicationic bipyridinium supported periodic mesoporous organosilica (PMO): Pd@Bipy–PMO as an efficient hybrid catalyst for Suzuki–Miyaura Cross-Coupling Reaction in Water. *Catalysts* **2019**, *9*, 140. [CrossRef]
15. Shokouhimehr, M.; Hong, K.; Lee, T.H.; Moon, C.H.; Hong, S.P.; Zhang, K.; Suh, J.M.; Choi, K.S.; Varma, R.S.; Jang, H.W. Magnetically retrievable nanocomposite adorned with Pd nanocatalysts: Efficient reduction of nitroaromatics in aqueous media. *Green Chem.* **2018**, *20*, 3809–3817. [CrossRef]
16. Shokouhimehr, M.; Shin, K.Y.; Lee, J.S.; Hackett, M.J.; Jun, S.W.; Oh, M.H.; Jang, J.; Hyeon, T. Magnetically recyclable core–shell nanocatalysts for efficient heterogeneous oxidation of alcohols. *J. Mater. Chem. A* **2014**, *2*, 7593–7599. [CrossRef]
17. Zhang, K.; Suh, J.M.; Choi, J.W.; Jang, H.W.; Shokouhimehr, M.; Varma, R.S. Recent Advances in the Nanocatalyst-Assisted NaBH4 Reduction of Nitroaromatics in Water. *ACS Omega* **2019**, *4*, 483–495. [CrossRef] [PubMed]
18. Shokouhimehr, M. Magnetically separable and sustainable nanostructured catalysts for heterogeneous reduction of nitroaromatics. *Catalysts* **2015**, *5*, 534–560. [CrossRef]
19. Siriworarat, K.; Deerattrakul, V.; Dittanet, P.; Kongkachuichay, P. Production of methanol from carbon dioxide using palladium-copperzinc loaded on MCM-41: Comparison of catalysts synthesized from flame spray pyrolysis and sol-gel method using silica source from rice husk ash. *J. Clean. Prod.* **2017**, *142*, 1234–1243. [CrossRef]
20. Yu, H.; Xue, X.; Huang, D. Synthesis of mesoporous silica materials (MCM-41) from iron ore tailings. *Mater. Res. Bull.* **2009**, *44*, 2112–2115. [CrossRef]
21. Sohrabnezhad, S.; Jafarzadeh, A.; Pourahmad, A. Synthesis and characterization of MCM-41 ropes. *Mater. Lett.* **2018**, *212*, 16–19. [CrossRef]
22. Santos, L.S.; Silva, J.A.B.; Urbina, M.M.; Silva, E.G.; Andrade, J.C.F.; Silva, V.L. Processo de Produção de Sílica Proveniente da Areia de Praia. BR Patent 1020140252835, 2 August 2016.
23. Carvalho, L.S.; Silva, E.; Andrade, J.C.; Silva, J.A.; Urbina, M.; Nascimento, P.F.; Carvalho, F.; Ruiz, J.A. Low-cost mesoporous adsorbents amines-impregnated for CO2 capture. *Adsorption* **2015**, *21*, 597–609. [CrossRef]
24. Dutov, V.V.; Mamontov, G.V.; Zaikovskii, V.I.; Liotta, L.F.; Vodyankina, O.V. Low-temperature CO oxidation over Ag/SiO2 catalysts: Effect of OH/Ag ratio. *Appl. Catal. B Environ.* **2018**, *221*, 598–609. [CrossRef]
25. Hernández-Maldonado, A.J.; Yang, F.H.; Qi, G.; Yang, R.T. Desulfurization of transportation fuels by π-complexation sorbents: Cu(I)-, Ni(II)-, and Zn(II)- zeolites. *Appl. Catal. B Environ.* **2005**, *56*, 111–126. [CrossRef]
26. Tang, H.; Li, W.; Zhang, T.; Li, Q.; Xing, J.; Liu, H. Improvement in diesel desulfurization capacity by equilibrium isotherms analysis. *Sep. Purif. Technol.* **2011**, *78*, 352–356. [CrossRef]
27. Hu, P.; Amghouz, Z.; Huang, Z.; Xu, F.; Chen, Y.; Tang, X. Surface-confined atomic silver centers catalyzing formaldehyde oxidation. *Environ. Sci. Technol.* **2015**, *49*, 2384–2390. [CrossRef] [PubMed]
28. Chen, Y.; Huang, Z.; Zhou, M.; Ma, Z.; Chen, J.; Tang, X. Single silver adatoms on nanostructured manganese oxide surfaces: Boosting oxygen activation for benzene abatement. *Environ. Sci. Technol.* **2017**, *51*, 2304–2311. [CrossRef]
29. Santos, V.P.; Pereira, M.F.R.; Órfão, J.J.M.; Figueiredo, J.L. The role of lattice oxygen on the activity of manganese oxides towards the oxidation of volatile organic compounds. *Appl. Catal. B Environ.* **2010**, *99*, 353–363. [CrossRef]
30. Losurdo, M.; Bergmair, I.; Dastmalchi, B.; Kim, T.H.; Giangregroio, M.M.; Jiao, W.; Bianco, G.V.; Brown, A.S.; Hingerl, K.; Bruno, G. Graphene as an electron shuttle for silver deoxidation: Removing a key barrier to plasmonics and metamaterials for SERS in the visible. *Adv. Funct. Mater.* **2014**, *24*, 1864–1878. [CrossRef]
31. Dutov, V.V.; Mamontov, G.V.; Zaikovskii, V.I.; Vodyankina, O.V. The effect of support pretreatment on activity of Ag/SiO2 catalysts in low-temperature CO oxidation. *Catal. Today* **2016**, *278*, 150–156. [CrossRef]
32. Padin, J.; Yang, R.T. New sorbents for olefin/paraffin separations by adsorption via π-complexation: Synthesis and effects of substrates. *Chem. Eng. Sci.* **2000**, *55*, 2607–2616. [CrossRef]
33. Beck, J.S.; Vartulli, J.C.; Roth, W.J.; Leonowicz, M.E.; Kresge, C.T.; Schmitt, K.D.; Chu, C.T.W.; Olson, D.H.; Sheppard, E.W.; McCullen, S.B.; et al. A new family of mesoporous molecular sieves prepared with liquid crystal templates. *J. Am. Chem. Soc.* **1992**, *114*, 10834–10843. [CrossRef]

34. Kowalczyk, A.; Borcuch, A.; Michalik, M.; Rutkowska, M.; Gil, B.; Sojka, Z.; Indyka, P.; Chmielarz, L. MCM-41 modified with transition metals by template ion-exchange method as catalysts for selective catalytic oxidation of ammonia to dinitrogen. *Microporous Mesoporous Mater.* **2017**, *240*, 9–21. [CrossRef]
35. Qin, J.; Li, B.; Zhang, W.; Lv, W.; Han, C.; Liu, J. Synthesis, characterization and catalytic performance of well-ordered mesoporous Ni-MCM-41 with high nickel content. *Microporous Mesoporous Mater.* **2015**, *208*, 181–187. [CrossRef]
36. Sing, K.S.W.; Everett, D.H.; Haul, R.A.W.; Moscou, L.; Pierotti, R.A.; Rouquerol, J.; Siemieniewska, T. Reporting physisorption data for gas/solid systems with special reference to the determination of surface area and porosity. *Pure Appl. Chem.* **1985**, *57*, 603–619. [CrossRef]
37. Carraro, P.M.; Blanco, A.A.G.; Soria, F.A.; Lener, G.; Sapag, K.; Eimer, G.A.; Oliva, M.I. Understanding the role of nickel on the hydrogen storage capacity of Ni/MCM-41 materials. *Microporous Mesoporous Mater.* **2016**, *231*, 31–39. [CrossRef]
38. Shen, S.; Chen, J.; Koodali, R.T.; Hu, Y.; Xiao, Q.; Zhou, J.; Wang, X.; Guo, L. Activation of MCM-41 mesoporous silica by transition-metal incorporation for photocatalytic hydrogen production. *Appl. Catal. B Environ.* **2014**, *150*, 138–146. [CrossRef]
39. Hoflund, G.B.; Weaver, J.F.; Epling, W.S. AgO XPS Spectra. *Surf. Sci. Spectra* **1995**, *3*, 163–168. [CrossRef]
40. Huang, Z.; Gu, X.; Cao, Q.; Hu, P.; Hao, J.; Li, J.; Tang, X. Catalytically active single-atom sites fabricated from silver particles. *Angew. Chem.* **2012**, *124*, 4274–4279. [CrossRef]
41. Cao, Y.; Dai, W.L.; Deng, J.F. The oxidative dehydrogenation of methanol over a novel Ag/SiO2 catalyst. *Appl. Catal. A Gen.* **1997**, *158*, L27–L34. [CrossRef]
42. Yang, R.T. *Adsorbents: Fundamentals and Applications*; Wiley: Hoboken, NJ, USA, 2003.
43. Battocchio, C.; Meneghini, C.; Fratoddi, I.; Venditti, I.; Russo, M.V.; Aquilanti, G.; Maurizio, C.; Bondino, F.; Matassa, R.; Rossi, M.; et al. Silver nanoparticles stabilized with thiols: A close look at the local chemistry and chemical structure. *J. Phys. Chem. C* **2012**, *116*, 19571–19578. [CrossRef]
44. Brobbey, K.J.; Haapanen, J.; Gunell, M.; Toivakka, M.; Mäkelä, J.M.; Eerola, E.; Ali, R.; Saleem, M.R.; Honkanen, S.; Bobacka, J.; et al. Controlled time release and leaching of silver nanoparticles using a thin immobilizing layer of aluminum oxide. *Thin Solid Films* **2018**, *645*, 166–172. [CrossRef]
45. Haro, N.K.; Vecchio, P.D.; Marcilio, N.R.; Féris, L.A. Removal of atenolol by adsorption—Study of kinetics and equilibrium. *J. Clean. Prod.* **2017**, *154*, 214–219. [CrossRef]
46. Habibi, A.; Belaroui, L.S.; Bengueddach, A.; Galindo, A.L.; Díaz, C.I.S.; Peña, A. Adsorption of metronidazole and spiramycin by an Algerian palygorskite. Effect of modification with tin. *Microporous Mesoporous Mater.* **2018**, *268*, 293–302. [CrossRef]
47. Choi, A.E.S.; Roces, S.; Dugos, N.; Arcega, A.; Wan, M. Adsorptive removal of dibenzothiophene sulfone from fuel oil using clay material adsorbents. *J. Clean. Prod.* **2017**, *161*, 267–276. [CrossRef]
48. Ho, Y.S.; McKay, G. Pseudo second-order model for sorption process. *Process Biochem.* **1999**, *34*, 451–465. [CrossRef]
49. Jeon, H.J.; Ko, C.H.; Kim, S.H.; Kim, J.N. Removal of Refractory Sulfur Compounds in Diesel Using Activated Carbon with Controlled Porosity. *Energy Fuels* **2009**, *23*, 2537–2543. [CrossRef]
50. Bu, J.; Loh, G.; Gwie, C.G.; Dewiyanti, S.; Tasrif, M.; Borgna, A. Desulfurization of diesel fuels by selective adsorption on activated carbons: Competitive adsorption of polycyclic aromatic sulfur heterocycles and polycyclic aromatic hydrocarbons. *Chem. Eng. J.* **2011**, *166*, 207–217. [CrossRef]
51. Oliveira, M.F.; Silva, M.G.C.; Vieira, M.G.A. Equilibrium and kinetic studies of caffeine adsorption from aqueous solutions on thermally modified Verde-lodo bentonite. *Appl. Clay Sci.* **2019**, *168*, 366–373. [CrossRef]
52. Meng, C.; Fang, Y.; Jin, L.; Hu, H. Deep desulfurization of model gasoline by selective adsorption on Ag+/Al-MSU-S. *Catal. Today* **2010**, *149*, 138–142. [CrossRef]
53. Muzic, M.; Sertic-Bionda, K.; Gomzi, Z.; Podolski, S.; Telen, S. Study of diesel fuel desulfurization by adsorption. *Chem. Eng. Res. Des.* **2010**, *88*, 487–495. [CrossRef]
54. Adelodun, A.A.; Ngila, J.C.; Kim, D.; Jo, Y.M. Isotherm, thermodynamic and kinetic studies of selective CO2 adsorption on chemically modified carbon surfaces. *Aerosol Air Qual. Res.* **2016**, *16*, 3312–3329. [CrossRef]
55. Raganati, F.; Alfe, M.; Gargiulo, V.; Chirone, R.; Ammendola, P. Isotherms and thermodynamics of CO2 adsorptionon a novel carbon-magnetite composite sorbent. *Chem. Eng. Res. Des.* **2018**, *134*, 540–552. [CrossRef]

56. Ben-Ali, S.; Jaouali, I.; Souissi-Najar, S.; Ouederni, A. Characterization and adsorption capacity of raw pomegranate peel biosorbent for copper removal. *J. Clean. Prod.* **2017**, *142*, 3809–3821. [CrossRef]
57. Sales, R.V.; Moura, H.O.M.A.; Silva, S.R.B.; Souza, M.A.F.; Campos, L.M.A.; Rodriguez-Castellon, E.; Carvalho, L.S. Experimental and theoretical study of adsorptive interactions in diesel fuel desulfurization over Ag/MCM-41 adsorbent. *Adsorption* **2019**, *25*, 1–13. [CrossRef]
58. Reichenberg, D. Properties of ion-exchange resins in relation to their structure. III. Kinetics of exchange. *J. Am. Chem. Soc.* **1953**, *75*, 589–597. [CrossRef]
59. Langmuir, I. The constitution and fundamental properties of solids and liquids. *Part I. Solids. J. Am. Chem. Soc.* **1916**, *38*, 2221–2295. [CrossRef]
60. Weber, T.W.; Chakravorti, R.K. Pore and solid diffusion models for fixed-bed adsorbers. *AIChE J.* **1974**, *20*, 228–238. [CrossRef]
61. Freundlich, H. Über die adsorption in Lösungen. *Z. Phys. Chem.* **1907**, *57*, 385–470. [CrossRef]
62. Pérez-Marín, A.B.; Zapata, V.M.; Ortuño, J.F.; Aguilar, M.; Sáez, J.; Lloréns, M. Removal of cadmium from aqueous solutions by adsorption onto orange waste. *J. Hazard. Mater. B* **2007**, *139*, 122–131. [CrossRef] [PubMed]
63. Dubinin, M. The potential theory of adsorption of gases and vapors for adsorbents with energetically nonuniform surfaces. *Chem. Rev.* **1960**, *60*, 235–241. [CrossRef]
64. Al-Anber, M.A. Thermodynamics approach in the adsorption of heavy metals. In *Thermodynamics—Interaction Studies—Solids, Liquids and Gases*; Moreno-Pirajan, J.C., Ed.; InTech: Rijeka, Croatia, 2011; pp. 737–764.

catalysts

MDPI

Article

A New Tool in the Quest for Biocompatible Phthalocyanines: Palladium Catalyzed Aminocarbonylation for Amide Substituted Phthalonitriles and Illustrative Phthalocyanines Thereof

Vanessa A. Tomé, Mário J. F. Calvete *, Carolina S. Vinagreiro, Rafael T. Aroso and Mariette M. Pereira *

Centro de Química de Coimbra (CQC), Department of Chemistry, University of Coimbra, Coimbra 3004-535, Portugal; vanessalmeida_97@hotmail.com (V.A.T.); carolina_svinagreiro@hotmail.com (C.S.V.); fortilap@gmail.com (R.T.A.)

* Correspondence: mcalvete@qui.uc.pt (M.J.F.C.); mmpereira@qui.uc.pt (M.M.P.); Tel.: +351-966-174744 (M.J.F.C.); +351-239-854474 (M.M.P.)

Received: 1 October 2018; Accepted: 17 October 2018; Published: 20 October 2018

Abstract: The amide peptide bond type linkage is one of the most natural conjugations available, present in many biological synthons and pharmaceutical drugs. Hence, aiming the direct conjugation of potentially biologically active compounds to phthalocyanines, herein we disclose a new strategy for direct modulation of phthalonitriles, inspired by an attractive synthetic strategy for the preparation of carboxamides based on palladium-catalyzed aminocarbonylation of aryl halides in the presence of carbon monoxide (CO) which, to our knowledge, has never been used to prepare amide-substituted phthalonitriles, the natural precursors for the synthesis of phthalocyanines. Some examples of phthalocyanines prepared thereof are also reported, along with their full spectroscopic characterization and photophysical properties initial assessment.

Keywords: peptide bond; phthalonitriles; phthalocyanines; aminocarbonylation; palladium catalysts

1. Introduction

Molecules of the tetrapyrrole family (e.g., porphyrin and phthalocyanine derivatives) are probably the most appealing chromophores for a vast array of photoactivated processes, such as phototherapy [1–3], photodiagnosis [4–6], photocatalysis [7,8], solar energy conversion [9,10], and also as photomaterials [11–13]. In particular, phthalocyanines largely fulfil a crucial optical requisite, necessary for photomedicinal applications, which is a strong absorption in near-infrared (NIR) spectral region (600–900 nm), as light in this region affords the deepest penetration in soft tissue. These highly stable compounds, possessing high molar absorptivity, high quantum yields of fluorescence and structural versatility [4,13,14], can be conveniently modified to grant suitable biological solubility [15–18], by introduction of hydrophilic moieties in the structure. The majority of the moieties used so far are negatively charged, such as sulfonates or carboxylates [15,19] or positively charged, like quaternized amines [15,20]. These ionic features present some important drawbacks like low cellular uptake and/or cellular internalization, due to the negatively-charged character of the cell plasma membranes, in the case of anionic phthalocyanines [21], or exaggerated phospholipid affinity, leading to phospholipidosis, in the case of cationic phthalocyanines [22]. Thus, in the search for biocompatibility, the amide peptide bond type linkage is one of the most natural conjugations available, present in many biological synthons, such as peptides, proteins, or amino acids [23], as well in pharmaceutical drugs [24,25]. Nevertheless, amide-substituted phthalocyanines are rare [26–35], when compared with other functionalities, including a few reports of phthalocyanines conjugated with amino acids [27–30] and peptides [31–35]. The main reason for the scarcity in phthalocyanine-amino acid conjugates arises from the difficult synthetic manipulation, which relies in troublesome transformations and purification procedures using highly polluting chemicals [25,27–30,36,37].

Apparently, phthalocyanine post-synthetic modulation would not be a very straightforward option, due to the chemical stability owned by phthalocyanines, which are quite stable against this type of structural variation. On the other hand, modification of precursory phthalonitriles bearing carboxylic acids is also demanding, given the sensitiveness of nitrile functions. It is worth mentioning that we have tested the strategies ourselves, to explain our points, and results were as described in the text. Whether in case of post-synthetic phthalocyanine modification or phthalonitrile modulation, no reproducible results could be obtained, always leading to cumbersome work-up approaches.

Palladium-catalyzed carbonylation reactions were first described by Heck almost 40 years ago [38]. Since then, many developments have been reported [6,39–46] and nowadays carbonylation has become an indispensable alternative to the classic organic synthesis of carbonyl compounds, including carboxylic acid derivatives (e.g., amides, esters) with valuable application in both industrial and fine chemistry. Among these reactions, aminocarbonylation [47–50], carried out using Ar–X substrates (X = I, Br, Cl, OTf, OTs, etc.), in the presence of *N*-nucleophiles, emerges as a sustainable, one-step synthetic approach, for the efficient, selective and mild synthesis of amides.

Herein we disclose a new strategy for direct modulation of phthalonitriles, inspired by an attractive synthetic strategy for the preparation of carboxamides based on optimized palladium-catalyzed aminocarbonylation of aryl halides in the presence of carbon monoxide (CO) [38,51] which, to our knowledge, has never been used to prepare amide-substituted phthalonitriles, the natural precursors for the synthesis of phthalocyanines. Furthermore, transformation thereof to the desired phthalocyanines is also described.

2. Results and Discussion

Modification of phthalonitriles is usually the chosen methodology when attempting to introduce significant changes at the phthalocyanine periphery, instead of phthalocyanine post-synthetic modulation [52], due to the known chemical stability owned by phthalocyanines. A conceivable example would be, for instance, to synthesize a phthalocyanine bearing peripheral four carboxylic acid groups, followed by acyl chloride formation, using a hazardous chlorinating agent, and then functionalization with an amine. The main issue regarding this strategy would be the proneness to form mixtures of mono-, di-, tri-, and tetra-amide substituted phthalocyanines requiring the use of excessive amounts of nucleophile, giving raise to cumbersome purification and low yields.

Our herein envisaged strategy uses 4-iodophthalonitrile (**1**) [53] as substrate and a range of amines as nucleophiles, in presence of a palladium catalyst formed in situ by addition of palladium(II) acetate to triphenylphosphine (in 1:2 molar ratio), together with Et_3N as base and carbon monoxide as reagent (Table 1) [54,55]. Our studies began with the aminocarbonylation of 4-iodophthalonitrile (**1**) using glycine methyl ester hydrochloride (**2a**) as model nucleophile, to optimize reaction conditions (temperature, pressure of CO and time reaction parameters) in the palladium-catalyzed aminocarbonylation reaction (Table 1).

The first reaction conditions employed (P_{CO} = 10 bar and T = 65 °C) (Table 1, entry 1) afforded only 25% substrate conversion, after 24 h. Then, we investigated the effect of increasing temperature, keeping the CO pressure at 10 bar, and it was found that, at 85 °C, the reaction proceeded faster, and full conversion of substrate **1** within 24 h was obtained (Table 1, entry 2). Keeping CO pressure at 10 bar and temperature at 85 °C, the reaction was not complete when reaction time was decreased to 12 h, reaching only 70% conversion of **1** (Table 1, entry 3). Conversely, when the reaction temperature was increased to 100 °C, keeping the pressure at 10 bar and reaction time at 12 h, substrate **1** was totally transformed into the desired amide (Table 1, entry 4). In addition, when the CO pressure was reduced to 5 bar, keeping the temperature at 100 °C, after 12 h, the conversion of substrate **1** was >97% (Table 1, entry 5). However, keeping CO pressure at 5 bar and reducing the temperature to 65 °C, it required 70 h until full conversion of substrate **1** was observed (Table 1, entry 6). Thus, this indicates that the temperature plays the most important role on the activity of the catalyst. To evaluate the effect of solvent, an additional experiment was performed using DMF instead of toluene and, regardless of the high conversion obtained using the same conditions, this reaction yielded a complex mixture of products (Table 1, entry 7), as checked and compared using thin layer chromatography-TLC, which may be attributed to decomposition of DMF. Summing, it was found that a temperature of 100 °C, a CO pressure of 5 bar and a reaction time of 12 h were the optimal reaction parameters selected to extend the scope of 4-iodophthalonitrile functionalization.

Table 1. Optimization of reaction conditions [a].

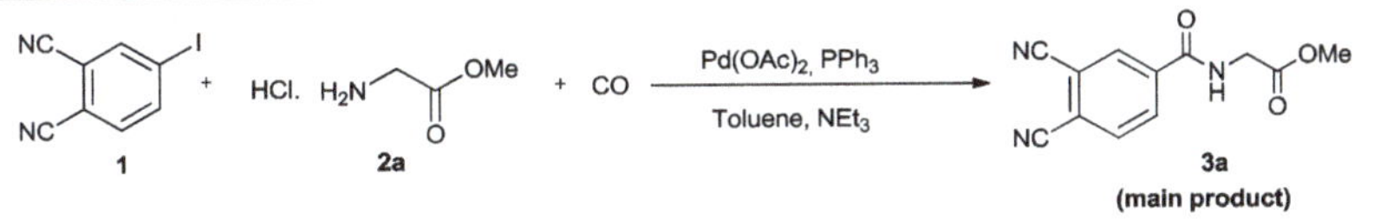

Entry	Solvent	CO Pressure (bar)	Temperature (°C)	Time (h)	Conversion (%) [b]
1	toluene	10	65	24	25
2	toluene	10	85	24	>98
3	toluene	10	85	12	70
4	toluene	10	100	12	>98
5	toluene	5	100	12	>98
6	toluene	5	65	70	>97
7 [c]	DMF	5	100	12	>98

[a] General reaction conditions: 2.5 mol % $Pd(OAc)_2$, 5 mol % PPh_3, 8 equiv. Et_3N, 1.1 equiv. **2a**. [b] Substrate conversion determined by ^{1}H-NMR on the reaction mixture obtained after evaporation of the solvent; [c] gave a complex mixture of products.

Hence, our next step was to promote the catalytic aminocarbonylation reaction between 4-iodophthalonitrile (**1**) with a wide range of amines as nucleophiles (**2a**–**2g**) to obtain the corresponding carboxamides (Table 2). Several structurally different amines as *N*-nucleophiles were used: three amino acid methyl esters (methyl glycinate (**2a**), methyl leucinate (**2b**) and methyl phenylalaninate (**2c**)), *tert*-butylamine (**2d**), *N*-BOC-ethylenediamine (**2e**), chalcone (*E*)-1-(4-aminophenyl)-3-(3,4,5-trimethoxyphenyl)prop-2-en-1-one (**2f**) [56], and piperazine (**2g**) (Table 2). Each reaction's progress was followed by TLC of aliquots taken from the reactor via *cannula*. After complete conversion of **1** to the corresponding carboxamides, the reaction mixture was then subjected to work-up and chromatographic purification procedures (See SI), yielding amide substituted phthalonitriles **3a**–**3g**, in good isolated yields (54–80%) at optimized reaction conditions (Table 2) (see also Figures S1–S21, SI).

When aminoesters were used as nucleophiles (**2a**–**2c**), 12 h were necessary for the complete conversion of the substrate, leading to carboxamides **3a**, **3b** and **3c** in 65%, 54%, and 59% isolated yields, respectively (Table 2, entries 1–3). It is worth mentioning that the yields obtained were higher than the ones reported for the model substrate iodobenzene using similar aminoesters as nucleophiles [57,58]. This may be attributed to the presence of cyano electron-withdrawing groups in 4-iodophthalonitrile, which enables an easier oxidative addition step in the catalytic cycle (A in Figure 1).

Table 2. Palladium catalyzed aminocarbonylation of 4-iodophthalonitrile using several amines as nucleophiles [a].

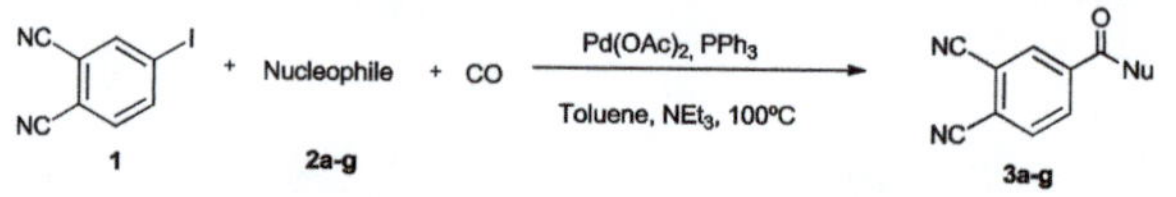

Entry	Nucleophile (NuH)	Amine/(1) Ratio (equiv.)	Time (h)	Product	Yield (%) [b]
1	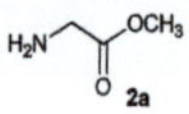 **2a**	1.1	12	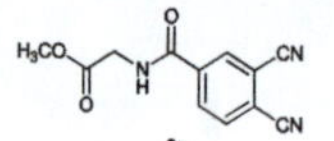 **3a**	65
2	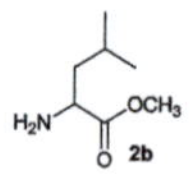 **2b**	1.1	12	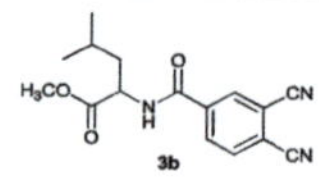 **3b**	54
3	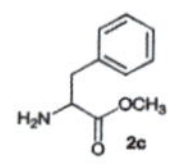 **2c**	1.1	12	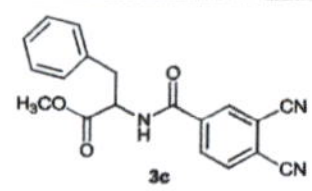 **3c**	59
4	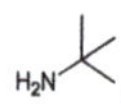 **2d**	3.3	4	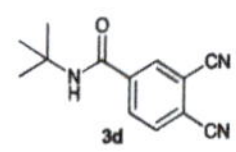 **3d**	74
5	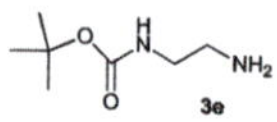 **3e**	1.2	3	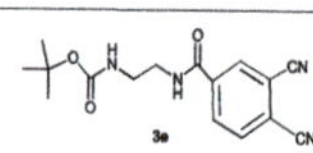 **3e**	80
6	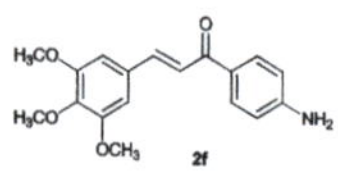 **2f**	1.2	25	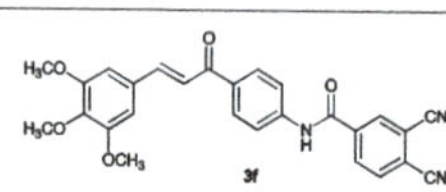 **3f**	70
7	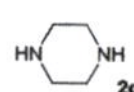 **2g**	6	7	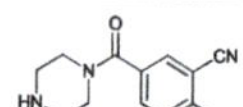 **3g**	77

[a] General reaction conditions: 5 bar (CO), 2.5 mol % $Pd(OAc)_2$, 5 mol % PPh_3, 8 equiv. Et_3N. Reactions were carried out in toluene (0.1 M, concerning substrate **1**). [b] Isolated yield.

Simpler aliphatic amines such as *tert*-butylamine (**2d**) were also used as nucleophiles. In this case, using 3.3 equivalents of **2d**, the aminocarbonylation reaction of **1** proceeded in the presence of palladium catalyst formed in situ by addition of palladium(II) acetate to triphenylphosphine (in 1:2 molar ratio), using toluene as solvent, under a CO pressure of 5 bar. Complete conversion of **1** was obtained in just 4 h, yielding carboxamide **3d** in 74% isolated yield (Table 2, entry 4).

Moreover, an *N*-mono-protected-ethylenediamine (**2e**) was also used. We had to prepare the mono-protected amine since, under the same reaction conditions, when unprotected ethylenediamine as nucleophile was used, a complex mixture of *N*-mono and *N,N'*-bis-substituted ethylenediamine, along with degradation products was formed, according to ^{1}H-NMR analysis. To overcome this problem, we then prepared **2e**, using *tert*-butyloxycarbonyl protecting group (BOC group) following a literature procedure [59]. Next, we promoted the aminocarbonylation reaction of **1** with nucleophile *N*-BOC-ethylenediamine (**2e**), yielding **3e** in 80% isolated yield after just 3 h (Table 2, entry 5). Next, the aminocarbonylation of **1** with chalcone **2f**, which is a potential anti-microbial agent [60], yielded carboxamide **3f** in 70% isolated yield, under standard reaction conditions, after 25 h (Table 2, entry 6). Since the chalcone is an aromatic amine, it is expected to be less nucleophilic and, consequently a prolonged period of time was necessary for the complete conversion of the substrate **1** into the corresponding carboxamide.

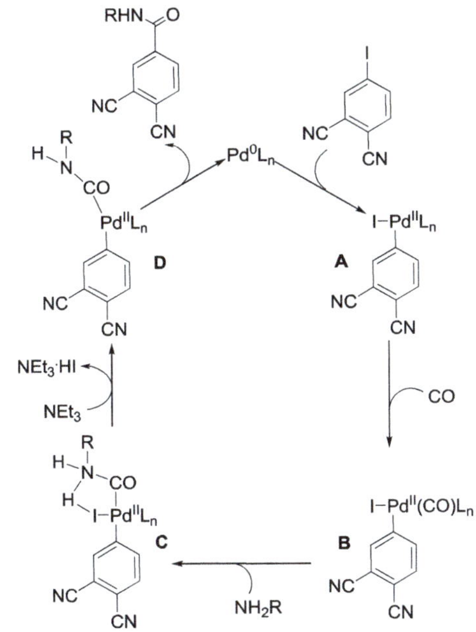

Figure 1. Simplified catalytic cycle describing the formation of 4-amide substituted phthalonitriles. L = PPh_3.

Using similar conditions, we also investigated the use of cyclic diamines in the palladium catalyzed aminocarbonylation reaction for the synthesis of *N*-mono-substituted diamines. Unprotected diamine piperazine (**2g**) is quite useful and interesting because the presence of two amine groups could enable the conjugation with bioactive molecules or functionalization with other relevant chemical groups. In order to attain the desired *N*-mono-substituted diamine we have selected an excess of 6 equiv. of the diamine **2g**. In the presence of palladium catalyst formed in situ by addition of palladium(II) acetate to triphenylphosphine (in 1:2 molar ratio), together with Et_3N as base in toluene solvent, under a CO pressure of 5 bar, complete conversion of **1** was obtained, after 7 h, yielding carboxamide **3g** after work-up and purification in 77% isolated yield (Table 2, entry 7).

According to previously described [49,51,61–64], a simplified mechanism for the formation of 4-amide substituted phthalonitrile is proposed in Figure 1. The catalytic cycle begins with the oxidative addition of the in situ formed Pd(0)L_n active species to the 4-iodophthalonitrile, resulting in an arylpalladium(II) intermediate **A**, which is able to coordinate to carbon monoxide, leading to intermediate **B**. Then, this complex undergoes a nucleophilic attack by the desired amine (*N*-nucleophile), affording **C**. Through HI elimination with the aid of Et_3N, intermediate **D** is formed, yielding the desired 4-amide substituted phthalonitrile, upon reductive elimination.

All carboxamide substituted phthalonitriles were characterized by ^{1}H, ^{13}C-NMR and mass spectrometry and their structures confirmed. It is worth mentioning that, under the reaction conditions employed (100 °C and 5 bar), 100% chemoselectivity toward the mono-carboxamide products was obtained, since no double carbon monoxide insertion product was observed, using these amines as nucleophiles [55,65].

Having established a methodology for the synthesis of several carboxamide-containing phthalonitriles **3a–3g**, we have then prepared, as selected examples, phthalocyanines **4a**, **4c**, and **4d**, starting from the corresponding phthalonitriles **3a**, **3c**, and **3d** (Table 3) (see also Figures S22–S27, SI). We have used an approach where the tetramerization of the phthalonitriles was carried out in pentan-1-ol at 140 °C, in the presence of zinc(II) acetate, for 20 h, with all reactions progress being followed by TLC and UV–VIS spectroscopy. Phthalocyanines **4a**, **4c** and **4d** were obtained, after purification and isolation by column chromatography on silica gel, in 58, 65 and 68% yields, respectively.

We have observed that the purification procedure for phthalocyanines **4a** and **4c** was considerably more demanding than for phthalocyanine **4d**. We have found that, even after repeated recrystallization from methanol/diethyl ether, pentan-1-ol remained coordinated with the waxy phthalocyanine molecules **4a** and **4c**, as observable on their corresponding ^{1}H-NMR spectra. On the other hand, solid phthalocyanine **4d**, bearing *tert*-butyl carboxamide groups, was easily recrystallized from methanol. We assume this occurrence to the nature of the carboxamide substituent, as amino acid derivatives are more prone to establish interactions with alcohol molecules, in our case pentan-1-ol [66,67]. This was also corroborated by the elemental analysis of **4a** and **4c**, which agreed with the presence of two molecules of pentan-1-ol per molecule of phthalocyanine. All the other typical metallophthalocyanine characteristics in terms of ^{1}H-NMR, mass spectrometry and UV–VIS spectroscopy were met, in agreement with the structures.

Table 3. Synthesis of zinc (II) metallophthalocyanines **4a**, **4c**, and **4d** and their spectral fundamental/excited state properties, studied in THF.

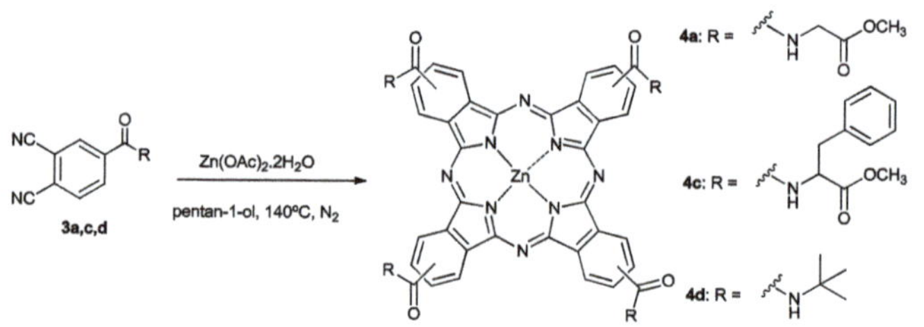

Compound	Isolated Yield (%)	λ_{max} (nm) (log ε)	Stokes Shift Δ_{stokes} (nm)	Emission λ_{max} (nm)	Φ_F [a]
4a	58	350 (4.52); 611 (4.22); 676 (4.87)	9	685	0.26
4c	65	350 (4.14); 610 (3.81); 675 (4.48)	11	686	0.31
4d	68	351 (4.41); 610 (4.49); 676 (5.10)	9	685	0.38

[a] Relative to unsubstituted ZnPc in DMSO (ΦF = 0.18) [68].

Initial photophysical assessment was carried out for the synthesized metallophthalocyanines. Absorption, emission and fluorescence quantum yields for the phthalocyanines **4a**, **4c**, and **4d** were recorded, using THF as solvent and the results are presented in Table 3.

The electronic absorption spectra of **4a**, **4c**, and **4d**, whose values of molar absorptivity coefficients (ε) are in the typical of range for zinc(II) metallophthalocyanines (Table 3), showed monomeric behavior evidenced by a single and sharp Q band, typical of non-aggregated metallated phthalocyanine complexes, with a maximum at respectively 676, 675 and 676 nm in THF, and a Soret band (the B-band) being observed at around 350 nm, as shown in Table 3 and Figure 2a. The B-bands are broad due to the superimposition of the B_1 and B_2 bands in the 350 nm region. Moreover, the absorption spectra, Figure 2a, shows that the introduction of the different substituents at the periphery of the phthalocyanine, does not disturb the UV–VIS spectrum, since the absorption bands maximum are similar.

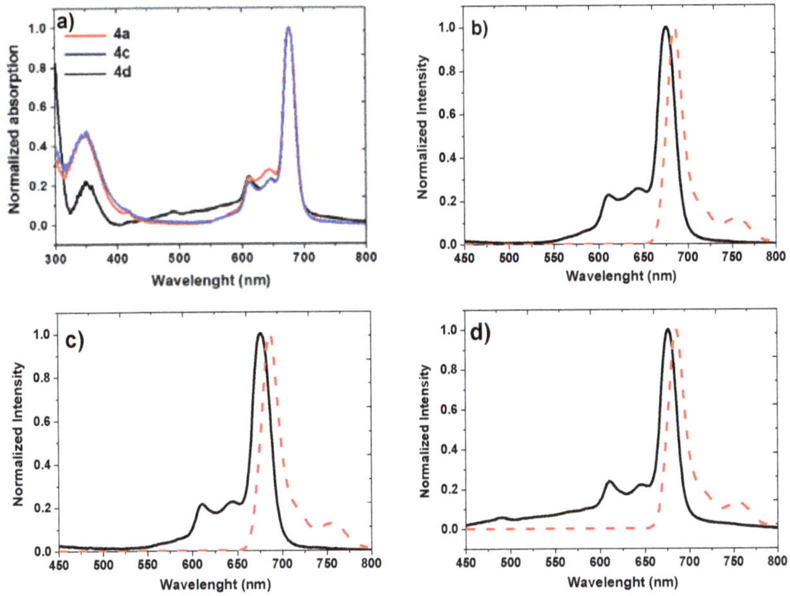

Figure 2. UV–VIS spectra of metallophthalocyanines **4a, 4c**, and **4d** in THF (**a**); normalized UV–Vis of studied phthalocyanines with absorption (black solid line) and emission spectra (red dashed line) in THF of: Zn(II)-**4a** (**b**); Zn(II)-**4c** (**c**); Zn(II)-**4d** (**d**). Fluorescence quantum yields (Φ_F) of the zinc phthalocyanines **4a** and **4c–d,** are presented in Table 3, were determined by the comparative method (Equation (1)) using the unsubstituted Zn phthalocyanine in DMSO as standard (Φ_F = 0.18) [68], and both the samples and the standard were excited at the same wavelength (640 nm). The Φ_F were calculated as 0.26, 0.31 and 0.38 for **4a**, **4c** and **4d**, respectively. The Φ_F value of zinc phthalocyanine complexes functionalized with the amino acid esters **4a** and **4c** have the same order of magnitude (Φ_F = 0.26–0.31) and are lower than non-biocompatible zinc phthalocyanine **4d** (Φ_F = 0.38).

The steady-state fluorescence emission spectra of the compounds in THF are shown in Figure 2 and the related data were listed with Stokes shifts in Table 3. Upon excitation at 640 nm, **4a**, **4c**, and **4d** showed fluorescence emission at 685, 686, and 685 nm, respectively. Again, and as expected, the fluorescence emission spectra of all phthalocyanines were similar, as all zinc metal complexes have maximum emission at the same wavelength (λ_{max} = 685–686 nm). It should be noted that the absorption spectra of all phthalocyanines were mirror images of the fluorescent spectra in THF, and that the emission is observed in the region of NIR, a pre-requisite for applications in fluorescence imaging within the important therapeutic window (λ = 650–900 nm) [2,5,69]. The observed Stokes shifts, were within the region ≈9–11 nm are typical of β-substituted phthalocyanines, which is a consequence of the rigidity of the macrocyclic ligand [70].

3. Experimental

3.1. Materials and Methods

Commercially available reagents were purchased from Aldrich (Lisbon, Portugal) and Fluorochem (Derbyshire, UK) and used as received. All solvents were pre-dried according to standard laboratory techniques. UV–VIS absorption spectra were recorded on *a Hitachi U-2010* (Hitachi Corporation, Tokyo, Japan) using quartz cells. The molar absorption coefficients were determined using THF as solvent. The fluorescence spectra for the determination of fluorescence quantum yields were acquired on a Spex Fluorolog 3 spectrofluorimeter (Horiba Instruments Incorporated, Edison, NJ, USA). ^{1}H and ^{13}C-NMR spectra were recorded on a *Bruker Advance III* spectrometer (Bruker, Karlsruhe, Germany) (400.13 for ^{1}H, and 100.61 MHz for ^{13}C). Chemical shifts for ^{1}H and ^{13}C are expressed in ppm, relatively to an internal pattern of TMS. The MALDI-TOF mass spectra were acquired using a Bruker Daltonics Flex Analysis apparatus (Bruker, Madrid, Spain). High-resolution mass spectrometry analysis was carried out with a Bruker Microtof apparatus (Bruker, Madrid, Spain), equipped with selective ESI detector. Elemental analyses were acquired using a FISONS model EA 1108 (Thermo Scientific, Waltham, MA, USA). Column chromatography was performed with silica gel grade 60, 70–230 mesh. 4-Iodophthalonitrile (**1**) was prepared according to the literature procedure [53] starting from 4-nitrophthalonitrile. The nucleophiles (*E*)-1-(4-aminophenyl)-3-(3,4,5-trimethoxyphenyl)prop-2-en-1-one (**2f**) [56] and *N*-BOC-ethylenediamine (**2e**) [59] were prepared as described in the literature.

Fluorescence quantum yields (Φ_F) were determined in DMSO using a comparative method with the Equation (1), using unsubstituted zinc(II) phthalocyanine (ZnPc) in DMSO (Φ = 0.18)[68] as standard:

$$\Phi_F = \Phi_F{}^{Std} \frac{F\, A_{Std}\, \eta^2}{F_{Std}\, A\, \eta_{Std}{}^2} \tag{1}$$

where F and F_{Std} are the areas under the fluorescence curves of the samples and the standard, respectively; A and A_{Std} are the corresponding absorbances of the samples and standard at the excitation wavelengths, respectively; η^2 and $\eta^2{}_{Std}$ are the refractive indices of solvents used for the sample and standard, respectively. The absorbance of the solutions at the excitation wavelength was around 0.1.

3.2. General Procedure for Synthesis of CARBOXAMIDE Substituted Phthalonitriles **3a–g**

In a typical aminocarbonylation reaction, the catalyst precursor $Pd(OAc)_2$, triphenylphosphine (PPh_3) ligand, substrate 4-iodophthalonitrile and the nucleophile were directly introduced in a high pressure reactor having a magnetic stirrer inside. The reactor was sealed and three vacuum/CO gas cycles were performed. Under vacuum, the reaction solvent was then added (toluene) via cannula, followed by triethylamine as base. The reactor was then pressurized using 5 bar CO and the reaction mixture maintained at 100 °C for the required period of time. After this period, the reactor was cooled to room temperature and depressurized. Palladium particles were filtered, the solvent rotary evaporated, and the crude product was then purified according to the corresponding procedure. All new compounds were characterized by means of ^{1}H-, ^{13}C-NMR, and mass spectrometry and presented in ESI.

Methyl 2-(3,4-dicyanobenzamido)acetate (glycine substituted phthalonitrile) (**3a**). Following the above described procedure, 6.75 mg (0.030 mmol) of $Pd(OAc)_2$, 15.74 mg (0.060 mmol) of PPh_3, 300 mg (1.18 mmol) of 4-iodophthalonitrile, 163.4 mg (1.30 mmol) glycine methyl ester hydrochloride (**2a**), and 1.1 mL Et_3N were dissolved in 10 mL of toluene. The reaction was pressurized and maintained at 100 °C for 12 h. The residue was dissolved in dichloromethane (20 mL), washed with brine (3 × 20 mL) and water (3 × 20 mL). The organic phase was dried with sodium sulfate and the solvent evaporated. The product was purified by recrystallization with ethyl acetate/*n*-hexane yielding **3a** in 65% yield (158 mg). ^{1}H-NMR (400.13 MHz, $CDCl_3$) δ 8.26 (s, 1H), 8.15 (d, J = 8.1 Hz, 1H), 7.93 (d, J = 8.1 Hz, 1H), 6.77 (s, 1H), 4.27 (d, J = 4.9 Hz, 2H), 3.84 (s, 9H). ^{13}C-NMR (100.61 MHz, $CDCl_3$) δ 170.0, 163.7, 138.4, 134.1, 132.4, 131.6, 118.6, 116.9, 114.8, 114.7, 53.0, 42.1. HRMS (ESI-TOF) m/z calcd for $[M + Na]^+$: $C_{12}H_9N_3NaO_3$ 266.0536; found 266.0532.

(S)-Methyl 2-(3,4-dicyanobenzamido)-4-methylpentanoate (leucine substituted phthalonitrile) (**3b**). Following the above described procedure, 6.75 mg (0.030 mmol) of $Pd(OAc)_2$, 15.74 mg (0.06 mmol) of PPh_3, 300 mg (1.18 mmol) of 4-iodophthalonitrile, 236.6 mg (1.30 mmol) leucine methyl ester hydrochloride (**2b**) and 1.1 mL Et_3N were dissolved in 10 mL toluene. The reaction was pressurized and maintained at 100 °C for 12 h. The residue was dissolved in dichloromethane (20 mL), washed with brine (3 × 20 mL) and water (3 × 20 mL).The organic phase was dried with sodium sulfate and the solvent evaporated. The product (**3b**) was purified by column chromatography on silica gel (stationary phase) first using chloroform and then a mixture of chloroform/ethyl acetate (20/1) and obtained in 54% yield (120 mg), after being washed with *n*-hexane. ^{1}H-NMR (400.13 MHz, $CDCl_3$) δ 8.24 (d, J = 8.1 Hz, 1H), 8.14 (dd, J = 1.7 Hz, J = 8.1 Hz, 1H), 7.91 (d, J = 8.1 Hz, 1H), 6.77 (br s, 1H), 4.86–4.80 (m, 1H), 3.79 (s, 3H), 1.80–1.66 (2 m, 3H), 1.00–0.97 (m, 6H). ^{13}C-NMR (100.61 MHz, $CDCl_3$) δ 173.4, 163.5, 138.6, 134.1, 132.4, 131.9, 118.3, 116.6, 114.9, 52.9, 51.8, 41.6, 25.1, 22.9, 22.0. HRMS (ESI-TOF) m/z calcd for $[M + Na]^+$: $C_{16}H_{17}N_3NaO_3$ 322.1162; found 322.1153.

(S)-Methyl 2-(3,4-dicyanobenzamido)-3-phenylpropanoate (phenyl alanine substituted phthalonitrile) (**3c**). Following the above described procedure, 6.75 mg (0.030 mmol) of $Pd(OAc)_2$, 15.74 mg (0.06 mmol) of PPh_3, 300 mg (1.18 mmol) of 4-iodophthalonitrile, 280.4 mg (1.30 mmol) phenyl alanine methyl ester hydrochloride (**2c**) and 1.1 mL Et_3N were dissolved in 10 mL toluene. The reaction was pressurized and maintained at 100 °C for 12 h. The residue was dissolved in dichloromethane (20 mL), washed with brine (3 × 20 mL) and water (3 × 20 mL). The organic phase was dried with sodium sulfate and the solvent evaporated. The product (**3c**) was purified by column chromatography on silica gel (stationary phase) first using chloroform and then a mixture of chloroform/ethyl acetate (10/1) and obtained in 59.0% yield (192 mg), after being washed with *n*-hexane. ^{1}H-NMR (400.13 MHz, $CDCl_3$) δ 8.12 (d, J = 1.7 Hz, 1H), 8.00 (dd, J = 8.1, 1.7 Hz, 1H), 7.88 (d, J = 8.1 Hz, 1H), 7.30–7.29 (m, 3H), 7.10–7.08 (m, 2H), 6.64 (br s, 1H), 5.09–5.04 (m, 1H), 3.82 (s, 3H), 3.34–3.21 (m, 2H). ^{13}C-NMR (100.61 MHz, $CDCl_3$) δ 171.7, 163.3, 138.6, 135.4, 134.1, 132.4, 131.6, 129.3, 128.9, 127.6, 118.4, 116.6, 114.8, 114.8, 54.0, 52.9, 37.7. HRMS (ESI-TOF) m/z calcd for $[M + Na]^+$: $C_{19}H_{15}N_3NaO_3$ 356.1003; found 356.1006.

N-Tert-butyl-3,4-dicyanobenzamide (**3d**). Following the above described procedure, 4.4 mg (0.020 mmol) of $Pd(OAc)_2$, 10.5 mg (0.040 mmol) of PPh_3, 200 mg (0.79 mmol) of 4-iodophthalonitrile, 0.28 mL (2.6 mmol) of *tert*-butyl amine (**2d**) and 0.8 mL Et_3N were dissolved in 6 mL of toluene. The reaction was pressurized and maintained at 100 °C for 4 h. The residue was dissolved in dichloromethane (20 mL), washed with brine (3 × 20 mL) and water (3 × 20 mL). The organic phase was dried with sodium sulfate and the solvent evaporated. The product (**3d**) was purified by column chromatography on silica gel (stationary phase) using a mixture of dichloromethane/ethyl acetate (20/1) and obtained in 74% yield (132.9 mg). ^{1}H-NMR (400.13 MHz, $CDCl_3$) δ 8.14 (bs, 1H), 8.06 (d, J = 8.1 Hz, 1H), 7.88 (d, J = 8.1 Hz, 1H), 5.94 (br s, 1H), 1.49 (s, 9H). ^{13}C-NMR (100.61 MHz, $CDCl_3$) δ 163.1, 140.6, 134.0, 132.1, 131.6, 117.8, 116.4, 115.0, 53.0, 28.8. HRMS (EI) m/z calcd for $[M]^+$: $C_{13}H_{13}N_3O$ 227.1059; found: 227.1060.

N-BOC-Ethylenediamine-3,4-dicyanobenzamide (**3e**). Following the above described procedure, 4.4 mg (0.020 mmol) of $Pd(OAc)_2$, 10.5 mg (0.040 mmol) of PPh_3, 200 mg (0.79 mmol) of 4-iodophthalonitrile, 151 mg (0.94 mmol) of *N*-BOC-ethylenediamine (**2e**) and 0.8 mL Et_3N were dissolved in 6 mL toluene. The reaction was pressurized and maintained at 100 °C for 3 h. The product (**3e**) precipitated in the middle of the reaction and then was washed with *n*-hexane and obtained in 80% yield (198.5 mg). ^{1}H-NMR (400.13 MHz, $CDCl_3$) δ 8.31 (sl, 1H), 8.21 (dd, J = 8.1, 1.1 Hz, 1H), 8.17 (br s, 1H), 7.89 (d, J = 8.1 Hz, 1H), 5.18 (br s, 1H), 3.56–3.53 (m, 2H), 3.43–3.41 (m, 2H), 1.43 (s, 9H). ^{13}C-NMR (100.61 MHz, $CDCl_3$) δ 163.6, 158.7, 138.9, 133.9, 132.4, 131.7, 117.9, 116.4, 115.0, 114.9, 80.9, 43.6, 39.7, 28.4. HRMS (ESI-TOF) m/z calcd for $[M + H]^+$: $C_{16}H_{18}N_4NaO_3$ 337.1271; found 337.1271.

(E)-3,4-Dicyano-N-(4-(3-(3,4,5-trimethoxyphenyl)acryloyl)phenyl)benzamide (**3f**). Following the above described procedure, 4.4 mg (0.020 mmol) of $Pd(OAc)_2$, 10.5 mg (0.040 mmol) of PPh_3, 200 mg (0.79 mmol) of 4-iodophthalonitrile, 296 mg (0.94 mmol) of (*E*)-1-(4-aminophenyl)-3-(3,4,5-trimethoxyphenyl)prop-2-en-1-one (**2f**), and 0.8 mL Et_3N were dissolved in 6 mL toluene. The reaction was pressurized and maintained at 100 °C for 25 h. The product (**3f**) precipitated in the middle of reaction and was washed with methanol and cyclohexane and obtained in 70% yield (256 mg). ^{1}H-NMR (400.13 MHz, acetone-d_6) δ 10.23 (s, 1H), 8.64 (d, J = 1.5 Hz, 1H), 8.51 (dd, J = 8.2, 1.6 Hz, 1H), 8.26 (d, J = 8.2 Hz, 1H), 8.18 (d, J = 8.7 Hz, 2H), 8.01 (d, J = 8.7 Hz, 2H), 7.84 (d, J = 15.5 Hz, 1H), 7.73 (d, J = 15.5 Hz, 1H), 7.19 (s, 2H), 3.91 (s, 6H), 3.79 (s, 3H). ^{13}C-NMR (100.61 MHz, acetone-d_6) δ188.7, 163.9,

154.9, 145.3, 143.8, 141.8, 140.6, 135.4, 133.9, 133.8, 131.8, 130.7, 122.1, 120.8, 120.7, 118.9, 117.0, 116.3, 107.4, 60.9, 56.8. HRMS (ESI-TOF) m/z calcd for $[M]^+$: $C_{27}H_{21}N_3O_5$ 468.1554; found 468.1555.

N-Piperazine-3,4-dicyanobenzamide (**3g**). Following the above described procedure, 4.4 mg (0.020 mmol) of $Pd(OAc)_2$, 10.5 mg (0.040 mmol) of PPh_3, 200 mg (0.79 mmol) of 4-iodophthalonitrile, 407 mg (4.72 mmol) piperazine (**2g**), and 0.8 mL Et_3N were dissolved in 6 mL toluene. The reaction was pressurized and maintained at 100 °C for 7 h. The product (**3g**) was purified by column chromatography on silica gel (stationary phase) using ethanol as eluent and obtained in 77% yield (146 mg). ^{1}H-NMR (400.13 MHz, $CDCl_3$) δ 7.87 (d, J = 8.0 Hz, 1H), 7.83 (d, J = 1.3 Hz, 1H), 7.75 (dd, J = 8.0, 1.3 Hz, 1H), 3.75 (s, 2H), 3.32 (s, 2H), 2.90 (d, J = 49.8 Hz, 4H). ^{13}C-NMR (100.61 MHz, $CDCl_3$) δ 166.1, 141.1, 134.0, 132.2, 131.7, 116.8, 116.7, 114.9, 114.8, 49.01, 46.6, 45.9, 43.6. HRMS (ESI-TOF) m/z calcd for $[M + H]^+$: $C_{13}H_{13}N_4O$ 241.1084; found 241.1081.

3.3. General Procedure for Synthesis of Carboxamide Substituted Phthalocyanines

In a typical experiment, the desired phthalonitrile and $Zn(OAc)_2{\cdot}2H_2O$ were dissolved in high boiling solvent pentan-1-ol and the mixture heated to reflux temperature for the required time for total consumption of the substrate (checked by TLC) under nitrogen atmosphere. After distilling off most of the solvent, the mixture was cooled to room temperature, and n-hexane was added to precipitate the crude compound. The solid was filtered, washed with water and purified according to the corresponding procedure. All new compounds were characterized by means of ^{1}H-NMR, UV–VIS, fluorescence and mass spectrometry and presented in ESI.

2(3)-Tetra-(keto-N-glycinyl) phthalocyaninato zinc(II) (**4a**). Following the procedure described above, 100 mg of phthalonitrile **3a** (0.41 mmol) and 29.7 mg $Zn(OAc)_2{\cdot}2H_2O$ (0.14 mmol) were dissolved in 1 mL of pentan-1-ol. The mixture was heated to 140 °C and stirred for 20 h. After workup procedure, the zinc(II) phthalocyanine complex **4a** was purified by column chromatography on silica gel first using dichloromethane/ethyl acetate (1/1) and then a mixture of dichloromethane/ethanol (20/1) as eluent to obtain 62 mg of **4a** (58% yield), as a waxy dark blue solid. UV–VIS (THF) λ_{max} (log ε) 350 (4.52), 611 (4.22), 676 (4.87). ^{1}H-NMR (400.13 MHz, acetone-d_6, 30 °C) δ 8.46 (br s, 4H), 8.34 (d, J = 7.7 Hz, 4H), 8.29 (s, 4H), 7.94 (d, J = 7.7 Hz, 4H), 4.04–3.94 (m, 8H), 2.76 (s, 12H). MS (MALDI-TOF-INFUSION) m/z calcd for $[M + Li]^+$: $C_{48}H_{36}N_{12}O_{12}LiZn$ 1043.2023; found 1043.2050. EA calcd for $C_{48}H_{36}N_{12}O_{12}Zn{\cdot}2C_5H_{12}O{\cdot}2H_2O$ C, 55.70; H, 5.16; N, 13.44; found C, 55.55; H, 5.35; N, 13.50.

(S,S,S,S)-2(3)-Tetra-(keto-N-phenyl alaninyl) phthalocyaninato zinc(II) (**4c**). Following the procedure described above, 45 mg of phthalonitrile **3c** (0.14 mmol) and 11 mg $Zn(OAc)_2{\cdot}2H_2O$ (0.05 mmol) were dissolved in 0.5 mL of pentan-1-ol. The mixture was heated to 140 °C and stirred for 20 h. After workup procedure, the zinc(II) phthalocyanine complex **4c** was purified by column chromatography on silica gel first using dichloromethane/ethyl acetate (5/1) and then a mixture of dichloromethane/ethanol (20/1) as eluent to obtain 32 mg of **4c** (65% yield), as a waxy dark blue solid. UV–VIS (THF) λ_{max} (log ε) 350 (4.14), 610 (3.81), 675 (4.48). ^{1}H-NMR (400.13 MHz, acetone-d_6) δ 8.34 (br s, 4H), 8.24 (d, J = 7.8 Hz 4H), 8.20 (s, 4H), 7.89 (d, J = 7.7 Hz, 4H), 7.27 (2m, 20H), 4.93 (m, 4H), 3.31 (m, 4H), 3.19 (m, 4H), 2.86 (s, 12H). MS (ESI-TOF-INFUSION) m/z calcd for $[M]^+$: $C_{76}H_{60}N_{12}O_{12}Zn$ 1396.3745; found 1396.3754. EA calcd for $C_{76}H_{60}N_{12}O_{12}Zn{\cdot}2C_5H_{12}O{\cdot}H_2O$ C, 64.84; H, 5.44; N, 10.55; found C, 64.59; H, 5.75; N, 10.83.

2(3)-Tetra-(tert-butyl-carboxamidyl) phthalocyaninato zinc(II) (**4d**). Following the procedure described above, 100 mg of phthalonitrile **3d** (0.44 mmol) and 32.9 mg $Zn(OAc)_2{\cdot}2H_2O$ (0.15 mmol) were dissolved in 0.5 mL of pentan-1-ol. The mixture was heated to 140 °C and stirred for 20 h. After workup procedure, the zinc(II) phthalocyanine complex (**4d**) was purified by column chromatography on silica gel using a mixture dichloromethane/methanol (20/1) as eluent to obtain 74 mg of (**4d**) (68% yield) as a dark blue solid, after recrystallization from methanol. UV–VIS (THF) λ_{max} (log ε) 351 (4.41), 610 (4.49), 676 (5.10). ^{1}H-NMR (400.13 MHz, acetone-d_6) δ 8.26–8.20 (br s, 8H), 7.88–7.86 (br s, 4H), 7.58 (s, 4H), 1.48 (sl, 36H). MS (MALDI-TOF) m/z calcd for $[M]^+$: $C_{52}H_{52}N_{12}O_4Zn$ 972.3; found 972.3;

$[M + Na]^+$, *m/z*: 995.3. EA calcd for $C_{52}H_{52}N_{12}O_4Zn{\cdot}2H_2O$ C, 61.81; H, 5.59; N, 16.63; found C, 62.06; H, 5.50; N, 16.43.

4. Conclusions

In conclusion, we have established a straightforward methodology to prepare carboxamide substituted phthalonitriles, using the well-known palladium-catalyzed aminocarbonylation of aryl halides in the presence of carbon monoxide (CO) to our advantage. In virtue of this direct modification of phthalonitriles, a more accessible preparation of biocompatible phthalocyanines is, hence, achieved. Current efforts are being devoted to extending the methodology to other phthalonitriles and phthalocyanines thereof. Initial assessment of the photophysical properties led us to conclude that this type of phthalocyanines may be usable in medicinal applications, namely optical fluorescence imaging, given the high fluorescence quantum yields ($\Phi_F = 0.31$ for biocompatible amino acid ester substituted phthalocyanine **4c**) and acceptable Stokes shifts.

Supplementary Materials: The following are available online at http://www.mdpi.com/2073-4344/8/10/480/s1, 1. Experimental procedures for the synthesis of phthalonitriles **3a–g** and copies of 1H, 13C NMR and Mass Spectra, 2. Experimental procedures for the synthesis of phthalocyanines and copies of 1H NMR and Mass Spectra.

Author Contributions: Conceptualization: M.J.F.C. and M.M.P.; investigation: V.A.T., M.J.F.C., C.S.V. and R.T.A.; methodology: M.M.P.; supervision: M.J.F.C.; validation: V.A.T., C.S.V. and R.T.A.; writing—original draft: V.A.T., M.J.F.C. and M.M.P.; writing—review and editing: M.J.F.C. and M.M.P.

Funding: This research was funded by FCT-Portugal (Portuguese Foundation for Science and Technology) and FEDER (ERDF)—European Regional Development Fund through the COMPETE Programme (Operational Programme for Competitiveness) with grants PEst-OE/QUI/UI0313/2014, to PT2020 POCI-01-0145-FEDER-027996, PTDC/QEQ-MED/3521/2014 and PTDC/QUI-OUT/27996/2017. M.J.F.C. post-doctoral grant SFRH/BPD/99698/2014 funded by FCT-Portugal. C.S.V., V.A.T. and R.T.A. PhD grants (PD/BD/128317/2017, PD/BD/128318/2017 and PD/BI/135341/2017, respectively) funded by FCT-Portugal. APC waived by MDPI.

Conflicts of Interest: There are no conflicts to declare.

References and Notes

1. Lovell, J.F.; Liu, T.W.B.; Chen, J.; Zheng, G. Activatable Photosensitizers for Imaging and Therapy. *Chem. Rev.* **2010**, *110*, 2839–2857. [CrossRef] [PubMed]
2. Josefsen, L.B.; Boyle, R.W. Unique Diagnostic and Therapeutic Roles of Porphyrins and Phthalocyanines in Photodynamic Therapy, Imaging and Theranostics. *Theranostics* **2012**, *2*, 916–966. [CrossRef] [PubMed]
3. Jiblaoui, A.; Leroy-Lhez, S.; Ouk, T.S.; Grenier, K.; Sol, V. Novel polycarboxylate porphyrins: Synthesis, characterization, photophysical properties and preliminary antimicrobial study against Gram-positive bacteria. *Bioorg. Med. Chem. Lett.* **2015**, *25*, 355–362. [CrossRef] [PubMed]
4. Calvete, M.J.F.; Pinto, S.M.A.; Pereira, M.M.; Geraldes, C.F.G.C. Metal coordinated pyrrole-based macrocycles as contrast agents for magnetic resonance imaging technologies: Synthesis and applications. *Coordin. Chem. Rev.* **2017**, *333*, 82–107. [CrossRef]
5. Calvete, M.J.F.; Simoes, A.V.C.; Henriques, C.A.; Pinto, S.M.A.; Pereira, M.M. Tetrapyrrolic Macrocycles: Potentialities in Medical Imaging Technologies. *Curr. Org. Synth.* **2014**, *11*, 127–140. [CrossRef]
6. Tekdas, D.A.; Garifullin, R.; Senturk, B.; Zorlu, Y.; Gundogdu, U.; Atalar, E.; Tekinay, A.B.; Chernonosov, A.A.; Yerli, Y.; Dumoulin, F.; et al. Design of a Gd-DOTA-Phthalocyanine Conjugate Combining MRI Contrast Imaging and Photosensitization Properties as a Potential Molecular Theranostic. *Photochem. Photobiol.* **2014**, *90*, 1376–1386. [CrossRef] [PubMed]
7. Sorokin, A.B. Phthalocyanine Metal Complexes in Catalysis. *Chem. Rev.* **2013**, *113*, 8152–8191. [CrossRef] [PubMed]
8. Calvete, M.J.F.; Silva, M.; Pereira, M.M.; Burrows, H.D. Inorganic helping organic: Recent advances in catalytic heterogeneous oxidations by immobilised tetrapyrrolic macrocycles in micro and mesoporous supports. *RSC Adv.* **2013**, *3*, 22774–22789. [CrossRef]
9. Li, L.L.; Diau, E.W.G. Porphyrin-sensitized solar cells. *Chem. Soc. Rev.* **2013**, *42*, 291–304. [CrossRef] [PubMed]

10. Urbani, M.; Gratzel, M.; Nazeeruddin, M.K.; Torres, T. Meso-Substituted Porphyrins for Dye-Sensitized Solar Cells. *Chem. Rev.* **2014**, *114*, 12330–12396. [CrossRef] [PubMed]
11. Bayda, M.; Dumoulin, F.; Hug, G.L.; Koput, J.; Gorniaka, R.; Wojcik, A. Fluorescent H-aggregates of an asymmetrically substituted mono-amino Zn(II) phthalocyanine. *Dalton Trans.* **2017**, *46*, 1914–1926. [CrossRef] [PubMed]
12. Jiblaoui, A.; Brevier, J.; Ducourthial, G.; Gonzalez-Nunez, H.; Baudequin, C.; Sol, V.; Leroy-Lhez, S. Modulation of intermolecular interactions in new pyrimidine-porphyrin system as two-photon absorbing photosensitizers. *Tetrahedron* **2015**, *71*, 2428–2434. [CrossRef]
13. Dini, D.; Calvete, M.J.F.; Hanack, M. Nonlinear Optical Materials for the Smart Filtering of Optical Radiation. *Chem. Rev.* **2016**, *116*, 13043–13233. [CrossRef] [PubMed]
14. Calvete, M.J.F.; Dini, D. Conjugated macrocyclic materials with photoactivated optical absorption for the control of energy transmission delivered by pulsed radiations. *J. Photochem. Photobiol. C Photochem. Rev.* **2018**, *35*, 56–73. [CrossRef]
15. Dumoulin, F.; Durmus, M.; Ahsen, V.; Nyokong, T. Synthetic pathways to water-soluble phthalocyanines and close analogs. *Coord. Chem. Rev.* **2010**, *254*, 2792–2847. [CrossRef]
16. Pinto, S.M.A.; Tome, V.A.; Calvete, M.J.F.; Pereira, M.M.; Burrows, H.D.; Cardoso, A.M.S.; Pallier, A.; Castro, M.M.C.A.; Toth, E.; Geraldes, C.F.G.C. The quest for biocompatible phthalocyanines for molecular imaging: Photophysics, relaxometry and cytotoxicity studies. *J. Inorg. Biochem.* **2016**, *154*, 50–59. [CrossRef] [PubMed]
17. Hanack, M.; Crucius, G.; Calvete, M.J.F.; Ziegler, T. Glycosylated Metal Phthalocyanines. *Curr. Org. Synth.* **2014**, *11*, 59–66. [CrossRef]
18. Lobo, A.C.S.; Silva, A.D.; Tome, V.A.; Pinto, S.M.A.; Silva, E.F.F.; Calvete, M.J.F.; Gomes, C.M.F.; Pereira, M.M.; Arnaut, L.G. Phthalocyanine Labels for Near-Infrared Fluorescence Imaging of Solid Tumors. *J. Med. Chem.* **2016**, *59*, 4688–4696. [CrossRef] [PubMed]
19. Simoes, A.V.C.; Adamowicz, A.; Dabrowski, J.M.; Calvete, M.J.F.; Abreu, A.R.; Stochel, G.; Arnaut, L.G.; Pereira, M.M. Amphiphilic meso(sulfonate ester fluoroaryl)porphyrins: Refining the substituents of porphyrin derivatives for phototherapy and diagnostics. *Tetrahedron* **2012**, *68*, 8767–8772. [CrossRef]
20. Roffey, S.J.; Obach, R.S.; Gedge, J.I.; Smith, D.A. What is the objective of the mass balance study? A retrospective analysis of data in animal and human excretion studies employing radiolabeled drugs. *Drug Metab. Rev.* **2007**, *39*, 17–43. [CrossRef] [PubMed]
21. Ball, D.J.; Mayhew, S.; Wood, S.R.; Griffiths, J.; Vernon, D.I.; Brown, S.B. A comparative study of the cellular uptake and photodynamic efficacy of three novel zinc phthalocyanines of differing charge. *Photochem. Photobiol.* **1999**, *69*, 390–396. [CrossRef]
22. Anderson, N.; Borlak, J. Drug-induced phospholipidosis. *FEBS Lett.* **2006**, *580*, 5533–5540. [CrossRef] [PubMed]
23. El-Faham, A.; Albericio, F. Peptide Coupling Reagents, More than a Letter Soup. *Chem. Rev.* **2011**, *111*, 6557–6602. [CrossRef] [PubMed]
24. Dunetz, J.R.; Magano, J.; Weisenburger, G.A. Large-Scale Applications of Amide Coupling Reagents for the Synthesis of Pharmaceuticals. *Org. Process. Res. Dev.* **2016**, *20*, 140–177. [CrossRef]
25. Pattabiraman, V.R.; Bode, J.W. Rethinking amide bond synthesis. *Nature* **2011**, *480*, 471–479. [CrossRef] [PubMed]
26. Osburn, E.J.; Chau, L.K.; Chen, S.Y.; Collins, N.; OBrien, D.F.; Armstrong, N.R. Novel amphiphilic phthalocyanines: Formation of Langmuir-Blodgett and cast thin films. *Langmuir* **1996**, *12*, 4784–4796. [CrossRef]
27. Dong, R.N.; Qiu, Y.; Tian, S.F.; Wang, S.Q.; Song, X.Q. Oxidative biomacromolecular damage from novel phthalocyanine. *Sci. China Ser. B* **1998**, *41*, 45–49. [CrossRef]
28. Mikhalenko, S.A.; Solov'eva, L.I.; Luk'yanets, E.A. Phthalocyanines and related compounds: XXXVII. Synthesis of covalent conjugates of carboxy-substituted phthalocyanines with alpha-amino acids. *Russ. J. Gen. Chem.* **2004**, *74*, 451–459. [CrossRef]
29. Haywood-Small, S.L.; Vernon, D.I.; Griffiths, J.; Schofield, J.; Brown, S.B. Phthalocyanine-mediated photodynamic therapy induces cell death and a G(0)/G(1) cell cycle arrest in cervical cancer cells. *Biochem. Biophys. Res. Commun.* **2006**, *339*, 569–576. [CrossRef] [PubMed]

30. Drechsler, U.; Pfaff, M.; Hanack, M. Synthesis of novel functionalised zinc phthalocyanines applicable in photodynamic therapy. *Eur. J. Org. Chem.* **1999**, 3441–3453. [CrossRef]
31. Sibrian-Vazquez, M.; Ortiz, J.; Nesterova, I.V.; Fernandez-Lazaro, F.; Sastre-Santos, A.; Soper, S.A.; Vicente, M.G.H. Synthesis and properties of cell-targeted Zn(II)-phthalocyanine-peptide conjugates. *Bioconjugate Chem.* **2007**, *18*, 410–420. [CrossRef] [PubMed]
32. Ongarora, B.G.; Fontenot, K.R.; Hu, X.K.; Sehgal, I.; Satyanarayana-Jois, S.D.; Vicente, D.G.H. Phthalocyanine-Peptide Conjugates for Epidermal Growth Factor Receptor Targeting. *J. Med. Chem.* **2012**, *55*, 3725–3738. [CrossRef] [PubMed]
33. Dubuc, C.; Langlois, R.; Benard, F.; Cauchon, N.; Klarskov, K.; Tone, P.; van Lier, J.E. Targeting gastrin-releasing peptide receptors of prostate cancer cells for photodynamic therapy with a phthalocyanine-bombesin conjugate. *Bioorg. Med. Chem. Lett.* **2008**, *18*, 2424–2427. [CrossRef] [PubMed]
34. Ali, H.; Ait-Mohand, S.; Gosselin, S.; van Lier, J.E.; Guerin, B. Phthalocyanine-Peptide Conjugates via Palladium-Catalyzed Cross-Coupling Reactions. *J. Org. Chem.* **2011**, *76*, 1887–1890. [CrossRef] [PubMed]
35. Chen, Z.; Zhou, S.Y.; Chen, J.C.; Deng, Y.C.; Luo, Z.P.; Chen, H.W.; Hamblin, M.R.; Huang, M.D. Pentalysine beta-Carbonylphthalocyanine Zinc: An Effective Tumor-Targeting Photosensitizer for Photodynamic Therapy. *Chemmedchem* **2010**, *5*, 890–898. [CrossRef] [PubMed]
36. Montalbetti, C.A.G.N.; Falque, V. Amide bond formation and peptide coupling. *Tetrahedron* **2005**, *61*, 10827–10852. [CrossRef]
37. Han, S.Y.; Kim, Y.A. Recent development of peptide coupling reagents in organic synthesis. *Tetrahedron* **2004**, *60*, 2447–2467. [CrossRef]
38. Schoenberg, A.; Heck, R.F. Palladium-Catalyzed Formylation of Aryl, Heterocyclic, and Vinylic Halides. *J. Am. Chem. Soc.* **1974**, *96*, 7761–7764. [CrossRef]
39. Gabriele, B.; Veltri, L.; Mancuso, R.; Carfagna, C. Cascade Reactions: A Multicomponent Approach to Functionalized Indane Derivatives by a Tandem Palladium- Catalyzed Carbamoylation/Carbocylization Process. *Adv. Synth. Catal.* **2014**, *356*, 2547–2558. [CrossRef]
40. Veltri, L.; Mancuso, R.; Altomare, A.; Gabriele, B. Divergent Multicomponent Tandem Palladium-Catalyzed Aminocarbonylation-Cyclization Approaches to Functionalized Imidazothiazinones and Imidazothiazoles. *ChemCatChem* **2015**, *7*, 2206–2213. [CrossRef]
41. Veltri, L.; Grasso, G.; Rizzi, R.; Mancuso, R.; Gabriele, B. Palladium-Catalyzed Carbonylative Multicomponent Synthesis of Functionalized Benzimidazothiazoles. *Asian J. Org. Chem.* **2016**, *5*, 560–567. [CrossRef]
42. Veltri, L.; Giofrè, S.V.; Devo, P.; Romeo, R.; Dobbs, A.P.; Gabriele, B. A Palladium Iodide-Catalyzed Oxidative Aminocarbonylation–Heterocyclization Approach to Functionalized Benzimidazoimidazoles. *J. Org. Chem.* **2018**, *83*, 1680–1685. [CrossRef] [PubMed]
43. Gabriele, B.; Mancuso, R.; Salerno, G. Oxidative Carbonylation as a Powerful Tool for the Direct Synthesis of Carbonylated Heterocycles. *Eur. J. Org. Chem.* **2012**, *2012*, 6825–6839. [CrossRef]
44. Wu, X.-F.; Neumann, H.; Beller, M. Synthesis of Heterocycles via Palladium-Catalyzed Carbonylations. *Chem. Rev.* **2013**, *113*, 1–35. [CrossRef] [PubMed]
45. Wu, X.-F.; Neumann, H.; Beller, M. Palladium-Catalyzed Oxidative Carbonylation Reactions. *Chemsuschem* **2013**, *6*, 229–241. [CrossRef] [PubMed]
46. Gadge, S.T.; Bhanage, B.M. Recent developments in palladium catalysed carbonylation reactions. *RSC Adv.* **2014**, *4*, 10367–10389. [CrossRef]
47. Wu, L.; Fang, X.; Liu, Q.; Jackstell, R.; Beller, M.; Wu, X.-F. Palladium-Catalyzed Carbonylative Transformation of C(sp3)–X Bonds. *ACS Catal.* **2014**, *4*, 2977–2989. [CrossRef]
48. Kalck, P.; Urrutigoïty, M. Recent improvements in the alkoxycarbonylation reaction catalyzed by transition metal complexes. *Inorg. Chim. Acta* **2015**, *431*, 110–121. [CrossRef]
49. Wu, X.-F. Palladium-catalyzed carbonylative transformation of aryl chlorides and aryl tosylates. *RSC Adv.* **2016**, *6*, 83831–83837. [CrossRef]
50. Shen, C.; Wu, X.-F. Palladium-Catalyzed Carbonylative Multicomponent Reactions. *Chem.A Eur. J.* **2017**, *23*, 2973–2987. [CrossRef] [PubMed]
51. Kollar, L. *Modern Carbonylation Methods*; Kollar, L., Ed.; Wiley VCH: Weinheim, Germany, 2008.
52. Nemykin, V.N.; Lukyanets, E.A. Synthesis of substituted phthalocyanines. *Arkivoc* **2010**, 136–208.

53. Yang, S.I.; Li, J.Z.; Cho, H.S.; Kim, D.; Bocian, D.F.; Holten, D.; Lindsey, J.S. Synthesis and excited-state photodynamics of phenylethyne-linked porphyrin-phthalocyanine dyads. *J. Mater. Chem.* **2000**, *10*, 283–296. [CrossRef]
54. Carrilho, R.M.B.; Heguaburu, V.; Schapiro, V.; Pandolfi, E.; Kollar, L.; Pereira, M.M. An efficient route for the synthesis of chiral conduritol-derivative carboxamides via palladium-catalyzed aminocarbonylation of bromocyclohexenetetraols. *Tetrahedron* **2012**, *68*, 6935–6940. [CrossRef]
55. Carrilho, R.M.B.; Pereira, M.M.; Takacs, A.; Kollar, L. Systematic study on the catalytic synthesis of unsaturated 2-ketocarboxamides: Palladium-catalyzed double carbonylation of 1-iodocyclohexene. *Tetrahedron* **2012**, *68*, 204–207. [CrossRef]
56. Fun, H.K.; Kobkeatthawin, T.; Ruanwas, P.; Chantrapromma, S. (E)-1-(4-Aminophenyl)-3-(2,4,5-trimethoxyphenyl)prop-2-en-1-one. *Acta Crystallogr. E* **2010**, *66*, O1973-U1802. [CrossRef] [PubMed]
57. Muller, E.; Peczely, G.; Skoda-Foldes, R.; Takacs, E.; Kokotos, G.; Bellis, E.; Kollar, L. Homogeneous catalytic aminocarbonylation of iodoalkenes and iodobenzene with amino acid esters under conventional conditions and in ionic liquids. *Tetrahedron* **2005**, *61*, 797–802. [CrossRef]
58. Takacs, A.; Abreu, A.R.; Peixoto, A.F.; Pereira, M.; Kollar, L. Synthesis of Ortho-alkoxy-aryl Carboxamides via Palladium-Catalyzed Aminocarbonylation. *Synth. Commun.* **2009**, *39*, 1534–1548. [CrossRef]
59. Li, B.Y.; Zhang, Y.M.; Ma, D.X.; Li, L.; Li, G.H.; Li, G.D.; Shi, Z.; Feng, S.H. A strategy toward constructing a bifunctionalized MOF catalyst: Post-synthetic modification of MOFs on organic ligands and coordinatively unsaturated metal sites. *Chem. Commun.* **2012**, *48*, 6151–6153. [CrossRef] [PubMed]
60. Prasad, Y.R.; Rao, A.S.; Rambabu, R. Synthesis of Some 4′-Amino Chalcones and their Antiinflammatory and Antimicrobial Activity. *Asian J. Chem.* **2009**, *21*, 907–914.
61. Fang, W.W.; Zhu, H.B.; Deng, Q.Y.; Liu, S.L.; Liu, X.Y.; Shen, Y.J.; Tu, T. Design and Development of Ligands for Palladium-Catalyzed Carbonylation Reactions. *Synthesis-Stuttgart* **2014**, *46*, 1689–1708. [CrossRef]
62. De la Fuente, V.; Godard, C.; Claver, C.; Castillon, S. Highly Selective Palladium-Catalysed Aminocarbonylation of Aryl Iodides using a Bulky Diphosphine Ligand. *Adv. Synth. Catal.* **2012**, *354*, 1971–1979. [CrossRef]
63. Brennfuhrer, A.; Neumann, H.; Beller, M. Palladium-Catalyzed Carbonylation Reactions of Aryl Halides and Related Compounds. *Angew. Chem. Int. Ed.* **2009**, *48*, 4114–4133. [CrossRef] [PubMed]
64. Wu, X.F.; Neumann, H.; Beller, M. Selective Palladium-Catalyzed Aminocarbonylation of Aryl Halides with CO and Ammonia. *Chem. Eur. J.* **2010**, *16*, 9750–9753. [CrossRef] [PubMed]
65. Carrilho, R.M.B.; Almeida, A.R.; Kiss, M.; Kollar, L.; Skoda-Foldes, R.; Dabrowski, J.M.; Moreno, M.J.S.M.; Pereira, M.M. One-Step Synthesis of Dicarboxamides through Pd-Catalysed Aminocarbonylation with Diamines as N-Nucleophiles. *Eur. J. Org. Chem.* **2015**, 1840–1847. [CrossRef]
66. Dwyer, D.S.; Bradley, R.J. Chemical properties of alcohols and their protein binding sites. *Cell. Mol. Life Sci.* **2000**, *57*, 265–275. [CrossRef] [PubMed]
67. Graton, J.; Besseau, F.; Brossard, A.M.; Charpentier, E.; Deroche, A.; Le Questel, J.Y. Hydrogen-Bond Acidity of OH Groups in Various Molecular Environments (Phenols, Alcohols, Steroid Derivatives, and Amino Acids Structures): Experimental Measurements and Density Functional Theory Calculations. *J. Phys. Chem. A* **2013**, *117*, 13184–13193. [CrossRef] [PubMed]
68. Atilla, D.; Durmus, M.; Gurek, A.G.; Ahsen, V.; Nyokong, T. Synthesis, photophysical and photochemical properties of poly(oxyethylene)-substituted zinc phthalocyanines. *Dalton Trans.* **2007**, 1235–1243. [CrossRef] [PubMed]
69. Sekkat, N.; van den Bergh, H.; Nyokong, T.; Lange, N. Like a Bolt from the Blue: Phthalocyanines in Biomedical Optics. *Molecules* **2012**, *17*, 98–144. [CrossRef] [PubMed]
70. Isago, H. *Optical Spectra of Phthalocyanines and Related Compounds: A Guide for Beginners*; Springer: Tokyo, Japan, 2015.

catalysts MDPI

Communication

Layered Double Hydroxides as Bifunctional Catalysts for the Aryl Borylation under Ligand-Free Conditions

Lorenna C. L. L. F. Silva [1], Vinícius A. Neves [1], Vitor S. Ramos [2], Raphael S. F. Silva [3], José B. de Campos [2], Alexsandro A. da Silva [4], Luiz F. B. Malta [1,*] and Jaqueline D. Senra [1,4,*]

1 Laboratório de Química Supramolecular e de Sólidos, Instituto de Química, Universidade Federal do Rio de Janeiro, Rio de Janeiro 21941-909, Brazil; lorennaconti97@gmail.com (L.C.L.L.F.S.); alevato.v@gmail.com (V.A.N.)

2 Faculdade de Engenharia, Universidade do Estado do Rio de Janeiro, Rio de Janeiro 20550-900, Brazil; vramos00@gmail.com (V.S.R.); brantjose@gmail.com (J.B.d.C.)

3 Instituto Federal do Rio de Janeiro, Campus Rio de Janeiro, Rio de Janeiro 20270-021, Brazil; raphael.silva@ifrj.edu.br

4 Instituto de Química, Universidade do Estado do Rio de Janeiro, Rio de Janeiro 20550-900, Brazil; alexsandro.silva@uerj.br

* Correspondence: lfbmalta@iq.ufrj.br (L.F.B.M.); jaqueline.senra@uerj.br (J.D.S.); Tel.: +55-21-3938-7738 (L.F.B.M.); +55-21-2334-0563 (J.D.S.)

Received: 29 January 2019; Accepted: 18 March 2019; Published: 27 March 2019

Abstract: Organic derivatives of boron, such as boronic esters and acids, are important precursors for a wide range of environmental, energy, and health applications. Several catalytic methods for their synthesis have been reported, even though with the use of toxic and structurally complex ligands. Herein, we demonstrate preliminary studies envisaging the synthesis of boronic esters from an inexpensive catalytic system based on Cu/Al layered double hydroxides (LDH) in the presence of Na_2PdCl_4. The Cu/ Al LDHs were prepared according to coprecipitation method and characterized by X-ray diffraction (XRD) (with Rietveld refinement) to evaluate the contamination with malachite and other phases. Preliminary catalytic results suggest that pure Cu/Al LDH has potential for the borylation of aryl iodides/ bromides in the absence of base. Indeed, a synergic effect between copper and palladium is possibly related to the catalytic efficiency.

Keywords: boronic esters; borylation; Suzuki–Miyaura; layered double hydroxides; copper; palladium

1. Introduction

Organoboron compounds have received great attention in last years due to significant impact in analytical [1], technological [2], and medicinal fields [3]. Particularly, low-weight compounds containing boronic acid/ester moieties have played an important role in the synthesis of new hybrid materials, sensors and complex organic molecules, through synthetic protocols such as Suzuki–Miyaura reactions. However, mild and selective synthesis of boronic acids and esters still represent a challenge. In general, the classic reactions involving the generation of organometallics (e.g., arylmagnesium or aryllithium) followed by the reaction with a borate have been substituted by catalytic protocols [4]. Recently, new methods involving metal-free conditions have also been discovered [5,6]. However, most of them suffer from low reactivity/selectivity along with relatively high costs of the boron reagents. Some advantages of the catalytic protocols are the improved reaction selectivity and the possibility of the whole system reuse. Since the general steps for the catalytic synthesis of boronic acids and esters involve a sequence of oxidative addition and reductive elimination, several semihomogenous or heterogeneous systems are described based on the employment of noble metals (e.g., Pd and Ir) as catalysts [7].

In 1995, Miyaura first reported the Pd-catalyzed aryl borylation protocol by using a phosphine-based catalyst—$Pd(dppf)Cl_2$—in the presence of KOAc for the activation of diboron reagent [8]. Since then, several efficient methodologies have been reported based on similar reaction conditions [9–11]. Recently, Ratniyom and coworkers [12] reported an efficient cooperative catalysis based on Pd(0)/Cu(I) in the presence of triphenylphosphine with good results towards aryl iodides and bis(pinacolate)diboron. However, the yields were very sensitive to the nature of the base and reduced drastically with the use of aryl bromides/chlorides.

Inorganic networks with basic properties represent one alternative to avoid the use of expensive ligands as well as strong bases in order to achieve the B–B bond activation. Mostly, bimetallic systems usually involve low loadings of a noble metal in combination with an early transition metal to allow a cost-effective, broader functional group tolerance, and scalable reaction condition. Layered double hydroxides (LDHs)—anionic clays also known as hydrotalcites—have a structure based on layers of bi- and trivalent metals hydroxides and interlayer spaces occupied by anions and neutral molecules [13], see Figure 1. They can be composed of a wide range of different metals and anions, which make them useful as anionic exchangers, flame retardants, catalysts, and supports for metallic nanoparticles [14], and have low-cost of synthesis and basic surface properties [15]. Silva and coworkers [16] previously reported a catalyst system based on Mg/Al LDHs, PdNPs, and cyclodextrins for the efficient Suzuki–Miyaura reaction between aryl bromides and arylboronic acids. Lately, Sreedhar [17] employed Cu/Al LDHs as catalysts in Ullmann reactions between aryl chlorides and amines. So far, cross-coupling with heterogeneous ligand-free Cu(II) based cocatalysts has not been reported to date in borylation reactions.

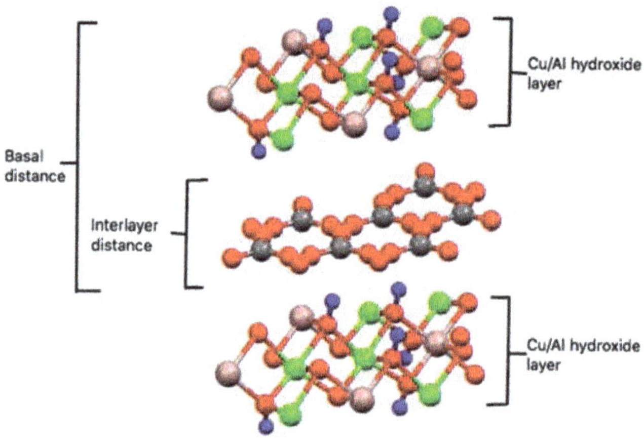

Figure 1. Representation of layered double hydroxide (LDH) hydrotalcite-like structure, with the Cu/Al hydroxide layers and the carbonate containing interlayer space. Cu → green, O → red, H → blue, C → grey, Al → light red.

In the present work, we disclosed Cu/Al LDHs as bifunctional cocatalysts in the presence of Na_2PdCl_4 for the borylation reactions between aryl halides and bis(pinacolato)diboron. The key points examined here are the influence of LDH purity and the catalytic conditions towards the reaction outcome.

2. Results

2.1. Synthesis and Characterization of Cu/Al LDH

A set of conditions were varied to accomplish the synthesis of the Cu/Al LDH. Figure 2 shows the X-ray diffraction (XRD) patterns of samples obtained using or not sodium carbonate as an additive. These diffractograms were refined using the Rietveld method in order to obtain the composition of phases. In addition, the refinement also afforded the unit cell parameters for LDH phase: these data are available in the supporting information file (Tables S1–S4). The LDH synthesized without sodium carbonate (Figure 2a) exhibited a XRD pattern composed of reflections from the mineral-like phases of gerhardtite ($Cu(OH)NO_3$, JCPDS-ICDD 01-082-1991), hydrotalcite ($Mg_{0.67}Al_{0.33}(OH)_2)(CO_3)_{0.165}(H_2O)_{0.48}$, JCPDS-ICDD 01-089-5434), and nitratine ($NaNO_3$, JCPDS-ICDD 00-036-1474). The LDH phase was obtained as a low crystallinity, minority phase with wide and low intensity peaks at 9.7° and 19.9°. The $Cu(OH)NO_3$ phase (gerhardtite) was obtained as the majority phase with the most intense peaks at 12.7°, 25.6°, 33.9°, 36.1°, 40.4°, and 42.9°. Finally, $NaNO_3$ (nitratine) was found as another minority phase, however showing a thin peak at 2θ = 22.8°, besides 29.4°, 31.8°, and 38.9°.

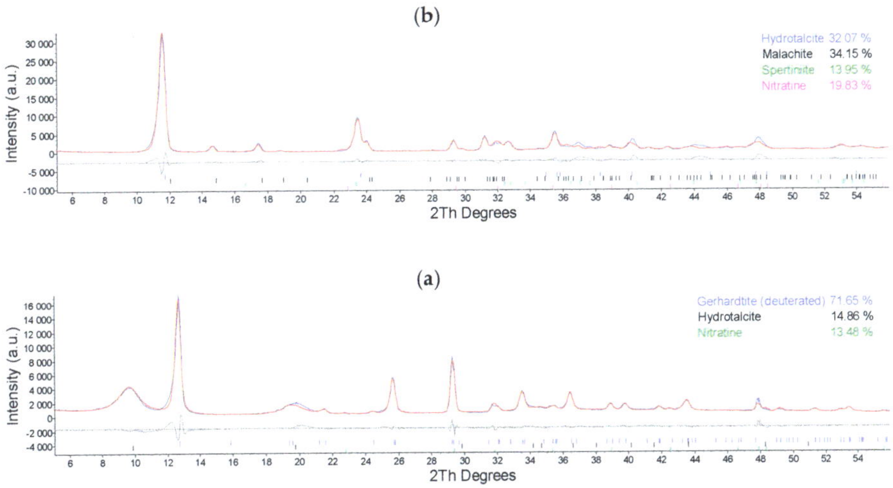

Figure 2. Rietveld-refined X-ray diffraction (XRD) patterns for Cu/Al LDH obtained (**a**) without and (**b**) with Na_2CO_3 dissolved in the precipitation agent solution.

In contrast, Figure 2b shows the XRD pattern for the sample synthesized using sodium carbonate. In this case, the majority phase was the layered double hydroxide with peaks at 2θ = 11.6°, 23.5°, 35.6°, 39.5°, and 47.0°. Another significant phase was malachite ($Cu_2(OH)_2CO_3$, JCPDS-ICDD 01-075-1163) related to the peaks 2θ = 11.9°, 14.7°, 17.5°, 24.0° 29.7°, 31.7°, 32.6°, and 35.4°. As minority phases, it was found that spertiniite ($Cu(OH)_2$ JCPDS-ICDD 01-080-0656) related to peaks at 2θ = 16.7°, 23.8°, 34.0°, 35.9°, and 39.8°, and nitratine with peaks at 2θ = 22.8°, 29.4°, 31.8° and 38.9°.

Regarding the LDH unit cell parameters (Table S1, supporting information), both XRD were refined as having a rhombohedral crystal system and belonging to the R-3m spatial group. However, the sample synthesized without carbonate presented the biggest unit cell dimensions (a = 3.039 Å and c = 26.921 Å against a = 2.979 Å and c = 22.519 Å for the carbonate Cu/Al LDH), which means that an expanded structural network was obtained.

The addition of sodium carbonate to the reaction medium allowed obtaining LDH as the main phase, but did not avoid forming malachite, a by-product. Still, it was decided to use the carbonate salt dissolved in the precipitation agent solution in the fore coming tests.

In the next set of experiments the LDH precipitation pH and postsynthesis work-up were evaluated.

The precipitation pH was varied between two values, 8 and 10, which are the most used in LDH synthesis. The XRD patterns for both samples are presented in Figure 3. Figure 3a evidences the refined XRD pattern for the LDH precipitated at pH = 8, which, compared to that of Figure 3b, related to LDH precipitated at pH = 10; thus pH is not a significant parameter in the synthesis. This is also supported by strong similarities of unit cell dimensions between these two samples (Table S2, supporting information). Therefore, a pH of 8 was chosen as the working pH in the next tests.

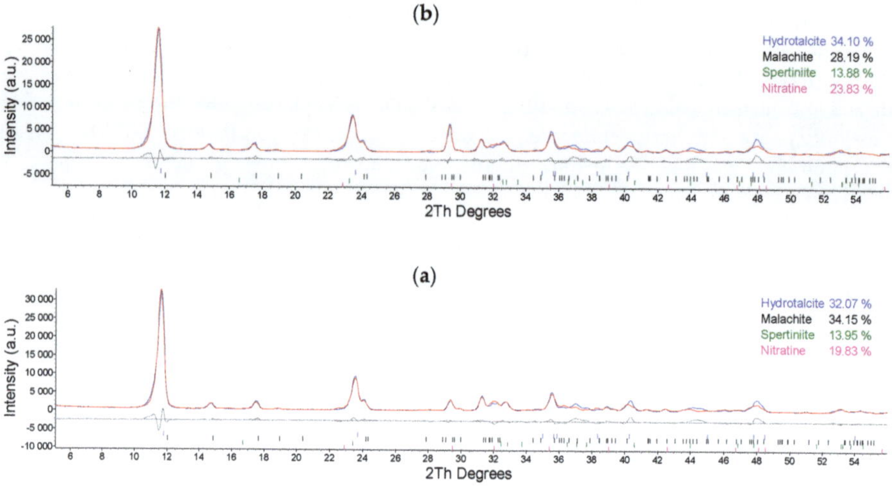

Figure 3. Rietveld-refined XRD patterns for Cu/Al LDH obtained at (**a**) pH = 8 and (**b**) pH = 10.

In all experiments the LDH postsynthesis work-up proceeded using centrifugation, but one sample was isolated using filtration. Their refined XRD patterns are shown in Figure 4. Concerning the unit cell dimensions, no significant differences between these two samples can be perceived (Table S3, supporting information) As evidenced when nitratine phase percentages are compared the centrifuged sample (Figure 4a) it exhibits 5-fold more $NaNO_3$ than the filtrated solid (Figure 4b), which implies that the postsynthesis work up is an important parameter to be aware of.

To reinforce that postsynthesis steps are very important one last sample was synthesized following the conditions established in the present study; however, this LDH was submitted to washing, using organic solvents (EtOH/Acetone). Its refined XRD measurements are presented in the Figure 5. The calculated unit cell parameters show no significant differences from those previously shown (Table S4, supporting information). In the phase distribution it is clearly observed that the malachite % dropped from 31.93% (Figure 4b) to 10.23% (Figure 5), increasing LDH % from 47.92% to 65.70%.

Besides, the XRD peaks shown in Figure 5 appear wider and less intense than those in Figure 4b, for instance. According to the Full Width Half the Maximum (FWHM) criterium used in the Scherrer formula, the mean crystal size evolved from 21 nm (for samples of Figures 3 and 4) to 14 nm, signaling a decrease in crystallinity upon postsynthesis treatment, such as washing with organic solvents.

To evaluate the effect on the borylation reaction, catalytic tests were carried out by using malachite and LDH as catalysts.

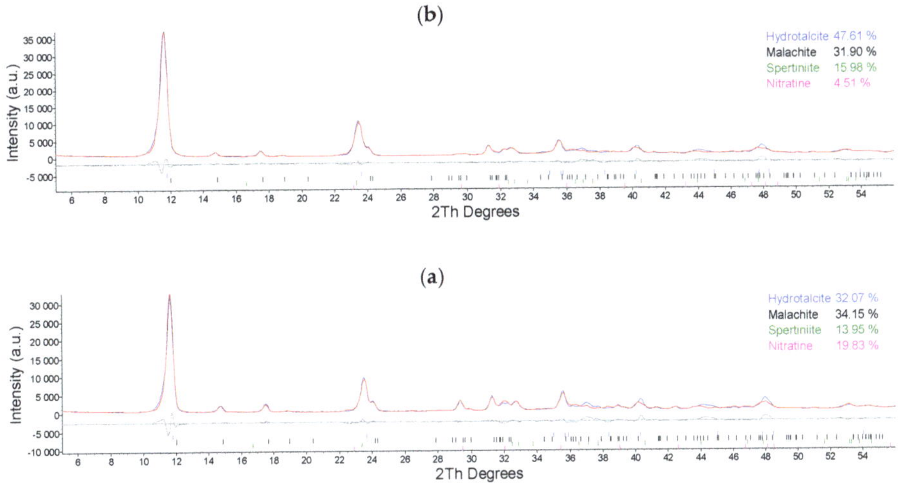

Figure 4. Rietveld-refined XRD patterns for Cu/Al LDH obtained after (**a**) centrifugation and (**b**) filtration.

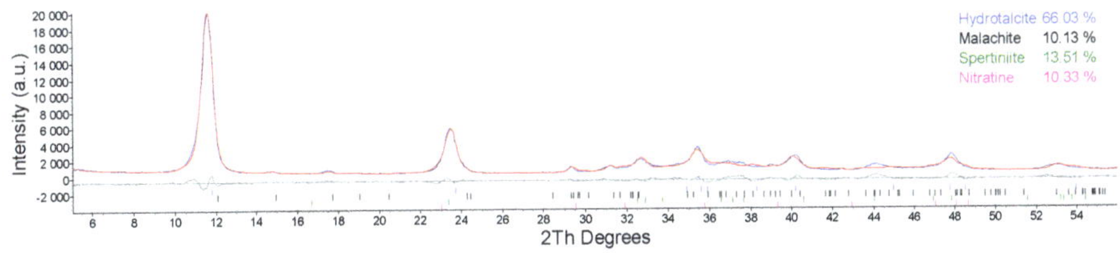

Figure 5. Rietveld-refined XRD pattern for Cu/Al LDH obtained after washing with ethanol and acetone.

2.2. *Synthesis of Aryl Boronic Esters Employing Cu/Al LDH Catalyst*

The reaction between bis(pinacolato)diboron, ($B(pin)_2$) and 1-iodo-4-nitrobenzene was taken as a model. Initially, purified Cu/Al LDH and malachite were evaluated as catalysts with acetonitrile as solvent. From these data, it is possible to confirm that malachite did not show catalytic activity towards the borylation reaction. To evaluate the copper loading, we used the LDH composition according to the method previously described by our group [18]. It was observed that both copper-based catalysts were not able to catalyze the reaction (Table 1, entries 1 and 2). Similarly, the use of Na_2PdCl_4 as the sole catalyst was ineffective (Table 1, entry 3). Indeed, addition of $CuSO_4$ in combination with Na_2PdCl_4 did not react under this condition (Table 1, entry 4). The comparison of the Pd precursor led us to evaluate a semihomogeneous system composed of palladium nanoparticles (PdNPs) stabilized by cyclodextrins (Table 1, entries 5 and 6). Our research group have already described this catalytic system to the carbon–carbon cross-coupling reactions [19]. However, it was not observed significant conversion even when using acetonitrile. Remarkably, the addition of 2 mol% Na_2PdCl_4 in the presence of LDH (30 mol% Cu) allowed a yield of 98% of the expected product (Table 1, entry 7). Since the positive effect could be related to the LDH structural and compositional properties, we also tested the most common Mg/Al LDH in the presence of $CuSO_4$ and Na_2PdCl_4, which rendered a good yield (Table 1, entry 8). Having in mind that the Cu/Al LDH is an anionic exchanger, it has been tested a $[PdCl_4]^{2-}$ exchanged Cu/Al LDH, and surprisingly, the yield obtained was significantly lower, 35% (Table 1, entry 8).

Table 1. Survey of reaction condition.

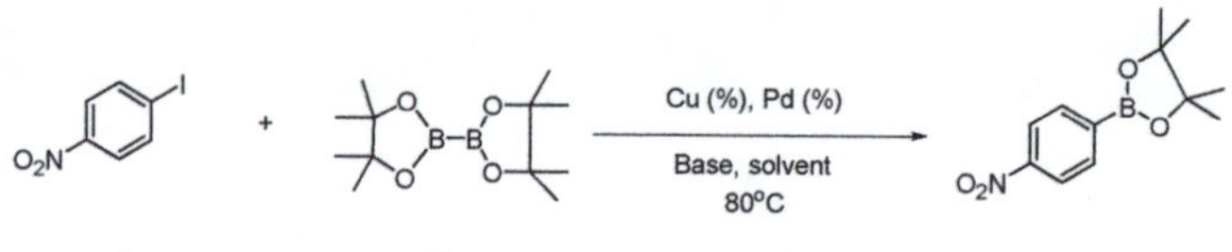

Entry	Catalyst System	Cu (mol%)	Pd (mol%)	Yield (%) [1] of 3
1	LDH	30	-	29%
2	Malachite	30	-	<5%
3	Na_2PdCl_4	-	2	<5%
4	PdNPs-CD	-	2	<5%
5	PdNPs-CD	-	2	<5%
6	$Na_2PdCl_4/CuSO_4$	30	2	<5%
7	Na_2PdCl_4/LDH	30	2	98%
8	$Na_2PdCl_4/CuSO_4$/LDH [2]	30	2	77%
9	Na_2PdCl_4/LDH [3]	30	2	35%
10	Na_2PdCl_4/LDH	15	2	54%
11	Na_2PdCl_4/LDH	7.5	2	42%
12	Na_2PdCl_4/LDH	3.75	2	26%
13	Na_2PdCl_4/LDH	30	1	46%
14	Na_2PdCl_4/LDH	30	0.5	42%
15	Na_2PdCl_4/LDH	30	0.05	23%
16	Na_2PdCl_4/LDH [4]	30	2	<10%
17	Na_2PdCl_4/LDH [5]	30	2	<10%
18	Na_2PdCl_4/LDH [6]	30	2	<5%

[1] Determined by gas chromatography–mass spectrometry (GC-MS); [2] Use of Mg/Al LDH, $CuSO_4$ and Na_2PdCl_4; [3] LDH intercalated with Na_2PdCl_4; [4] solvent = THF; [5] solvent = dioxane; [6] Addition of Cs_2CO_3.

In order to see the effect of copper in the reaction, we have tested different copper loadings (Table 1, entries 10–12). According to the results, a relatively high loading of copper has shown to be the most adequate for a high conversion (Table 1, entry 7). Similarly, to verify the influence of the Pd loading it was varied from 2 to 0.05% under the same conditions (Table 1, entries 13–15). In this case, 2 mol% Pd was necessary to keep an acceptable turnover. Analogously to Cu%, it was observed a direct relationship between the conversion rate and the Pd%. With the aim of evaluating the effect of noncoordinating solvents, THF and dioxane (Table 1, entries 16 and 17) were also tested under the conditions described in Table 1, entry 3. Surprisingly, no appreciable yield of the product was observed in both cases. Additionally, the use of base hampered the reaction (Table 1, entry 18).

Since the above results demonstrated that the catalytic system based on Cu/Al LDH and Pd(II) can be efficiently used for the borylation of an aryl iodide under ligand-free conditions, we carried out preliminary reactions with some aryl halides in order to examine its applicability. In this case, we have tested the influence of electron-withdrawing/donating groups substituted in the aromatic core. By considering their comparatively low-cost and availability, we have mainly focused on the evaluation of the reactivity of aryl bromides.

In general, it was possible to note an influence of the electron-withdrawing/ donating capabilities of the substitutional groups. Remarkably, it was evident that the strongly electron-donating amine group did not favor the reaction in both cases (Table 2, entries 1 and 5). In general, the absence of a substituent led to small yields.

Table 2. Preliminary scope for the borylation reaction catalyzed by Cu/Al LDH and Na_2PdCl_4 [1].

Entry	Aryl Halide	Product	Yield (%)
1	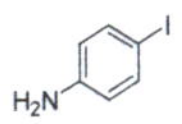	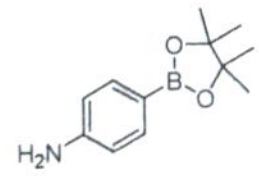 **4**	<10
2	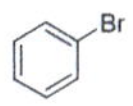	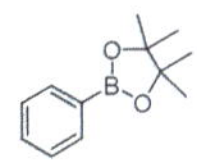 **5**	<10
3	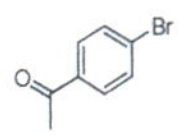	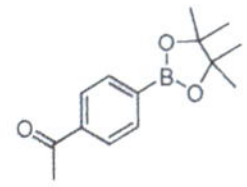 **6**	55
4	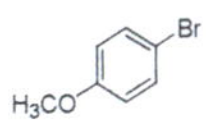	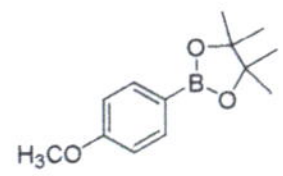 **7**	98
5	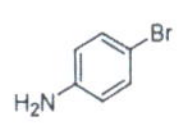	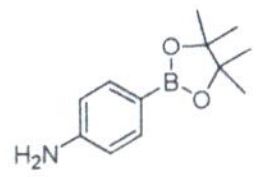 **4**	<10
6	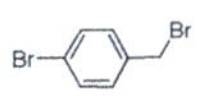	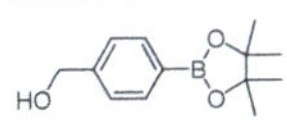 **8**	30
7	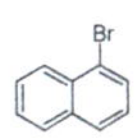	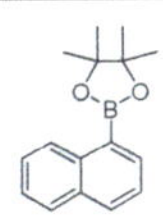 **9**	45

[1] Determined by GC-MS.

Encouraged by the preliminary results, we next investigated whether this catalyst could have recycling potential. In such case, it could shed a light on the behavior of Cu/Al LDH as a reservoir or as real catalytically active species. However, attempts to recover the material after reaction work up failed.

3. Discussion

Malachite is a common second phase present in Cu/Al LDH synthetic samples [20–26], and it is known to affect the general properties of the LDH, such as its catalytic properties [23–25]. In most cases, it arises from the need to use sodium carbonate as a precipitation co-agent, otherwise LDH is not formed as the majority phase, as observed in Figure 2a. The way the unit cell dimensions obtained for this phase (Table S1, supporting material) evidenced an expanded network in comparison to the carbonate containing LDH phases. This can be understood in terms of attraction electrostatic forces acting between layer and interlayer parts of the material that permit to pack more efficiently when CO_3^{2-} instead of NO_3^- is used as the intercalated ion.

Some authors in the literature have tried to eliminate the malachite impurity by a number of different approaches.

Muñoz et al. [23] sought for pure Cu/Al LDH in order to obtain its calcined-oxide derivative to catalyze the reduction of NO and CO gases. Through coprecipitation method synthesis they were able to verify that precursors solutions with lower concentrations prevented the formation of malachite, and that the oxide derived from malachite-free LDH showed better catalytic results than the derived from impure LDH. Gao et al. [24] synthetized Cu/Zn/Al/Zr LDH in order to obtain its calcined oxide for the catalysis of CO_2 hydrogenation, and also found that catalysts with malachite prior to the calcination presented worse catalytic activity than the ones without the contamination. Both Muñoz and Gao agree that higher ratios of Cu/Al induce the formation of malachite. Ichikawa et al. [25] tried a different way to obtain noncontaminated Cu/Al LDH: they used a coprecipitation method followed by an aging process at 90 °C under air bubbling for 1 h. They stated that this process helped to smooth the crystallization process removing the excess of CO_3^{2-} ions. They also found that the calcined oxide derived from the pure Cu/Al LDH was a better catalyst to the conversion of acrylonitrile to acrylamide than the contaminated one. Recently, Qu et al. [26] reported a new mechanochemical synthesis method for Cu/Al LDH, in which they dry milled $Cu_2(OH)_2CO_3$ and $Al(OH)_3$ at a planetary ball mill for 2 h at 600 rpm, producing an amorphous solid mixture that was treated in aqueous medium at room temperature under magnetic stirring for 4 h, producing a pure Cu/Al LDH.

Another important feature is the presence of nitratine ($NaNO_3$) in the phase mixture composition after LDH synthesis. Firstly, this phase arises from the combination of sodium of NaOH solution and nitrate from salt precursors of Cu^{2+} and Al^{3+}. However, considering sodium nitrate is soluble in aqueous medium, it is not clear how this phase coprecipitates with LDH. Figure 4 gives a hint: the centrifugation process leads to 5-fold increase of nitratine% in the phase mixture composition. In addition, Figure 5 points to a decrease by half of this phase percentage guaranteed by washing with EtOH/Acetone. Therefore, the combined use of filtration and washing in the postsynthesis work-up of LDH is beneficial for this catalyst.

The catalytic conditions studied for the borylation model reaction were initiated with an activated aryl iodide since electron-withdrawing groups are known for the accelerated effect on the reaction rate [7]. Under the basic conditions of the classical Pd cycle, it is assumed that the higher the Ar-Pd(II)-X electrophilicity, the faster the transmetallation with bisorganodiboron. In fact, the results pointed to a high catalytic conversion by using 4-iodonitrobenzene as an electrophile. To evidence the clean and selective conditions for the borylation reaction, the crude ^1H NMR spectrum of **3** is presented in the Supporting Information. A relatively high copper loading was, however, necessary to increase the substrate conversion. As pointed out, this fact has already been observed by Ratnyom [12] and other groups [13] but involving a typically homogeneous catalytic system based on Cu or Pd/Cu combined with phosphines. Regarding the usual organometallic mechanism, it is interesting to note that the LDH surface is possibly responsible for the presence of basic species involved in the preactivation step. The addition of base, however, was detrimental to the reaction, suggesting a partial decomposition of the LDH structure.

Even though the cooperative catalysis of Pd and Cu was already reported in some recent works, the ligand-free condition has been scarcely explored mostly because of the difficult electronic tuning towards the transmetallation step even though the soft coordination of acetonitrile cannot be ruled out. However, the surprising fact here is related to the superior activity of a heterogeneous layered Cu matrix, which allowed high substrate conversion and product selectivity in the reaction model. In detriment, common copper salts, such as the basic copper carbonate (malachite), were ineffective under the same conditions, indicating that the chemical environmental around copper is decisive to the reaction outcome even in the presence of palladium. This fact could be reinforced when testing Mg/Al LDH in the presence of $CuSO_4$ and Na_2PdCl_4 that resulted in an inferior product yield compared to the same catalytic conditions by using the Cu/Al matrix.

In addition, the preliminary scope of the method showed an unclear electronic effect of substituents: apparently, the reaction is favored by both electron-donating and -withdrawing groups but, intriguingly, it was not the case for the amine substituent. It is conceivable that a hybrid mechanism is operative with the moderate to low yields of α-naphtyl, 4-bromobenzyl and phenyl bromides arising from a predominant oxidative addition modulation. However, the low reactivity of bromobenzene was intriguing. We presume the occurrence of possible alternative pathways (e.g., Ullmann-type reaction and hydrodehalogenation) in minor extension but the solubility factor cannot be ruled out. Indeed, the role of LDH matrix as a catalytic reservoir of active Cu species can be considered based on experiments with coordinating solvents [27]. According to the frustrated catalyst recycling attempts, it is also possible that acetonitrile act as a moderate LDH exfoliation agent. On the other hand, the effect of $PdCl_4^{2-}$ intercalation in LDH seems to slow down the possible Pd(II) reduction. Lastly, $(Bpin)_2$ seems to be involved in the reduction of Pd(II) species, as suggested by the solution darkening according to visual inspection.

Taken together, these results suggest that a synergic effect between the Cu(II) and Pd(0) species can arise from a combination of factors related to the electronic effects on substrate along with the chemical environment of both metals.

A schematic view containing some of these effects is shown in Scheme 1.

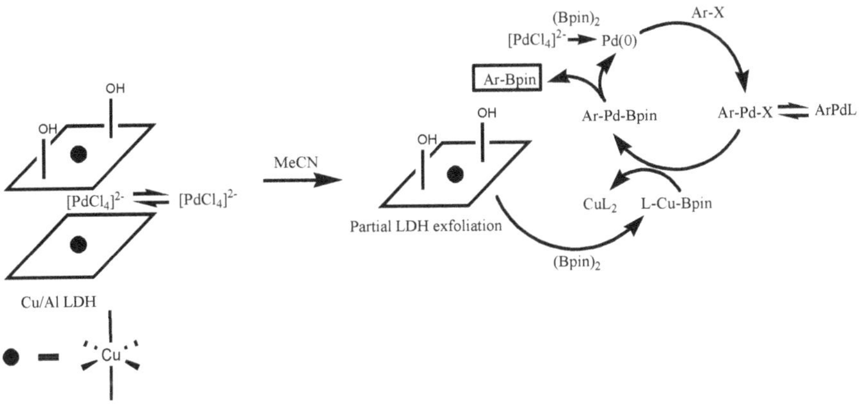

Scheme 1. Proposed mechanism for the Pd and Cu-catalyzed aryl borylation in the presence of LDH.

4. Materials and Methods

All the reactants and solvents in this work were commercially acquired and used without previous purification.

The powder X-ray diffractograms were recorded by a diffractometer Panalytical X'Pert PRO MPD (Malvern Panalytical Ltd., Royston, United Kingdom), with generator of Cu X-ray (λ = 1.5418 Å), line focus (1.8 kW), universal Theta-2Theta goniometer with 240 mm radius, divergence slit fixed $\frac{1}{2}^{\circ}$, flat-diffracted beam monochromator, and proportional detector Xe with 40 kV of tension and 40 mA of current in the X-ray tube. The samples were analyzed self-supported in aluminum sample holder and used under the Bragg–Brentano geometry; it used ranges between $5^{\circ} < 2\theta < 70^{\circ}$, with 0.05° step and 2.5 s per step.

The calculation based on the Rietveld Refinement applied in this work, through the program TOPAS academic V5.0 (Coelho Software, Brisbane, Australia), was based on the Fundamental Parameters Approach, based on the instrumental parameters with background correction. If necessary, the following parameters were refined; unit cell dimensions, sample height displacement, zero-shift, weight fraction (scaling), preferred orientation, atomic species/substitutions, atomic coordinates, site occupancies, thermal displacement parameters, crystallite size, and lattice strain.

All the organic products were characterized by ^1H and ^{13}C NMR spectroscopy (Bruker Analytics, Berlin, Germany).

4.1. Synthesis of the Cu/Al Layered Double Hydroxides

The Cu/Al layered double hydroxides (Cu/Al LDH) and $Cu_4Al_2(OH)_{12}CO_3{\cdot}4H_2O$, were synthetized using a constant pH method: one aqueous solution of $Cu(NO_3)_2{\cdot}3H_2O$ (0.225 $mol{\cdot}L^{-1}$) and $Al(NO_3)_3{\cdot}9H_2O$ (0.075 $mol{\cdot}L^{-1}$) and other of NaOH (0.50 $mol{\cdot}L^{-1}$) and, alternatively, Na_2CO_3 (0.15 $mol{\cdot}L^{-1}$), were added (40 mL each) simultaneously by continuous dropping to 40 mL of a NaOH solution (pH = 8 or 10), under magnetic stirring at room temperature. The pH was maintained during the process. A blue slurry was formed, which was then filtrated or centrifuged and dried at room temperature. The filtrated material was alternatively washed with ethanol:acetone 1:1 volume mixture.

4.2. Synthesis of Boronic Esters Employing Palladium Nanoparticles

The synthesis of PdNPs followed method described by Senra et al. [19], Scheme 2. At first, 1 mL of a Na_2PdCl_4 0.005 $mol{\cdot}L^{-1}$ aqueous solution was mixed with 69 mg of β-cyclodextrin in a 10 mL glass flask and the mixture was kept at 70 °C for 1 h under magnetic stirring. Then, it was added to the flask 0.25 mmol of 1-bromo-4-methoxybenzene, 0.25 mmol de bis(pinacolato)diboron, 0.375 mmol of K_2CO_3, and 2 mL of solvent (ethanol/H_2O 50% or acetonitrile). The system was kept at 70 °C for 24 h, and then the reaction mixture was extracted with dichloromethane and NaCl aqueous saturated solution (1:1). After the extraction, the mixture was dried with anhydrous sodium sulfate and evaporated under reduced pressure.

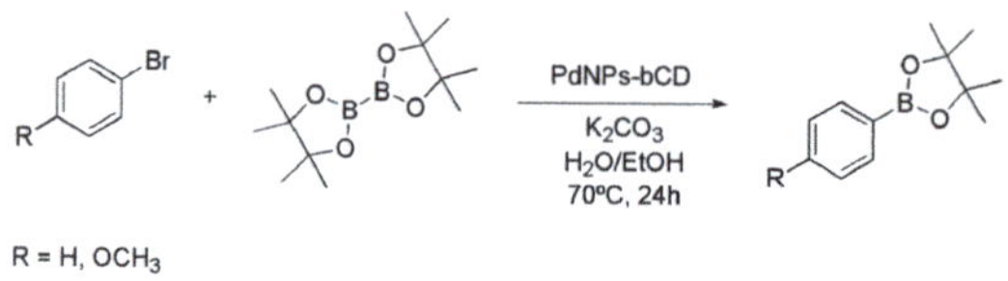

Scheme 2. Preliminary reaction.

4.3. Synthesis of Boronic Esters Employing Na_2PdCl_4/Cu/Al LDH

Into a 5-mL screw cap flask was added 0.9 mmol de bis(pinacolato)diboron, 0.6 mmol de 1-iodo-4-nitrobenzene, 2 mL of solvent, and 0.15 mmol (Cu) of LDH/Pd or 0.15 mmol (Cu) of Cu/Al LDH and 2 mL of Na_2PdCl_4 0.005 $mol{\cdot}L^{-1}$ aqueous solution, Scheme 3. In the latter, the percentage of palladium varied from 2% to 0.5%. The system was kept at 70 °C for 24 h under magnetic stirring, and its extraction was done as described in Supporting Information.

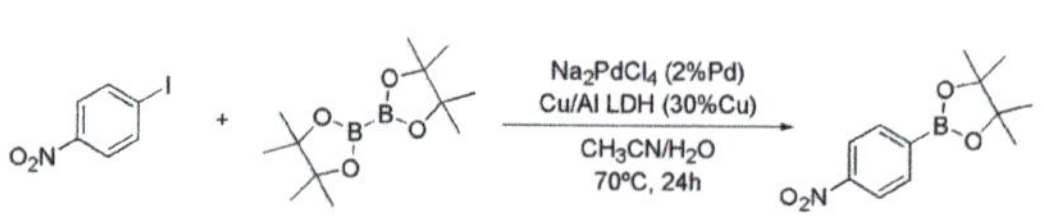

Scheme 3. Model reaction.

5. Conclusions

Cu/Al LDH had some synthesis parameters evaluated in order to produce the least quantity of by-products. For such material, the list of contaminants includes the mineral-like phases malachite, spertiniite, and nitratine. The catalytic system based on the least-contaminated LDH heterogeneous matrix and aqueous Na_2PdCl_4, under ligand-free conditions, showed efficient catalytic properties in the borylation reactions by using bis(pinacolato)diboron as boron precursor. This catalyst system can be a promising low-cost alternative for use in classical homogeneous phosphine-based systems.

Additional experiments to investigate the possible reaction intermediates and their correlation with the LDH structure are under investigation and will be reported in due course.

Supplementary Materials: The following are available online at http://www.mdpi.com/2073-4344/9/4/302/s1: Figure S1: ^{1}H NMR of **3**, Figure S2: ^{13}C NMR of **3**, Figure S3: ^{1}H NMR of **4**, Figure S4: ^{13}C NMR of **4**, Figure S5: ^{1}H NMR of **6**, Figure S6: ^{13}C NMR of **6**, Figure S7: ^{1}H NMR of **7**, Figure S8: ^{13}C NMR of **7**, Figure S9: ^{1}H NMR of **8**, Figure S10: ^{13}C NMR of **8**, Figure S11: ^{1}H NMR of **9**, Figure S12: ^{13}C NMR of **9**).

Author Contributions: L.C.L.L.F.S. conducted all the borylation reactions. V.A.N. prepared the catalysts. V.S.R. and J.B.d.C. conducted the XRD analysis and the Rietveld refinements. R.S.F.S. and A.A.d.S. conducted the product characterization. L.F.B.M. and J.D.S. supervised the work and wrote the paper.

Funding: This study was financed in part by the Coordenação de Aperfeiçoamento de Pessoal de Nível Superior—Brasil (CAPES), Finance Code 001. All authors acknowledge the financial support by CNPq and FAPERJ. Specifically, L.F.B.M. would like to thank the funding from CNPq, Universal project code 425613/2016-0, and from FAPERJ, Jovem Cientista do Nosso Estado project code E-26/203.212/2017.

Acknowledgments: The authors thank CBPF (Centro Brasileiro de Pesquisas Físicas), UFRJ and UERJ for the analytical support.

Conflicts of Interest: The authors declare no conflicts of interest.

References

1. Wu, X.; Chen, X.X.; Jiang, Y.B. Recent advances in boronic acid-based optical chemosensors. *Analyst* **2017**, *142*, 1403–1414. [CrossRef] [PubMed]
2. Roll, M.F. Ionic borohydride clusters for the next generation of boron thin-films: Nano-building blocks for electrochemical and refractory materials. *J. Mater. Res.* **2016**, *31*, 2736–2748. [CrossRef]
3. Pisarev, M.A.; Dagrosa, M.A.; Juvenal, G.J. Boron neutron capture therapy in cancer: Past, present and future. *Arq. Bras. Endocrinol. Metab.* **2007**, *51*, 852–856. [CrossRef]
4. Hall, D.G. *Boronic Acids*; Wiley: Weinheim, Germany, 2005.
5. Cheng, Y.; Mgck-Lichtenfeld, C.; Studer, A. Metal-Free Radical Borylation of Alkyl and Aryl Iodides. *Angew. Chem. Int. Ed.* **2018**, *57*, 16832–16836. [CrossRef]
6. Zhang, L.; Jiao, L. Pyridine-Catalyzed Radical Borylation of Aryl Halides. *J. Am. Chem. Soc.* **2017**, *139*, 607–610. [CrossRef]
7. Chow, W.K.; Yuen, O.Y.; Choy, P.Y.; So, C.M.; Lau, C.P.; Wong, W.T.; Kwong, F.Y. A decade advancement of transition metal-catalyzed borylation of aryl halides and sulfonates. *RSC Adv.* **2013**, *3*, 12518–12539. [CrossRef]
8. Ishiyama, T.; Murata, M.; Miyaura, N. Palladium(0)-Catalyzed Cross-Coupling Reaction of Alkoxydiboron with Haloarenes: A Direct Procedure for Arylboronic Esters. *J. Org. Chem.* **1995**, *60*, 7508–7510. [CrossRef]
9. Avitia, B.; MacIntosh, E.; Muhia, S.; Kelson, E. Single-flask preparation of polyazatriaryl ligands by a sequential borylation/Suzuki-Miyaura coupling. *Tetrahedron Lett.* **2011**, *52*, 1631–1634. [CrossRef]
10. Xie, D.; Rong, L.; Zhang, D.; Hu, J.; Xiao, D.; Li, X.; Xiang, Y.; Jin, W. Palladium-catalyzed borylation of m-dibromobenzene derivative and its applications in one-pot tandem Suzuki-Miyaura arenes synthesis. *Tetrahedron* **2015**, *71*, 8871–8875. [CrossRef]
11. Hemming, D.; Fritzemeier, R.; Westcott, S.A.; Santos, W.L.; Steel, P.G. Copper-boryl mediated organic synthesis. *Chem. Soc. Rev.* **2018**, *47*, 7477–7494. [CrossRef]
12. Ratniyom, J.; Dechnarong, N.; Yotphan, S.; Kiatisevi, S. Convenient Synthesis of Arylboronates through a Synergistic Pd/Cu-Catalyzed Miyaura Borylation Reaction under Atmospheric Conditions. *Eur. J. Org. Chem.* **2014**, *7*, 1381–1385. [CrossRef]
13. Mills, S.J.; Christy, A.G.; Génin, J.-M.R.; Kameda, T.; Colombo, F. Nomenclature of the hydrotalcite supergroup: Natural layered double hydroxides. *Miner. Mag.* **2012**, *75*, 1289–1336. [CrossRef]
14. Vaccari, A. Preparation and catalytic properties of cationic and anionic clays. *Catal. Today* **1998**, *41*, 53–71. [CrossRef]
15. Béres, A.; Pálinkó, I.; Kiricsi, I.; Nagy, J.B.; Kiyozumi, Y.; Mizukami, F. Layered double hydroxides and their pillared derivatives—Materials for solid base catalysis; synthesis and characterization. *App. Catal. A Gen.* **1999**, *182*, 237–247. [CrossRef]

16. Silva, A.C.; Senra, J.D.; de Souza, A.L.F.; Malta, L.F.B. A Ternary Catalytic System for the Room Temperature Suzuki-Miyaura Reaction in Water. *Sci. World J.* **2013**. [CrossRef]
17. Sreedhar, B.; Arundhathi, R.; Reddy, P.L.; Reddy, M.A.; Kantam, M.L. Cu-Al Hydrotalcite: An efficient and reusable ligant-free catalyst for the coupling of aryl chlorides with aliphatic, aromatic, and N(H)-heterocyclic amines. *Synthesis* **2009**, *15*, 2517–2522. [CrossRef]
18. Neves, V.A.; Costa, M.V.; Senra, J.D.; Aguiar, L.C.S.; Malta, L.F.B. Thermal behavior of LDH 2CuAl.CO3 and 2CuAl.CO3/Pd. *J. Therm. Anal. Calorim.* **2017**, *130*, 689–694. [CrossRef]
19. Senra, J.D.; Malta, L.F.B.; Michel, R.C.; Cordeiro, Y.; Simão, R.A.; Simas, A.B.C.; Aguiar, L.C.S. Hydrophilic cyclodextrin protected Pd nanoclusters: Insights into their size control and host–guest behavior. *J. Mater. Chem.* **2011**, *21*, 13516–13523. [CrossRef]
20. Segal, S.R.; Carrado, K.A.; Marshall, C.L.; Anderson, K.B. Catalytic decomposition of alcohols, including ethanol, for in situ H_2 generation in a fuel stream using a layered double hydroxide-derived catalyst. *Appl. Catal. A Gen.* **2003**, *248*, 33–45. [CrossRef]
21. Manivannan, R.; Pandurangan, A. Formation of ethyl benzene and styrene by side chain methylation of toluene over calcined LDHs. *Appl. Clay Sci.* **2009**, *44*, 137–143. [CrossRef]
22. Lwin, Y.; Yarmo, M.A.; Yaakob, Z.; Mohamad, A.B.; Daud, W.R.W. Synthesis and characterization of Cu/Al layered double hydroxides. *Mater. Res. Bull.* **2001**, *36*, 193–198. [CrossRef]
23. Muñoz, V.; Zotin, F.M.Z.; Palacio, L.A. Copper-aluminum hydrotalcite type precursors for NO_x abatement. *Catal. Today* **2015**, *250*, 173–179. [CrossRef]
24. Gao, P.; Xie, R.; Wang, H.; Zhong, L.; Xia, L.; Zhang, Z.; Wei, W.; Sun, Y. Cu/Zn/Al/Zr catalyst via phase-pure hydrotalcite-like compounds for methanol synthesis from carbon dioxide. *J. CO_2 Util.* **2015**, *11*, 41–48. [CrossRef]
25. Ichikawa, S.; Miyazoe, S.; Matsuoka, O. A highly efficient Cu/Al(OH)$_3$ catalyst for the hydration of acrylonitrile to acrylamide. *Chem. Lett.* **2011**, *40*, 512–514. [CrossRef]
26. Qu, J.; He, X.; Chen, M.; Hu, H.; Zhang, Q.; Liu, X. Mechanochemical synthesis of Cu/Al and methyl orange intercalated Cu/Al layered double hydroxides. *Mater. Chem. Phys.* **2017**, *191*, 173–180. [CrossRef]
27. Neves, V.A. Nanostructured Materials Based on Cu/Al Layered Double Hydroxides, Pd and Cyclodextrin as Catalysts for Cross-Coupling Reactions. Master's Thesis, Federal University of Rio de Janeiro, Rio de Janeiro, Brazil, 2018.

catalysts

MDPI

Article

Aldol Condensation of Furfural with Acetone Over Mg/Al Mixed Oxides. Influence of Water and Synthesis Method

Almudena Parejas, Daniel Cosano, Jesús Hidalgo-Carrillo *, José Rafael Ruiz, Alberto Marinas, César Jiménez-Sanchidrián and Francisco J. Urbano

Departamento de Química Orgánica, Instituto Universitario de Investigación en Química Fina y Nanoquímica IUIQFN, Universidad de Córdoba, Campus de Rabanales, Edificio Marie Curie, E-14071 Córdoba, Spain; q12pabaa@uco.es (A.P.); q92cohid@uco.es (D.C.); qo1ruarj@uco.es (J.R.R.); alberto.marinas@uco.es (A.M.); qo1jisac@uco.es (C.J.-S.); fj.urbano@uco.es (F.J.U.)

* Correspondence: jesus.hidalgo@uco.es; Tel.:+34-957-218-638

Received: 23 January 2019; Accepted: 20 February 2019; Published: 23 February 2019

Abstract: Aldol condensation of furfural and acetone (an important initial step to obtain diesel from biomass) was studied over MgAl mixed oxides. The influence of the utilization of microwaves and/or a surfactant (Pluronic 123) during the synthesis as well as the use of water (either pre-hydrating the solids before catalytic studies or in water/toluene mixtures as the reaction medium) is discussed. The combined use of Pluronic 123 and microwaves led to solids with bigger pore sizes, exhibiting lower basicity and higher acidity than the conventional synthetic method, thus resulting in an increase in the yield of the desired product of condensation, comprising two molecules of furfural and one of acetone (F2Ac). As for the influence of water, re-hydration of the mixed oxides was detrimental to activity, probably as a result of the partial blocking (solvation) of active sites. On the contrary, the increase in water percentage in the reaction medium resulted in higher conversions, though selectivity to F2Ac decreased. The weakening of the C=O bond of furfural in the presence of water as well as the higher solubility of the first condensation product (FAc) in toluene, as compared to water, could account for that. A 44.5% yield of F2Ac (66% conversion) after 16 h was obtained with the most active solid, which maintained the activity for three consecutive reactions.

Keywords: aldol condensation; biomass valorization; Mg/Al mixed oxides; surfactant; microwaves; influence of water

1. Introduction

Fossil fuel depletion and environmental concern have boosted the search for renewable energies, one of the possible sources being biomass [1,2]. Furfural is a so-called platform molecule from biomass obtained through xylose dehydration [3,4] and can be transformed into a wide range of chemicals via hydrogenation, oxidation, decarbonylation, nitration, or condensation processes, just to cite some of them [5]. For instance, aldol condensation and subsequent hydrogenation and hydrodeoxygenation can lead to liquid hydrocarbons for use as diesel [6–8].

Aldol condensation is a well-known C–C bond formation process which can occur in acidic or basic sites, the latter being more frequently reported in the literature [6–15]. It requires the existence of a reactive hydrogen in alpha position, with respect to a carbonyl compound able to form an enol, which reacts with another carbonyl compound, and after dehydration, yields a conjugated enone. Focusing on aldol condensation between furfural and acetone (Figure 1), it can initially lead to 4-(2-furanyl)-3-buten-2-one (FAc), a subsequent aldol condensation with another furfural molecule, forming 1,5-bis-(2-furanyl)-1,4-pentadien-3-one (F2Ac) (Figure 1a) [16,17]. Some side reactions include acetone self-condensation to form diacetone-alcohol and mesityl oxide (Figure 1b), condensation between FAc and acetone (Figure 1c), and multiple aldol condensations between different carbonyl compounds, thus forming polymers [18,19] (Figure 1d).

Aldol condensations have been traditionally performed in organic media, using base catalysts such as sodium or calcium hydroxides. Nevertheless, the existence of corrosion problems and the difficult reutilization have led to the use of some other base heterogeneous catalysts, such as hydrotalcites and hydrotalcite-derived mixed oxides [20,21], amorphous aluminophosphate [15], and diamine-functionalized MCM-41 [22], just to cite some examples.

In the present work, different AlMg mixed oxides were obtained through calcination of layered double hydroxides (LDHs) and tested for aldol condensation with acetone to form F2Ac. The influence on the catalytic results of two synthetic variables (conventional or microwave heating with the presence or absence of Pluronic 123 as the surfactant) was explored. Furthermore, the effect of water (either pre-hydrating the solids before catalytic studies or in water/toluene mixtures as the reaction medium) is discussed.

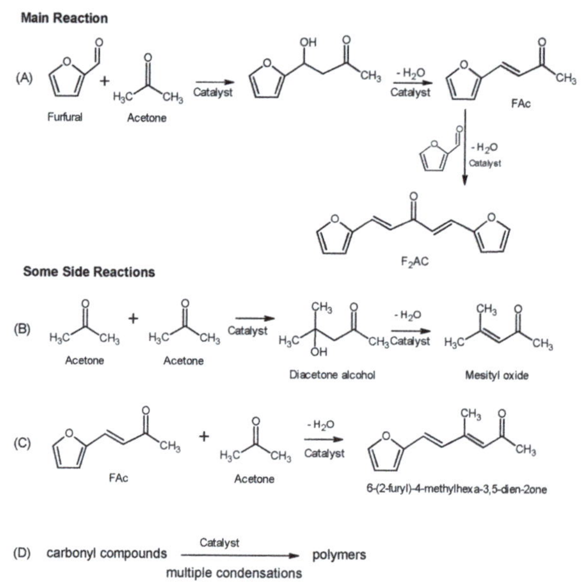

Figure 1. Reaction scheme for aldol condensation of furfural and acetone. Some side reactions have also been included. (**A**) aldol condensation between furfural and acetone, (**B**) acetone self-condensation, (**C**) condensation between FAc and acetone and (**D**) multiple aldol condensations between different carbonyl compounds.

2. Results and Discussion

2.1. Textural, Structural and Acid–Base Characterization of the Solids

X-ray diffractograms of uncalcined and calcined hydrotalcites are shown in Figure 2. As can be seen, uncalcined solids exhibit a typical hydrotalcite crystallinity profile (JCPDS 22-700), with symmetric reflections at $2\theta = 11°$, 22°, 36°, 37°, 45°, 60°, and 62°. Therefore, sharper peaks corresponded to (003), (006), (010), and (013) reflections, whereas broader signals were obtained for (009), (015), and (016) reflections, all of them representative of layered materials. As for calcined solids (Figure 2b), diffraction patterns are very similar to each other, exhibiting (111), (200), and (220) reflections ascribed to periclase. In a previous paper, Aramendia et al. [23], using ^{27}Al NMR-MAS, demonstrated that the coordination of Al^{3+} changed from octahedral to tetrahedral upon calcination of hydrotalcites, Al^{3+} ions thus isomorphically substituting Mg^{2+} ions, forming MgAlOx periclase.

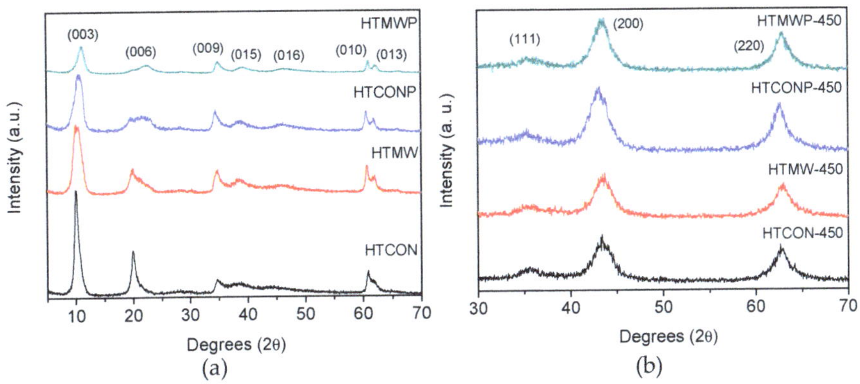

Figure 2. X-ray diffractograms of the different solids synthesized in the present work. (**a**) Uncalcined solids. (**b**) Solids calcined at 450 °C.

Thermal stability of hydrotalcites was determined by TG-DTA (Figure 3). In all cases, weight loss percentage is in the 42–45% range (Figure 3a). HTCON and HTMW thermogravimetric profiles are consistent with those reported in the literature for hydrotalcites [24,25]. Therefore, two main weight losses are observed. The first one at 100–200 °C is ascribed to the loss of intercalated water molecules, whereas nitrate coming from both the precursor and hydroxyl groups can account for the second loss at higher temperatures (250–500 °C). For the solids synthesized using the surfactant (HTCONP and HTMWP), the second weight loss seems to be produced quicker (i.e., at lower temperatures), thus suggesting that for those systems, re-structuration to form periclase is somehow favored by Pluronic 123. Heat flow profiles (Figure 3b) seem to confirm this hypothesis, the exothermal peak centered at 450 °C in HTCON and HTMW being shifted to lower temperatures (300–350 °C) for HTCONP and HTMWP.

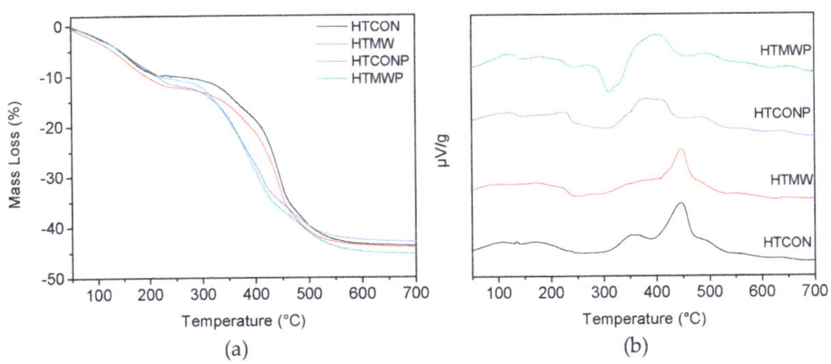

Figure 3. TG analyses (**a**) and heat flow (**b**) of hydrotalcites.

Raman spectra of uncalcined solids (i.e., hydrotalcites) are represented in Figure 4. The band appearing at 557 cm^{-1} can be assigned to vibrations of brucite-like octahedral layers, Al-O-Mg, which are present in all Mg-Al hydrotalcites [26]. Moreover, the spectra also exhibit bands at 710 and 1055 cm^{-1}, corresponding to nitrate vibrations [27] and bands at ca. 3500 cm^{-1}, due to surface hydroxyl groups. In the case of HTCONP and HTMWP solids, there are also some intense bands of C-H stretching Pluronic 123 at 2986, 2941, and 2933 cm^{-1} [28].

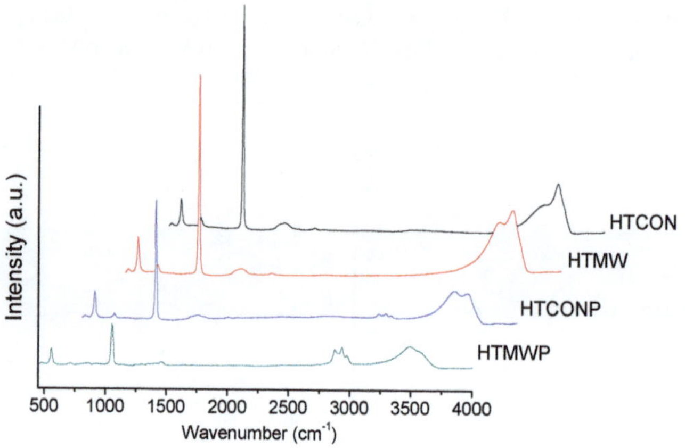

Figure 4. Raman spectra of the uncalcined solids (hydrotalcites).

N_2 adsorption–desorption isotherms of calcined solids are shown in Figure 5. In all cases, type IV isotherms corresponding to mesoporous materials were obtained. BET surface areas, pore volume and average pore diameter values are given in Table 1. With regards to the BET areas, they are in the 160–210 $m^2 \cdot g^{-1}$ range, the highest value corresponding to HTCONP-450. Modification of conventional synthesis by using microwave irradiation and/or in the presence of the surfactant (Pluronic 123) led in all cases to an increase in BET area. Solids that aged under microwave irradiation exhibit smaller pore diameters than their conventionally-heated counterparts (compare HTMW-450 vs. HTCON-450 or HTMWP-450 vs. HTCONP-450). Systems synthesized in the presence of the surfactant present bigger pores (compare HTCONP-450 vs. HTCON-450 and HTMWP-450 vs. HTMW-450). Therefore, the effect of microwaves and the presence of a surfactant on pore volume is the opposite. However, if both variables are changed simultaneously, the influence of the surfactant is more important, thus resulting in the pore diameter increasing (compare HTCON-450 and HTMWP-450 with pore diameters of 6.8 and 8.4 nm, respectively).

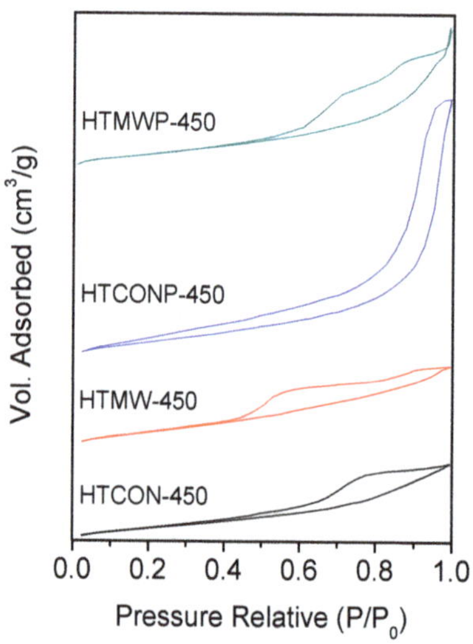

Figure 5. Nitrogen adsorption–desorption isotherms corresponding to the mixed oxides.

Table 1. Summary of the main features of the mixed oxides synthesized in this work.

Catalyst	Textural Properties			Base Sites (CO_2-TPD) (μmol/g)	Acid Sites (Py-TPD) (μmol/g)	Mg/Al Ratio	
	S_{BET} (m^2/g)	V_P (cm^3/g)	D_P (nm)			Nominal	XRF
HTCON-450	160	0.26	6.8	1105	871	2.00	2.01
HTMW-450	201	0.29	4.2	880	1030	2.00	1.94
HTCONP-450	210	0.86	16.6	932	960	2.00	2.04
HTMWP-450	183	0.40	8.4	775	1124	2.00	1.97

X-Ray fluorescence results (Table 1) evidenced a good incorporation of Mg and Al to the solids, with Mg/Al ratios very similar to the nominal value (Mg/Al = 2).

Base characterization of the solids was performed using thermal programmed desorption of pre-adsorbed CO_2 (CO_2-TPD) and the results are given in Tables 1 and 2, and in Figure 6. In all cases, signals were deconvoluted in peaks, which, depending on the desorption temperature, were ascribed to weak (80–200 °C), medium (200–300 °C), or strong (>300 °C) basic sites, respectively. Taking HTCON-450 as the reference, the use of microwave irradiation and/or the presence of the surfactant in the synthesis results in a drop in total basicity. Interestingly, as far as the base site distribution is concerned, the effect of microwave irradiation, the presence of the surfactant, or both variables simultaneously considered is different. Therefore, in the absence of Pluronic 123, microwave irradiation does not vary base site distribution. On the contrary, the presence of Pluronic 123 results in an increase in the strong base sites' percentage, to the detriment of weak ones. Finally, simultaneous use of microwaves and Pluronic 123 leads to an increase in the percentage of weak base sites.

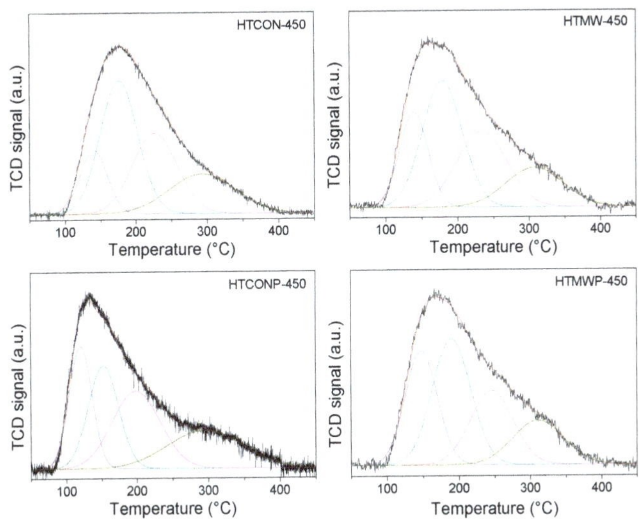

Figure 6. Temperature-programmed desorption profiles of CO_2 for the mixed oxides synthesized in this work.

Table 2. Base site distribution of the solids expressed as μmol CO_2/g. Values in brackets represent the percentage of the total basicity.

Catalyst	Base Sites Distribution, μmol CO_2/g		
	Weak (80–200 °C)	Medium (200–300 °C)	Strong (>300 °C)
HTCON-450	580.34 (52.53%)	304.89 (27.60%)	219.51 (19.87%)
HTMW-450	468.05 (53.20%)	241.95 (27.50%)	169.80 (19.30%)
HTCONP-450	428.62 (45.97%)	276.64 (29.67%)	227.13 (24.36%)
HTMWP-450	470.96 (60.80%)	177.38 (22.90%)	126.26 (16.30%)

Acid characterization of the solids was performed by thermal programmed desorption of pre-adsorbed pyridine (Py-TPD), results being given in Tables 1 and 3, and in Figure 7. Taking HTCON-450 as the reference, contrary to basicity, total acidity increases when the solids are synthesized utilizing a microwave and/or in the presence of Pluronic 123. Furthermore, with regards to acid site distribution, only microwave irradiation has some effect (acid strength decreases), whereas the presence of Pluronic 123, either under conventional heating or microwave irradiation, does not vary acid site distribution.

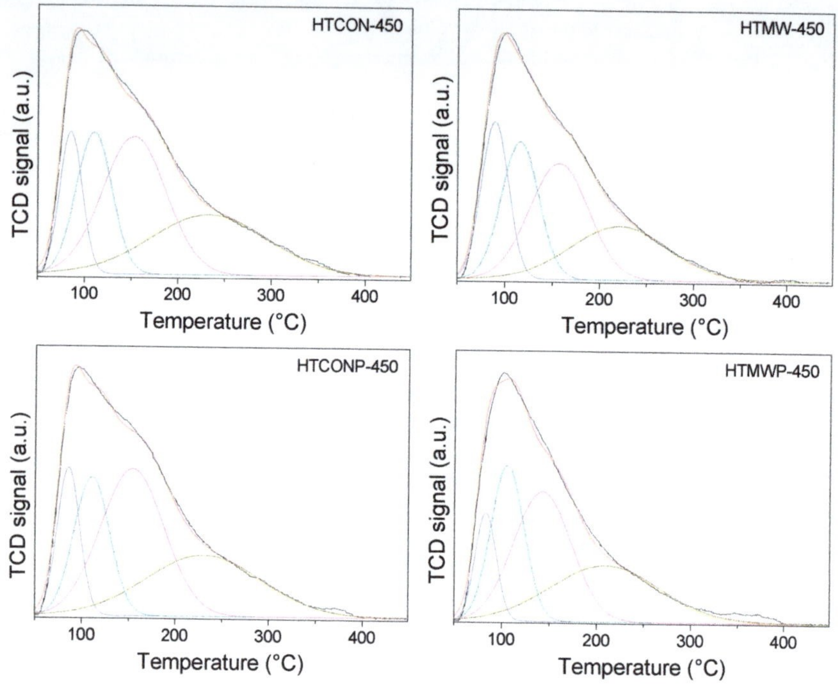

Figure 7. Temperature-programmed desorption profiles of pyridine for the solids synthesized in this work.

Table 3. Acid site distribution of the solids expressed as μmol Py/g. Values in brackets represent the percentage of the total basicity.

Catalyst	Acid Site Distribution, μmol Py/g		
	Weak (80–200 °C)	Medium (200–300 °C)	Strong (>300 °C)
HTCON-450	296.36 (34.01%)	320.14 (36.74%)	254.88 (29.25%)
HTMW-450	447.56 (43.44%)	329.59 (31.99%)	253.14 (24.57%)
HTCONP-450	320.58 (33.39%)	365.71 (38.09%)	273.82 (28.52%)
HTMWP-450	397.24 (35.33%)	398.70 (35.46%)	328.43 (29.21%)

All in all, microwave irradiation and/or the use of Pluronic 123 as the surfactant result in an increase in total acidity and a decrease in total basicity together with an increase in BET areas.

2.2. Catalytic Activity

The solid synthesized under conventional heating and in the absence of the surfactant (HTCON-450) was used as the reference material in order to study the influence of reaction temperature and the presence of water in the reaction medium on catalytic activity.

2.2.1. Influence of Reaction Temperature

Table 4 summarizes catalytic results obtained for t = 3 h at different temperatures. From that table, it is evident that the increase in temperature results in the increase in conversion, whereas selectivity to F2Ac, the desired product, hardly changes. Therefore, from then on, 100 °C was selected as the reaction temperature.

Table 4. Aldol condensation of furfural and acetone on HTCON-450: Influence of temperature on catalytic activity. Reaction conditions: Reactor pressurized to 5 bar with N_2; 10 mmol of furfural, 20 mmol acetone, 20 mL toluene, and 400 mg catalyst, t = 3 h.

Temperature (°C)	Conv. (%)	Sel. FAc (%)	Sel. F2Ac (%)	Yield FAc (%)	Yield F2Ac (%)
60	3.8	45.5	54.5	1.7	2.1
80	14.4	50.3	49.7	7.3	7.2
100	35.0	42.5	57.6	14.8	20.1

2.2.2. Influence of Water

Two different approaches were made to study the influence of water on catalytic activity of HTCON-450. On the one hand, the solid was re-hydrated using a N_2 flow saturated in water. On the other hand, reactions were performed in water/toluene mixtures (0%, 5%, 10%, 50%). Regarding the former approach (Table 5), the pretreatment of HTCON-450 with a flow of nitrogen saturated in water results in a drop in conversion (35.0% and 23.7% for HTCON-450 and HTCON-450-rehydrated, respectively, at t = 3 h), whereas selectivity values to F2Ac are quite similar. This suggests that rehydration results in the elimination of active sites to a certain extent, which could be ascribed to solvation. Results obtained for the uncalcined HTCON solid (which exhibits very low catalytic activity) are also given for the sake of comparison.

Table 5. Aldol condensation of furfural and acetone on HTCON-450: Influence of calcination and rehydration of the solid. Reaction conditions: Reactor pressurized to 5 bar with N_2; 10 mmol of furfural, 20 mmol acetone, 20 mL toluene, and 400 mg catalyst, t = 3 h, 100 °C.

Catalyst	Conv. (%)	Sel. FAc (%)	Sel. F2Ac (%)	Yield FAc (%)	Yield F2Ac (%)
HTCON	0.7	85.0	14.9	0.6	0.1
HTCON-450	35.0	42.5	57.6	14.8	20.1
HTCON-450-rehidrated	23.7	40.1	59.8	9.5	14.2

Table 6 summarizes the results obtained for the study conducted using different water/toluene ratios as the reaction medium. As can be seen, the higher the percentage of water, the higher the conversion, but in general, the lower the selectivity to the desired product, F2Ac. In a previous paper [29] on chemoselective hydrogenation of alpha,beta-unsaturated carbonyl compounds, our research group found evidence by Raman spectroscopy that water interacted with the carbonyl group and made the double bond weaker and thus more reactive (carbonyl band shifted to lower wavelength values). The same could occur in the C=O group in furfural and account for its higher conversion in the presence of water. With regards to the change in selectivity, one should consider that we are working in a biphasic media and thus the catalyst hydrophilic character, as well as the relative solubility of reactants and products both in toluene and water, is important. Active sites in the catalyst will probably interact better with water than with toluene. Furfural is partially soluble in water (50–100 $mg \cdot mL^{-1}$). Its condensation with one molecule of acetone will produce FAc, whose solubility in water is much lower (1–10 $mg \cdot mL^{-1}$). Therefore, once formed, FAc will pass to the organic phase (toluene) and will not be able to undergo subsequent condensation with another acetone molecule to produce F2Ac. This results in the increase in selectivity to FAc and probably in conversion, since FAc is retired of the aqueous phase as the reaction proceeds. All in all, the highest yield to F2Ac, the desired product, is achieved with pure toluene. Therefore, this reaction medium was selected for subsequent studies on other catalysts.

Table 6. Aldol condensation of furfural and acetone on HTCON-450: Influence of the presence of water in water/toluene mixtures. Reaction conditions: Reactor pressurized to 5 bar with N_2; 10 mmol of furfural, 20 mmol acetone, 20 mL (toluene + water) and 400 mg catalyst, t = 3 h, 100 °C.

Water (%)	Conv. (%)	Sel. FAc (%)	Sel. F2Ac (%)	Yield FAc (%)	Yield F2Ac (%)
0	35.0	42.5	57.6	14.8	20.1
5	37.6	60.8	39.2	22.9	14.7
10	53.2	70.3	29.6	37.4	15.8
50	80.4	88.1	11.9	70.8	9.6

2.3. Catalytic Activity of the Other Mixed Oxides

Once pure toluene and 100 °C had been selected as the reaction conditions for aldol condensation of furfural and acetone in order to obtain F2Ac, the study was extended to the other mixed oxides. Reactions were conducted at 3 h and 16 h, the main catalytic results being summarized in Table 7.

A first conclusion from that table is that the lowest conversion values correspond to HTMW-450. It is important to note that this solid was the one exhibiting the lowest pore diameter (4.2 nm, Table 1), which could account for that. Focusing on the other solids, the highest conversion value at 3 h is achieved for HTCON-450, whereas as the reaction proceeds, the rate is higher for the systems synthesized using Pluronic 123, which together with their higher selectivity to F2Ac results in F2Ac yields of 24.6%, 28.3%, and 44.5% for HTCON-450, HTCONP-450, and HTMWP-450, respectively, at t = 16 h. As evidenced by thermal-programmed desorption of pyridine and CO_2, the use of microwave irradiation and/or the presence of Pluronic 123 in the reaction medium during the synthesis resulted in an increase in total acidity and a decrease in total basicity. In a previous paper, Climent et al. [15] described the cooperative effect of weak acid and base sites of an amorphous aluminophosphate in aldol condensation, thus resulting in higher selectivities than those presented by stronger acid or base catalysts. This effect together with the increase in pore size could explain the higher yields obtained for the solids synthesized using the surfactant.

Table 7. Results obtained for aldol condensation of furfural and acetone on the different solids. Reaction conditions: Reactor pressurized to 5 bar with N_2; 10 mmol of furfural, 20 mmol acetone, 20 mL toluene and 400 mg catalyst, 100 °C.

Time (h)	Catalyst	Conv. (%)	Sel. FAc (%)	Sel. F2Ac (%)	Yield FAc (%)	Yield F2Ac (%)
3	HTCON-450	35.0	42.5	57.6	14.8	20.1
	HTMW-450	19.5	63.2	36.8	12.3	7.2
	HTCONP-450	28.4	47.6	52.4	13.5	14.9
	HTMWP-450	32.9	32.9	67.1	10.8	22.1
16	HTCON-450	46.9	47.6	52.4	22.4	24.6
	HTMW-450	27.4	42.9	57.1	11.8	15.7
	HTCONP-450	45.2	37.4	62.6	18.9	28.3
	HTMWP-450	66.2	32.8	67.2	21.7	44.5

2.4. Reutilization of HTCON-450 and HTMWP-450

Finally, some reutilization studies were conducted on HTCON-450 and HTMWP-450 solids, results being summarized in Table 8. In all cases, the Mg/Al ratio of the solids was quite similar to the nominal value (Mg/Al = 2). Moreover, after the reactions, the reaction medium was analyzed by inductively coupled plasma mass spectrometry (ICP-MS). No Mg or Al was detected which is evidence of the stability of the solids, which do not undergo leaching.

Table 8. Results obtained for reutilization studies. Reaction conditions: Reactor pressurized to 5 bar with N_2; 10 mmol of furfural, 20 mmol acetone, 20 mL toluene and 400 mg catalyst, 100 °C.

Catalyst	Conv. (%)	Sel. FAc (%)	Sel. F2Ac (%)	Yield FAc (%)	Yield F2Ac (%)	Mg/Al Ratio (XRF)
HTCON-450	35.0	42.5	57.6	14.8	20.1	2.01
HTCON-450-R	37.4	41.2	58.8	15.4	22.0	2.03
HTMWP-450	32.9	32.9	67.1	10.8	22.1	1.97
HTMWP-450-R	31.1	41.5	58.5	12.9	18.2	1.98
HTMWP-450-R2	34.4	43.5	56.5	15.0	19.4	2.02

As far as the catalytic activity is concerned, neither HTCON-450 nor HTMWP-450 exhibited any remarkable deactivation keeping F2Ac yield in the ca. 20% order after three hours. In the case of the most active solid at long reaction times (HTMWP-450), its activity and selectivity only decreased slightly (from 67.1 to 58.5%) after three consecutive uses.

3. Materials and Methods

Hydrotalcites were synthesized by a co-precipitation method, starting from two solutions containing 0.2 mol $Mg(NO_3)_2 \cdot 6H_2O$ and 0.1 mol $Al(NO_3)_3 \cdot 9H_2O$ in 25 mL deionized water, respectively (Mg/Al = 2). The mixture was slowly added to a pH 10 aqueous solution under continuous stirring and an inert atmosphere (N_2), with temperature maintained at 60 °C. During precipitation, the pH value was maintained, adding NaOH 1M. The suspension was divided into four portions for further treatment. One part was kept under conventional heating at 80 °C for 24 h, followed by filtration and washing with deionized water, thus obtaining the solid called HTCON. A second portion was aged under microwave heating at 80 °C for 1 h, thus leading, after filtration and washing, to the solid termed as HTMW. A flexiWave platform for microwave synthesis (22 V, 50 Hz) with an IR temperature sensor (p/n IRT0500) was used. The other two portions were submitted to the same conventional or microwave heating while performing the synthesis in the presence of surfactant Pluronic 123 (2% by weight), thus leading to the solids named HTCONP and HTMWP, respectively. Finally, all four solids were calcined at 450 °C in the air for 8 h (1 $°C \cdot min^{-1}$ ramp). Nomenclature of these solids include the suffix 450, referring to calcination temperature (HTCON-450, HTMW-450, HTCONP-450, and HTMWP-450). Subsequent treatment of HTCON-450 for 2 h at 450 °C in the presence of a N_2 flow (50 $mL \cdot min^{-1}$) saturated in water at 20 °C led to a solid called HTCON-450-rehydrated.

A Setaram SetSys 12 instrument (SETARAM Instrumentation, Caluire, France) was used for thermogravimetric analyses (TGA). Experiments were performed on 20 mg samples placed in an alumina crucible and heated in the 30–600 °C range (10 $°C \cdot min^{-1}$, 50 $mL \cdot min^{-1}$ air stream).

Textural properties (BET surface area, cumulative pore volume, and average pore diameter) were measured in a Micromeritics ASAP-2010 instrument (Micromeritics, Norcross, GA, USA.). Samples were heated at 120 °C and degassed to 0.1 Pa before measurement.

The measure of magnesium or aluminium leaching (presence in filtered reaction medium) was performed by inductively coupled plasma mass spectrometry (ICP-MS) on a Perkin–Elmer ELAN DRC-e instrument.

The Mg/Al ratio of solids was measured by X-ray fluorescence (XRF) spectroscopy (Rigaku ZSK PrimusIV wavelength X-ray spectrometer (Rigaku, The Woodlands, TX, USA). Further details are given elsewhere [30].

Raman spectra were recorded on a Renishaw spectrometer (InVia Raman Microscope, Renishaw, Gloucestershire, UK), equipped with a Leica microscope with various lenses, monochromators, filters, and a CCD detector. Spectra were recorded over the 150–4000 cm^{-1} range, using green laser light excitation (532 nm) and gathering 32 scans.

X-ray diffraction (XRD) analysis was performed on a Siemens D-5000 diffractometer (Bruker Corporation, Billerica, MA, USA) using CuKα radiation over the range 5–80°.

Surface acidity of samples was measured by thermal programmed desorption of pre-adsorbed pyridine (Py-TPD) using TC detection. Samples (30 mg) were cleaned by heating to 450 °C (10 $°C \cdot min^{-1}$

ramp) under He flow (75 mL·min^{-1}) and then cooled down to 50 °C. The catalysts were subsequently saturated with pyridine for 30 min, cleaned for 60 min with He and TPD monitored from 50 to 450 °C (10 °C·min^{-1}), the final temperature being held for 45 min.

Surface basicity of the catalysts was determined on a Micromeritics Autochem II instrument by thermal programmed desorption of pre-absorbed CO_2 (CO_2-TPD) with TCD detection. Samples (100 mg) were cleaned in an Air stream (20 mL·min^{-1} Ar, heating at 450 °C at a rate of 10 °C·min^{-1} for 60 min and then cooled down to 40 °C). Then, solids were saturated with CO_2 (5% CO_2/Ar flow at 20 mL·min^{-1} for 60 min), physisorbed CO_2 removed with Ar flow (20 mL·min^{-1} for 30min) and TPD monitored from 50 to 450 °C (5 °C·min^{-1}), the final temperature being held for 60 min.

The solids were tested for aldol condensation of furfural using a Berghof HR-100 stainless steel high-pressure autoclave equipped with a 75 mL PTFE insert vessel. Under standard conditions, 10 mmol of furfural, 20 mmol acetone, 20 mL toluene, and a 400 mg catalyst were introduced in the vessel. Reactor was purged with nitrogen and pressurized to 5 bar of N_2. The reaction temperature was set to 100 °C and started by switching on the stirring at 750 rpm. To stop the reaction, the vessel was submerged in an ice bath. The choice of toluene as the organic medium was motivated by a previous paper [3] on xylose dehydration to furfural where toluene was found to give the highest yield to furfural. The final strategy would be to make the one-pot transformation of xylose to furfural and then F2Ac.

Experiments to evaluate the influence of the presence of water in the reaction medium were conducted varying the water/toluene ratio (0%, 5%, 10%, and 50% volume) while keeping the total solvent volume constant (20 mL).

Once the reactions were finished, the products were analyzed by gas chromatography (Agilent 7890) with a flame ionization detector (GC-FID), using a Supelco Nukol™ capillary column. In the case of using biphasic media (toluene/water mixtures), products were extracted from the aqueous phase with dichloromethane before GC-FID analysis. Quantification of furfural and condensation products was performed using the appropriate calibration curves. In all cases, mass balance considering unreacted furfural, FAc, and F2Ac was over 95%.

For reutilization experiments, after the reaction, the solids were filtered, washed with ethanol, and dried at 100 °C, followed by calcination at 450 °C under the same conditions as described in the synthesis. Nomenclature of reused catalysts include the suffix R (one reuse) or R2 (two reuses).

Furfural conversion and FAc and F2Ac selectivity were defined by Equations (1)–(3):

$$\text{Furfural conversion (\%)}: \frac{\text{initial furfural concentration} - \text{final furfural concentration}}{\text{initial furfural concentration}} \times 100 \quad (1)$$

$$\text{FAc selectivity (\%)}: \frac{\text{FAc concentration}}{\text{FAc concentration} + 2\cdot\text{F2Ac concentration}} \times 100 \quad (2)$$

$$\text{F2Ac selectivity (\%)}: \frac{2\cdot\text{F2Ac concentration}}{\text{FAc concentration} + 2\cdot\text{F2Ac concentration}} \times 100 \quad (3)$$

4. Conclusions

The synthesis of hydrotalcites in the presence of Pluronic 123 led, after calcination, to MgAl mixed oxides with bigger pore sizes than untreated solids. On the other hand, microwave irradiation led to smaller pore sizes as compared to conventional thermal treatment. As far as acid–base characteristics are concerned, the use of both microwave irradiation and Pluronic 123 during the synthesis resulted in a decrease of total basicity and an increase in total acidity.

Rehydration of mixed oxides by treating them with a nitrogen flow saturated with water led to solids exhibiting lower catalytic activity in aldol condensation of furfural, probably as a result of the partial blocking (solvation) of active sites. By contrast, the increase in the percentage of water in water/toluene biphasic media resulted in an increase in conversion values, though selectivity to FAc also increased at the expense of the desired product F2Ac. A plausible explanation is that water weakens the C=O bond in furfural, thus favoring its transformation. Moreover, once FAc is produced, its higher solubility in toluene, as compared to water, favors its transfer to the organic medium, thus avoiding its subsequent reaction with another furfural molecule to yield F2Ac. The fact that the produced FAc is retired to the organic phase could also account for the observed increase in conversion.

A comparison of catalytic activity of the reference material (HTCON-450) with that of the other solids allows us to conclude that the use of Pluronic 123 during synthesis (especially in combination with microwave irradiation) resulted in solids exhibiting higher F2Ac yields at long reaction times. This could be the result of the combination of two factors: The above-mentioned larger pore size achieved with the surfactant and the increase in total acidity which could favor aldol condensation.

HTMWP-450 exhibited a good stability without any significant loss of activity after three uses.

Author Contributions: Conceptualization, C.J.-S. and F.J.U.; methodology, A.M. and J.R.R.; validation, F.J.U., J.R.R. and J.H.-C.; formal analysis, A.M. and J.R.R.; investigation, A.P. and D.C.; data curation, A.M., J.H.-C. and A.P.; writing—original draft preparation, A.P. and D.C.; writing—review and editing, J.H.-C and A.M.; supervision, C.J.-S. and F.J.U.

Funding: This research was funded by Ramón Areces Foundation.

Acknowledgments: The scientific support from the Central Service for Research Support (SCAI) at the University of Cordoba is acknowledged.

Conflicts of Interest: The authors declare no conflict of interest.

References

1. Field, C.B.; Campbell, J.E.; Lobell, D.B. Biomass energy: The scale of the potential resource. *Trends Ecol. Evol.* **2008**, *23*, 65–72. [CrossRef] [PubMed]
2. Chheda, J.N.; Huber, G.W.; Dumesic, J.A. Liquid-phase catalytic processing of biomass-derived oxygenated hydrocarbons to fuels and chemicals. *Angew. Chem. Int. Ed.* **2007**, *46*, 7164–7183. [CrossRef] [PubMed]
3. Parejas, A.; Montes, V.; Hidalgo-Carrillo, J.; Sanchez-Lopez, E.; Marinas, A.; Urbano, F.J. Microemulsion and Sol-Gel Synthesized ZrO2-MgO Catalysts for the Liquid-Phase Dehydration of Xylose to Furfural. *Molecules* **2017**, *22*, 2257. [CrossRef] [PubMed]
4. Sheldon, R.A. Green and sustainable manufacture of chemicals from biomass: State of the art. *Green Chem.* **2014**, *16*, 950–963. [CrossRef]
5. O'Neill, R.E.; Vanoye, L.; De Bellefon, C.; Aiouache, F. Aldol-condensation of furfural by activated dolomite catalyst. *Appl. Catal. B Environ.* **2014**, *144*, 46–56. [CrossRef]
6. Smolakova, L.; Frolich, K.; Kocik, J.; Kikhtyanin, O.; Capek, L. Surface Properties of Hydrotalcite-Based Zn(Mg)Al Oxides and Their Catalytic Activity in Aldol Condensation of Furfural with Acetone. *Ind. Eng. Chem. Res.* **2017**, *56*, 4638–4648. [CrossRef]
7. Faba, L.; Díaz, E.; Ordóñez, S. Aqueous-phase furfural-acetone aldol condensation over mixed oxides. *Appl. Catal. B Environ.* **2012**, *113–114*, 201–211. [CrossRef]
8. Chheda, J.N.; Dumesic, J.A. An overview of dehydration, aldol-condensation and hydrogenation processes for production of liquid alkanes from biomass-derived carbohydrates. *Catal. Today* **2007**, *123*, 59–70. [CrossRef]
9. Climent, M.J.; Corma, A.; Iborra, S. Conversion of biomass platform molecules into fuel additives and liquid hydrocarbon fuels. *Green Chem.* **2014**, *16*, 516–547. [CrossRef]
10. Shen, W.Q.; Tompsett, G.A.; Hammond, K.D.; Xing, R.; Dogan, F.; Grey, C.P.; Conner, W.C.; Auerbach, S.M.; Huber, G.W. Liquid phase aldol condensation reactions with MgO-ZrO2 and shape-selective nitrogen-substituted NaY. *Appl. Catal. A Gener.* **2011**, *392*, 57–68. [CrossRef]
11. Cota, I.; Ramirez, E.; Medina, F.; Sueiras, J.E.; Layrac, G.; Tichit, D. New synthesis route of hydrocalumite-type materials and their application as basic catalysts for aldol condensation. *Appl. Clay Sci.* **2010**, *50*, 498–502. [CrossRef]
12. West, R.M.; Liu, Z.Y.; Peter, M.; Gaertner, C.A.; Dumesic, J.A. Carbon-carbon bond formation for biomass-derived furfurals and ketones by aldol condensation in a biphasic system. *J. Mol. Catal. A Chem.* **2008**, *296*, 18–27. [CrossRef]

13. Daniel, E.; Resasco, S.S.; Faria, J.; Prasomsri, T.; Ruiz, A.M.P. Furfurals as chemical platform forbiofuels production. In *Heterogeneous Catalysis in Biomass to Chemicals and Fuels*; D.K.A.I., Ed.; Stanford Court: Irvine, CA, USA, 2011.
14. Bao, Q.; Qi, H.; Zhang, C.; Ning, C.; Zhang, Y.; Wu, Y.; Gui, W.; Wang, Z. Highly Catalytic Activity of Ba/γ-Ti–Al2O3 Catalyst for Aldol Condensation of Methyl Acetate with Formaldehyde. *Catal. Lett.* **2018**, *148*, 3402–3412. [CrossRef]
15. Climent, M.J.; Corma, A.; Fornés, V.; Guil-Lopez, R.; Iborra, S. Aldol condensations on solid catalysts: A cooperative effect between weak acid and base catalysts. *Adv. Synth. Catal.* **2002**, *344*, 1090–1096. [CrossRef]
16. Fakhfakh, N.; Cognet, P.; Cabassud, M.; Lucchese, Y.; Rios, M.D.D.L. Stoichio-kinetic modeling and optimization of chemical synthesis: Application to the aldolic condensation of furfural on acetone. *Chem. Eng. Process.* **2008**, *47*, 349–362. [CrossRef]
17. Xing, R.; Subrahmanyam, A.V.; Olcay, H.; Qi, W.; van Walsum, G.P.; Pendse, H.; Huber, G.W. Production of jet and diesel fuel range alkanes from waste hemicellulose-derived aqueous solutions. *Green Chem.* **2010**, *12*, 1933–1946. [CrossRef]
18. Patel, A.A.; Patel, S.R. Synthesis and characterization of furfural-acetone polymers. *Eur. Polym. J.* **1983**, *19*, 231–234. [CrossRef]
19. Gandini, A.; Belgacem, M.N. Furans in polymer chemistry. *Prog. Polym. Sci.* **1997**, *22*, 1203–1379. [CrossRef]
20. Hora, L.; Kelbichova, V.; Kikhtyanin, O.; Bortnovskiy, O.; Kubicka, D. Aldol condensation of furfural and acetone over Mg-Al layered double hydroxides and mixed oxides. *Catal. Today* **2014**, *223*, 138–147. [CrossRef]
21. Ordonez, S.; Diaz, E.; Leon, M.; Faba, L. Hydrotalcite-derived mixed oxides as catalysts for different C-C bond formation reactions from bioorganic materials. *Catal. Today* **2011**, *167*, 71–76. [CrossRef]
22. Choudary, B.M.; Kantam, M.L.; Sreekanth, P.; Bandopadhyay, T.; Figueras, F.; Tuel, A. Knoevenagel and aldol condensations catalysed by a new diamino-functionalised mesoporous material. *J. Mol. Cataly. A Chem.* **1999**, *142*, 361–365. [CrossRef]
23. Aramendia, M.A.; Aviles, Y.; Borau, V.; Luque, J.M.; Marinas, J.M.; Ruiz, J.R.; Urbano, F.J. Thermal decomposition of Mg Al and Mg Ga layered-double hydroxides: A spectroscopic study. *J. Mater. Chem.* **1999**, *9*, 1603–1607. [CrossRef]
24. Morato, A.; Alonso, C.; Medina, F.; Cesteros, Y.; Salagre, P.; Sueiras, J.E.; Tichit, D.; Coq, B. Palladium hydrotalcites as precursors for the catalytic hydroconversion of CCl2F2 (CFC-12) and CHClF2 (HCFC-22). *Appl. Catal. B Environ.* **2001**, *32*, 167–179. [CrossRef]
25. Xu, C.; Gao, Y.; Liu, X.; Xin, R.; Wang, Z. Hydrotalcite reconstructed by in situ rehydration as a highly active solid base catalyst and its application in aldol condensations. *RSC Adv.* **2013**, *3*, 793–801. [CrossRef]
26. Frost, R.L.; Palmer, S.J.; Theiss, F. Synthesis and Raman spectroscopic characterisation of hydrotalcites based on the formula $Ca_6Al_2(CO_3)(OH)_{16}\cdot 4H_2O$. *J. Raman Spectrosc.* **2010**, *42*, 1163–1167. [CrossRef]
27. Frost, R.L.; Erickson, K.L. Vibrational spectroscopic study of the nitrate containing hydrotalcite Mbobomkulite, Spectrochim. *Acta Part A* **2005**, *61*, 2919–2925. [CrossRef] [PubMed]
28. Mahtab, P.; Seyed-Rasulzade, S.K.; Nikzad-Kojanag, B. Effect of Preparation Methods and Pluronic Template on the Catalytic Activity of Ca/SBA-15. *Iran. J. Chem. Chem. Eng.* **2018**, *37*, 53–60.
29. Hidalgo-Carrillo, J.; Marinas, A.; Marinas, J.M.; Delgado, J.J.; Raya-Miranda, R.; Urbano, F.J. Water as solvent in the liquid-phase selective hydrogenation of crotonaldehyde to crotyl alcohol over Pt/ZnO: A factorial design approach. *Appl. Catal. B Environ.* **2014**, *154*, 369–378. [CrossRef]
30. Cosano, D.; Esquivel, D.; Mateos, L.D.; Quesada, E.; Jiménez-Sanchidrián, C.; Ruiz, J.R. Spectroscopic analysis of corrosion products in a bronze cauldron from the Late Iberian Iron Age. *Spectrochim. Acta Part A Mol. Biomol. Spectrosc.* **2018**, *205*, 489–496. [CrossRef] [PubMed]

MDPI
St. Alban-Anlage 66
4052 Basel
Switzerland
Tel. +41 61 683 77 34
Fax +41 61 302 89 18
www.mdpi.com

Catalysts Editorial Office
E-mail: catalysts@mdpi.com
www.mdpi.com/journal/catalysts

www.ingramcontent.com/pod-product-compliance
Lightning Source LLC
Chambersburg PA
CBHW061759210926
51850CB00006B/597
* 9 7 8 3 0 3 9 2 8 8 9 1 5 *